Latin Names Explained

Latin Names Explained

A Guide to the Scientific Classification of Reptiles, Birds & Mammals

A. F. Gotch

 Facts On File®

AN INFOBASE HOLDINGS COMPANY

Contents

Preface

There is an increasing tendency for books about animals to give their Latin names; this must be welcomed for it enables a species to be identified as *that particular species*, regardless of its English or vernacular name, and it is applicable throughout the world. However, no attempt is made to 'translate' the Latin names into English or to explain the reason for these names.

Some years ago I was studying animals which I had seen in Uganda while on a photographic safari. I was annoyed, because not being a classical scholar, many of the Latin names meant little or nothing to me. After making exhaustive enquiries in bookshops and libraries, including university and zoological libraries, it became apparent that there was no book available that would really solve my problem. Thus I decided, as there was no such book, I had better write one. Actually one or two books of a dictionary type have been published and these might be considered to fulfil the need, but they do not; the names do not refer to any particular species and no attempt is made to explain the *reason* for the names.

There are about 4,300 known species of mammals and I have included over 1,000 of these; most have been selected because they are well known, some because they are of particular interest, and some because they are rare or even on the point of extinction. Some subspecies of particular interest have been included, as they may be ranked as full species by certain authors. Thousands of mammal subspecies, which are sometimes called races, have been named by zoologists. One large mammal, the Virginian Deer *Odocoileus virginianus*, has been divided into nearly forty subspecies, and among the small mammals much greater numbers than these have sometimes been given.

Obviously it would be impossible to include all species of all animals in one book, since such an undertaking would run to many

7

volumes and the research would constitute a lifetime's work for a team of classicists. A Cambridge zoologist carrying out research on beetles worked out that if he spent his entire life studying just one family, the Staphylinidae, he would only be able to devote about twenty minutes to any one species! Each chapter in Parts Two to Four of this book refers to one Order, and includes all the Families in that Order, and a diagram shows clearly what animals are included in each Family and how they are classified. Throughout Parts Two to Four, the Order, followed by the Suborder where relevant, is shown on the running headline of the left-hand page, and the Family on the right-hand page.

In Part One of the book (Chapters 1 to 4), an explanation is given of the system of classification started by the Swedish naturalist Carl von Linné during the eighteenth century, and which became known as the Binominal System. The basic principles of this system as in use today are explained, avoiding unnecessary detail as far as possible, and throughout the book running headlines at the top of each page give the classification of the animals listed on that page. This makes for quick and easy reference, but the main feature is that with every animal the translation and interpretation of the Latin name is given —and when necessary an explanation of the reason for giving the species that particular name.

In a number of cases reference has been made to T. S. Palmer's standard work *Index Generum Mammalium* 1904 (see page 468); the period from 1904 to 1951 is covered by L. R. Conisbee's British Museum publication *Genera and Subgenera of Recent Mammals*, and from 1951 to 1971 by his four Supplementary Notes in the *Journal of Mammalogy*.

Any reader who is interested in further study of classification and nomenclature should consult the Bibliography given on page 657 at the end of this book

A. F. Gotch

PART ONE
The Animal Kingdom

1 Generic and Specific Names

The student or amateur naturalist who is interested in the study of taxonomy is faced with a formidable array of Latin and Greek words and a system of grouping the animals into phyla, classes, orders and families which have little or no meaning to the layman. In this book my aim is to explain this arrangement and give a meaning to the names by translating them into English; no attempt is made to describe the animal other than the features and associations that explain the reason for the scientific name.

Carl von Linné, the Swedish botanist, born in the year 1707, was responsible for the first real attempts to classify and name living organisms—although Aristotle had done some work on a simple form of classification as long ago as 384 B.C. Linnaeus, the Latin form of the name von Linné, is normally used in classification and his system of nomenclature is now in use throughout the world. The authoritative tenth edition of his *Systema Naturae* was published in 1758, and this is an all-important date, January 1st of that year being the starting point; all Latin names given before that date are considered invalid.

The Linnaean classification is known as the binominal system, i.e. 'two names'; every plant and animal must be given two names. The first is the generic name (from *genus*, which is Latin for birth or origin) and a genus comprises a group of closely related animals (or plants); the second is the specific name, the name of the actual species, which distinguishes it from any other animal in that group, and so from any other animal in the world.

The generic name should be a noun, and written with a capital letter, and the specific name should be an adjective, though sometimes it is a noun; it should be written with a small letter even though it is a proper name of a place or person; and both names should be printed in *italics*. To be completely authentic, it should be followed by the name of the zoologist who first gave the animal that name, and the

date of naming; for example the European Hedgehog would be
Erinaceus europaeus Linnaeus, 1758.

The author's name and date should not be in italics; if there has been
a change of the generic name, as sometimes happens, the original
author's name will be given in parentheses. This information about
the author and date is not always necessary in zoological publications
and is not given in this book.

PRIORITY

The saying goes 'Priority is the basic principle of zoological nomenclature'; the first scientific name given to an animal after January 1st
1758 stands, even though it is not descriptively accurate. However,
by its plenary powers the International Commission on Zoological
Nomenclature can moderate the application of this rule to preserve
long established names, in order to avoid inconvenience and confusion. A recent ruling has discouraged zoologists from digging out
and reviving forgotten names; they must not displace names that
have been accepted for fifty years or more.

HOMONYMS

A new name for an animal is considered invalid if it has been used
previously for some other form in the *animal* kingdom, as this would
result in the same name for two different animals; it is known as a
homonym. A difference of one letter is sufficient distinction, for
example *Apis*, *Apos* and *Apus* are not homonyms. *Platypus* was given
as the Latin name for the Duckbill in 1799, but this was a homonym
because the same name had been used for a beetle some years before;
it was therefore invalid and a new name had to be given; this was
Ornithorhynchus.

SYNONYMS

In some cases, alternative names are used and recognised. For
example, in the primates, an author may use Prosimii and Simiae
for the names of the suborders: *pro* (L) before; *simia* (L) an ape, a
monkey, i.e. 'early monkeys' or 'primitive monkeys', and 'modern
monkeys'; another author may use Lemuroidea and Anthropoidea:
anthrōpos (Gr) man; *-oides* (New L) from *eidos* (Gr) shape, resemblance,

i.e. 'lemur-like' and 'man-like'. These are known as synonyms, and where there are alternative names these will be shown thus: Lemuroidea (or Prosimii), or if there are more than two, Anthropoidea (or Simiae, Pithecoidea): *pithēkos* (Gr) an ape, a monkey. Similarly, for the Alligator Snapper, in the family Chelydridae, some authors may use *Macrochelys temminckii* and others *Macroclemys temminckii;* for Swainson's Francolin, a bird of the pheasant family, some authors may use *Pteristis swainsoni* and others *Francolinus swainsoni*. These, too, are synonyms and are shown as *Macrochelys temminckii* (= *Macroclemys*) or *Pternistis swainsoni* (or *Francolinus*). If a generic name has been rejected and considered invalid, but still appears in some publications, it will be shown thus: (*Aotus* formerly *Nyctipithecus*).

TAUTONYMS

It will be seen that in some cases the Latin name of an animal has the generic name repeated for the specific name, for example the Pine Marten is *Martes martes*. This is known as a tautonym, from *tauton* (Gr) the same, and *onoma* (Gr) a name; it has no particular significance and is due to a change of the generic name. For example, a Linnaean specific name later used as a generic name so that the full name is a repeated word; a rule states that the specific name *may not be changed*, even though it results in a tautonym. The ruling has been modified for botanical names and tautonyms are forbidden.

I would like to quote an extract from an article by Mr Michael Tweedie which appeared in the magazine *Animals* (now *Wildlife*):

'The wren was among the birds that Linnaeus himself named, and he called it *Motacilla troglodytes*. Under his genus *Motacilla* he included a number of small birds which ornithologists have now split up into several genera, reserving *Motacilla* for the wagtails. In naming the American wren-like birds in the first decade of the 19th century, the French authority Vieillot chose the Linnaean specific name *troglodytes* (troglodyte or cave-dweller, by reference to the form of the nest) as a generic name. Later it was found that the European wren was so closely allied to these that it must go in Vieillot's genus, so it became *Troglodytes troglodytes*.'

A rule states that the name of the nominate subspecies must repeat the specific name of the species (see Chapter 4). Thus, when subspecies were named, the nominate subspecies of the wren group had to be

Troglodytes troglodytes troglodytes. The veteran zoologists, who were classical scholars, thoroughly disliked this and there were heated interchanges of correspondence, and they refused to use these 'monstrosities'; but as they passed from the scene the modernists prevailed, the rules were accepted, and so some stability was achieved.

Some authors state that a tautonym indicates the typical or common species in a genus, but this is not correct; it is most unusual for an animal to have a tautonym for its original name. It is true to say that quite often the common species, after a change in its generic name, does then have such a name; for example, the Common Toad *Bufo bufo* and the Common Bat *Pipistrellus pipistrellus*. On the other hand, the names of many common species do not have the tautonymous form; for example, the Common Seal *Phoca vitulina* and the Common Dormouse *Muscardinus avellanarius*.

TYPE SPECIES

The first species to be named in a particular genus is usually, though not always, the type species. Other species in that genus will resemble it more than those in a different genus and they will be distinguished by their specific names. For example, *Martes flavigula* the Yellow-throated Marten, and *Martes americana* the American Marten.

ORIGIN OF NAMES

The Latin name of an animal might originate from the naturalist who first discovered it, but it is more likely that it would originate from a zoologist working in a laboratory, and studying the anatomy of the animal; the specific name is quite often given in honour of the person who discovered it. The Kiwi has been given the generic name *Apteryx*, which is derived from the Greek *a-*, a prefix meaning 'not', or 'there is not', and *pterux* 'a wing'. The kiwi is a flightless bird, the wings being very small, hidden under the body feathers, and useless for flight; the Great Spotted Kiwi *A. haasti* was named in honour of Sir Julius von Haast, the New Zealand explorer.

USE OF LATIN AND GREEK

The advantages of using languages such as Latin and classical Greek are obvious; if Linnaeus had used the Swedish language, then his system would not have been accepted internationally, and in any case other countries would not have understood it. In those days Latin was the international language of European scholars and Linnaeus wrote most of his scientific work in Latin to make it more widely read and understood. Even today, there are classical scholars throughout the world who are familiar with Latin and classical Greek and who understand the meaning of the words. It is true they would find some of the words wrongly construed and incorrectly spelt, and even hybridised, combining a mixture of Latin and Greek to form one word; anathema to the purist. Sometimes they are obscure native words which are known as 'barbarisms'. By international agreement, once these names appear in print and are accepted by the International Commission on Zoological Nomenclature, any mistakes must remain. However, the 1961-63 revised Code permits correction of spelling; at the same time it forbids the use of hyphens,* diacritic marks such as the apostrophe, the diaeresis, and the umlaut —so that *Hyperoödon* becomes simply *Hyperoodon* and *mülleri* becomes *muelleri*.

The classical significance of the names is not of first importance, but some knowledge of Latin and Greek is a help in remembering the names, and will often tell you something about the animal. For instance, *Felis nigripes* must surely mean a cat with black feet; in fact it is the Black-footed Cat of South Africa. The real importance of the names is that they are international fixed labels, identifying a particular species for zoologists throughout the world.

PECULIAR NAMES

Sometimes the names are just invented; the zoologist who named the Kookaburra, or Laughing Jackass as it is sometimes called, was so hard put to it in deciding on a generic name that would distinguish it from other kingfishers, that he used the word *Dacelo*; the generic name of the Common Kingfisher is *Alcedo* which is Latin for a kingfisher; *Dacelo* is simply an anagram of *Alcedo*.

There are a number of other peculiar names apart from anagrams, for instance *Ia io* for a bat, named by Oldfield Thomas; the shortest

---

*The Comma butterfly *Polygonia c-album,* meaning 'many-angled with a white 'c' ', has the honour of being allowed to use the hyphen; probably the only exception! The name refers to a white letter 'C' which is plainly marked on the under surface of the wing.

name ever given to a mammal. Ia was a young woman of classical times and Thomas says: 'Like many women of those times a bat is essentially flighty'; *io* is a Latin exclamation of joy, like hurrah! He obviously had a sense of humour—even if it was a rather chauvinistic one. Another name he invented was *Zyzomys* for an Australian mouse but he never gave an explanation; perhaps he wanted to make sure that it would appear last in any index!

LOCAL AND PERSONAL NAMES

Many zoologists have used the names of little-known localities, which can make interpretation difficult, and there are references to ancient Greek mythology which may have no special significance. Sometimes the naturalist or collector who discovered the animal is honoured by using his name. The French missionary Père Armand David discovered a strange Asiatic deer near Peking; this has been given the name *Elaphurus davidianus*. Commemorative names usually have -*i*, -*ii*, or -*iana* added to the complete name to form the genitive if the person is a man, and -*ae*, or -*iae* if a woman. The pheasant *Chrysolophos amherstiae* was named in honour of Lady Amherst; note that in such cases the specific name is written with a small letter in accordance with the rules of nomenclature, although it is a personal name.

In many cases the names are derived from Greek words, but they are 'Latinised', that is to say they are put in a Latin form, and so are referred to as 'Latin names'; it is the popular term but rather disparaged by zoologists who prefer to call them 'scientific names'; purists even suggest that the correct term is 'Linnaean names'.

MISLEADING NAMES

Some animals have been given a name that is misleading when translated; this may have come about because they were named when knowledge of them was incomplete. Here are some examples: *galē* (Gr) a weasel or marten-cat, used for a wallaby, which is a marsupial (see page 472); *meles* (L) a badger, used for the bandicoot, another marsupial (see page 467); *kapros* (Gr) a wild boar, used for the hutiacouga, which is a rodent (see page 565); *galē*, and some other names such as *kuōn* (Gr) a dog, have become conventional final elements in nomenclature for the names of small mammals quite remote from weasel, marten-cat, or dog. The Greek word *mus* (genitive

muos) a mouse, is used for all kinds of small mammals. Consider, for ·
instance, *Cynomys ludovicianus* the Prairie Dog (see page 544); *Cynomys*
means 'dog-mouse', but it is not a mouse and not even in the same
family as the rats and mice, and it certainly is not a dog! It will be
seen that it is spelt -*mys*, which is the Latin System of transliteration
from the Greek; in the Modern System, as used in this book, it is
spelt -*mus*. This is the same as the Latin word *mus* (genitive *muris*)
which is derived from the Greek and has the same meaning. The two
systems of transliteration will be found in the Appendix.

2 Separating the Animals into Groups

A division or group is known as a taxon (plural taxa) from the Greek word *taxis*, an arrangement, though the term is not often used for groups higher than classes. The divisions are named as shown in the following list, the main divisions in the left-hand column:

KINGDOM
    SUBKINGDOM
PHYLUM
    SUBPHYLUM
    SUPERCLASS
CLASS
    SUBCLASS
    SUPERORDER
ORDER
    SUBORDER
    SUPERFAMILY
FAMILY
    SUBFAMILY
    Genus
      Subgenus
    Species
      Subspecies (or *Race*)

These categories do not all have to be used in the classification of any particular group or species; each division or subdivision is given its own name, as will be seen in Chapter 4, and as shown in the classification key on pp. 524–5. A subspecies will have a subspecific name, for example the Helmeted Guinea Fowl *Numida meleagris mitrata*; this is known as trinominal, i.e. three names. It is a subspecies of the Tufted Guinea Fowl *Numida meleagris*. A rule made by the International Commission on Zoological Nomenclature states that the subspecific name of the species on which a group is founded must be the same, i.e. it must *repeat* the specific name. Thus, in this case it becomes *Numida meleagris meleagris*. Therefore it is easily distinguished and is known as the nominate subspecies.

A subspecies is always basically the same as the species, but there is some slight difference; it usually has a different geographical distribution and develops different characteristics. For instance, the crest of the Helmeted Guinea Fowl is larger than that of the Tufted Guinea Fowl and the colour of the body feathers is different. Sub-genera are also sometimes recognised, and like subspecies the animal is basically the same as that in the genus; a subgenus is shown thus: *Nycticeius (Scotoecus) falabae* the Falaba House Bat. The subgeneric name is given in parentheses, and does not count as one of the words in the name—so it is not a trinominal name. Some zoologists might consider this bat worthy of the status of a full genus, in which case the name would be *Scotoecus falabae*. Examples of suborders will be found on page 539, and subclasses on page 460 (see also the classification key on pages 524 and 525).

In deciding to which particular taxon an animal belongs, the zoologists must study its anatomy. During the early days of classification not enough attention was paid to this aspect and many animals were placed in the wrong taxon simply because of their outward appearance. These mistakes have gradually been corrected over the years, which accounts for the tautonymic names (see page 13). However, there is always likely to be a difference of opinion among the scientists concerned in this work. One of the most important parts of the anatomy which must be studied is the teeth; during a discussion on the classification of mammals Baron Cuvier, the famous French zoologist, is reputed to have said:

'Show me your teeth and I will tell you what you are.'

Apart from the anatomical structure some notice is taken of the habits of an animal: where does it live, what does it eat? As already mentioned, it is no use looking at an animal and deciding from its appearance to which group it belongs. For instance, take the lizard known in Britain as the 'slow-worm'; the uninitiated on seeing this animal would cry, 'Look, a snake'! Further, many might probably get a stick and beat the poor thing to death, though a more harmless and docile little creature it would be difficult to find (this could well be the reason for the decrease in numbers of the slow-worm in the

British Isles!).

What then makes this creature a lizard and not a snake? For one thing, it has eyelids, and snakes do not have eyelids—though lizards do; and another thing, if caught by the tail the slow-worm, like the lizard, can break it off. This is known as autotomy, from *auto-* (Gr) 'self', and *tomē* (Gr) 'a cutting off'. There is a muscular mechanism in the tail which breaks it off and then actuates to prevent loss of blood; the tail then grows again. This leaves the surprised predator with only the broken-off tail, while the slow-worm makes good its escape. Indeed the Latin name *Anguis fragilis* suggests this, though it also shows that the naturalist who named it thought it was a snake: *anguis* (L) 'a snake', *fragilis* (L) 'brittle'—hence 'the brittle snake'. One can imagine his surprise if he grabbed it by the tail, and was left with it wriggling in his hand; there is no doubt that the dismembered piece does wriggle, as I have seen myself, and this probably adds to the general surprise and confusion and gives more time for the main body to escape.

The main divisions of the Animal Kingdom are the phyla (singular, phylum) from *phulon* (Gr) 'a stock, race or kind'. There is not complete agreement among zoologists as to the number of phyla, as some have separated a group and classified it as a subphylum, and it can take many years before international agreement is reached. As a result a number of different systems have become more or less established and generally recognised, and although they are basically the same there are certain differences which can be confusing; for example one well-known system is based on 22 phyla, and another, as given in this book in Chapter 3, is based on 27 phyla.

In any particular phylum there will be assembled all the animals having a common basic plan. Let us take as an example the phylum Chordata. The distinctive character of this phylum is the notochord; *nōton* (Gr) 'the back' and *khordē* (Gr) 'gut, string', giving rise to *chordata* (New L) meaning 'provided with a back-string'. This is a cord running along the back, made of a special tough elastic tissue, and present in all animals in this phylum. In the humblest members of the group, such as the lancelets (small marine creatures about 5 cm or 2 in long) the notochord is retained throughout life. In the higher forms, the true vertebrates, it is present in the embryo, but is replaced more or less completely by the stronger and yet flexible spinal column of jointed vertebrae. However, as all the animals in the phylum Chordata do not develop a spinal column, and there are

other differences, the group is divided into four subphyla, the subphylum Vertebrata being one of these; this will be considered in more
detail in Chapter 4.

3　The Phyla

As already mentioned, there is not complete agreement about the number of phyla needed to classify the animal kingdom, but the system used in this book comprises 27 phyla, one of which, Echinodermata, has two subphyla, and another, Chordata, four subphyla. Considering that the phylum Chordata includes such diverse animals as the acorn worm, the mouse, the whale and humans, it is no wonder that some subphyla are needed. Even so, the one subphylum Vertebrata contains all the animals best known to us: birds, fishes and reptiles; amphibians, such as the frogs and newts; and mammals, such as dogs, horses, antelopes, lions, whales and ourselves.

I do not advise anyone attempting to learn, like a parrot, the Latin and Greek names that now follow; in the course of study they will become familiar without any special effort. The sequence is not fixed; it begins with what the compiler considers to be the most primitive forms of life and ends with the most advanced. The number of species given for each phylum is approximate; taxonomists do not always agree about the number of species in a particular group, and it can never be finally settled as new species may be discovered at any time.

The word phylum itself is derived from *phulon* (Gr) plural *phula*, a race or kind. The phyla are as listed below.

PROTOZOA　30,000 species. Amoeba, mycetozoa, etc.
prōtos (Gr) first; *zōon* (Gr) an animal, a living thing.
amoibē (Gr) a change, an alteration. The amoeba, a tiny single-celled animal, continually changes its shape.
mukēs (Gr), genitive *mukētos*, a fungus; any knobbed body shaped like a fungus.

22

PORIFERA (or PARAZOA, SPONGIDA)   4,500 species. The sponges.
poros (Gr) a way through, a passage; *gero* (L) I bear, I carry.

COELENTERATA (or CNIDARIA)   9,000 species. Jellyfish, corals, etc.
koilos (Gr) hollow; *enteron* (Gr) bowel, intestine.

CTENOPHORA   80 species. Comb jelly, sea gooseberries.
kteis (Gr), genitive *ktenos*, a comb; *phora* (Gr) carrying, bearing.

MESOZOA   7 species. Minute worms, parasites in the kidneys of squids and octopuses.
mesos (Gr) middle; *zōon* (Gr) an animal, a living thing.

PLATYHELMINTHES   9,000 species. Flatworms, liver flukes, bilharzia, etc.
platus (Gr) flat; *helmins* (Gr), genitive *helminthos*, a worm.

NEMERTEA   570 species. Ribbon worms, bootlace worms.
nēma (Gr) genitive *nēmatos*, a thread.

NEMATODA (or NEMATA)   10,500 species. Roundworms, vinegar eelworms, etc.
nēma (Gr), genitive *nēmatos*, a thread; *-oda* (New L) from *eidos* (Gr) form, like.

ROTIFERA (or ROTATORIA)   1,200 species. Wheel animalcules, the smallest many-celled animals about 1 mm (less than $\frac{1}{16}$ in) long, mostly living in fresh water.
rota (L) a wheel; *fero* (L) I bear, I carry. They do not, of course, carry a wheel, but the first ones discovered and examined under a microscope showed tiny hairs round the mouth; these wave in a circular motion that gives the impression of a turning wheel.

GASTROTRICHA   100 species. Gastrotrichs, tiny transparent creatures less than 0·5 mm ($\frac{1}{50}$ in) long, mostly living in fresh water.
gaster (Gr) the belly, stomach; *thrix* (Gr), genitive *trichos*, hair; they have hairs on their underparts.

KINORHYNCHA (or ECHINODERA, ECHINODERIDA)   30 species. Kinorhynchs, marine animals about 1 mm (less than $\frac{1}{16}$ in) long.
kineō (Gr) I move; *rhunkhos* (Gr) a snout, a beak; they pull themselves along by a kind of snout.

PRIAPULIDA   6 species. Priapulids, wormlike marine animals 7·6 cm (about 3 in) long, living on muddy bottoms and sometimes at a great depth. Priapos was the god of gardens and vineyards and Priapus, in Roman mythology, meant a representation or symbol of the male generative organ, or phallus. The name priapulida probably refers to the shape of the animal, which is not unlike the human penis; -*ida* (New L) from *idea* (Gr) species, sort.

NEMATOMORPHA (or GORDIACEA)   80 species. Horsehair worms, not usually marine, and very variable in length from 10 cm (4 in) upwards, and from 0·3 to 3 mm ($\frac{1}{80}$–$\frac{1}{8}$ in) in diameter. *nēma* (Gr) genitive nēmatos, thread; *morphē* (Gr) form shape. Sometimes found in horse drinking troughs, hence the English name.

ACANTHOCEPHALA   400 species. Spiny-headed worms; parasites in the intestines of vertebrates. *akantha* (Gr) a thorn, prickle; *kephalē* (Gr) the head.

ENTOPROCTA (or ENDOPROCTA, CALYSSOZOA, KAMPTOZOA, POLYZOA ENDOPROCTA, POLYZOA ENTOPROCTA)   60 species. Entoprocts; a flower-like body on a stalk less than 6 mm ($\frac{1}{4}$ in) high. The anus opens within a circlet of tentacles, hence the name: *entos* (Gr) within; *prōktos* (Gr) anus, hinder parts.

CHAETOGNATHA   30 species. Arrow worms; common in sea water near the shore or in the depths, from 2 to 10 cm ($\frac{3}{4}$–4 in) in length. *khaitē* (Gr) hair, mane; *gnathos* (Gr) jaw; it has short bristles surrounding the mouth.

POGONOPHORA (or BRACHIATA)   22 species. Beard worms; a deep sea animal about 3 mm ($\frac{1}{10}$ in) in diameter and up to 33 cm (13 in) long. It has no digestive system and how it obtains nourishment is something of a mystery. *pōgōn* (Gr) a beard; *phora* (Gr) carrying.

PHORONIDA (or PHORONIDEA)   15 species. Phoronids; small marine animals that range from 2 to 30 cm (1 – 12 in) in length. They build themselves a tube in which to live, and catch their food by means of tentacles projecting from the end of the tube. The reason for their name is obscure, but is thought to originate from Phoronis, surname of Io, the daughter of Inachus; there is a strange legend which involves her wandering all over the earth.

BRYOZOA (or POLYZOA, POLYZOA ECTOPROCTA, ECTOPROCTA) 6,000 species. Moss animals; tiny animals, about 0·4 mm ($\frac{1}{64}$ in) long, which live in sea water and fresh water. You might find them attached to the bottom of your boat when you pull it ashore, and though they look rather like a mossy plant they are probably these tiny animals, the bryozoans.
bruo (Gr) I swell, I sprout, giving rise to bryon, lichen or tree moss; *zōon* (Gr) an animal.

BRACHIOPODA 260 species. Lamp shells; the brachiopod shell is shaped somewhat like the oil lamps used by the Greeks and Romans in ancient times.
brakhus (Gr) short; *pous* (Gr) genitive podos, a foot; they do not have feet, but this refers to a short stalk by which they attach themselves to some support.

SIPUNCULA (or SIPUNCULOIDEA) 250 species. Peanut worms; marine animals about 20 to 40 cm (8–18 in) long and 12 mm ($\frac{1}{2}$ in) in diameter. They can change their shape, and sometimes take the shape of a peanut. A peculiar feature is a pump-like action with one part of the body sliding up and down inside the posterior cylindrical part.
siphōn (Gr) a sucker, as of a pump; *-culus* (L) suffix meaning small: 'a little pump'.

ECHIURA (or ECHIUROIDEA, ECHIURIDA) 60 species. Curious sausage-shaped marine animals, up to about 30 cm (12 in) long, and having a proboscis which in some species may be 1 m (3 ft) long. This is used for gathering food: *ekhis* (Gr) an adder, a serpent: *oura* (Gr) the tail; the 'serpent' part is more a nose than a tail.

MOLLUSCA 50,000 species. Oysters, octopuses, slugs, snails, squids, etc.
mollusca (L), neuter plural of *molluscus*, soft.

ANNELIDA (or ANNULATA) 6,000 species. Leeches, earthworms, ragworms, etc.
anellus (L) a little ring; *-ida* (New L) from *idea* (Gr) species, sort; the rings that mark the body of an earthworm or a ragworm give this phylum its name.

ARTHROPODA 815,000 species, including the insects which number about 750,000 species, possibly many more. Crustaceans, spiders, insects, etc.
arthron (Gr) a joint; *pous* (Gr) genitive podos, a foot; in this case it is taken to mean leg, as the arthropods have jointed legs. Some also have a hard external skeleton; *crusta* (L) a shell, crust.

ECHINODERMATA 5,500 species.
ekhinos (Gr) a hedgehog; *derma* (Gr) the skin; a reference to some that have 'spiky' skins.

Subphylum Pelmatozoa Sea lilies, feather stars.
pelma (Gr) genitive *pelmatos*, the sole of the foot, a stalk; *zōon* (Gr) an animal. The sea lily has a flower-like body supported on a stalk, the feather star begins life on a stalk but later breaks away and is free to swim about.

Subphylum Eleutherozoa Sea urchins, sea cucumbers, star-fishes etc.
eleutheros (Gr) free, not bound as on a stalk; *zōon* (Gr) an animal: the sea urchin's body is equipped with movable spikes.

CHORDATA about 44,750 species.
nōtos (Gr) the back; *khordē* (Gr) gut, string; *-ata* (New L) suffix used for divisions of zoological names; the notochord, or 'back-string', is a rod-like structure made of tough elastic tissue which is present in all early embryos in the phylum Chordata.

Subphylum Hemichordata (or Stomochordata, Branchio-tremata) 90 species. The acorn worms and their kin.
hēmi- (Gr) prefix meaning half; suggesting halfway between primitive chordates and the next stage. The notochord is found only in the proboscis.

Subphylum Urochordata (or Tunicata) 1,600 species. Sea squirts, salps, etc.
oura (Gr) the tail; *khordē* (Gr) gut, string. The notochord extends into the tail.

Subphylum Cephalochordata (or Acrania, Leptocardii) 13 species. Lancelets; small marine animals about 5 cm (2 in) long and pointed at both ends.
lancea (L) a small spear; *kephalē* (Gr) the head: *khordē* (Gr) gut, string. The notochord extends into the head.

Subphylum Vertebrata   About 43,000 species. The vertebrates; fishes, amphibians, reptiles, birds, and mammals, including humans (see page 28).

vertebra (L) a joint, specially a joint of the back, derived from *verto* (L) I turn. With the exception of certain fishes, such as the cartilage fishes, during development from the embryo the notochord is replaced by the bony spinal column.

4 The Vertebrates*

In the subphylum Vertebrata there is a great variety of animals, and so they are divided into groups, or taxa, called classes. There are certain recognised sequences for compiling lists of animals, but they are arbitrary, and different authors adopt different plans. It is not possible to make a linear series that is scientifically correct from the point of view of evolution, because mammals, or any other forms of life, have not descended one from another in a long line. The classes are as listed below.

MARSIPOBRANCHII (or AGNATHA)   Lampreys and hagfishes.

SELACHII (or CHONDROPTERYGII, CHONDRICHTHYES, ELASMOBRANCHII)   Sharks, dogfishes, rays, i.e. the cartilage fishes.

BRADYODONTI   Rabbit fishes.

PISCES (or OSTEICHTHYES)   Bony fishes.

AMPHIBIA   Frogs, toads, newts, and their kin.

REPTILIA   Tortoises, lizards, snakes, crocodiles.

AVES   Birds.

MAMMALIA   Dogs, cats, horses, whales, man, and their kin.

The reptiles are divided into 4 taxa called orders, with 5 suborders, and these are again divided into 82 families and subfamilies; the family names have the ending -idae, and the subfamilies -inae; the complete list

* Including the lampreys and cartilage fishes.

can be seen in Chapter 6. The names are formed by adding the suffix to the stem of the name of the type-genus, for example the family name of the crocodiles is formed from the generic name *Crocodylus* and so becomes Crocodylidae. The type-genus is that genus whose structure and characteristics are most representative of the group as a whole, although in some cases it may have been selected because it is the largest, best-known, or earliest-described genus.

To bring us up to date so far, let us work through an example, the skinks. We will list the divisions, starting with the phylum and going through to the species:

Phylum	**CHORDATA**
Subphylum	**VERTEBRATA**
Class	**REPTILIA**
Order	**SQUAMATA**
Suborder	**SAURIA**
Family	**SCINCIDAE**
Subfamily	**SCINCINAE**
Genus	
and species	*Scincus conirostris* Eastern Skink
	S. muscatensis Arabian Skink
	S. philbyi Philby's Arabian Skink
	Ophiomorus brevipes Short-footed Sand Skink
	O. persicus Persian Snake Skink
	(others)

It is considered sufficient, where the genus is the same as the one mentioned immediately before, simply to put the initial capital letter, as will be seen in the list above. The English interpretation of all the foregoing Latin names will be found in the section dealing with the skinks (see p. 96).

Now let us look at another example, a well-known snake subspecies, the Burmese Python *Python molurus bivittatus*, with translations into English of each stage of its classification:

Phylum   CHORDATA

khordē (Gr) gut, string; giving rise to *chordata* (New L) having a notochord or backstring; *nōtos* (Gr) the back. The notochord is made of a special tough elastic tissue, and is present in the embyro of all animals in this phylum.

Subphylum   VERTEBRATA

vertebra (L) a joint, specially a joint of the back, derived from *verto* (L) I turn; *vertebrata* (New L) having a jointed back. In this subphylum the notochord develops into the spinal column.

Class   REPTILIA
repto (L) I crawl; *reptile* (L) a crawling animal, a reptile.

Order   SQUAMATA
squama (L) scale.

Suborder   SERPENTES
serpens (L) genitive *serpentis*, a serpent.

Family   BOIDAE
boa (L) a kind of snake; *-idae* (New L) suffix added to generic names to indicate the name of a family.

Subfamily   PYTHONINAE
puthōn (Gr) the serpent slain by Apollo, thence surnamed the Pythian.

Genus   *Python*.

Species   *molurus moluros* (Gr) a kind of serpent.

Subspecies   *molurus bivittatus bi-* (L) prefix meaning two; *vitta* (L) a ribbon, a band; *-atus* (L) suffix meaning provided with; *bivittatus* is known as the subspecific name, so this snake is the Two-banded or Burmese Python *Python molurus bivittatus*.

The birds are divided into 27 taxa called orders, and can be distinguished by the ending -iformes. These orders are divided into a very large number of families, totalling 161, which have the ending -idae, and in some cases these are again divided into subfamilies which have the ending -inae. The complete list can be seen in Chapter 13. The names are formed by adding the suffix to the stem of the name of the type-genus. For example, the family name of the starlings is formed from the generic name *Sturnus* (L) a starling, and so becomes Sturnidae. The type-genus is that genus whose structure and characteristics are most representative of the larger group as a whole, although in some cases it may have been selected because it is the largest, best-known, or earliest-described genus.

To bring us up to date so far, let us work through an example, the owls. We will list the divisions, starting with the phylum and going through to the species:

Phylum   **CHORDATA**
Subphylum   **VERTEBRATA**
Class   **AVES**

Order   **STRIGIFORMES**

Family TYTONIDAE   Barn Owls
Genus and species   *Tyto alba*   Barn Owl
  T. tenebricosa   Sooty Owl
  Phodilus prigoginei   Tanzanian Bay Owl
  (others)

Family STRIGIDAE   Typical Owls
Genus and species   *Otus albogularis*   White-throated Screech-Owl

Lophostrix cristata   Crested Owl
Uroglaux dimorpha   Papuan Hawk-Owl
(others)

It is considered sufficient, where the genus is the same as the one mentioned immediately before, simply to put the initial capital letter, as will be seen in the list above. The English interpretation of all the foregoing Latin names will be found in the chapter dealing with the owls.

Now let us look at another example, that of a small well-known British bird, the Chaffinch, with translations into English of each stage of its classification:

Phylum   CHORDATA

khordē (Gr) gut, string; giving rise to *chordata* (New L) having a notochord or backstring   *nōtos* (Gr) the back. The notochord is made of a special tough elastic tissue, and is present in the embyro of all animals in this phylum.

Subphylum   VERTEBRATA

vertebra (L) a joint, specially a joint of the back, derived from *verto* (L) I turn   *vertebrata* (New L) having a jointed back; in this subphylum the notochord develops into the spinal column.

Class   AVES

avis (L) a bird.

Order   PASSERIFORMES
passer (L) a sparrow, or other small bird   *-iformes* (New L) suffix added to indicate the name of an order.

Family FRINGILLIDAE

fringilla (L) a small bird, according to some authors the chaffinch   *-idae* (New L) suffix added to generic names to indicate the name of a family.

Subfamily FRINGILLINAE

-inae (New L) suffix added to generic names to indicate a subfamily.

Genus *Fringilla*

Species *coelebs* = *caelebs* (L) unmarried, single, whether a bachelor or a widower; sometimes chaffinches gather in flocks of one sex only, and at one time it was thought that the females migrated south leaving the males behind, and they became known as 'bachelor birds'. So the name of the Chaffinch is *Fringilla coelebs* —a small bird, a bachelor.

The mammals are divided into 19 taxa called orders. They show the most amazing variation in form, and have adapted themselves to live in the air, in the water, on the ground, and under the ground.

The orders are:

Monotremata	Duck-billed platypus, and echidna or spiny ant-eater
Marsupialia	Kangaroos, opossums, and their kin
Insectivora	Hedgehogs, shrews, moles, and their kin
Chiroptera	Bats
Dermoptera	Colugos or flying lemurs
Primates	Tree shrews, lemurs, monkeys, apes and man
Edentata	Sloths, armadillos, anteaters
Pholidota	Pangolins
Lagomorpha	Rabbits, hares, pikas
Rodentia	Mice, rats, squirrels, porcupines, and their kin
Cetacea	Whales, dolphins, porpoises
Carnivora	Lions, dogs, mongooses, weasels, and their kin

Pinnipedia	Seals, sea-lions, walrus
Tubulidentata	Aardvark
Proboscidea	Elephants
Hyracoidea	Hyraxes
Sirenia	Manatees, dugong
Perissodactyla	Horses, tapirs, rhinoceroses
Artiodactyla	Deer, camels, giraffes, hippopotamuses, and their kin

The orders are divided into families and the family taxon is easy to distinguish because the scientific name ends with -idae and a subfamily with -inae. It is formed by adding the suffix -idae, or -inae, to the stem of the name of the type-genus. For example, the family name of the squirrels is formed from the squirrel genus *Sciurus*, and so becomes Sciuridae. The type-genus is that genus whose structure and characteristics are most representative of the larger group as a whole, although in some cases it may have been selected because it is the largest, best-known, or earliest-described genus.

To bring us up to date so far we will list the divisions, starting with the phylum, which apply to the Perissodactyla:

Phylum   **CHORDATA**
Subphylum   **VERTEBRATA**
Class   **MAMMALIA**

Order   PERISSODACTYLA

Families   **EQUIDAE**   Horses
               TAPIRIDAE   Tapirs
               RHINOCEROTIDAE   Rhinoceroses

Now we come to the genera and species, which will give each animal its own name.

Family EQUIDAE
Genus and species   *Equus przewalskii*   Przewalski's horse

E. caballus	Horse
E. hemionus	Wild ass
E. asinus	North African wild ass (donkey)
E. zebra	Mountain zebra
E. burchelli	Burchell's or common zebra
E. grevyi	Grèvy's zebra

Family TAPIRIDAE

Genus and species	*Tapirus indicus*	Malayan tapir
	T. terrestris	Brazilian tapir
	T. pinchaque	Mountain tapir
	T. bairdi	Baird's tapir

Family RHINOCEROTIDAE

Genus and species	*Rhinoceros unicornis*	Indian rhinoceros
	R. sondaicus	Javan rhinoceros
	Didermocerus sumatrensis	Sumatran rhinoceros
	Diceros simus	White rhinoceros
	D. bicornis	Black rhinoceros

It is standard practice, where the genus is the same as the one mentioned immediately before, simply to put the initial capital letter. This practice has been followed in the lists above. The English interpretation of all the foregoing Latin names will be given in the Chapter dealing with each particular genus.

Phylum   Chordata
Subphylum   Vertebrata

AN EXAMPLE

Now let us look at an example of a small mammal, the House Mouse, with translations into English of each stage of its classification.

khordē (Gr) gut, string, giving rise to chordata (New L) having a notochord, or backstring: *nōtos* (Gr) the back. The notochord is made of a special tough elastic tissue, and is present in the embyro of all animals in this Phylum. In the Subphylum Vertebrata it develops into the spinal column, as in this case.

vertebra (L) a joint, specially a joint of the back, derived from *verto* (L) I turn.

Class   Mammalia

mamma (L) a breast. All animals which feed their young from the breast are in this class, so it includes the whales, seals and other sea-living mammals.

rodo (L) I gnaw, eat away.

Order   RODENTIA

Family Muridae *mus* (L) genitive muris, a mouse or a rat: *-idae* (L) suffix added to generic names to form family names.

Genus   *Mus*

Species *musculus -culus* (L) suffix added to form diminutive. So the House Mouse is known as *Mus musculus*—'*Mouse, little mouse*'.

PART TWO
The Reptile Orders

5   Reptiles in General – The Class Reptilia

The reptiles, which have descended from the amphibians and which are the ancestors of the birds and mammals, are classed as Reptilia. They number about 5,400 species and are one of the three groups of cold-blooded vertebrates, the other two being the fishes (Pisces) and the amphibians (Amphibia); with the two warm-blooded vertebrate groups, the birds (Aves) and the mammals (Mammalia), they constitute the subphylum Vertebrata (see Chapter 4). The four orders of the Reptilia are given below, the Squamata being divided into two suborders:

Order	**CHELONIA**	
	Turtles, tortoises and terrapins	220 species
Order	**CROCODYLIA**	
	Crocodiles, alligators and gavials	21 species
Order	**RHYNCHOCEPHALIA**	
	The Tuatara	1 species
Order	**SQUAMATA**	
Suborder	**SAURIA** Lizards	2,800 species
Suborder	**SERPENTES** Snakes	2,200 species

Nowadays zoologists prefer not to use the term 'cold-blooded' when referring to fish, reptiles and amphibians, because the temperature of the blood varies with the surroundings; the condition is known as poikilothermic, from the Greek *poikilos*, variable, and *thermē*, heat. In some cases the blood can become so hot in the sun that a lizard, for instance, must seek the shade or it would perish. It is now known that the body temperature of some 'cold-blooded' animals, for instance

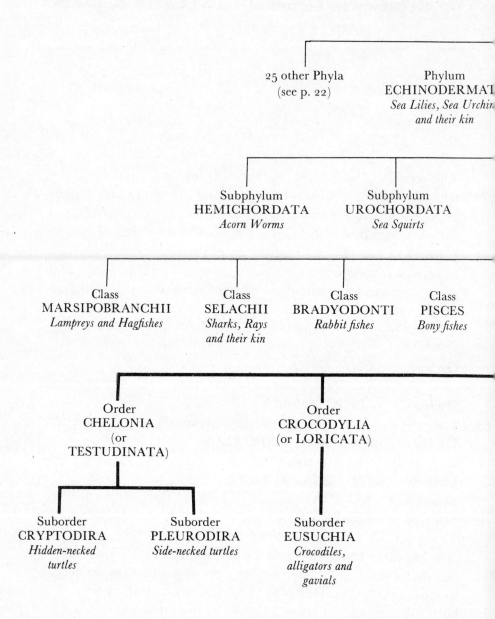

25 other Phyla
(see p. 22)

Phylum
ECHINODERMAT
*Sea Lilies, Sea Urchir
and their kin*

Subphylum
HEMICHORDATA
Acorn Worms

Subphylum
UROCHORDATA
Sea Squirts

Class
MARSIPOBRANCHII
Lampreys and Hagfishes

Class
SELACHII
*Sharks, Rays
and their kin*

Class
BRADYODONTI
Rabbit fishes

Class
PISCES
Bony fishes

Order
CHELONIA
(or
TESTUDINATA)

Order
CROCODYLIA
(or LORICATA)

Suborder
CRYPTODIRA
*Hidden-necked
turtles*

Suborder
PLEURODIRA
Side-necked turtles

Suborder
EUSUCHIA
*Crocodiles,
alligators and
gavials*

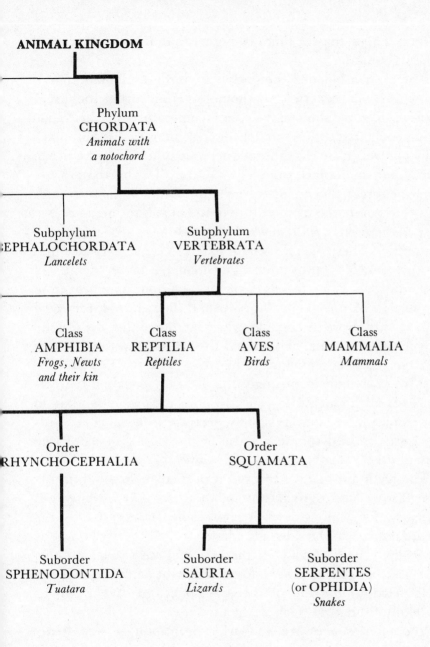

ANIMAL KINGDOM

Phylum
CHORDATA
*Animals with
a notochord*

Subphylum
CEPHALOCHORDATA
Lancelets

Subphylum
VERTEBRATA
Vertebrates

Class
AMPHIBIA
*Frogs, Newts
and their kin*

Class
REPTILIA
Reptiles

Class
AVES
Birds

Class
MAMMALIA
Mammals

Order
RHYNCHOCEPHALIA

Order
SQUAMATA

Suborder
SPHENODONTIDA
Tuatara

Suborder
SAURIA
Lizards

Suborder
SERPENTES
(or **OPHIDIA**)
Snakes

certain snakes, may at times be higher than the temperature of the surroundings.

The 'warm-blooded' condition in birds and mammals is known as homeothermic (or homoiothermic) from the Greek *homoios*, like or similar; i.e. the temperature of the blood remains about the same under all conditions and is automatically controlled by the animal's nervous system. It is higher in birds than mammals, about 41°C (106°F). The reptiles are almost entirely dependent on the temperature of the surroundings to regulate the temperature of the blood, so there are two areas where they will never be found: the Arctic and the Antarctic. There is one small lizard, *Lacerta vivipara*, that ventures north as far as the Arctic Circle, but no reptiles live in the Antarctic. In many cases, they depend on some heat from the sun; their movements are slow and they are not interested in food until the body becomes warm, and then they are usually very active and agile, particularly the lizards. Most reptiles are at their best at a temperature between 25°C (77°F) and 38°C (100°F), although surprisingly the Tuatara prefers a temperature of only about 12°C (53°F). Reptiles are most abundant in hot tropical areas and gradually decrease in number towards the northern and southern parts of the world; in temperate climates they usually hibernate in winter.

The teeth are modified to suit a particular diet and, in the case of some snakes, to accommodate the poison fangs, and in all snakes for swallowing their prey whole. In most lizards, the teeth are fixed to the side of the jaw bone. These are known as pleurodont, from *pleura* (Gr) the side, and *odous* (Gr), genitive *odontos*, a tooth; some lizards, for example the agamids, have the teeth fixed to the top of the jaw bone and are known as acrodont, from *akros* (Gr), at the top.

Venomous snakes are known as 'front-fanged' and 'rear-fanged', depending on whether the poison fangs are situated at the front part of the mouth or at the rear; the scientific names are Proteroglypha and Opisthoglypha, from *proteros* (Gr) first, in space or time, and *gluphē* (Gr) anything carved, can mean a whole cut; and *opisthe* (Gr) after, behind. This refers to the grooves on the fangs that conduct the poison to the wound. Snakes with fangs that are actually hollow, like a hypodermic needle, are termed Solenoglypha from *sōlēn* (Gr) a pipe or channel, and the non-venomous snakes Aglypha, from the Greek prefix *a* meaning not, or there is not.

The rear-fanged Opisthoglypha, are not usually dangerous to Man; as the fangs are at the back of the mouth they are not easily brought into action on a large animal; an exception to this is the Boomslang *Dispholidus typus*, which is known to have caused several fatalities, including the tragic death of the well-known American herpetologist Karl Schmidt.

In most reptiles, the sense of smell is augmented by an anatomical structure known as Jacobson's Organ, which consists of two pits in the roof of the mouth lined with a sensitive membrane similar to that of the nose. It is well developed in snakes and most lizards, and is present in tortoises and turtles and the Tuatara, but is only vestigial in crocodiles. To make use of this organ, the tongue picks up tiny particles from the air, almost as though tasting the air; the tongue is then withdrawn into the mouth and the tips of the tongue are inserted into the organ where the sensitive membrane records the smell. It is very noticeable in snakes, who constantly flick the tongue in and out, which means they are smelling the air, the ground, or their prey, and it is probably used in courtship. Monitor lizards also have this habit, which suggests that Jacobson's Organ is particularly important in their case.

6 Reptile Orders, Suborders, Families and Subfamilies

Order	**CHELONIA** (or **TESTUDINATA**) Turtles, tortoises and terrapins
Suborder	**CRYPTODIRA** Hidden-necked Turtles
Families	**DERMATEMYDIDAE** River Turtle
	CHELYDRIDAE Snapping Turtles
	KINOSTERNIDAE Musk Turtles
	PLATYSTERNIDAE Big-headed Turtle
	EMYDIDAE Pond Turtles, Terrapins
	TESTUDINIDAE True Tortoises
	CHELONIIDAE Marine Turtles
	DERMOCHELYIDAE Leatherback Turtle
	CARETTOCHELYIDAE Papuan Turtle
	TRIONYCHIDAE Soft-shelled Turtles
Subfamilies	**CYCLANORBINAE** Soft-shelled Turtles
	TRIONYCHINAE Soft-shelled Turtles
Suborder	**PLEURODIRA** Side-necked Turtles
Families	**PELOMEDUSIDAE** Side-necked Turtles
	CHELIDAE Snake-necked Turtles
Order	**CROCODYLIA** (or **LORICATA**) Crocodiles and their kin
Suborder	**EUSUCHIA**

Families	**ALLIGATORIDAE** Alligators	
	CROCODYLIDAE Crocodiles	
	GAVIALIDAE Gavial or Gharial	

Order	**RHYNCHOCEPHALIA** The Tuatara	
Suborder	**SPHENODONTIDA** The Tuatara	
Family	**SPHENODONTIDAE** The Tuatara	

Order	**SQUAMATA** Lizards and Snakes	
Suborder	**SAURIA** (or **LACERTILIA**) Lizards	
Families	**GEKKONIDAE** Geckos	
	PYGOPODIDAE Scaly-footed Lizards	
	DIBAMIDAE Two-legged Lizards	
	IGUANIDAE Iguanas	
Subfamilies	**SCELOPORINAE** Spiny Lizards and their kin	
	TROPIDURINAE Keel-tailed Lizards and their kin	
	IGUANINAE Iguanas	
	BASILISCINAE Basilisks	
	ANOLINAE Anoles, Long-legged Lizards	

Families	**AGAMIDAE** Agamas and their kin	
	CHAMAELEONIDAE Chameleons	
	SCINCIDAE Skinks	
Subfamilies	**TILIQUINAE** Blue-tongued Skinks and their kin	
	SCINCINAE Skinks	
	LYGOSOMINAE Slender Skinks and their kin	

Families	**FEYLINIIDAE** Feylin's Lizards	
	ANELYTROPSIDAE Mexican Blind Lizard	
	CORDYLIDAE Girdle-tailed Lizards	
Subfamilies	**CORDYLINAE** Girdle-tailed Lizards	
	GERRHOSAURINAE Plated Lizards and their kin	

Families	**XANTUSIIDAE** Night Lizards
	TEIIDAE Teiids, Racerunners, and their kin
	LACERTIDAE True Lizards or Typical Lizards
	ANGUIDAE Lateral Fold Lizards
Subfamilies	**DIPLOGLOSSINAE** Folded-tongue Lizards and their kin
	GERRHONOTINAE Alligator Lizards
	ANGUINAE Slow-worms
Families	**ANNIELLIDAE** Legless Lizards
	XENOSAURIDAE Crocodile Lizards and their kin
	HELODERMATIDAE Beaded Lizards
	VARANIDAE Monitors
	LANTHANOTIDAE Earless Monitor
	BIPEDIDAE Two-legged Worm Lizards
	AMPHISBAENIDAE Ringed or Worm Lizards
	TROGONOPHIDAE Sharp-tailed Worm Lizards
Suborder	**SERPENTES** (or **OPHIDIA**)
Families	**TYPHLOPIDAE** Blind Snakes
	LEPTOTYPHLOPIDAE Slender Blind Snakes
	ANILIIDAE Pipe Snakes
	UROPELTIDAE Shield-tailed Snakes
	XENOPELTIDAE Sunbeam Snake
	ACROCHORDIDAE Wart Snakes
	BOIDAE Boas and Pythons
Subfamilies	**LOXOCEMINAE** Mexican Python
	PYTHONINAE Pythons
	BOINAE Boas
	BOLYERIINAE Round Island Boas

Family	**COLUBRIDAE** Typical Snakes
Subfamilies	**XENODERMINAE** Xenodermin Snakes
	SIBYNOPHINAE Spear Snakes
	XENODONTINAE Hog-nosed Snakes and their kin
	NATRICINAE Water Snakes
	COLUBRINAE Typical Snakes
	CALAMARINAE Reed Snakes
	LYCODONTINAE Wolf Snakes
	DIPSADINAE Thirst Snakes
	DASYPELTINAE Egg-eating Snakes
	HOMALOPSINAE Rear-fanged Water Snakes
	BOIGINAE Boigine Snakes
Families	**ELAPIDAE** Cobras
	HYDROPHIIDAE Sea Snakes
Subfamilies	**LATICAUDINAE** Sea Kraits
	HYDROPHIINAE Sea Snakes
Families	**VIPERIDAE** Vipers
	CROTALIDAE Pit Vipers and Rattlesnakes

7 Turtles, Tortoises and Terrapins
CHELONIA (or TESTUDINATA)

Galapagos Giant Tortoise
Testudo elephantopus

Turtles, tortoises, and terrapins, with about 220 species, all belong to the order Chelonia, the main differences being anatomical; while some live on the land, others live mainly in water, either sea water or fresh water. They are the oldest type of living reptiles, older than the ancient fossil dinosaurs and other reptiles that are now extinct. The layman may tend to think that frogs, newts and their kin are reptiles, but they cannot be so classified; they are fundamentally different creatures, starting life in water and breathing by means of gills, as for example tadpoles. Even when their gills disappear and they venture onto the land, breathing by means of their lungs, for many species the skin plays an important part in breathing and must never be allowed to dry out; they are classified in the order Amphibia.

The Chelonia can be divided into two suborders, the Cryptodira and the Pleurodira; *kruptos* (Gr) secret, hidden; *deirē* (Gr) neck, throat; the neck, when withdrawn, is bent in a vertical plane and the head and neck are then not visible and thus protected; *pleura* (Gr) a rib, can mean the side; the neck is bent sideways when withdrawn and is not completely hidden.

In the USA, those living on land are called tortoises and we tend to follow this ruling in Great Britain, although the European Pond Tortoise *Emys orbicularis* lives partly in water and partly on land; turtles live in water although some spend part of their time on land. The word terrapin is derived from an American-Indian word *terrapene* a small turtle, and some American herpetologists consider the name terrapin should only be used for the Diamondback Terrapin *Malaclemys terrapin*. In Great Britain, we tend to think of terrapins as the small fresh water turtles kept as pets and often seen in pet shops in small aquaria. Exchange of books on natural history between English-speaking countries seems to have added to the confusion of the names and the Latin names do not help as some turtles have a Latin name (though actually a Greek word) that translates as a tortoise, e.g. the Spotted Turtle *Clemmys guttata*, where *klemmus* (Gr) means a tortoise.

The order is divided into 12 families and 2 subfamilies which now follow under their various headings.

Order   CHELONIA

khelōnē (Gr) a tortoise.

Suborder   CRYPTODIRA

kruptos (Gr) secret, hidden; *deirē* (Gr) neck, throat, the neck, when withdrawn, is bent in a vertical plane; the head and neck are then not visible and are thus protected.

Family   DERMATEMYDIDAE   1 species

derma (Gr) genitive *dermatos* skin, leather; *emus* (Gr) genitive *emudos* a freshwater turtle.

American River Turtle   *Dermatemys mawii*
Dermatemys see Family Dermatemydidae above. Occupying a family on its own because the anatomical features do not show that it belongs to any of the larger groups; the Latin name is

strange, as the shell is bony rather than leathery. It is quite a large turtle, growing to a length of about 40 cm (16 in). Inhabiting large rivers and ranging from Mexico to Guatemala.

Family CHELYDRIDAE 2 species

kheludros (Gr) an amphibious serpent, a kind of tortoise.

Snapping Turtle *Chelydra serpentis*
The translation of *kheludros* given above is as given in Liddell and Scott's *Greek-English Lexicon;* *serpens* (L) genitive *serpentis* a serpent; the name does not seem suitable, although it has a vicious bite and will even attack human beings. It is widespread in North America.

Alligator Snapper *Macrochelys temminckii* (= *Macroclemys*)
makros (Gr) long, large; *khelus* (Gr) a tortoise; Professor C. J. Temminck (1778–1858) was a Dutch zoologist and at one time Director of the Natural History Museum at Leiden in the Netherlands. This is one of the largest freshwater turtles known and may weigh up to 90 kg (200 lb), and grow to a length of 75 cm (29 in); it lives in rivers in the southern USA, particularly the Mississippi Basin.

Family KINOSTERNIDAE 9 species, possibly more

kineō (Gr) I move; *sternon* (Gr) the breast, chest; the central part of the plastron, i.e. the lower part of the shell, the chest, is integral with the upper part, the carapace; however two parts of the plastron, a forward part and a rear part, are hinged to it and can move up and down, so enclosing the head and legs when necessary.

Common Musk Turtle *Sternotherus odoratus*
sternon (Gr) the breast, chest; *therus* is derived from *thairos* (Gr) the hinge of a door; the plastron is hinged (see Family Kinosternidae above); *odor* (L) a smell; *-atus* (L) suffix mean-

ing provided with; it has a strong musky smell and is sometimes known as the 'stinkpot'! Inhabiting a large area of north-eastern USA, from Maine and southern Ontario, southwest to the Gulf States and Mexico.

Keel-backed Musk Turtle *S. carinatus*
carina (L) a keel; *atus* (L) provided with; a reference to the ridge along the central line of the back. Inhabiting the central southern area of the USA bordering on the Gulf of Mexico.

Mud Turtle *Kinosternon subrubrum*
Kinosternon see Family Kinosternidae, p. 40; *sub-* (L) under, almost; *rubrum* (L) red; a reference to the brown shell. Inhabiting the eastern area of the USA.

Family PLATYSTERNIDAE 1 species
platus (Gr) flat; *sternon* (Gr) the chest.

Big-headed Turtle *Platysternon megacephalus*
Platysternon see Family Platysternidae above; *megas* (Gr) wide, big; *kephalē* (Gr) the head; 'a flat-chested big-head'! A more correct interpretation would be 'big-headed flat-chest', as in strict nomenclature the generic name is a noun and the specific name is an adjective. A most peculiar animal, unlike any other turtle, so has to be classified on its own in its special family; the head is so big and wide that, at its base, it appears to be almost as wide as the body; it cannot be completely withdrawn into the shell and is heavily armoured; even more extraordinary, with its sharp claws, it climbs well and is reported to have been seen up trees! It inhabits southern Burma, ranging south to Vietnam, east along a coastal belt of southern China, and the Philippines.

Family EMYDIDAE 76 species
emus (Gr) genitive *emudos* a freshwater turtle.

Blanding's Turtle   *Emys blandingii*
Emys see Family Emydidae above. Named after William
Blanding, a nineteenth-century American herpetologist. It
inhabits the area of the Great Lakes of North America.

European Pond Tortoise   *E. orbicularis*
orbis (L) a circle; *orbiculus* a little circle or disc; *-aris* (L)
pertaining to; a small tortoise (sometimes called a Swamp
Turtle) with a length up to 25 cm (9½ in); it is 'at home' in
water or on the land. Inhabiting Europe, but not the northern
parts, and a small area in North Africa.

Painted Turtle   *Chrysemys picta picta*
khrusos (Gr) gold; *emus* (Gr) a freshwater turtle; a reference to
the unmarked golden-yellow plastron; *pictor* (L) a painter; a
reference to the brightly coloured markings on the carapace,
head, and legs. Inhabiting the Atlantic coastal area of the
USA; this is the nominate subspecies (see pp. 13 and 19).

(No recognised English name)   *C. p. belli*
The subspecies inhabit distinctive areas and will not necessarily
all be known as Painted Turtles; they are likely to have local
names. Dr J. G. Bell (1812–1889) was an American physician
and taxidermist; he accompanied Audubon on his Missouri
River expedition in 1843; J. J. L. Audubon (1785–1851) was
the famous American ornithologist and artist. This turtle
inhabits the eastern side of the Rockies and ranges south to
northern Mexico.

(No recognised English name)   *C. p. marginata*
margo (L) genitive *marginis* edge, border; *-atus* (L) provided
with; the carapace has bright red markings round the edge.
Ranging from the Alleghenies to the Great Lakes, North
America. This turtle and the two above are subspecies (see pp.
13 and 19).

Diamondback Terrapin   *Malaclemys terrapin*
malakos (Gr) soft, gentle; *klemmus* (Gr) a tortoise; *terrapene*
(American-Indian) a small turtle. This is a good example of the

confusion of names in the order Chelonia: here we have the English word terrapin, coined from the American-Indian *terrapene*, which translates as a small turtle, and the Greek word *klemmus* which means a tortoise. If any animal can be called a terrapin this is it; indeed some American herpetologists say that this is the one and only terrapin. 'Diamondback' is a reference to the shape of the plates of the carapace; it has a wide distribution along the Atlantic coast of the USA from Massachusetts, south to Florida, west to Texas and south to northern Central America.

Eastern Box Turtle   *Terrapene carolina*
terrapene see Diamondback Terrapin above. Named from Carolina, it is not confined to that part of the USA. The name Box Turtle refers to the two hinged lobes of the plastron (see Family Kinosternidae, p. 50) which can be closed to form an almost airtight box. It inhabits a large area of the USA from the east coast ranging west as far as Texas.

Spotted Turtle   *Clemmys guttata*
klemmus (Gr) a tortoise; *gutta* (L) a drop, a spot; *-atus* (L) provided with; a reference to the yellow and orange spots on the shell. Inhabiting the eastern part of the USA and the area of the Great Lakes.

Pacific Pond Turtle   *C. marmorata*
marmor (L) marble; *-atus* (L) provided with; having the appearance of marble. It is becoming rare as it has been extensively hunted and sold in markets for the decorative carapace. Inhabiting the border of the Pacific Ocean from Vancouver south to the border of Mexico.

Muhlenberg's Turtle   *C. muhlenbergii*
Named after G. H. E. Muhlenberg (1753–1815), an American botanist, in whose honour a plant genus was also named. This little turtle, only about 10 cm (4 in) long, spends a lot of its time on land in swampy areas, eating plant food; it is sometimes known as the Bog Turtle. Inhabiting New Jersey and surrounding areas on the eastern part of the USA.

Black Pond Turtle   *Geoclemys hamiltonii*
gē (Gr) the earth; *klemmus* (GR) a tortoise; the only species in
this genus, and probably more often seen on land than in the
water. Sir Ian Hamilton (1853–1947), was an officer in the
British Army who served in India and other places in the Far
East. Inhabiting a broad belt of country from north-western
India across to the River Ganges, where it is most common in
the Lower Ganges area.

(No recognised English name)   *Geoemyda spinosa*
gē (Gr) the earth; *emus* (Gr) genitive *emudos*, a freshwater turtle;
the turtles in this genus are mainly terrestrial; *spina* (L) a
prickle, or spine of certain animals; -*osus* (L) full of, augmented;
an allusion to the peculiar spines which arise from the plates of
the carapace. Inhabiting the Malay Peninsula, Sumatra and
Borneo.

(No recognised English name)   *G. pulcherrima*
pulcher (L) beautiful; *pulcherrima* very beautiful; 'A very beauti-
fully coloured turtle' (Grzimek). Inhabiting southern California, Mexico and the northern part of Central America.

Family   TESTUDINIDAE   35 species

testudo (L) genitive *testudinis* a tortoise; members of this family
are essentially land tortoises, visiting the water only for
drinking and bathing. Their feet are shaped for walking, quite
different from the paddle-shaped feet of the turtles.

West African Hinge-backed Tortoise   *Kinixys erosa*
kineō (Gr) I move; *ixus* (Gr) the back, the waist; an allusion to
the rear part of the carapace, which is hinged and can move up
and down; *erosus* (L) gnawed off, eaten into; an allusion to the
anterior plates of the plastron which have a 'scooped-out'
shape. Inhabiting a wide belt of country in central Africa from
Senegal to Lake Victoria.

Spider Tortoise   *Pyxis arachnoides*
puxis (Gr) a box made from box-wood, also any box; a reference

to the carapace and plastron forming a box; *arakhnēs* (Gr) a
spider; *-oides* (New L) from *eidos* (Gr) form, like; it is not the
shape of a spider; this is an allusion to the furrows on the
carapace which have a radiating pattern that looks something
like a spider's web. It is found only in Madagascar and inhabits
a small area on the western side.

Galapagos Giant Tortoise   *Testudo elephantopus*

testudo see Family Testudinidae, p. 44; *elephas* (Gr) genitive
elephantos an elephant; *pous* (Gr) the foot; 'elephant-footed'.
This enormous tortoise can grow to a length of 1.1 m
(3 ft 6 in), and live to a great age; it is difficult to establish
authentic figures because they can certainly live for well over
100 years, so no man is ever likely to observe one individual
from birth to death; even small tortoises can live to a great age.
Inhabiting the Galapagos Islands.

Starred Tortoise   *T. elegans*

elegans (L) neat, elegant; a reference to the very decorative
carapace with yellowish stars on a brown background. Owing
to the great heat at mid-day in its habitat, it hides in shady
places at this time; even in captivity in a temperate climate it is
determined to have its 'siesta' and crawls to a secluded place for
some pleasant inactivity! Inhabiting the southern part of the
Malay Peninsula, Sumatra, Java and Borneo.

Indian Ocean Giant Tortoise   *T. gigantea*

gigas (Gr) genitive *gigantos* a giant; another huge tortoise living
in quite a different place, in fact the other side of the world. It
may grow to a length of 1.2 m (nearly 4 ft), slightly larger even
than *elephantopus* (see above) and also living to a great age. It
inhabits the Seychelles and other islands in the Indian Ocean.

Common European Tortoise   *T. greaca*

greacus (L) of Greece; the specific name is misleading as it
inhabits quite large areas of southern Europe and northern
Africa, ranging east to the Caspian Sea. This species is often to
be seen for sale in pet shops.

Mediterranean Tortoise   *T. hermanni*
Professor J. Hermann (1738–1800) was a Professor of Natural
History in Strasburg. This tortoise, and *T. greaca*, above, are
very similar in habits and appearance, the difference being that
greaca has spurs on the thighs and *hermanni* has similar spurs on
the tail. Both species are sold as pets and can usually be found
in pet shops; if you have one, close examination will tell you its
species. It inhabits Greece, southern Italy, Corsica, Sardinia
and southern France.

Leopard Tortoise   *T. pardalis*
pardus (L) a leopard; *-alis* (L) relating to, similar to; an allusion
to the brightly coloured, almost orange carapace, covered with
dark spots. This is quite a large tortoise and can grow to a
length of about 65 cm (just over 2 ft); in the hot summer period
it usually aestivates. Widespread in Africa south of the Sahara.

African Spurred Tortoise   *T. sulcata*
sulcus (L) a furrow; *-atus* (L) provided with; the name is a
reference to the concentric furrows on the carapace. The
English name is an allusion to the horny spines on the legs;
usually slightly bigger than *pardalis* (above); it inhabits a large
area of the central part of Africa.

Desert Gopher Tortoise   *Gopherus agassizi*
The name gopher is used for various burrowing animals and is
said to derive from *gaufre* (Fr) a honeycomb, on account of the
many holes made by these animals; this is a burrowing tortoise.
It is named in honour of Professor J. L. R. Agassiz
(1807–1873), a Swiss zoologist and famous as a teacher of
zoology; he founded the Zoological Museum at Harvard
University. It inhabits the plains and deserts of southern USA
and northern Mexico; to escape the terrible heat of these areas
during the day, it digs long tunnels and only appears in the cool
of the evening and night to look for food, probably mostly cacti
as these contain moisture.

Texas Tortoise   *G. berlandieri*
J. L. Berlandier (1805–1851), was a Belgian botanist who was

employed by the Mexican government from 1827 to 1834; he was first collecting in Texas in 1827 and a genus of plants was named in his honour. Inhabiting southern USA and Texas.

Mexican Tortoise   *G. flavomarginatus*
flavus (L) yellow, golden; *margo* (L) genitive *marginis* a border, an edge; a reference to the yellow border of the carapace. This tortoise is now very rare and may even be extinct; it lives in southern USA and Mexico.

Family   CHELONIIDAE   5 species

khelōnē (Gr) a tortoise. This family consists of true marine turtles, built for swimming with powerful flippers in front; they only come ashore to lay their eggs.

Green Turtle   *Chelonia mydas*
Chelonia see Family Cheloniidae above. This is the edible turtle, which has been used as food for centuries and the one from which turtle soup is made, although some other species are eaten. One hears accounts of how they are turned on their backs in the hold of a ship so that they cannot move about or escape, and remain alive and therefore fresh. There is possibly a second reason for this procedure: when out of water the enormous weight of the shell restricts the breathing, so if not turned on their backs they would eventually die from suffocation. Inhabiting tropical seas, but becoming rare on account of being hunted for food, and now in danger of extinction.

Hawksbill   *Eretmochelys imbricata*
eretmos (Gr) an oak; *khelus* (Gr) a tortoise; a reference to the paddle-shaped flippers; *imbrex* (L) genitive *imbricis* a hollow tile; *-atus* (L) provided with. Imbricate means overlapping like the tiles on a roof; when young, the scales of the Hawksbill are noticeably overlapping, but become less so as it grows older. The hooked shape of the mouth gives rise to the English name. This is the turtle used for 'tortoise-shell', the translucent plates

of the carapace are very decorative; here is another instance of the confusing names in this order; a turtle is used to make tortoise-shell; Inhabiting tropical seas and in danger of extinction.

Loggerhead   *Caretta caretta*
caretta is a name coined from *carey* (Sp) meaning tortoise-shell. Turtles in the family Cheloniidae have big heads, too big to be drawn back into the shell, and the Loggerhead's head is even bigger than the other heads in this family. They have a wide distribution, ranging both north and south of the tropical seas.

Ridley   *Lepidochelys olivacea*
lepidus (L) elegant, neat; *chelys* (L) a tortoise; from *khelus* (Gr) a tortoise; pure Latin would be *testudo;* *oliva* (L) an olive; *-aceus* (L) similar to, like; 'olive-coloured'. This is the smallest sea turtle, less than 1 m (about 3 ft), and lives in the gulf of Mexico and other tropical seas. It is named after H. N. Ridley FRS who was in Brazil and also on the island of Fernando de Noronha in 1887.

Family   DERMOCHELYIDAE   1 species

derma (Gr) skin, leather; *khelus* (Gr) a tortoise; an allusion to the smooth skin which lacks any external shell, but has small bony plates embedded in the skin. There are several species of turtles which for purposes of classification have been put in a family on their own; this is because their general habits and behaviour, and particularly their anatomical structure, do not indicate that they belong to any of the larger groups.

Leatherback Turtle   *Dermochelys coriacea*
Dermochelys see Family Dermochelyidae above; *corium* (L) leather; *-aceus* (L) similar to, like. This turtle can grow to an enormous size, even bigger than the giant tortoises of the genus *Testudo;* a weight of 1 tonne has been suggested, but this is probably an exaggeration. It inhabits tropical seas, though it has been known to range as far north as the coast of England and Maine in the USA.

Family CARETTOCHELYIDAE   1 species

caretta is a name coined from *carey* (Sp) meaning tortoise-shell; *khelus* (Gr) a tortoise.

Papuan Turtle   *Carettochelys insculpta*
Carettochelys see Family Carettochelyidae above; *insculptus* (L) carved; the plates are pitted, which gives the appearance of being carved; it is sometimes known as the Pitted-shell Turtle. Inhabiting south-eastern New Guinea.

Family TRIONYCHIDAE   25 species, possibly more.

tria (Gr) three; *onux* (Gr) genitive *onukhos* a claw; see also Subfamily Trionychinae below. The carapace and the plastron are covered by a tough tissue of skin, hence the name 'soft-shelled'; they are freshwater turtles. For purposes of classification the family can be divided into two subfamilies; the reader should be reminded, as first mentioned in Chapter 4, that the suffix *-idae* denotes a family and *-inae* a subfamily.

Subfamily CYCLANORBINAE   6 species

kuklas (Gr) encircling; *orbis* (L) anything of a circular shape, a disk.

Indian Flap-shelled Turtle   *Lissemys punctata*
lissos (Gr) smooth; *emus* (Gr) a freshwater turtle; *punctum* (L) a puncture; *-atus* (L) provided with; spotted as with punctures; a reference to the yellow spots on the carapace and head. This turtle has two flaps on the plastron that close over the back legs when they are withdrawn. Inhabiting India, Sri Lanka and Burma.

Senegal Soft-shelled Turtle   *Cyclanorbis senegalensis*
Cyclanorbis see Subfamily Cyclanorbinae above; a reference to the flat and almost circular shell; *-ensis* (L) belonging to, usually a place; 'of Senegal'. It ranges along a broad belt of country in Africa from Senegal to the Nile.

Subfamily TRIONYCHINAE   19 species, possibly more

tria (Gr) three; *onux* (Gr) genitive *onukhos* a claw; they have five toes, but only three claws; the fourth and fifth toes are clawless and webbed to form paddles.

Long-headed Soft-shelled Turtle   *Chitra indica*
chitra is from *chitraka* (Sanskrit) having a speckled body; a reference to the spotted carapace; cf. cheetah, from its spotted coat; *indicus* (L) Indian. It has an unusually long and narrow head and neck. Inhabiting northern India, the Burma-Vietnam area and Malaysia.

Malayan Soft-shelled Mud Turtle   *Pelochelys bibroni*
pelos (or *pellos*) (Gr) dark-coloured, dusky; *khelus* (Gr) a tortoise. Inhabiting Malaysia.

Euphrates Soft-shelled Turtle   *Trionyx euphraticus*
Trionyx see Subfamily Trionychinae above; *-icus* (L) belonging to. Inhabiting Asian Turkey, Syria, Israel, Iraq and the River Euphrates to the Persian Gulf.

Spiny Soft-shelled Turtle   *T. spiniferus*
spina (L) a thorn, a spine; *fero* (L) I bear; an allusion to the spiny tubercles on the front of the shell. Inhabiting the St Lawrence and Mississippi Rivers in North America.

African Soft-shelled Turtle   *T. triunguis*
tri (L) three; *unguis* (L) a claw, a hoof; here we have the name first in Greek and then in Latin: 'a three-clawed three-claw'. Inhabiting a large area of central Africa from the Red Sea to the west coast.

Suborder PLEURODIRA

pleura (Gr) side; *deira* (Gr) neck, throat; the head and neck, when withdrawn, are bent sideways and part of the neck remains visible.

Family **PELOMEDUSIDAE** 10 species

pelos (or *pellos*) (Gr) dark coloured, dusky; Medusa was one of the three Gorgons, winged females with their heads covered in snakes instead of hair. In 1758, Linnaeus used several names from classical mythology for the names of certain species and it is probable that they were given without thought of any physical significance. Other zoologists, such as Lacépède, have followed his example; the genus *Pelomedusa*, from which the family takes its name, is the only one in the family and was named by Lacépède in 1788.

African Water-tortoise or Turtle   *Pelomedusa subrufa*
Pelomedusa see Family Pelomedusidae above; *sub-* (L) under, somewhat; *rufus* (L) red, ruddy; it is brownish rather than red. This turtle, although mainly aquatic, spends a certain amount of time roaming about on land looking for food, both plants and animals. When water holes dry up in the hot season, it buries itself in the mud at the bottom and sleeps; an example of aestivation, a 'summer hibernation'. The word is derived from *aestas* (L) summer, and *aestivo* (L) I spend the summer. Sometimes known as the Helmeted Terrapin, so now we have an animal that has, by various authors, been called a tortoise, a turtle and a terrapin, but there is no need to allow these English names to confuse you; simply remember that turtles and terrapins are essentially swimming animals and tortoises are walking animals; this rule has very few exceptions, when certain species are 'at home' in water and on land. As already mentioned, even the Latin names, when the Latin or Greek words are translated, can be confusing; the essential thing to remember is that these names, whether 'right or wrong', are *fixed labels*, internationally recognised throughout the world, and apply to only *one particular species*. *P. subrufa* is widespread in Africa south of the Sahara, including the western part of Madagascar, but not the very southern part of the mainland.

Adanson's Mud Turtle   *Pelusios adansonii*
pelusios is a coined name derived from *pēlos* (Gr) earth, mud; an

allusion to its habit of aestivating in a mud wallow (see African Water-tortoise above). Named after N. Adanson (1727–1806) a well-known French naturalist. 'His explorations in Senegal and collections and descriptions of its plants and animals were noteworthy' (E. C. Jaeger). Inhabiting Senegal and a wide belt of country stretching east as far as Uganda.

Madagascar Turtle *Podocnemis madagascariensis*
pous (Gr) genitive *podos*, the foot; *knēmis* (Gr) legging, a greave, armour between knee and ankle; an allusion to the enlarged scales at the back of each leg; *-ensis* (L) suffix meaning belonging to, usually a place. Surprisingly, although not known on the mainland of Africa it inhabits quite a large area on the western side of Madagascar.

Family CHELIDAE 23 species

khelus (Gr) a tortoise. The turtles in this family are known as Snake-necked Turtles; they are in the suborder Pleurodira; they bend the neck sideways to conceal the head under the front edge of the carapace, but the neck still remains visible.

Matamata *Chelys fimbriatus*
Chelys see Family Chelidae above; *fimbriatus* (L) fringed; a reference to the fleshy fringes of skin on the head and neck. Matamata is a Portuguese name derived from the Tupi language of a people of South America, especially those living in the Amazon Basin in Brazil. This turtle is largely crepuscular and nocturnal and inhabits a large area of northern South America, including the Amazon Basin.

Otter Turtle *Hydromedusa tectifera*
hudōr (Gr) water; when used as a prefix, *hudrō-;* Medusa was one of the three Gorgons (see Family Pelomedusidae, p. 61); *tectum* (L) a roof, a covering; *fero* (L) I bear, carry; a reference to the carapace. It inhabits the Argentine and Paraguay area of South America.

Brazilian Turtle *Phrynops rufipes*
phrunos (Gr) a species of toad; *ops* (Gr) the eye, face, can also

mean appearance; 'toad-like'; an allusion to the broad head, a deeply split mouth and almost no neck; *rufus* (L) red; *pes* (L) the foot; the legs and throat are a brownish red. Inhabiting a large area of Brazil based on the upper Amazon Basin.

Colombian Turtle   *Batrachemys dahli*
batrakhos (Gr) a frog; *emus* (Gr) a freshwater turtle; 'frog-like turtle'. Professor K. Dahl (1871–1951) was a Norwegian zoologist. Some herpetologists may consider that this species, only recently discovered, and named in 1958, should be in the genus *Phrynops* (above), together with other species in *Batrachemys*. Apart from the frog-like head there are other anatomical features that suggest this move. It is found only in a small area in Colombia, South America.

(No recognised English name)   *Platemys platycephala*
platus (Gr) wide, flat; *emus* (Gr) a freshwater turtle; the carapace and head are unusually flat; *kephalē* (Gr) the head. It inhabits the northern half of the Amazon Basin and ranges west as far as the Andes.

Australian Snake-necked Turtle   *Chelodina longicollis*
khelus (Gr) a tortoise; *dina* is derived from *deinos* (Gr) terrible; 'a terrible tortoise'; quite apart from the confusion of 'turtle' and 'tortoise', this is a misleading name as they are not usually vicious; 'These creatures are quite inoffensive and are not known to bite, however roughly handled' (Schmidt & Inger). It is quite a small turtle, only about 13 cm (5 in) long; *longus* (L) long; *collum* (L) the neck. Inhabiting a large area in the south-eastern part of Australia.

Krefft's Turtle   *Emydura krefftii*
emus (Gr) genitive *emudos* a freshwater turtle; *oura* (Gr) the tail; J. L. G. Krefft (1830–1881) was an Australian zoologist at the Natural History Museum in Sydney. This turtle inhabits a large area in north-eastern Australia, ranging from the Joseph Bonaparte Gulf to the Mackenzie River in Queensland.

8 Alligators, Crocodiles and the Gavial
CROCODYLIA (OR LORICATA)

Gavial
Gavialis gangeticus

These large reptiles all belong to the order Crocodylia, with only 21 known species. Many people want to know how one can tell the difference between a crocodile and an alligator; the two are, in fact, very similar. As a general rule the alligators are smaller, but this cannot be taken as a safe guide; the American Alligator has been known to grow to a length of 6 m (over 19 ft), which is about the same as a big crocodile, though the normal length of any species in the order Crocodylia probably does not exceed 5 m (16 ft). The most obvious difference is in the teeth; put in simple terms, in the crocodile the enlarged fourth tooth of the lower jaw can be seen when the jaws are closed; in the alligators this tooth is covered by the upper jaw.

Crocodiles spend much of their time floating at the surface and, though almost completely submerged, their anatomical structure allows them to continue to breathe. The ears, eyes and nostrils are the highest points on the head and project just above the surface, but a curious feature is that they have no lips covering the jaw bone, so the mouth when closed is not completely watertight. There is, however, a valve to close the throat and the air passages from the nostrils open behind this valve. The nostrils can also be closed by valves and the eyes have upper and lower lids and a nictitating membrane.

The only remaining species of gavial lives in India, mostly in the Ganges, the Indus and the Brahmaputra Rivers. It is sometimes known as a gharial and the two names have probably come about through a misspelling and there is some disagreement as to which is correct. In India, the name gharial is used but, as the family is recognised by scientists throughout the world as Gavialidae, it seems reasonable to use the name gavial. Although classified in the order Crocodylia it is not as closely related to the crocodiles and alligators as these two families are to each other. The food of the gavial consists almost entirely of fish and it has a long slender snout, quite different in appearance to that of the crocodiles; this enables it to catch fish by a sideways 'swipe' of the head and jaws.

The newly hatched crocodile, like the cobra, is an aggressive little creature. While helping Dr Hugh Cott with his research into crocodiles on the Nile in Uganda, we found that handling them required considerable caution to avoid a bite; this would not be serious, but with their sharp little teeth they could quickly draw blood, and if they did not attack us they attacked each other! The population of the Nile Crocodile is becoming much reduced through poaching, for the value of the skin, and if they become extinct it would be a sad blow for conservation, and for the Kabalega Falls National Park (formerly Murchison Falls). We did our best to assist their preservation by digging out crocodile nests that we knew were ready for hatching and, after taking measurements of the newly hatched young crocodiles, and other statistics, we put them in the river. In the normal course of events many must be lost through predators like the Nile Monitors, who dig up the nests and eat the eggs, or other predators who seize the youngsters on their way to the water.

Some crocodiles are known to guard the spot where the eggs are buried, and even take some care of the young after hatching, which is unusual behaviour for a reptile; with great delicacy, the mother crocodile will pick up the young in her huge jaws and carry them to the water, which keeps them safe during this perilous journey. It is estimated that only about

10% of baby turtles reach the water before being seized by predators, as they receive no parental care after the eggs are laid and buried.

Order   CROCODYLIA (or LORICATA)

crocodilus (L) a crocodile.

Suborder   EUSUCHIA

eu- (Gr) prefix meaning well, nicely, sometimes used to mean typical; *soukhos* (Gr) the name for the crocodile in one part of Egypt.

Family   ALLIGATORIDAE   7 species

alligator is derived from *el* (Sp) the, and *lagarto* (Sp) a lizard.

American Alligator   *Alligator mississippiensis*
Alligator see Family Alligatoridae above; *-ensis* (L) suffix meaning belonging to; inhabiting the River Mississippi and other rivers in the south-eastern part of the USA. The scientific name was originally misspelt as *mississipiensis* and the Code (Rules) of the International Commission on Zoological Nomenclature did not allow misspelling to be corrected when a name had appeared in print and become established, but this ruling has been modified recently and corrections of spelling are permitted.

Chinese Alligator   *A. sinensis*
sinae (Late L) Chinese; *-ensis* (L) suffix meaning belong to. This alligator was not known in the western world until late in the nineteenth century; it inhabits the lower reaches of the Yangtze River.

Spectacled Caiman   *Caiman crocodilus*
caiman (Sp) an alligator; *crocodilus* (L) a crocodile; it inhabits
Central America ranging south to the central part of South
America, particularly the valleys of the Orinoco and Amazon
Rivers. The English name refers to a ridge of bone on the
forehead connecting the eyes like the bridge of a pair of
spectacles.

Broad-nosed Caiman   *C. latirostris*
latus (L) wide, broad; *rostrum* (L) genitive *rostris* the snout.
Inhabiting eastern Brazil.

Black Caiman   *Melanosuchus niger*
melas (Gr) genitive *melanos* black; *soukhos* (Gr) the name for the
crocodile in one part of Egypt; *niger* (L) black. Inhabiting the
valleys of the Orinoco and the Amazon Rivers and ranging
north to the Guiana region.

Dwarf Crocodile   *Palaeosuchus palpebrosus*
palaios (Gr) ancient; *soukhos* see Black Caiman above; the
crocodiles and their kin are the sole survivors of the ancient
reptiles like the dinosaurs; *palpebra* (L) the eyelid; *-osus* (L)
suffix meaning full of; a reference to the prominent bony scales
above the eyes. This small crocodile is only about 1.5 m (4 ft 10
in) in length. Inhabiting northern and central South America.

Smooth-fronted Dwarf Crocodile   *P. trigonatus*
trigōnos (Gr) three-cornered; *-atus* (L) suffix meaning provided
with; a reference to the shape of the head. Inhabiting north and
central South America, particularly the Amazon Basin. The
name 'smooth-fronted' refers to the absence of the bony ridge
on the forehead connecting the eyes, as in the Spectacled
Caiman.

Family   CROCODYLIDAE   15 species
krokodeilus (Gr) a kind of lizard = *crocodilus* (L) a crocodile.

American Crocodile   *Crocodylus acutus*
Crocodylus see Family Crocodylidae above; *acutus* (L) sharp, pointed; a reference to the snout which tapers more to a point than that of other crocodiles. It has a wide distribution from Florida to Central America, north-western South America and the West Indies.

Slender-snouted Crocodile   *C. cataphractus*
kataphraktos (Gr) clad in armour, decked in. Inhabiting the Congo River valley.

Morelet's Crocodile   *C. moreletii*
Named after P. M. A. Morelet (1809–1892), the French naturalist and explorer who discovered this crocodile in Mexico in 1850; then followed a long period during which no more were seen, and zoologists were in some doubt as to whether this was a genuine new species or had now become extinct; however, several more were discovered and collected in 1923 near Belize. It inhabits south-eastern Mexico and Central America as far east as Guatemala.

Nile Crocodile   *C. niloticus*
-icus (L) suffix meaning belonging to; one of the big crocodiles and becoming scarce now in Africa due to poaching for the value of the skin; however, many can still be seen on the Nile below the Kabalega Falls (formerly Murchison Falls); it also inhabits Madagascar.

New Guinean Crocodile   *C. novaeguineae*
Distinguished by a rather narrow snout, it inhabits New Guinea and neighbouring islands of Indonesia, such as Sula, and some of the Philippine Islands.

Mugger   *C. palustris*
palustris (L) marshy, swampy; it inhabits India and Sri Lanka and is particularly well known in the River Ganges. The English name is from *magar* (Hindi) a water monster.

Salt-water Crocodile   *C. porosus*
porosis (Gr) a callosity; *-osus* (L) suffix meaning full of; a

reference to the double row of scaly humps on the upper surface of the snout. This is probably the largest crocodile known, with a length of about 7 m (23 ft), considerably longer than the Nile Crocodile which has always tended to have its length exaggerated by African explorers. This crocodile, being able to swim in the open sea, has a very wide distribution in south-east Asia, ranging from southern India, through the Philippines, the Moluccas and New Guinea, to the New Hebrides and south to northern Australia.

African Dwarf Crocodile   *Osteolaemus tetraspis*
osteon (Gr) a bone; *laimos* (Gr) the throat; a reference to the bony shields on the throat and belly; *tetra* (Gr) four; *aspis* (Gr) a shield; it has four large horny plates arranged in a square on the nape of the neck, known as 'nuchal shields'. This is one of the smallest crocodiles with a length of about 1.9 m (5 ft 10 in); it inhabits the tropical part of central and western Africa.

False Gavial   *Tomistoma schlegelii*
tomos (Gr) cutting, sharp; *stoma* (Gr) the mouth; it has a long narrow snout with many sharp teeth and thus *resembles* a true gavial, but in fact it is closely related to the crocodiles and so became known as the False Gavial; Professor H. Schlegel (1804–1884) was a Dutch zoologist and Director of Leiden Museum from 1858 until his death in 1884. It inhabits the Malayan Peninsula, Sumatra and Borneo.

Family   GAVIALIDAE   only 1 species
gavial is derived from *ghariyāl* (Hindi) a name for the crocodile.

Indian Gavial or Gharial   *Gavialis gangeticus*
Gavialis see Family Gavialidae above; *-alis* (L) suffix meaning pertaining to; *icus* (L) suffix meaning belonging to; in addition to the Ganges it also inhabits the Mahanadi and Brahmaputra Rivers. It is largely aquatic, the legs being weak and little used and the tail large and powerful and used in swimming; the long narrow snout enables it to make a sideways 'swipe' for catching fish, its main diet.

9 The Tuatara RHYNCHOCEPHALIA

Tuatara
Sphenodon punctatus

This strange reptile, although in outward appearance a lizard, has to be classified in an order and family of its own for it is not a lizard. It has been called 'a living fossil', being the only survivor of a very ancient order of reptiles, the Rhynchocephalia; it roamed the earth, along with the ancient turtles, even before the rise of the dinosaurs. There appears to have been no real change or modification in its anatomy for about 200 million years and it possesses a number of peculiar features. The structure of the skull is most unusual; it has abdominal ribs; it has no copulatory organ, so fertilization has to be achieved by the male pressing its cloaca to that of the female; it cannot tolerate much heat and, although most reptiles are active with a body temperature of between 25°C and 38°C (77°F and 100°F), the Tuatara prefers a temperature of about 12°C (54°F); the arrangement of the teeth is unlike that of any other animal.

Order RHYNCHOCEPHALIA

rhunkhos (Gr) the beak, snout; *kephalē* (Gr) the head; the front part of the skull is elongated into a snout and sometimes the animal is known as a 'beak-head'.

Suborder **SPHENODONTIDA**

sphen (Gr) a wedge; *odous* (Gr) genitive *odontos* a fang, tooth; *-ida* (New L) suffix used for zoological group names; a reference to the two wedged-shaped teeth situated in front of the maxillary bone; the teeth as a whole have a very unusual formation.

Family **SPHENODONTIDAE**   1 species

Tuatara   *Sphenodon punctatus*
Sphenodon see Suborder Sphenodontida above; *punctum* (L) a hole, a prick; *punctatus* (L) spotted as with punctures; the young Tuatara has light spots on the throat, body, and legs, especially noticeable after shedding the skin, but these spots gradually fade with age. Originally inhabiting New Zealand, it is no longer found on the two main islands, but inhabits a number of small islands, mostly off the northern coast of North Island. The English name Tuatara arises from the native name for the animal and refers to the spiny crest along its back; *tua* (Maori) the back; *tara* (Maori) a spine.

10 Lizards SQUAMATA: SAURIA

Jackson's Chameleon
Chameleo jacksonii

The lizards, of the order Squamata, suborder Sauria, with a total of about 3,000 species, come in a great variety of shapes and sizes; a number of species have no legs and thus have the appearance of a snake. They range in size from the tiny Caribbean Gecko *Sphaerodactylus elegans*, about 4 cm ($1\frac{1}{2}$ in) long, to the huge Komodo Dragon *Varanus komodoensis* at 3 m (almost 10 ft). They have a world-wide distribution, although mostly in the warm regions; no reptiles inhabit the Arctic or Antarctic areas.

As already mentioned in Chapter 2, there are a number of anatomical features, some easily seen or demonstrated, that prove the creatures to be lizards although they may look like snakes: they have eyelids (except the geckos) and most have the ability to break off their tail (autotomy) as a means of escape, which then grows again. If lizards are kept in captivity, and the smaller species make most interesting, active and quite intelligent pets, they soon become tame; I do not advise an experiment to see whether the tail breaks off! Although the tail will grow again, it will not be so long and decorative as originally.

I was interested to see, on one occasion, an experienced

herpetologist pick up a lizard by its tail when he was catching some for me to take home; he said they seem to know when they are in danger and, if you hold the tail lightly, they will not shed it. I have seen two male lizards fighting over territory and one shed its tail, which continued to wriggle vigorously and arch about; this might serve to distract the attention of a predator so that the lizard has time to escape. In captivity, lizards should, if possible, be kept in an outdoor vivarium (which I prefer to call a 'herpetarium' although this is not a recognised word), where they have plenty of room to run about and where they can get as much sun as they need; this is essential if they are going to remain in good health. However, there must be rocks and plants under which they can hide to escape the heat of the mid-day sun if necessary, or the blood may become too hot and this would be fatal (see Chapter 5 about blood-heat control in reptiles).

Order SQUAMATA

squama (L) a scale.

Suborder SAURIA lizards

saura (Gr) genitive *sauros* a lizard.

Family GEKKONIDAE about 550 species

gekoq is a Malayan word; the name gecko, or tokay, is derived from the sound made by the Asian species; their ability to make little chirps and croaks is unique among lizards. Furthermore, they are able to crawl up very smooth surfaces and even hang upside down on ceilings. It is now established that this is not done by means of suction pads on the feet but by pads furnished with thousands of minute hooks, too small to be seen with the naked eye, which cling to the tiny irregularities of any surface, even glass.

Common House Gecko   *Hemidactylus mabouia*
hemi (Gr) half; *daktulos* (Gr) a finger or toe; a reference to the
clinging pads on the under surface of the toes being divided into
two halves, separated longitudinally along the centre line; they
are sometimes known as 'half-toes'. *mabuya* (New L) from an
American-Spanish word for a type of lizard. Inhabiting Central America and northern South America and Africa south of
the Sahara, including Madagascar.

Turkish Gecko   *H. turcicus*
-icus (L) suffix meaning belonging to; it inhabits the eastern
Mediterranian coast, the Red Sea area and other parts of the
Near East.

European Leaf-fingered Gecko   *Phyllodactylus europaeus*
phullon (Gr) a leaf; *daktulos* (Gr) a finger or toe; an allusion to
the broad leaf-like clinging pads; *europaeus* (L) of Europe.
Inhabiting Corsica, Sardinia and neighbouring islands.

Naked-toed Gecko   *Gymnodactylus pulchellus*
gumnos (Gr) naked; *daktulos* (Gr) a finger or toe; an allusion to
the toes which do not have the clinging pads; *pulcher* (L)
beautiful; *-ellus* (L) diminutive suffix: 'a little beauty'. Inhabiting the southern part of eastern Europe and ranging east to
northern Arabia and Iran.

Dwarf Gecko   *Lygodactylus capensis*
lugos (Gr) a pliant twig, a willow twig, i.e. flexible; *daktulos* (Gr)
a finger or toe; an allusion to the curling movement of the toes
which enables the clinging pads to grip and release; *-ensis* (L)
belonging to; named from Cape Province, it inhabits the
western part of Africa. Although named Dwarf Gecko, it is not
the smallest gecko and is about 8 cm (3 in) long.

Spotted Gecko   *Pachydactylus maculatus*
pakhus (Gr) thick, fat; *daktulos* (Gr) a finger or toe; a reference to
the tips of the toes being expanded like clubs; *macula* (L) a spot,
or mark; *-atus* (L) provided with; it has dark spots on a

brownish background. Inhabiting the eastern part of South Africa.

Fan-footed or House Gecko *Ptyodactylus hasselquistii*

ptuon (Gr) a winnowing shovel or fan; *daktulos* (Gr) a finger or toe; an allusion to the clinging pads which are wide and expanded like fans. F. Hasselquist (1722–1752) was a Swedish naturalist living in Palestine from 1747 to 1752, where doubtless he died from one of the diseases of the East for which no cure was known in those days. This gecko ranges from Algeria across the northern part of Africa to Iraq and Iran, where it often makes its home in houses.

Lizard-toed Gecko *Saurodactylus fasciata*

saura (Gr) a lizard; *daktulos* (Gr) a finger or toe; as the English name suggests, it has lizard-like toes without the usual clinging pads, and so will not be seen on smooth walls or ceilings of houses; it is a ground dweller; *fascia* (L) a band or girdle; *-atus* (L) provided with; a reference to the dark marking on the back. Inhabiting Algeria and Morocco.

Day Gecko *Phelsuma quadriocellata*

Named by Gray in 1831 after 'a Mr Phelsum, but herpetological records do not give any details about this naturalist; *quadruus* (L) fourfold; *ocellus* (L) a little eye; *-atus* (L) suffix meaning provided with; 'having four little eyes'; this is an allusion to four black spots outlined with blue on the sides of the body, one behind each fore leg and one just in front of each hind leg. Dr J. E. Gray FRS (1800–1875) was a zoologist at the British Museum (Natural History) in London from 1840 to 1874. This gecko is often seen out hunting during the daytime; it inhabits central and southern Madagascar.

Leaf-tailed Gecko *Uroplatus fimbriatus*

oura (Gr) the tail; *platus* (Gr) flat, wide; a reference to the peculiar tail which from a normal base spreads out flat and wide, 'leaf-tailed'; *fimbriatus* (L) fringed; the sides of the body and legs have a scaly flange or fringe which is pressed down when the animal is at rest, thus reducing any shadow and making it very inconspicuous. Inhabiting Madagascar.

Tokay or Common Gecko   *Gekko gekko*

gekko the name is onomatopoeic (see Family Gekkonidae, p. 73); the whole family Gekkonidae takes its name from this genus, although many species do not make the peculiar croaking noises. The double name, i.e. using the generic name again as the specific name, is known as a tautonym; for an explanation of this strange feature of nomenclature see TAUTONYMS, p. 13. This gecko is widespread and frequently seen is south-east Asia, including India, the Burma-Vietnam area, Malaysia, Indonesia and the Philippines. Visitors to this part of the world are likely to be surprised and startled in the middle of the night by a loud bark, apparently coming from immediately overhead! It is indeed overhead and made by this gecko crawling about on the ceiling hunting for insects.

Emerald Gecko   *G. smaragdinus*

smaragdos (Gr) a precious stone; *-inus* (L) like, belonging to; according to Liddell and Scott's *Greek–English Lexicon*: 'A precious stone of a light green colour; probably not the same as our emerald, but a semi-transparent stone like the *aqua marina*'; a reference to the colour of this gecko. Inhabiting the Philippines.

Kuhl's Gecko   *Ptychozoon kuhli*

ptux (Gr) genitive *ptukhos* a fold, a layer; *zōon* (Gr) a living being, an animal; a reference to the fold of skin which extends along the sides of the head, body, thighs, and tail; this can be stretched out and may assist the gecko when jumping or falling, like a glider; it may also assist in camouflage (see *Uroplatus fimbriatus*, p. 75). Dr Heinrich Kuhl (1796–1821), a German naturalist, was in the Dutch East Indies (now known as Indonesia) in 1820 and 1821; a family of fishes, Kuhliidae, is also named after him. Inhabiting south-east Asia including Indonesia.

Panther Gecko *Eublepharis macularius*
eu- (Gr) prefix meaning well, nicely; sometimes used to mean typical; *blepharon* (Gr) the eyelid; 'well-made eyelid'; 'typical eyelid'; most geckos have eyelids that are fused and transparent, forming a permanent covering over the eye; this gecko has movable eyelids; *macula* (L) a spot, a mark; *-arius* (L) suffix meaning pertaining to; a reference to the black spots on the head, body and tail, and this also gives it the English name. In fact the name panther is frequently used for the unspotted type of leopard, but a panther and a leopard are the same animal. Inhabiting Afghanistan, Pakistan and western India.

Caribbean Gecko *Sphaerodactylus elegans*
sphaira (Gr) a ball, a sphere; *daktulos* (Gr) a finger or toe; a reference to the round clinging pads on the toes; *elegans* (L) neat, elegant; possibly the smallest living reptile, it measures only 4 cm ($1\frac{1}{2}$ in) total length, although the Nossi Bé Dwarf Chameleon has been recorded as measuring 3.2 cm ($\frac{1}{4}$ in). This gecko inhabits The Antilles.

Kidney-tailed Gecko *Nephrurus laevis*
nephros (Gr) the kidneys; *oura* (Gr) the tail; the peculiar tail is not only bulbous but has a small knob at the end and is really not much like a kidney; the function of the knob is not known; *laevis* (L) = *levis* (L) light (in weight); can also mean quick, nimble. Inhabiting desert areas in the western part of Australia.

Fat-tailed Gecko *Oedura marmorata*
oidos (Gr) a swelling; *oura* (Gr) the tail; the tail is probably used as a store of fat when food is scarce; *marmor* (L) marble; *-atus* (L) provided with; 'marbled'; a reference to the colour pattern of white marking on a brown background. One of the larger geckos with a length up to 18 cm (7 in). Inhabiting Australia.

Horned Gecko *Phyllurus cornutus*
phullon (Gr) a leaf; *oura* (Gr) the tail; a very unusual tail that is sometimes an almost flat rectangular shape spreading out immediately behind the narrow base; *cornu* (L) the horn of an

animal; *cornutus* (L) horned; it has two small horns, one on each side of the head. Inhabiting the north-eastern part of Australia.

Family PYGOPODIDAE   13 species
pugē (Gr) the rump, buttocks; *pous* (Gr) genitive *podos* the foot; 'rump-footed'; this small family of lizards has no fore legs and the hind legs consist only of small scaly flaps at the rear end of the body, at about the level of the cloaca. Peculiar creatures, they have been given various English names: Scaly-footed Lizards, Flap-footed Lizards and Snake Lizards. They are said to behave like snakes as part of a defensive behaviour, but two things 'give the show away': they have small external ear openings and small flaps instead of hind legs; they are also able to shed their tails but this attribute cannot be demonstrated without damaging the small harmless creature. For anatomical reasons, they appear to be closely related to the geckos.

Common Scaly-foot   *Pygopus lepidopodus*
Pygopus see Family Pygopodidae above; *lepis* (Gr) a scale; *lepidōtos* (Gr) covered with scales; *pous* (Gr) genitive *podos* a foot. This is one of the largest species and may be 65 cm (26 in) long; it is widespread in Australia except the most eastern part.

Western Scaly-foot   *P. nigriceps*
niger (L) black; *ceps* (New L) from *caput* (L) the head; a reference to the black marking round the neck. Inhabiting western Australia.

(No recognised English name)   *Aprasia pulchella*
aprasia (Gr) is a word meaning 'no sale', which seems a very odd name for a genus of lizards; it appears that, when Gray gave this name to the genus in 1839, in his description of the animal he gave no indication of the derivation of the name; *pulcher* (L) beautiful; *-ellus* (L) a diminutive suffix: 'a little beauty'. It inhabits western Australia. Dr J. E. Gray FRS (1800–1875) was an English zoologist and at one time keeper at the British Museum (Natural History) in London.

Family DIBAMIDAE   3 species

dibamos (Gr) on two legs; the females have no legs and the males have hind legs that are only vestigial, consisting of fin-like flaps.

(No recognised English name)   *Dibamus novaeguineae*
Dibamus see Family Dibamidae above; *novaeguineae* (New L) New Guinea. This strange creature is blind and wormlike and yet, like a lizard, it can shed its tail to escape from predators if necessary. It inhabits New Guinea and the Sumatra, Java, Flores group of islands.

Family IGUANIDAE   about 650 species

iguana is derived from *iwana*, meaning a lizard; it is Spanish-Arawak, the language of a people of the West Indies. For purposes of classification, the Iguanidae can be divided into five subfamilies; the reader should be reminded, as first mentioned in Chapter 4, that the suffix *-idae* denotes a family and *-inae* a subfamily.

Subfamily SCELOPORINAE

skelos (Gr) the leg; *poros* (Gr) a passage through the skin; a reference to the large femoral pores.

Western Fence Lizard   *Sceloporus occidentalis*
Sceloporus see Subfamily Sceloporinae above; *occidentalis* (L) western. Like most lizards it has a habit of basking in the sun and is often seen on the top of fence posts; it inhabits large areas of western and southern North America, ranging south into Central America.

Tree Lizard   *Urosaurus ornatus*
oura (Gr) the tail; *saura* (Gr) a lizard; an allusion to the typical lizard tail; *ornatus* (L) ornament, decoration; it has dark spots and bars on a grey-brown background, which is good camouflage; it is very difficult to see if it remains still on a tree trunk or branch; the male is decorated with blue or green spots

on the belly. Inhabiting southern California, Arizona and northern Mexico.

Banded Rock Lizard   *Petrosaurus mearnsi*
petra (Gr) a rock; *saura* (Gr) a lizard; the usual habitat is amongst rocks where it basks in the sun and can easily hide from predators; Dr E. A. Mearns (1856–1916) was a Lieutenant Colonel in the US Medical Corps in Mexico from 1892 to 1894; he was a keen naturalist and student of wildlife and at least two birds have been named in his honour. The English name of this lizard refers to a black band across the neck; it inhabits a restricted area from southern California to northern Mexico.

Greater Earless Lizard   *Holbrookia texana*
Girard named this lizard in 1851 in honour of Dr J. E. Holbrook (1794–1871), a well-known herpetologist and the author of a number of books on American reptiles. (It is unusual for a naturalist to be commemorated by having his name used for a genus in zoology, being more common in botany, for example the well-known Butterfly Bush *Buddleia davidii* named after Adam Buddle, the English botanist.) The English name is misleading as it is only about 8 cm (3 in) long; furthermore it has ears but there are no visible external ear openings; it inhabits the south-western part of the USA. Charles Girard (1822–1895) was another well-known American herpetologist; a subspecies of the Striped Whipsnake *Coluber taeniatus girardi* was named in his honour.

Zebra-tailed Lizard   *Callisaurus draconoides*
kallos (Gr) beauty; *kalli-* (Gr) is used as a prefix to denote beauty; *saura* (Gr) a lizard; probably not more attractive than other lizards in this family except for the tail which is decorated with black bands; *drakōn* (Gr) a dragon; *-oides* (New L) from *eidos* (Gr) apparent shape, resemblance. Inhabiting south-western USA.

Texas Horned Lizard   *Phrynosoma cornutum*
phrunos (Gr) a species of toad; *sōma* (Gr) the body; sometimes

known as the Horned Toad, it has a toad-like body; *cornu* (L) the horn of animals; *cornutus* (L) horned; a reference to the spiky horns on the rear part of the head. A most peculiar lizard in appearance and yet quite common in Texas where it can sometimes be seen even in towns and gardens, and other parts of the western USA and Mexico. They have an extraordinary defensive weapon when angry or frightened which consists of squirting blood from the eyes! It is not known how this can be of any use for protection or attack.

Flat-tailed Horned Toad *P. macalli*
G. A. McCall (1802–1868) was a Brigadier General in the US Army; the name was given by Cope in recognition of his keen interest in wildlife and his extensive collections for the Smithsonian Institution. It is also known as the Flat-tailed Horned Lizard and the Latin name has been given as *P. m'calli*, but the apostrophe is forbidden in modern usage (see Chapter 1 under USE OF LATIN AND GREEK). E. D. Cope (1840–1897) was a well-known American zoologist and at one time was the editor of *The American Naturalist*. This Horned Lizard inhabits the western part of the USA and Mexico.

Collared Lizard *Crotaphytus collaris collaris*
krotōn (Gr) a louse, a tick; *phuton* (Gr) a creature, an animal; an allusion to the insectivorous diet; *collum* (L) the neck; *collaris* (L) pertaining to the neck, a collar; a reference to two black bands across the back of the neck; it inhabits the country bordering the Mississippi River. This is the nominate subspecies (see Chapter 2).

Bailey's Collared Lizard *C. c. baileyi*
This subspecies was named after V. O. Bailey (1864–1942), at one time chief field naturalist with the US Biological Survey and author of several books on the wildlife of America. Inhabiting an area to the west of that occupied by *C. c. collaris* above.

Subfamily TROPIDURINAE

tropis (Gr) genitive *tropios* which later became *tropidos* a ship's keel; *oura* (Gr) the tail; a reference to a ridge along the upper surface of the tail.

(No recognised English name) *Tropidurus torquatus*
Tropidurus see Subfamily Tropidurinae above; *torquatus* (L) wearing a collar; a reference to the fold of skin around the neck. Inhabiting the Amazon area.

(No recognised English name) *Strobilurus torquatus*
strobilos (Gr) anything twisted, that turns round; can mean a fir cone; *oura* (Gr) the tail; the tail is unusually round and very spiny, giving the appearance of a fir cone; *torquatus* (L) wearing a collar; see *Tropidurus torquatus* above. Inhabiting a small area in the eastern part of Brazil, south of the Amazon.

Spiny-tailed Iguanid *Uracentron azureum*
oura (Gr) the tail; *kentron* (Gr) a spike; *azureus* (New L) blue, derived from *azul* (Sp) blue; in life this lizard is a rather bright green, but specimens preserved in museums change to a more bluish colour. Inhabiting the Amazon Basin area in Brazil.

(No recognised English name) *Plica umbra*
plico (L) I fold; the skin at the side of the neck is folded and can be distended when the lizard is threatened or angry; *umbra* (L) a shade, can mean a shady place; a reference to its habitat, as it seems to prefer shady places in the rain forest rather than bright sunlight. Most lizards require some heat from the sun before they become active and feel the need for food, but this lizard and some related species can feed at lower temperatures. Inhabiting the rain forests of the lower Amazon Basin.

Santiago Smooth-throated Lizard *Liolaemus altissimus*
leios (Gr) smooth; *laimos* (Gr) the throat; *altus* (L) high; *altissimus* (L) very high; the habitat of this lizard is unusually high, just below the snow line, but it is still able to find insects at that height. Inhabiting the mountains to the east of Santiago.

Magellan Smooth-throated Lizard   *L. magellanicus*
The Magellan Straits separate Tierra del Fuego from the
mainland at the southern end of South America; this lizard
inhabits Tierra del Fuego and it is safe to assume that no lizard
lives nearer to the South Pole.

Narrow-tailed Lizard   *Stenocercus crassicaudatus*
stenos (Gr) narrow; *kerkos* (Gr) the tail; *crassus* (L) thick, heavy;
caudatus (L) having a tail; this apparently contradictory name
was caused by a change of the generic name: originally named
Scelotrema crassicaudatum, for anatomical reasons it was transferred to the genus *Stenocercus;* by observing the Rules (now known
as the Code) of Zoological Nomenclature, the contradiction
could not be avoided by changing the specific name as this is
forbidden (see p. 12). Most *Stenocerus* species do have a narrow
tail but that of *S. crassicaudatus* is robust and rather spiny.
Inhabiting Ecuador and Peru.

Weapon-tailed Lizard   *Hoplocercus spinosus*
hoplon (Gr) a tool, a weapon, usually a weapon of war; *kerkos*
(Gr) the tail; *spina* (L) a prickle or spine of animals; *-osus* (L)
suffix meaning full of; the tail is short and heavy and the scales
are very spiny. Inhabiting southern Brazil.

Subfamily   IGUANINAE

iguana is derived from *iwana*, meaning a lizard; it is Spanish-
Arawak, the language of a people of the West Indies.

West Indian Iguana   *Iguana delicatissima*
Iguana see Subfamily Iguaninae above; *delicatus* (L) giving
pleasure, delicate; *delicatissimus* (L) very delicate; a reference to
the tail which easily breaks off if seized by a predator, a
characteristic known as autotomy. Inhabiting the islands of the
West Indies.

Common Iguana   *I. iguana*
iguana see Subfamily Iguaninae above: This is a big lizard, with
a high crest, and is quite an impressive creature; it can grow to

a length of 2.2 m (about 7ft 6 in). It is widespread and inhabits Central America, the islands of the West Indies and the northern half of South America. For an explanation of the name *Iguana iguana* see TAUTONYMS, p. 13.

Rhinoceros Iguana　*Cyclura cornuta*

kuklos (Gr) round, circular; *oura* (Gr) the tail; an allusion to the tail which is circular in form, quite heavy and powerful, and can inflict considerable damage with a 'side-swipe' action; it is a fairly large lizard with an overall length of about 70 cm (2 ft 3 in); *cornu* (L) the horn of animals; *cornutus* (L) horned; it has horns on the nose like a rhinoceros. Inhabiting Haiti.

Marine Iguana　*Amblyrhynchus cristatus*

amblus (Gr) blunt; *rhunkhos* (Gr) the snout, beak; *crista* (L) a tuft or crest, usually used for the crest on the head of animals; *cristatus* (L) crested; it has an unusually short blunt nose and a spiky crest on the top of the head and extending along the back. This lizard is a good swimmer, with a tail that is flattened like the blade of an oar, in contrast to the rounded tail of *Cyclura*, and the legs are not used for swimming; it is herbivorous and searches in the sea for seaweed and algae. Inhabiting the Galapagos Islands.

Land Iguana　*Conolophus subcristatus*

kōnos (Gr) a cone, or peak of a helmet; *lophos* (Gr) a crest; *sub-* (L) a prefix meaning under, or somewhat; *crista* (L) a tuft or crest, usually used for the crest on the head of animals; *cristatus* (L) crested; it has a small raised cone on the head and the crest along the back is smaller than that of *Amblyrhynchus*. This iguana inhabits the Galapagos Islands, and Darwin studied its habits while he was making a special study of the finches of these islands, which are sometimes known as 'Darwin's Finches'. C. R. Darwin FRS (1809–1882) was the zoologist who became famous for his work on the theory of evolution by natural selection; today there are some zoologists who tend to disagree with his theory.

Fijian Iguana   *Brachylophus fasciatus*
brachus (Gr) short; *lophos* (Gr) a crest; a reference to the short spines that form the crest along the back, much smaller than those of other iguanas; *fascia* (L) a band or girdle; *fasciatus* (L) banded; it has broad dark bands round the green body and tail. Inhabiting the Fiji and Tonga Islands in the South Pacific, it is rare and now considered to be an endangered species.

Desert Iguana   *Dipsosaurus dorsalis*
dipsa (Gr) thist; *saura* (Gr) a lizard; 'a thirsty lizard'; as it lives in desert areas, Baird and Girard, who named this iguana in 1852, evidently thought that drinking water would be a problem for it; eating plants, probably cacti which contain water, would help to solve this problem; it also eats insects. Records show that this remarkable lizard can survive in temperatures of over 40°C (about 100°F) which would certainly kill most reptiles, as already mentioned in Chapter 5. Professor S. F. Baird (1823–1887) was a very well-known and hard-working American ornithologist; several birds have been named after him. C. Girard (1822–1895) was an American herpetologist; a subspecies of snake is named in his honour. This iguana inhabits the desert areas of south-western USA.

Chuckwalla   *Sauromalus ater*
saura (Gr) a lizard; *omalos* (Gr) level, flat; a reference to the broad flat body; *ater* (L) black; in the male, the head, neck, legs and front part of the body are dark or black, and the rear part is reddish grey with a yellow tail. The name chuckwalla is derived from a Mexican-Spanish name, *chacahuala*; it inhabits the desert areas of the south-western USA.

Subfamily   BASILISCINAE
basiliskos (Gr) a little king, a chieftain.

Common Basilisk   *Basiliscus basiliscus*
Basiliscus see Subfamily Basiliscinae above. Known as a basilisk, it takes its name from the mythical creature which was portrayed with a crown-like crest on the head and whose

glance could kill, as could its fiery breath; a strange name for a harmless lizard but it has a casque on the head which no doubt was the reason that Linnaeus, in 1758, gave it this name.

Helmeted Lizard   *Corytophanes cristatus*
korus (Gr) genitive *koruthos* a helmet; *phainō* (Gr) I show; *crista* (L) usually a tuft on the head of animals; *cristatus* (L) crested; it has a bony casque or helmet on the head and a crest of skin on the neck; the size of these can be increased to form a display when the animal is angry or threatened. It inhabits Honduras in Central America and ranges south to the north-western part of Colombia, South America.

Subfamily   ANOLINAE

anoli (West Indian) a lizard.

Long-legged Lizard   *Polychrus marmoratus*
polukhroos (Gr) many-coloured; a reference to its ability to change colour; *marmor* (L) marble; *marmoratus* (L) marbled; the brown back with pale markings gives an impression of marble. The unusually long hind legs enable it to leap rather like a frog and so to jump from branch to branch of a tree; sometimes known as the Bush Lizard, it is widespread in tropical South America.

Brazilian Tree Lizard   *Enyalius catenatus*
enualios (Gr) warlike, ferocious; a reference to its behaviour when meeting other male lizards; normally a dark brown colour, when angry it changes to green; *catena* (L) a chain; can mean a chain of gold or silver worn by women as an ornament; *-atus* (L) provided with; an allusion to the 'chains' of white spots running in various directions across the back. Inhabiting south-eastern Brazil.

Patagonian Lizard   *Diplolaemus darwinii*
diploē (Gr) a fold, doubling; *laimos* (Gr) the throat, gullet; an allusion to the folds of skin round the neck; named after Charles R. Darwin (1809–1882), the English zoologist and author of

the theory of organic evolution. Patagonia is the name given to the southern part of Argentina; this lizard was discovered there by Darwin.

(No recognised English name)   *Anisolepis iheringii*
anisos (Gr) unequal, uneven; *lepis* (Gr) a scale; a reference to the type of scales on the body, some large and some small. Professor H. von Ihering (1850–1930) was Director of the Museum in São Paulo, Brazil from 1893 to 1916. This lizard inhabits Brazil.

Keel-toed Lizard   *Tropidodactylus onca*
tropis (Gr) genitive *tropios* which later became *tropidos* a keel; *daktulos* (Gr) a finger or toe; a reference to the lack of adhesive pads on the toes; *onkos* (Gr) (= *ogkos*) mass, bulk; can mean a tumour; a reference to the unusually large dewlap on the throat of the male. Inhabiting a belt of country along the north-eastern coast of South America from Venezuela to Guinea.

False Chameleon   *Chamaeliolis chamaeleontides*
khamaileon (Gr) a chameleon; *Chamaeliolis* is a name coined by Duméril in 1837 when this lizard was discovered and named; *chamaeleontides* has the suffix -*ides* (L) a relationship; the 't' is evidently inserted for the sake of euphony. It is not a chameleon, but similar in appearance; its slow deliberate movements, quite unlike a normal lizard, the prehensile tail and the fact that it lives in trees all add to the illusion. It inhabits Cuba. Professor A. M. C. Duméril (1774–1860) was a well-known French herpetologist and author.

Carolina or Green Anole   *Anolis carolinensis*
anoli (West Indian) a lizard; -*ensis* (L) suffix meaning belonging to; it is not confined to Carolina and inhabits other south-eastern states of the USA. Like other anoles, it can change colour and is sometimes known as the American Chameleon, but it is not a chameleon; due to anatomical reasons, the change of colour is much slower in the anoles than it is in the chameleons; in the former, the change is controlled by chemical changes in the blood and in the latter by the nerves in the skin.

Knight Anole   *A. equestris*
equestris (L) relating to horsemen; can mean relating to the
knights; a reference to the head, which is casque-like and bears
some resemblance to an armoured helmet. Anoles are widespread in Central America, the Greater and Lesser Antilles and
the northern half of South America.

Leaf-nosed Anole   *A. phyllorhinus*
phullon (Gr) a leaf; *rhis* (Gr) genitive *rhinos* the nose; it has a
peculiar flat formation of scaly tissue that extends beyond the
nose but the function of this is not known. Inhabiting Brazil.

Sagré's Anole   *A. sagrei*
R. de la Sagré (1801–1871) was a Cuban historian and at one
time Director of the Botanical Gardens in Havana; this anole
was named in his honour by Duméril in 1837. It inhabits Cuba.

Family   AGAMIDAE   about 300 species

agama (New L) derived from the Dutch-Guianan language.

Common Agama   *Agama agama*
Agama see Family Agamidae above. The repetition of the
generic name for the specific name is caused by a change of the
generic name when some revision of the classification took
place; it does not necessarily mean that it will be known as the
'common species'. (For further information see TAUTONYMS, p.
13) This lizard has a brightly coloured head and neck and, to a
limited extent, can change colour; it is widespread in the
central area of Africa and can frequently be seen sunning itself
on walls and houses in populated areas.

Asian Agama   *A. agilis*
agilis (L) agile, busy; an unusual agama with eyes that are
independently movable, like those of a chameleon. The range is
from Arabia to Pakistan.

Black-necked Agama   *A. atricollis*
ater (L) black; *collum* (L) the neck; quite a big lizard with a

length up to 30 cm (12 in). It inhabits southern and eastern Africa.

Nigerian Agama   *A. benuensis*
-*ensis* (L) suffix meaning belonging to; it takes the name from the River Benue in Nigeria. A brightly coloured lizard with white and green spots and dark stripes. It inhabits Nigeria and Cameroon.

African Spiny-tailed Lizard   *Uromastyx acanthinurus*
oura (Gr) the tail; *mastix* (Gr) a whip; *akantha* (Gr) a thorn, a prickle; *oura* (Gr) the tail; 'a thorny-tailed whip-tail'; it has a short heavy tail which it lashes vigorously from side to side as a means of defence. Inhabiting desert areas in the northern part of Africa.

Egyptian Spiny-tailed Lizard   *U. aegypticus*
-*icus* (L) suffix meaning belonging to; one of the largest lizards in this family it can grow to a length of 75 cm ($29\frac{1}{2}$ in). Named from Egypt, it ranges across northern Africa from Algeria to north-western Arabia.

Toad-headed Agama   *Phrynocephalus mystaceus*
phrunos (Gr) a species of toad; *kephalē* (Gr) the head; *mustax* (Gr) genitive *mustakos* a moustache; it has a strange-looking head which somewhat resembles that of a toad; 'moustache' refers to folds of skin with spiky scales on either side of the mouth which can be distended during a threat display and which appear to increase the size of the gaping mouth. Inhabiting south-eastern Russia and ranging to central Asia.

Moloch   *Moloch horridus*
Moloch was the tribal god of the Ammonites, who demanded the human sacrifice of children by fire; *horridus* (L) rough, bristly; it can also mean frightful, horrible. In spite of its name, the Moloch is a harmless inoffensive lizard, but grotesque in appearance, its body being covered with large spikes. The name was given by Gray in 1841; he was well known for inventing many apparently meaningless scientific names for

animals or, as a friend of mine put it, 'an inveterate coiner'! Doubtless the formidable appearance of this lizard suggested the name. Ants are its chief food and it inhabits the deserts of central Australia. Dr J. E. Gray FRS (1800–1875) was Keeper of Zoology at the British Museum (Natural History) from 1840 to 1874.

Bearded Lizard   *Amphibolurus barbatus*
amphibolos (Gr) thrown round, attacked on both sides; *oura* (Gr) the tail; the reason for this name is obscure and not explained in the herpetological records in the British Museum (Natural History); *barbatus* (L) bearded; the beard consists of a fold of skin beneath the chin equipped with sharp spines which can be expanded as a threat display. Widespread in Australia except the north.

(No recognised English name)   *Diporiphora bilineata*
di- (Gr) prefix meaning two; *poros* (Gr) a passage through the skin, pores; *phora* (Gr) a carrying, a bearing; an allusion to the two small pores in the scales just forward of the anus; *bilineatus* (L) having two lines; a reference to the two white lines along the body. Inhabiting the northern part of Australia and New Guinea.

Frilled Lizard   *Chlamydosaurus kingi*
khlamus (Gr) genitive *khlamudos*, a short cloak, or mantle; *saura* (Gr) a lizard; an allusion to the cloak-like frill of skin round the shoulders which can be spread out as a gesture of defiance and is also used in courtship display; named after Rear Admiral Philip P. King FRS (1791–1856), the author of books on Australia. After serving on *HMS Mermaid* and *HMS Adventure* he settled in Sydney. Inhabiting New Guinea and northern Australia.

Water Lizard   *Hydrosaurus amboinensis*
hudor (Gr) water; as a prefix it becomes *hudro-*; *saura* (Gr) a lizard; it takes the specific name from Amboina (=Ambon Island) in the Moluccas, Indonesia, but it is known throughout

that area, including Celebes (now known as Sulawesi), the Moluccas and New Guinea. A big lizard with a length up to 110 cm (40 in); the tail is flattened in a vertical plane and the toes have lobes of skin, all aids to swimming. The local name is soa-soa.

Angle-headed Lizard   *Gonocephalus godeffroyi*
gōnia (Gr) an angle, a corner; *kephalē* (Gr) the head; the name is descriptive, as the angle of the snout, seen from the side, appears more obtuse than in other lizards, so that the whole head is seen to be almost an equilateral triangle; J. C. Godeffroy (1813–1885) established the Godeffroy Museum in Hamburg and organised many journeys by naturalists to the South Pacific to collect specimens. Inhabiting north-eastern Australia, New Guinea and many islands in that area ranging to the Solomon Islands.

(No recognised English name)   *Acanthosaura lepidogaster*
akantha (Gr) a thorn, a prickle; *saura* (Gr) a lizard; *lepis* (Gr) genitive *lepidos*, a scale; *gaster* (Gr) the belly; it has a row of spines along the neck and a spine on each side of the head behind the eyes; 'a scaly-bellied thorn-lizard'. Inhabiting southern China and the northern part of Thailand and Vietnam.

Lyre-headed Lizard   *Lyriocephalus scutatus*
lura (Gr) a lyre; *kephalē* (Gr) the head; it has some peculiar scale formations and marks on the head but to liken it to a lyre requires some imagination; *scutum* (L) a shield; *scutatus* (L) armed with a shield; a reference to a hump on the snout covered with smooth scales; the function is not known. It inhabits Sri Lanka (formerly Ceylon).

Indian Bloodsucker   *Calotes versicolor*
kalotēs (Gr) (= *kallos*) beauty; a rare word formed by Chrysippus; *versicolor* (L) changing colour, of various colours; an

allusion to its ability to change colour, which it does very quickly; this means that the change is made by nerves in the skin, like the chameleons, and not by a change in the blood hormones, like the anoles. It is sometimes, incorrectly, called a chameleon. The name 'bloodsucker' does not mean literally what it says, but refers to the scarlet colour of the head and neck, and this is particularly bright and spectacular when the mood is aggressive or during courtship display. Inhabiting Iran and ranging east and south through India to southern China and Malaysia. Chrysippus (*c.* 280–207 BC) was a Greek Stoic born at Soli, Cilicia, and later studied at the Academy in Athens.

Bornean Bloodsucker   *C. cristatellus*
crista (L) a crest; usually means a tuft on the head of animals; *cristatus* (L) having a crest; *-ellus* (L) diminutive suffix: having a little crest; it has a small crest of scales from the rear of the head extending a short distance along the back. Named from Borneo, it also inhabits the southern Thailand to Vietnam area and the Philippines.

Ceylon Deaf Agama   *Cophotis ceylanica* (= *ceylonica*)
kōphos (Gr) blunt, obtuse, dumb, deaf; *ous* (Gr) genitive *ōtos* the ear; it is not deaf, but has no exposed tympanum such as is found in most lizards; *-icus* (L) suffix meaning belonging to; Ceylon is now known as Sri Lanka. This agama is unusual as it has a prehensile tail.

Horned Agama   *Ceratophora stoddarti*
keras (Gr) genitive *keratos* the horn of an animal; *phora* (Gr) a bearing, carrying; it has a peculiar small horn on the tip of the nose. The specific name is given in honour of Colonel Stoddart, of the Royal Staff college, who collected this lizard for Gray in 1831. It inhabits Sri Lanka (Ceylon).

(No recognised English name)   *Ptyctolaemus gularis*
ptux (Gr) genitive *ptukhos* a fold, layer; *ptuktos* (Gr) folded; *laimos* (Gr) the throat, gullet; *gularis* (L) pertaining to the throat; an

allusion to the folds of skin round the neck and throat. Inhabiting Assam, Asia.

Butterfly or Smooth-scaled Lizard   *Leiolepis belliana*
leios (Gr) smooth; *lepis* (Gr) a scale, particularly the scales of a lizard; Thomas Bell (1792–1880) was a naturalist who helped Dr Gray with his work. The name butterfly is an allusion to the brilliant colouring which includes shades of orange, black, brown, green and yellow, and may have an irridescent sheen. Inhabiting the Thailand-Kampuchea area, Malaysia and Sumatra.

Flying Dragon   *Draco volans*
draco (L) genitive *draconis* from *drakōn* (Gr) a dragon; *volo* (L) I fly; *volans* (L) flying; it cannot fly in a true sense, but has membranes of skin on the sides of the body which it can distend and is thus able to make quite long gliding flights of about 50 m (55 yds). The lizards in this genus are the only ones that have this ability. Inhabiting Malaysia, the Philippines and Indonesia.

Black-bearded Dragon   *D. melanopogon*
melas (Gr) genitive *melanos* black; *pōgōn* (Gr) the beard; a reference to the black dewlap which, when distended, could be likened to a beard. Inhabiting Sumatra and Borneo.

Family   CHAMAELEONIDAE   about 80 species (probably more)

The Liddell and Scott *Greek-English Lexicon* gives the complete word *khamai-leōn*, and simply translates it as the chameleon, but the name is derived from two Greek words: *khamai* (Gr) on the ground, near the ground, or dwarf; *leōn* (Gr) a lion, hence 'a lion near the ground', a little lion; a peculiar name as the chameleon can hardly be said to resemble a lion!

The chameleon is well known for its ability to change colour, although a number of other lizards can also do this. Equally remarkable are the eyes, which swivel independently of each other, and the tongue which can be shot out with incredible speed to trap an insect; it can be extended to a length equal to that of the head and body together. Although chameleons normally move with slow, deliberate steps, they can scuttle away quite fast on the ground if frightened. The peculiar hesitant movement typical of chameleons, when they seem to sway to and fro several times with each step, is considered to be an imitation of a leaf swaying in the wind, to deceive predators, and also to deceive their prey as they move up for the fatal strike with the tongue.

The colour change can be effected by light or by an emotional state, and can be quite fast (see *Anolis carolinensis*, p. 87); it does not necessarily match the background. I once watched two chameleons meet on a picture rail in a house in Uganda and neither would give way; they both turned black with rage! On another occasion, I brought one into my camp, sitting on my hand and clutching my fingers with its little feet; my African servants fled in terror, as they think that if touched by a chameleon you become sterile.

African Chameleon   *Chamaeleo africanus*
Chamaeleo see Family Chamaeleonidae above. This is one of the big chameleons, measuring up to a length of 37 cm (about $14\frac{1}{2}$ in). It inhabits the northern part of Africa south of the Sahara, ranging from the west coast across to Ethiopia and Somalia.

Two-lined Chameleon   *C. bitaeniatus*
bis (L) two, double; *taenia* (L) a band, a ribbon; *-atus* (L) suffix meaning provided with; it means striped and refers to two light stripes on a dark background along the sides of the body. Inhabiting a large area on the eastern side of Africa including Uganda, Kenya, Tanzania and the northern part of Mozambique.

Short-horned Chameleon   *C. brevicornis*
brevis (L) short; *cornu* (L) the horn of animals; the 80 (or more) species of chameleons display a variety of horns, some being flexible and hardly worthy of the name; this species has a very

short single horn in the male, but absent in the female. Inhabiting Madagascar.

European or Mediterranean Chameleon *C. chamaeleon chamaeleon*

Although the name suggests that it lives in Europe, southern Spain and Crete are probably the only places in Europe where it can be found; apart from this it inhabits the borders of the southern Mediterranean from Israel ranging west to Morocco. It is the nominate subspecies.

Zeylan Chameleon *C. c. zeylanicus*
This subspecies of *Chamaeleo chamaeleon* inhabits India and Sri Lanka and is probably the only chameleon to be found in that part of Asia.

Common Chameleon *C. dilepis*
dis (Gr) twice, or two; *lepis* (Gr) a scale, particularly the scales of a lizard; a reference to two lobes at the back part of the head which are erected when it meets another chameleon; this might be accompanied by a change of colour. Inhabiting the southern part of Africa from the area of the equator ranging south.

Fork-horned or Fork-nosed Chameleon *C. furcifer*
furca (L) a fork; *fero* (L) I bear, carry; the name 'fork-nosed' is misleading as it is the horn on the nose that is forked and this gives it a peculiar appearance. Inhabiting Madagascar.

Jackson's Chameleon *C. jacksonii*
Sir F. J. Jackson (1860–1929) was a naturalist and author and was Governor of Uganda from 1911 to 1917. This chameleon has three horns, one on the snout and two slightly further back on the head; it inhabits Kenya, Uganda and Tanzania. I have seen these chameleons, and other species, on many occasions in Kenya and Uganda without especially looking for them; they are quite happy to live more or less permanently on the rose trees and bushes in private gardens.

(No recognised English name)   *C. lateralis*
latus (L) genitive *lateris* the side, flank; *-alis* (L) suffix meaning
relating to; the body is normally a dull green with a light stripe
along the flank but, as an example of how chameleons defy
being described accurately by colour, this species, when angry,
shows a network of various colours including blue and yellow.
It is widespread in Madagascar but seems to prefer the
highlands in the central area.

Dwarf Chameleon   *C. pumilus*
pumilus (L) a dwarf, a pigmy; a very small chameleon with a
length of only about 14 cm ($5\frac{1}{2}$ in). Inhabiting a small area
round the southern coast of South Africa.

Nossi Bé Dwarf Chameleon   *Brookesia minima*
The genus *Brookesia* was given by Gray in 1827; he decided on
this name when examining the private museum of a Mr Joshua
Brookes and gave it in recognition of his services to zoology;
further information about Mr Brookes is not recorded; *minimus*
(L) smallest; only one other species of chameleon is known to be
so small, with a length of 3.2 cm ($1\frac{1}{4}$ in). Found only on Nossi
Bé, a small island off the north-west coast of Madagascar.

Armoured Chameleon   *B. perarmata*
per- (L) prefix meaning well, very; *armatus* (L) armed; a
reference to the double row of thorn-like spines along the back.
Inhabiting the western part of Madagascar.

Flat-headed Chameleon   *B. platyceps*
platus (Gr) flat or wide; *ceps* (New L) from *caput* (L) the head; it
has a flat top to the head. Inhabiting Malawi, East Africa.

Family   SCINCIDAE   about 700 species

scincus (L) a kind of lizard. The skinks are one of the largest
families of lizards and almost worldwide in distribution in the
tropical and warm areas, although most are found in Africa,
southern Asia and Australasia. Many species look just like
ordinary lizards, for example the pretty little Five-lined Skink

Mabuya quinquetaeniata, while others have degenerate legs or none at all, for example the South African Plain Skink *Scelotes inornatus*. Many species are good burrowers and spend much of their time underground.

Subfamily TILIQUINAE

Tiliqua is an obscure name given by Gray in 1825; herpetological records at the British Museum (Natural History) do not give any explanation.

(No recognised English name) *Tiliqua nigrolutea*
Tiliqua see Subfamily Tiliquinae above; *niger* (L) black; *luteus* (L) saffron yellow; it has a blackish pattern on the back merging into shades of dark to light brown. Inhabiting the southern part of Australia and Tasmania.

Stump-tailed Skink *T. rugosa*
ruga (L) a wrinkle; *-osus* (L) suffix meaning full of, prone to; wrinkled; a reference to the plates on the head and back which are thick and rough. The tail is very short and fat, so much so that it looks very like the animal's head and, to a casual observer, gives the impression of a two-headed lizard. Inhabiting western and south- western Australia.

Blue-tongued Skink *T. scincoides scincoides*
scincus (L) a kind of lizard; *-oides* (New L) from *eidos* (Gr) apparent shape, form. Inhabiting Western Australia. This is the nominate subspecies; two other subspecies are given below.

Giant Blue-tongued Skink *T. s. gigas*
gigas (L) a giant; one of the largest skinks with a length of about 50 cm (19½ in). Inhabiting New Guinea, the Moluccas and other islands in that area.

Intermediate Blue-tongued Skink *T. s. intermedia*
intermedius (L) between; another large skink coming between *gigas* and *scincoides* in length. Inhabiting the northern part of Western Australia.

Subfamily SCINCINAE

scincus (L) a kind of lizard.

Eastern Skink   *Scincus conirostris*
Scincus see Subfamily Scincinae above; *conus* (L) a cone; *rostrum* (L) the beak, snout; a reference to the shape of the nose which is adapted for burrowing through the sand. Inhabiting Iran.

Arabian Skink   *S. muscatensis*
-ensis (L) suffix meaning belonging to, usually a place; named from Muscat, it inhabits Arabian desert areas.

Philby's Arabian Skink   *S. philbyi*
H. St John Bridger Philby (1885–1960) was an explorer, author and orientalist and became a friend of Ibn Sa'ud. He crossed the 'Empty Quarter' (Rub al' Khalé) in southern Arabia in 1932. This skink inhabits the southern part of Arabia.

Common Skink   *S. scincus*
For explanation of the generic name being repeated for the specific, see TAUTONYMS, p. 13. Inhabiting desert areas ranging across northern Africa from Senegal to Egypt and Israel.

Short-footed Sand Skink   *Ophiomorus brevipes*
ophis (Gr) a snake; *moros* (= *phobos*) (Gr) fear, fright; some of the skinks in this genus have no visible legs and their method of moving is very snake-like; *brevis* (L) short; *pes* (L) the foot. This skink inhabits desert or sandy areas of southern Iran where it can easily bury itself.

Persian Snake Skink   *O. persicus*
-icus (L) suffix meaning belonging to; Persia is now known as Iran; it inhabits western Iran. This species has no legs and has the appearance of a snake.

Speckled Snake Skink   *O. punctatissimus*
punctum (L) a hole, a puncture; *punctatus* (New L) spotted as with punctures; *-issimus* (L) suffix meaning very, much; it has a

pale body covered with very small spots and, having no legs, looks very much like a snake. Inhabiting Greece and Turkey.

Three-toed Sand Skink   *O. tridactylus*
trias (Gr) the number three; *daktulos* (Gr) a finger, can mean a toe; although a true skink such as *Scinctus scinctus* has five toes and normal legs, a gradual degeneration has brought about a reduction in the number of toes and the length of the legs in many species; this species has already been reduced to three toes on each foot. Inhabiting West Pakistan, Afghanistan and the adjoining part of eastern Iran.

Algerian Skink   *Eumeces algeriensis*
eu- (Gr) a prefix meaning well, nicely; sometimes used to mean the typical or most advanced animals in a group; *mēkos* (Gr) length, stature; a large skink, it can be 42 cm (16 $\frac{1}{2}$ in) in total length, with a long tail and five toes on each foot, in fact a typical lizard. Furthermore, lizards in this genus are among the very few reptiles that show any care for the young after the eggs have hatched, guarding them and licking them clean, and indeed will guard the eggs and turn them regularly before hatching like the birds. Considering that the birds evolved from the more advanced lizards this is not so surprising. *-ensis* (L) belonging to, usually a place; taking its name from Algeria, it is also found in Morocco and Tunisia.

American Five-lined Skink   *E. fasciatus*
fascia (L) a band, girdle; *fasciatus*, banded; a misleading name as the lizard has five stripes or lines along the body and not bands. Inhabiting the eastern part of North America.

Broad-headed Skink   *E. laticeps*
latus (L) broad, wide; *ceps* (New L) from *caput* (L) the head. Sometimes known as the Greater Five-lined Skink, it inhabits North America.

(No recognised English name)   *Sphenops sepsoides*
sphen (Gr) a wedge; *ops* (Gr) eye, face; can mean appearance; a reference to the wedge-shaped snout used for burrowing

through the sand; *seps* (Gr) a serpent, the bite of which causes putrefaction; can also mean a type of lizard; *oides* (New L) from *eidos* (Gr) apparent shape, form. Inhabiting Egypt and Israel.

Cylindrical Skink   *Chalcides chalcides*
khalkis (Gr) genitive *khalkidos* a lizard with copper-coloured stripes on the back; this species is a bronze colour with several dark lines along the back; the body is more cylindrical than typical lizards and is snake-like in appearance. When in a hurry, the legs are not used and it adopts a snake-like movement. Inhabiting eastern Mediterraean countries.

Eyed Cylindrical Skink   *C. ocellatus*
ocellus (L) a little eye; *-atus* (L) suffix meaning provided with; it has a thick cylindrical body and tail, and is marked with small black spots and streaks of white. Inhabiting Sudan, Somalia, the Mediterranean area of northern Africa and ranging east to southern Asia.

Green Cylindrical Skink   *C. viridanus*
viridis (L) green, can mean all shades of green; *-anus* (L) belonging to; it is similar to other skinks in this genus but with an olive green back. Inhabiting western Canary Islands.

Subfamily   LYGOSOMINAE

lugos (Gr) a pliant twig, especially a twig for wickerwork; *soma* (Gr) the body; the subfamily name is taken from the genus *Lygosoma*, a skink with a long slender body.

Indian Keeled Skink   *Mabuya carinata*
mabuya (New L) from American-Spanish, a lizard; the name originates from the Brazilian species; *carina* (L) a keel; *-atus* (L) a suffix meaning provided with; a reference to the form of the toes. The Mabuya Skinks are widespread, being found in southern Asia, Africa and the Americas, but are not known in New Guinea and Australia. This skink inhabits Sri Lanka (formerly Ceylon).

Eastern Smooth Skink   *M. laevis*
laevis is from *levis* (L) smooth. Inhabiting the small islands in south-eastern Asia.

Common Mabuya   *M. mabuya*
A brown skink widespread in Mexico and ranging south through Central America to Brazil and Bolivia.

African Five-lined Skink   *M. quinquetaeniata*
quinque (L) five; *taenia* (L) a band, a ribbon; -*atus* (L) suffix meaning provided with; the five lines are not bands but stripes running lengthways. On one occasion in Uganda, Dr Cott and I were resting near the Nile just below the Kabalega Falls (formerly Murchison Falls) drinking our flask of tea (laced with limejuice from limes picked from a nearby tree!), when one of these pretty little lizards scampered across the rocks at our feet; they are widespread in northern Africa and down the Nile valley.

(No recognised English name   *Sphenomorphus florensis*
sphen (Gr) a wedge; *morphē* (Gr) form, shape; a reference to the short wedge-shaped head; -*ensis* (L) suffix meaning belonging to, usually a place; inhabiting Flores Island, Indonesia.

Indian Skink   *S. indicus*
-*icus* (L) suffix meaning belonging to. Inhabiting the lower slopes of the Himalayas in India and Tibet.

Slender Skink   *Lygosoma novaeguineae*
lugos (Gr) a pliant twig, specially a twig for wickerwork; *soma* (Gr) the body; a reference to the long slender body. Inhabiting New Guinea.

Kashmir Smooth Skink   *Leiolopisma ladacensis*
leios (Gr) smooth, soft; *lopos* (Gr) shell, bark; *lopisma* (Gr) a rare word used by Eustathius in 1160 AD with the same meaning; Ladakh is a district in Kashmir, India; -*ensis* (L) suffix meaning belonging to, usually a place. This skink has been found in the mountains of Kashmir at a height of over 4,500 m (15,000 ft). Eustathius (died *c.* 1193) was a Greek scholar who studied the

ancient classic authors; the unusual word *lopisma* was first published seven centuries later in 1863.

Himalayan Smooth Skink   *L. himalayana*

-anus (L) suffix meaning belonging to. In this species, the male can be distinguished by the bright red marking on the sides of the body during the mating season; it can be found living at an altitude of over 3,500 m (about 12,000 ft) in the Himalayas.

The Emo   *Emoia cyanura*

Gray named this species The Emo in 1845, but does not give an explanation; it seems likely that the name refers to a person, or the people, of one of the remote islands where it lives; *kuaneos* (Gr) blue; *oura* (Gr) the tail; there are about 40 species in this genus. Dr J. E. Gray FRS (1800–1875) was a zoologist at the British Museum (Natural History) from 1840 to 1874. These lizards are widespread in the Indo-Pacific island region.

Black-sided Emo   *E. atrocostata*

ater (L) black; *costa* (L) a rib; can mean the side; *atus* (L) provided with; 'having black sides'. Inhabiting the coasts of islands near Singapore, it has been seen swimming and hunting for prey in the sea.

Keeled Skink   *Tropidophorus sinicus*

tropis (Gr) genitive *tropios* which later became *tropidos* a ship's keel; *phora* (Gr) a carrying, bearing; a reference to the pronounced keels on the scales which probably assist them when swimming; they are one of the few skinks which are good swimmers. *sinae* (New L) a people of south-east Asia, usually taken to mean the Chinese; *-icus* (L) suffix meaning belonging to; this can be interpreted in the wider sense of south-east Asia as the skinks in this genus are also found in the Malay Peninsula, Borneo, Celebes (now known as Sulawesi) and the Philippines.

Lidless Skink   *Ablepharus pannonicus*

a- (Gr) prefix meaning not, or there is not; *blepharon* (Gr) an eyelid; an allusion to the eyelids which are fixed permanently shut and transparent; -*icus* (L) suffix meaning belong to; named from Pannonia, an ancient country not shown on modern maps; it is an area now known as Yugoslavia and included neighbouring countries such as Hungary. One of the very few skinks found in Europe it inhabits Hungary and Romania.

Wahlberg's Skink *Panaspis wahlbergi*
pan (Gr) all; *aspis* (Gr) a viper, an asp; the eyelids are permanently shut, as in *Ablepharus* above, and for this reason the lizards in this genus are sometimes known as 'snake-eyes'; J. A. Wahlberg (1810–1856) was a Swedish naturalist who was in Africa from 1839 to 1845. This skink inhabits Kenya and ranges south through Tanzania.

Two-legged Skink *Scelotes bipes*
skelos (Gr) the leg; -*tes* (Gr) suffix meaning pertaining to; *bis* (L) two, double; *pes* (L) the foot; the skinks in this genus show a complete range of reduced legs and toes, from *Scelotes bojeri*, which has quite normal legs and the usual five toes, to *S. inornatus* which has no legs at all. This skink, *bipes*, as the name tells us, has only two legs, the hind legs, and only two toes on each leg. Inhabiting Madagascar.

Bojer's Skink *S. bojeri*
Wenzel Bojer (1800–1856) was a Czechoslovakian naturalist from Prague; he collected in Madagascar and Zanzibar and was Curator of the Mauritius Natural History Museum. This skink has four perfectly good legs with five toes on each foot; it inhabits Mauritius.

Plain Skink *S. inornatus*
It is not surprising that this skink has acquired the name 'plain'; it has no legs at all; the Latin name has the same meaning: *ornatus* (L) ornament, decoration; *inornatus* (L) plain. It inhabits South Africa.

Black-sided Skink *S. melanopleura*
melas (Gr) genitive *melanos* black; *pleura* (Gr) a rib; usually plural, the ribs or sides of a man or any other animal. This

skink has five toes but the legs are much reduced and very short; it inhabits Madagascar.

Dart Skink   *Acontias plumbeus*
akōn (Gr) a javelin, a dart; *akontias* (Gr) a quick-darting serpent; it is not, of course, a serpent but is a limbless lizard, so it may well have been mistaken for a snake when the name was given; *plumbeus* (L) lead-coloured. Inhabiting Africa and Madagascar.

(No recognised English name)   *Sepsina angolensis*
sēps (Gr) a putrefying sore; a snake whose bite causes putrefaction; a kind of lizard; *-inus* (L) suffix meaning like; it has a snake-like body and very small legs and in ancient times it was thought to have a poisonous bite; *-ensis* (L) suffix meaning belonging to; it inhabits the Angolan area.

Family   FEYLINIIDAE   4 species

Feylinia appears to be named after a naturalist or collector named Feylin, but research in the Department of Zoology at the British Museum (Natural History) gives no explanation of the name.

Curror's Skink   *Feylinia currori*
Feylinia see Family Feyliniidae above. Little is known about J. Curror, an officer in the Royal Navy, except that he supplied the information that enabled Gray to describe the animal. It is limbless and lives underground in tropical Africa. Dr J. E. Gray FRS (1800–1875) was a zoologist at the British Museum (Natural History) from 1840 to 1874.

Family   ANELYTROPSIDAE   1 species

anelytros (Gr) without sheath, or covering; *opsis* (Gr) aspect, appearance; or *ops* (Gr) the eye; Cope describes this lizard as having the epidermis absolutely continuous over the eye, the eye being scarcely visible through the single ocular plate, so the generic name seems contradictory; furthermore if the eye is visible one assumes it is not blind!

Mexican Blind Lizard   *Anelytropsis papillosus*
Anelytropsis see Family Anelytropsidae above; *papilla* (L) a pimple; *-osus* (L) suffix meaning full of; 'very pimply'; a reference to the rough scales on the body. Very rare and only found in Mexico. E. D. Cope (1840–1897) was an American zoologist who took a special interest in reptiles and wrote articles for the herpetological magazine *Copeia*; it was named in his honour (E. C. Jaeger).

Family   CORDYLIDAE   33 species

kordulē (Gr) a club, a cudgel; not a reference to the shape of the tail but to its use as a weapon; there are large spines on the head, body and tail, and the latter can be lashed vigorously from side to side; the lizards in this genus have heavy protective armour.

Subfamily   CORDYLINAE

Blue-spotted Girdle-tailed Lizard   *Cordylus caeruleo-punctatus*
Cordylus see Family Cordylidae above; *caeruleus* (L) dark blue; *punctum* (L) a hole, a prick; *punctatus* (L) spotted as with punctures; it has very marked bright blue spots on the sides of the body. Inhabiting the mountains in the southern part of Cape Province, South Africa.

Armadillo Lizard   *C. cataphractus*
kataphraktos (Gr) covered, decked in; horses clad in full armour; also *cataphractus* (L) mail-clad; a reference to the heavy protective armour; the English name is from the mammal, Armadillo, which has protective bony scales. Inhabiting the western part of Cape Province, South Africa.

Common Girdle-tailed Lizard   *C. cordylus*
The name 'girdle-tailed' is a reference to the large spiked scales
that encircle the tail and enhance its use as a weapon; it lashes
vigorously with its tail if threatened. This lizard has a wider
range than the other 22 species in the genus, inhabiting
Ethiopia, Kenya and ranging south to Cape Province, South
Africa.

Giant Girdle-tailed Lizard   *C. giganteus*
gigas (L) a giant; *giganteus* (L) gigantic; the largest lizard in this
genus, it can grow to a length of about 38 cm (15 in).
Inhabiting South Africa.

Spotted or Red-tailed Flat Lizard   *Platysaurus guttatus*
platus (Gr) flat; *saura* (Gr) a lizard; the head and body are
unusually flat which enables it to hide in very narrow cracks in
the rocky terrain where it lives; *gutta* (L) a drop (of a fluid);
guttae (L) spots or marks on animals; *guttatus* (L) spotted,
speckled. Inhabiting the Transvaal area of South Africa.

(No recognised English name)   *P. imperator*
imperator (L) a commander, a leader; an allusion to its size,
being the largest of the 10 species in this genus, at 32 cm ($12\frac{1}{2}$
in). Inhabiting the north-east part of Zimbabwe and the
neighbouring part of Mozambique.

Cape Snake Lizard   *Chamaesaura anguina*
khamai (Gr) on the ground, on the earth; *saura* (Gr) a lizard; the
legs are much reduced, useless for crawling and barely visible,
so the body must slide along on the ground like a snake; *anguis*
(L) a snake; *-inus* (L) a suffix meaning like; not only is the
movement snake-like but the body tapers down gradually to
the tail which is more than twice the length of the body; only
close examination would reveal where the tail actually starts.
An inquisitive person, thinking it was a small snake and
grabbing it by the tail, would quickly discover it was not a
snake as the tail would break off and remain wriggling in his
hand! Snakes are not capable of autotomy. Inhabiting South
Africa.

Large-scaled Snake Lizard *C. macrolepis*
makros (Gr) long; the word is not really suitable as strictly
speaking it means long, not large; *lepis* (Gr) the scale of an
animal; the legs are even more reduced than *C. anguina* (above)
and are only vestigial. Inhabiting South Africa.

Subfamily GERRHOSAURINAE

gerrhon (Gr) anything made of wicker work, especially an
oblong shield; *saura* (Gr) a lizard; known as 'plated lizards'
because of the special head shields, the bony cores in the body
scales and their arrangement in longitudinal and cross rows
giving the appearance of wicker work; this encases the lizard in
a protective shell.

Yellow-throated Plated Lizard *Gerrhosaurus flavigularis*
Gerrhosaurus see Subfamily Gerrhosaurinae above; *flavus* (L)
yellow; *gula* (L) the throat; *-aris* (L) suffix meaning pertaining
to. It inhabits Sudan, Ethiopia and ranges south through East
Africa to the Cape.

Sudan Plated Lizard *G. major*
major (L) greater; although indicating a large lizard, it does not
mean the largest: *maximus* (L) greatest, would mean the largest
lizard. At 56 cm (22 in) it is not so big as *G. validus* (below). It
inhabits Sudan and Eritrea and ranges south through East
Africa to South Africa.

South African Plated Lizard *G. validus*
validus (L) strong, robust; at present, records show that this is
the largest species in the genus at 69 cm (27 in). It inhabits
Mozambique and ranges south to the eastern part of South
Africa.

(No recognised English name) *Cordylosaurus subtessellatus*
kordulē (Gr) see Family Cordylidae p. 95; *saura* (Gr) a lizard,
sub- (L) under, can mean less than usual; *tessella* (L) a cube-
shaped paving stone; *tessellatus* (L) set with small cubes; an
allusion to the shields on the front part of the head, there being

less than those of other related species. Inhabiting Angola and ranging south to the western part of South Africa.

African Seps　*Tetradactylus africanus*
tetras (Gr) four; *daktulos* (Gr) a finger, can mean a toe; this genus shows a great variation in the number of toes on the feet so the generic name may be misleading; in this species, the fore legs are very small with only one toe on each foot. Inhabiting South Africa.

Short-legged Seps　*T. seps*
seps (Gr) a putrefying sore; a snake whose bite causes putrefaction; a kind of lizard; although the legs are very much reduced there are still five toes on each foot, so the generic name is again misleading. Inhabiting South Africa.

Long-tailed Seps　*T. tetradactylus*
tetradaktulos (Gr) four toes (see above under *T. africanus*). This species actually does have four toes, so the generic and specific names are correct; for an explanation of the tautonymous name see p. 13 under TAUTONYMS. Inhabiting Table Mountain, South Africa.

Family　XANTUSIIDAE　12 species

Named in honour of L. Jason Xantus (1825–1894), a Hungarian zoologist; he was chiefly known for his interest in birds and Xantus's Murrelet *Brachyramphus hypoleuca* is named after him, but he was an all-round naturalist and became Director of the Zoological Gardens in Budapest in 1866. The Night Lizards, as their English name suggests, are nocturnal and spend the day hidden under rocks, only appearing at night to hunt for food, mostly insects.

Cuban Night Lizard　*Cricosaura typica*
krikos (Gr) a ring, circular; *saura* (Gr) a lizard; a reference to the rounded shape of the body; *tupos* (Gr) a blow, a mark; can mean the general character of a thing, its kind, sort, e.g.

typical; it is typical of this family. A very rare lizard found only in Cuba and threatened with extinction.

Island Night Lizard   *Klauberina riversiana*
Named after Dr L. M. Klauber (born 1883), an American who was by profession an engineer but had a keen interest in reptiles and wrote several books on snakes; he was a member of the Zoological Society of San Diego; *riversiana*: this lizard was described by Cope in 1883, but the description was first published in the *American Naturalist* in 1879 by an unnamed zoologist who said that he wished to commemorate Mr Rivers of the University of California; further details about Mr Rivers are not recorded. It has less nocturnal habits than other lizards in this small family; it inhabits several small islands off the coast of California. E. D. Cope (1840–1897) was an American zoologist with a particular interest in herpetology; the journal *Copeia* was named in his honour (E. C. Jaeger).

Desert or Yucca Night Lizard   *Xantusia vigilis*
Xantusia see Family Xantusiidae, p. 108; *vigil* (L) genitive *vigilis* alert, watchful; an allusion to the eyes which have no eyelids but are covered by a transparent scale so they are never shut. Another lizard which can change colour; it is paler during the day than at night. Inhabiting desert areas of California, Nevada, Utah and Arizona; it is often found in Yucca trees or hiding amongst the litter beneath.

Family   TEIIDAE   200 species

teju (Portuguese), from the Tupi language, a lizard. These are typical lizards and in many ways resemble the Lacertidae of Europe and Africa, although they vary in size from about 7.5 cm (3 in) up to 140 cm (4 ft 8 in) and some have the legs much reduced.

Seven-lined Racerunner   *Cnemidophorus deppei*
knēmis (Gr) genitive *knēmidos* a greave, a legging, armour between knee and ankle; *phor* (Gr) a carrying, a bearing;

knēmidophorus (Gr) wearing greaves to protect legs; an allusion to the thick scales on the legs; F. Deppe (1794–1860) was a naturalist who collected in Mexico and Hawaii during the early part of the nineteenth century. The lizards in this genus are often known as 'racerunners' on account of their great speed when seeking safety; during a chase they can stop suddenly which may cause the predator to over-run; they then dash off at lightning speed in a completely different direction. Speeds of about 30 kph (18 mph) have been estimated. Inhabiting Mexico and ranging south to Costa Rica.

Six-lined Racerunner   *C. sexlineatus*
sex (L) six; *linea* (L) a linen thread, string; can mean a line drawn; *-atus* (L) suffix meaning provided with; an allusion to the pale stripes along the back and extending to the tail. Inhabiting a large area of southern North America in Nevada and Utah, and ranging south and east to include Florida and west to part of Mexico.

Chequered or Tessellated Racerunner   *C. tessellatus*
tessella (L) a cube-shaped paving stone; *tessellatus* (L) set with small cubes; an allusion to the shape of the scales on the back. Inhabiting the southern part of North America.

Common or Yellow-throated Runner   *Ameiva chrysolaema*
ameiva, a native name for a lizard from the Tupi language; *khrusos* (Gr) gold; *khruseos* (Gr) golden-yellow; *laimos* (Gr) the throat. Inhabiting rocky barren areas in Haiti.

Dwarf Runner   *A. lineolata*
linea (L) a linen thread, string; can mean a line drawn; *lineola* (L) a little line; *-atus* (L) suffix meaning provided with; a reference to the stripes on the back. A small lizard and, although called a 'dwarf runner', at 15 cm (6 in) it is not the smallest teiid. Inhabiting Haiti.

Striped Racerunner   *Kentropyx striatus*
kentron (Gr) a spike, spur; *pux* (Gr) the rump, buttocks (a later form of *pugē*, as used by Aristotle); an allusion to the large spiky

keeled scales on the belly; *stria* (L) a channel, a furrow; *striatus* (New L) striped; it has small granular scales which form stripes along the flanks. Inhabiting north-eastern Brazil.

(No recognised English name)   *Tupinambis teguixin*
The Tupinambas were a group of Tupian peoples, now extinct, of the Brazilian coast, from the mouth of the Amazon to São Paulo; *tegu* = *teju* (Portuguese) from the Tupi language, a lizard; *ixia* (Gr) the chamaeleon plant; a reference to its liking for juicy fruits and leaves, although its diet is very varied, including small mammals and birds, their eggs, amphibians, insects and worms. This is a big lizard and can grow to a length of 140 cm (about 4 ft 6 in); it inhabits a large area of eastern South America from north of the Amazon to Uruguay.

Red Tegu   *T. rufescens*
rufus (L) red; *-escens* (L) suffix meaning nearly, approaching, i.e. 'reddish'; this is a big lizard but not attaining the size of *T. teguixin* (above). Inhabiting Argentina.

False Monitor   *Tejovaranus flavipunctatus*
tejo = *teju* (Portuguese) from the Tupi language, a lizard; *varanus* (New L) from *waran* (Ar) a monitor lizard; *flavus* (L) yellow; *punctum* (L) a hole, a prick; *punctatus* (L) spotted as with punctures; a reference to the yellow spots on the body. A big lizard measuring about 100 cm (3 ft 4 in) and certainly having the appearance of a monitor lizard, but known as the False Monitor because it is not a true monitor of the family Varanidae. It is a voracious carnivore and something of a cannibal as it often eats small lizards. Inhabiting Peru.

Caiman Lizard   *Dracaena guianensis*
drakōn (Gr) fem. *drakaina* a dragon; *-ensis* (L) suffix meaning belonging to; named from Guiana, it inhabits a large area of north-eastern South America. It is known as a Caiman from the Spanish *caiman*, an alligator; it has a vertically flattened tail for swimming and spends most of the day in the water, but it is not an alligator of the family Alligatoridae. Rather bigger, at about 125 cm (4 ft 2 in), than the False Monitor (above).

Dragon Lizardet   *Crocodilurus lacertinus*
krokodeilos (Gr) a kind of lizard; actually an Ionic Greek word
and can mean the Nile crocodile; *oura* (Gr) the tail; it has a
flattened tail similar to a crocodile; *lacerta* (L) a lizard; *-inus* (L)
suffix meaning like. A good swimmer, spending much of the
day in the water, it eats fish and frogs. Inhabiting Central
America and northern South America.

Water Teiid   *Neusticurus bicarinatus*
neustikos (Gr) able to swim; *oura* (Gr) the tail; *bi-* (L) prefix
meaning two; *carina* (L) the keel of a ship; *-atus* (L) suffix
meaning provided with; it has two parallel keels of spiny scales
along the top of the tail, which is flattened and well adapted for
swimming. It eats mostly tadpoles and small fish. Inhabiting
Venezuela and ranging east to north-eastern Brazil.

Spectacled Teiid   *Gymnophthalmus speciosus*
gumnos (Gr) naked; *ophthalmos* (Gr) the eye; the lower eyelid is
transparent and so forms a sort of spectacle when closed;
speciosus (L) showy, brilliant; a reference to the brilliant colours;
it has a blue line on the side of the head, a brownish head and
body with a metallic sheen, and an orange tail. A small lizard,
only 15 cm (6 in), inhabiting Mexico and ranging south to
Colombia.

Earless Teiid   *Bachia cophias*
bachia; Gray listed this name in his 1945 *Catalogue of Lizards* but,
as frequently happens regarding his names, he did not explain
it; one assumes that it commemorates a naturalist named Bach;
cŏphos (Gr) blunt, obtuse; can mean deaf; the lizard is not deaf
but does not have a visible tympanum like most lizards. It is a
strange-looking creature, worm-like, with tiny legs that are
almost useless for locomotion, so it moves about with the action
of a snake. It inhabits northern South America, ranging south
to Bolivia and Chile. Dr J. E. Gray FRS (1800–1875) was a
zoologist at the British Museum (Natural History) in 1840.

Rough Teiid   *Echinosaura horrida*
ekhinos (Gr) a hedgehog; *saura* (Gr) a lizard; *horridus* (L) rough,

prickly; a reference to the rows of large spiky scales on the back. Inhabiting Panama, Colombia and Ecuador.

Family **LACERTIDAE**   over 200 species

lacerta (L) a lizard. It is quite reasonable to consider the lizards in this family as 'typical lizards'; they do not have any special features to distinguish them like the geckos with their toes specially adapted for climbing smooth surfaces, the anoles with the expanding dewlap, or the chameleons with their unusual eyes, feet and ability to change colour, not to mention the extraordinary tongue.

Chameleons are not the only lizards that can change colour and some lizards can do this as quickly as the chameleons, but not the lacertids. They do all have the ability to shed their tails, known as autotomy (see p. 20), but this is not the case with all the Sauria.

Among the 200 or so species there are many small variations in diet, habitat, colour and scale formation, and this has caused taxonomists to divide them into many genera, species and subspecies; opinions about these divisions are always changing and have been doing so since the late nineteenth century. Herpetologists have spent many hours in the field and in the laboratory trying to establish satisfactory divisions and have even divided the genera into subgenera in an attempt to solve the problem; I am not including these as the whole subject is still under discussion and anything that appears now would soon be out of date.

Sand Lizard   *Lacerta agilis*

lacerta see Family Lacertidae above; *agilis* (L) agile, busy; it seems hardly necessary to explain these Latin names; almost all the lacertids are very active and agile when their body temperature is raised by the warmth of the sun, although *L. agilis* is not actually the most agile. In spite of the English name it is by no means always found in sandy areas. It is widespread in Eurasia, including southern England, southern Sweden and Spain in the west, and ranging east as far as Lake Baikal.

Lagarto   *L. atlantica*
atlantica (L) of the Atlantic; this lizard inhabits two of the islands in the Canary Islands group in the Atlantic. Lanzarote and Fuerteventura. The diet is unusual for a lacertid as instead of the usual insects it seems to prefer snails, worms, carrion and even the droppings of some animals; this is not surprising if they are rabbit droppings as rabbits always evacuate a certain amount of pellets, which differ from ordinary pellets in being rich in protein and which are eaten and redigested (R. M. Lockley). *Lagarto* is Spanish for a lizard.

Derjugin's Lizard   *L. derjugini*
Named by Nikolski in 1898 after Dr K. N. Derjugin (1878–1938) who was a Professor at the Leningrad State University. It inhabits the Krym (Crimea) Peninsula and the border of northern Georgia and southern Russia in the Caucasus Mountains. A. M. Nikolski (1858–1942) was a Russian herpetologist and Professor at Kharkov University.

Eyed or Jewelled Lizard   *L. lepida* (or *ocellata*)
lepida (L) graceful; at the opposite end of the scale to the little Dwarf Lizard (below), this is the largest in the family, sometimes reaching a length of 80 cm (2 ft 8 in). The English name is an allusion to the blue eyelike spots, usually bordered by a black ring, along the flanks; the body is a brilliant green. Inhabiting Italy, southern France, Spain, Portugal and northwest Africa.

Wall Lizard   *L. muralis muralis*
murus (L) a wall; usually meaning a city wall; *-alis* (L) suffix meaning relating to; this is the nominate subspecies (see Chapter 1). There are many subspecies and three are given here; this one inhabits a large area in central, southern and eastern Europe.

Albanian Lizard   *L. m. albanica*
-*icus* (L) suffix meaning belonging to; although named from
Albania, it is also found in Yugoslavia, Macedonia and
Peloponnesos. It lives in mountains up to a height of about
1,800 m (6,000 ft).

(No recognised English name)   *L. m. maculiventris*
macula (L) a mark, a spot; *venter* (L) genitive *ventris* the belly,
stomach; it has dark spots on the belly. Inhabiting northern
Italy and an area along the east coast.

Dwarf Lizard   *L. parva*
parvus (L) small; usually not more than 15 cm (6 in) in length.
Inhabiting Asian Turkey.

Rock Lizard   *L. saxicola*
saxum (L) a rock, a large stone; *colo* (L) I inhabit; it lives on the
rocky slopes of the Caucasus Mountains, which form the border
between Russia and Georgia.

Ruin Lizard   *L. sicula sicula*
siculus (L) belonging to Sicily; this is the nominate subspecies
(see Chapter 1). Although named from Sicily, it has a wide
distribution in Italy south of Rome, in Corsica and Sardinia,
and in the eastern part of Sicily. There are even more subspecies
named than with *L. muralis*, and 5 of these are given here; they
inhabit various islands in the Adriatic and Tyrrhenian Seas and
each island seems to have developed its own form and colour
patterns; often found living in the ruins of old buildings.

(No recognised English name)   *L. s. adriatica*
Considering the enormous number of species and subspecies of
Lacerta it is not surprising that many of them have no
vernacular name; to most Europeans, and certainly to most
visitors to Europe, they are just 'lizards'; they can be seen
almost everywhere in southern Europe scuttling away to hide
as one approaches. Inhabiting Pelagosa Picciola in the Adriatic; this species is inclined to be melanistic.

(No recognised English name) *L. s. coerulea*
coeruleus = caeruleus (L) blue, sky-blue, sometimes dark blue; it is greyish black on the back changing to blue flanks and sky-blue throat and belly. Inhabiting Capri.

Yellow-throated Lacertid *L. s. flavigula*
flavus (L) yellow; *gula* (L) the gullet, throat; it lives on San Giovanni Faro in the Adriatic Sea.

San Stephano Lacertid *L. s. sanctistephani* (formerly *sancti-stephani*)
Bolder markings distinguish it from other similar subspecies. Inhabiting San Stephano in the Tyrrhenian Sea. Although originally named *sancti-stephani*, the Code now forbids the use of hyphens in scientific names.

Simony's Lizard *L. simonyi*
Living on Hierro, one of the Canary Islands (though possibly not today), is this lizard named by Steindachner in 1889, after a Mr Simony who discovered it; further details of this naturalist are not known. It has not been seen and recorded since 1930, and later searches have proved fruitless so it is probably extinct. Dr F. Steindachner (1834–1919) was a German zoologist and Director of the Vienna Museum in 1882.

Three-lined Green Lizard *L. trilineata*
tria (L) three; *linea* (L) a linen thread, string; can mean a line drawn; *-atus* (L) suffix meaning provided with; in some cases there are five lines but there is always a central one along the backbone. Inhabiting Bulgaria, Greece and Turkey.

Green Lizard *L. viridis*
viridis (L) green; one of the larger lizards in this family it may grow to a length of 45 cm (17½ in). Inhabiting the warmer parts of Europe, from Spain ranging east to Asia Minor.

Common or Viviparous Lizard *L. vivipara*
vivus (L) alive; *pareo* (L) I bring forth, I beget; this lizard should really be known as ovoviviparous, from *ovum* (L) an egg; it has no placenta and the young are born alive *in the egg*; they have

received no nourishment from the mother. It has the advantage that the eggs are kept to a moderate temperature even in Arctic areas where this lizard is sometimes found; it can make use of the sun and move the eggs, still in its body, from place to place; it could not do this if the eggs were laid and buried. The young, normally black, soon break out of the egg membrane, which is very thin, and are immediately independent, there being no parental care.

This lizard has a wider distribution than any other lizard, inhabiting north and central Europe, including the British Isles, and ranging north in Sweden to the Arctic Circle and east to Mongolia. It is seldom seen, if ever, in southern Spain, Italy and Greece, but this depends on the local climate; preferring rather damp areas near water, it is a good swimmer and even hunts for prey in the water.

African Sand Lizard   *Psammodromus algirus*
psammos (Gr) sand; *dromos* (Gr) running, escape; *algirus* is a coined word and supposed to mean 'of Algiers'; it is a much larger lizard than *hispanicus* (below) at about 30 cm (12 in) in length. In addition to north-west Africa, it inhabits Spain and Portugal and can often be seen in parks and gardens.

Spanish Sand Lizard   *Ps. hispanicus*
hispanicus (L) Spanish; in a case like this the scientific name becomes very important; if the English name is Sand Lizard we must be able to distinguish it from the Sand Lizard inhabiting southern England, Europe and ranging east to Lake Baikal, which is named *Lacerta agilis*. They must be related because they are in the same family, but they are not in the same genus. *Ps. hispanicus* inhabits the French and Spanish Mediterranean coasts. The abbreviation of the generic name to *Ps.* is better than the initial *P.* because the Greek letter *psi* combines the two (see transliteration on p. 656).

Snake-eyed Lacertid   *Ophisops elegans*
ophis (L) a snake; *ops* (Gr) the eye; the eyelids are permanently closed and have a transparent window like that of snakes;

elegans (L) neat, elegant. Inhabiting north-eastern Africa and Asian Turkey.

Desert Racerunner   *Eremias arguta*
erēmia (Gr) desert, wilderness; *argutus* (L) lively; inhabiting mostly semi-desert areas in southern Ukraine, Moldavia and Romania.

(No recognised English name)   *E. velox*
velox (L) rapid, swift; inhabiting semi-desert areas of southern Kazakhstan bordering the Caspian Sea and ranging east to central Asia.

Fringe-toed Lizard   *Acanthodactylus erythrurus*
akantha (Gr) a thorn, a prickle; *daktulos* (Gr) a finger or a toe; an allusion to the scales on the borders of the toes, which increase the surface area; the feet are thus better adapted for running on loose sand; *eruthros* (Gr) red; *oura* (Gr) the tail; strangely, the red colour at the tip of the tail gradually fades in the male as he reaches maturity and finally disappears. Inhabiting sandy areas in Spain and Portugal.

African Fringe-toed Lizard   *Aporosaura anchietae*
a- (Gr) prefix meaning not, or there is not; *poros* (Gr) a passage through the skin, a pore; *saura* (Gr) a lizard; a reference to the absence in the male of the femoral pores; some doubt exists regarding the function of the pores on the under surface of the male femur, but herpetologists think that they may excrete a substance that gives the male a better grip of the female during copulation. This lizard appears to have been named after a comparatively unknown Portuguese naturalist, J. de Anchieta, who died in 1897; presumably a woman because of the feminine genitive ending *-ae*; at one time she was resident in Angola (Portuguese West Africa), where this lizard is found living in sandy desert areas. The English name 'fringe-toed' is explained under *Acanthodactylus* (above).

Long-tailed or Six-lined Grass Lizard *Takydromus sexlineatus*

takhus (Gr) quick, swift; *dromos* (Gr) running, escape; a very nimble small lizard measuring about 36 cm (14 in), but most of this is tail; probably the lizard with the longest tail in proportion to the length of the body; of the total length only about 6 or 7 cm ($2\frac{1}{4}$ to $2\frac{3}{4}$ in) consists of the body, about one fifth of the total length: *sex* (L) six; *linea* (L) a linen thread, string; can mean a line drawn; *-atus* (L) suffix meaning provided with. It lives among grasses and small plants in the Vietnam–Kampuchea area, Malaysia and Indonesia.

(No recognised English name) *Tropidosaura montana*

tropis (Gr) genitive *tropios* later to become *tropidos* a keel; *saura* (Gr) a lizard; an allusion to the noticeably keeled scales of the back which are also imbricated; *mons* (L) genitive *montis* a mountain; *montanus* of a mountain. Inhabiting the mountains of Cape Province, South Africa, up to an altitude of about 2,000 m (6,500 ft).

(No recognised English name) *Ichnotropis capensis*

ikhnos (Gr) a track, trace; *tropos* (Gr) a turn, direction; a reference to this lizard's peculiar method of escape from danger, running in wide zig-zags and making use of any available clumps of grass to hide in, at each change of direction; *capensis* (L) of the Cape. It inhabits Kalahari, Transvaal and Mozambique.

Family ANGUIDAE about 70 species

anguis (L) a snake; although many of the lizards in this family (known as Lateral Fold Lizards) have no legs and so are snake-like, some have four perfectly good legs; also they do not all have the lateral fold, but are grouped together for other anatomical reasons, chiefly characteristics of the teeth, tongue, and skull.

Subfamily   DIPLOGLOSSINAE

diploos (Gr) two-fold, double; *glōssa* (Gr) the tongue; a reference to the peculiar form of the tongue, the end being retractable into the basal portion, which forms a sort of sheath.

West Indies Galliwasp   *Diploglossus costatus*
Diploglossus see Family Diploglossinae above; *costa* (L) a rib; *costatus* (L) having ribs; can mean having rib-like lines; this refers to lines that run across the body and partly down the sides, as though indicating ribs. The lizards in this genus do not have the lateral fold. Inhabiting Hispaniola, one of the larger islands of the Greater Antilles. The origin of the name galliwasp is obscure.

Galliwasp   *D. tenuifasciatus*
tenuis (L) thin, slender; *fascia* (L) a band or girdle; *fasciatus* (L) banded, can mean striped; an allusion to the thin dark lines round the body similar to those of *costatus*, above. Inhabiting the eastern part of Central America.

Worm Lizard   *Ophiodes striatus*
ophis (Gr) a snake; *-odes* (New L) derived from *eidos* (Gr) apparent shape, form; in appearance it is similar to the Slow-worm, but has a pair of diminutive hind legs that are useless for crawling, but retain some muscular movement; *striatus* (New L) striped; an allusion to two brown stripes that run the length of the body and the tail. It inhabits Brazil and Argentina.

Subfamily   GERRHONOTINAE

gerrhon (Gr) a type of shield, anything made of wicker work; *nōton* (Gr) the back; a reference to the pattern of the scales; this, and the apparent stiffness of the body, gave rise to the name Alligator Lizards. This subfamily have the lateral fold, a kind of pleat along the flanks which compensates for the rigidity of the hard scales; it assists movement of the body, expansion for breathing and for pregnant females.

Scheltopusik or Pallas's Glass Lizard   *Ophisaurus apodus*

ophis (Gr) a snake; *saura* (Gr) a lizard; *a-*(Gr) prefix meaning not, or there is not; *pous* (Gr) genitive *podos*, the foot; 'snake-like lizard with no feet'; it has in fact two tiny vestigial hind legs only about ($\frac{1}{10}$ in) long, little more than pimples. It is quite a big lizard and may grow to a length of 1.3 m (over 4 ft). The name scheltopusik is from the Russian word zheltopuzik, which means 'yellow belly'. Inhabiting a large area ranging from Greece and the Adriatic coast, through Iraq and Iran, to central Russia. Named after Peter Simon Pallas (1741–1811), a German zoologist and explorer and an authority on Russian wildlife. He was a Professor at St Petersburg University in Russia and made contributions to most of the natural sciences. The reason for the name Glass Lizard is given below under *O. gracilis*.

Slender Glass Lizard   *O. attenuatus*

attenuo (L) I make thin, weak; *attenuatus* (L) weakened, enfeebled; this is not a well-chosen specific name, as it is supposed to mean slender and not enfeebled. It was named by Professor S. F. Baird (1823–1887), the American zoologist: 'Baird did more than any other man of his time to advance the study of ornithology and zoology' (T. S. Palmer). This lizard has a wide distribution in the River Mississippi area, USA.

Burman Glass Lizard   *O. gracilis*

gracilis (L) thin, slender; can mean plain, unadorned; like all the Glass Lizards it has no legs and the layman would almost certainly think it was a snake; even if he saw it blink its eyes he would not realise that it was not a snake. It can break off its tail (autotomy) to escape from a predator and the tail constitutes a large proportion of the whole animal; this may break into several pieces, which gives the impression that the whole animal has fallen to bits! Hence, a 'glass lizard'. Named from

Burma, it ranges from north-eastern India to south-western China.

Northern Alligator Lizard *Gerrhonotus coeruleus*
Gerrhonotus see Subfamily Gerrhonotinae, p. 110. *coeruleus* = *caeruleus* (L) dark blue, can mean dark-coloured. This is one of the species in this family that has four normal legs and five toes; it inhabits the north-western part of the USA, ranging south through California.

Popocatepetl Alligator Lizard *G. imbricatus*
imbrico (L) I cover with tiles; *imbricatus* (L) tiled; it usually means overlapping tiles like those on a roof; a reference to the type of scales. It lives high up on the volcanic mountain Popocatepetl, Mexico, at about 4,000 m (13,000 ft). There has not been a serious eruption since 1664; the name is an Aztec word meaning 'smoking mountain'.

Subfamily ANGUINAE
anguis (L) a snake; the Slow-worm is really a lizard, but Linnaeus gave it this name because it has no legs and he thought it was a snake (for details see Chapter 2, pp. 19–20).

Blue-spotted Slow-worm *Anguis fragilis colchicus*
Anguis see Subfamily Anguinae above; *fragilis* (L) fragile; an allusion to the tail, which easily breaks off if seized by a predator (autotomy); *-icus* (L) suffix meaning belonging to; Colchis is an ancient territory in Asia situated between the Black Sea and the Caspian Sea; it does not appear on modern maps and only exists as a Greek legend, once famous as a home of sorcery. This subspecies has small blue spots on the back; it inhabits eastern Europe and parts of Asia.

Slow-worm *A. f. fragilis*
Anguis fragilis (L) see above. It is widespread in Europe including the British Isles, but not northern Scandinavia; it can also be found in part of northern Africa and in Asia Minor. This is the nominate subspecies (see Chapter 2, p. 19).

Peloponnesos Slow-worm   *A. f. peloponnesiacus*
-*acus* (L) suffix meaning pertaining to; this subspecies inhabits southern Greece; Morea forms a large part of southern Greece and is sometimes considered to be an island, but it is joined to the mainland by the Isthmus of Corinth. Another name for Morea is Peloponnesos.

Family   ANNIELLIDAE   2 species

anniella (New L) from *anillo* (Sp) a ring, and *anellus* (L) a little ring; an allusion to its likeness to a worm (e.g. an earthworm) which has rings that mark the body and divide it into segments. A legless lizard, it has the appearance of a worm and is related to the Slow-worm.

Geronimo Legless Lizard   *Anniella geronimensis*
Anniella see Family Anniellidae above; -*ensis* (L) suffix meaning belonging to; it inhabits the island of San Geronimo and the adjacent coastal strip of Lower California, USA.

California Legless Lizard   *A. pulchra pulchra*
pulchrum (L) beautiful; although a worm-like creature could hardly be considered beautiful, this little lizard usually has a silvery back, yellow flanks, and dark stripes along the body. It inhabits California ranging south as far as Mexico. This is the nominate subspecies (see Chapter 2, p. 19).

Black Legless Lizard   *A. p. nigra*
niger (L) black; this subspecies is black or dark brown and inhabits the coastal area of California just south of San Francisco.

Family   XENOSAURIDAE   4 species

xenos (Gr) a stranger; *saura* (Gr) a lizard; a reference to this inconspicuous shy lizard being seldom seen; it hides in a hole or rock crevice for long periods with just the head peeping out.

Chinese Crocodile Lizard   *Shinisaurus crocodilurus*
shin does not mean China; the *shin* spelling may be used in one
of the lesser known European languages; *krokodeilos* (Gr) a kind
of lizard, a crocodile; *oura* (Gr) the tail; there is a double row of
prominent scales along the top of the tail as in crocodiles, and it
is a good swimmer and diver. It inhabits Kwangsi Province in
south-western China.

(No recognised English name)   *Xenosaurus newmanorum*
Xenosaurus see Family Xenosauridae above; the unusual ending
of the specific name, given in 1949, is the genitive plural, and is
in honour of R. J. Newman and his wife Marcella; he was a
zoologist at the Museum of Louisiana State University. They
found the lizard in San Luis Potosi State, Mexico, in 1946. It
inhabits the southern half of Mexico.

Flat-headed Xenosaur   *X. platyceps*
platus (Gr) flat; *ceps* (New L) from *caput* (L) the head; hybrid
words like this, combining Latin and Greek, are frowned on by
nomenclaturists; all xenosaurs have rather flat heads and
bodies. Inhabiting Mexico.

Family   HELODERMATIDAE   2 species

hēlos (Gr) a nail or stud, particularly as used for decoration;
derma (Gr) the skin; the scales of these lizards abut each other
and form rows of little studs, which also accounts for the name
Beaded Lizards. The two species are the only lizards that are
known to be poisonous; the bite is dangerous but not usually
fatal.

Beaded Lizard   *Heloderma horridum*
Heloderma see Family Helodermatidae above; *horridum* (L)
rough, bristly; can also mean frightful, horrible; it has a
poisonous bite. The reason for 'beaded lizard' is explained
above. Inhabiting western Mexico.

Gila Monster   *H. suspectum*
suspectus (L) distrusted, suspicious; it was not known for certain

when this lizard was first discovered and named whether it had a poisonous bite, but Cope, who named it in 1865, for some reason thought it had. One cannot blame him for not trying it out on his own finger, but two cases have since been recorded of men who were bitten on the finger; one died but the other recovered. The English name is from the Gila River in Arizona, where this lizard lives; it is also found in the neighbouring part of Mexico. E. D. Cope (1840–1897) was an American zoologist with a particular interest in herpetology (E. C. Jaeger).

Family VARANIDAE 31 species

varanus (New L) derived from *waran* (Ar) a monitor lizard; the name Monitor arises from a belief that these lizards give a warning when there are crocodiles about, but there is probably no truth in this belief.

(No recognised English name) *Varanus acanthurus*
Varanus see Family Varanidae above; *akantha* (Gr) a thorn, a prickle; *oura* (Gr) the tail; it has very spiky scales on the tail. Inhabiting Western Australia, the central area and the Northern Territory.

Bengal Monitor *V. bengalensis*
-ensis (L) suffix meaning belonging to; named from Bengal, it is widespread in southern Asia. Since 1947, the original Bengal Province no longer exists; it has been divided into West Bengal and Bangladesh.

Short-tailed Monitor *V. brevicauda*
brevis (L) short; *cauda* (L) the tail; it is the smallest monitor, at about 20 cm (8 in), and the tail is even shorter than the body, an exception in this genus. Inhabiting Western Australia.

(No recognised English name) *V. caudolineatus*
cauda (L) the tail; *linea* (L) a linen thread, string; can mean a line drawn; *-atus* (L) suffix meaning provided with; a reference to the four stripes along the tail. Inhabiting Western Australia.

Duméril's Monitor   *V. dumerilii*
Named after Professor A. M. C. Duméril (1774–1860), a well-known French zoologist; he studied at the Paris University and specialised in herpetology. This monitor inhabits Burma, Malaysia, Sumatra, Banka Island and neighbouring small islands.

Yellow Monitor   *V. flavescens*
flavus (L) yellow; *-escens* (L) suffix meaning approaching, beginning to; 'yellowish'; it is a brownish yellow with dark spots. Inhabiting Pakistan and ranging to West Bengal.

Giant Monitor   *V. giganteus*
gigas (L) a giant; *giganteus* (L) very big, gigantic; a big lizard which can grow to a length of 2.4 m (8 ft), but not so big as the Komodo Dragon, p. 127. Inhabiting desert areas of central Australia.

Desert Monitor   *V. griseus*
griseus (New L) grey, derived from *greis* (G) grey; variable in colour, it has been described as brownish yellow. Able to live in desert areas, it ranges across the Sahara through Arabia and Iran to Pakistan.

Indian or Pacific Monitor   *V. indicus*
-icus (L) suffix meaning belonging to; although named from India, it does not live there; the name refers to the East Indies. Inhabiting Sulawesi (formerly Celebes) and Timor and ranging east through various small islands to New Guinea and the Marshall Islands.

Schmidt's Monitor   *V. karlschmidti*
A very rare lizard and considered to be an endangered species; it is named after Karl P. Schmidt (1880–1957), Curator of Reptiles at the Natural History Museum in Chicago; a well-known herpetologist, he came to a tragic end when he died from the bite of a Boomslang in 1957. This lizard inhabits the northern part of New Guinea but very few have been seen and recorded.

Komodo Dragon   *V. komodoensis*
-ensis (L) suffix meaning belonging to, usually a place; it inhabits Komodo Island which lies between Sumbawa and Flores in the East Indies, also the western part of Flores and two other small islands nearby. The name Dragon is well suited; it is an enormous lizard measuring up to 3 m (nearly 10 ft) and considered to be the largest lizard in existence. It is rare and listed as an endangered species.

Nile Monitor   *V. niloticus*
-icus (L) suffix meaning belonging to; although named from the Nile, it is widespread in Africa south of the Sahara, particularly near rivers, and ranging south to South Africa. A big lizard it can grow to a length of 2 m (6 ft 6 in), is a good swimmer and eats frogs, fishes and other aquatic animals, and crocodile eggs when it can get them.

Family   LANTHANOTIDAE   1 species

lanthanō (Gr) I am unseen; *ous* (Gr) genitive *ōtos*, the ear; it has eardrums but they are covered. A rather drab ordinary lizard, but of interest to herpetologists because it has anatomical features similar to snakes; for example the snake-like tongue and the inability to break off the tail, common to most lizards (autotomy, see Chapter 2, p. 20). Only a few have been seen and collected; the first one was discovered in 1878 and, since then, there have been continued differences of opinion among herpetologists as to its proper place in the scheme of classification. It now seems to have been settled by giving it a family on its own with one genus and one species.

Earless Monitor   *Lanthanotus borneensis*
Lanthanotus see Family Lanthanotidae above; *-ensis* (L) suffix meaning belonging to, usually a place; it has been found only in Sarawak, Borneo.

Family   BIPEDIDAE   3 species

bi- (L) prefix meaning two; *pes* (L) genitive *pedis*, the foot. These are known as the Two-legged Worm Lizards.

Common Two-legged Worm Lizard   *Bipes biporus*
Bipes see Family Bipedidae above; *porus* (L) a passage through the skin, a pore; a reference to two pores situated forward of the cloacal opening. This is a peculiar little creature having only the front legs, but these have adequate claws which can be used for digging. Inhabiting Lower California and Mexico.

Family   AMPHISBAENIDAE   about 130 species

amphis (Gr) on or at both sides; *bainō* (Gr) I step, go; 'going at both sides', which is supposed to mean going at both ends; the Liddell and Scott *Lexicon* gives a word *amphisbainō*, and translates it as 'a kind of serpent that can go both ways'. Known as the Ringed or Worm Lizards, they have a cylindrical worm-like body and the loose skin is marked with rings; their appearance certainly does not suggest that they are lizards. It is true, however, that they can 'go both ways', as in their tunnels they have been seen to move both forwards and backwards, but they are not serpents; neither end is obviously the head and, to add to the confusion, they sometimes raise the tail and wave it about, rather like a snake raising its head. Locally they are often called 'two-headed snakes'; like most lizards the tail breaks off (autotomy) if seized by a predator, but it does not regenerate. Herpetologists are still in some doubt about their correct place in the classification of animals; to quote: '.... this is the section on lizards, yet under these headings we are including a group of animals we are not positive are reptiles, let alone lizards' (Schmidt & Inger, *Living Reptiles of the World*). Sad to relate, Karl Schmidt died from the bite of a poisonous snake shortly after the publication of their book.

White-bellied Worm Lizard   *Amphisbaena alba*
Amphisbaena see Family Amphisbaenidae above; *albus* (L) white; the body is usually a yellowish red, with a white belly; it inhabits South America except the most southern part. For explanation of 'worm lizard' see above.

Darwin's Ringed Lizard   *A. darwini*
Named after Charles R. Darwin FRS (1809–1882); he has become famous for his work on the theory of evolution by natural selection; the validity of his theory is still in dispute, even today. This lizard inhabits Argentina, Paraguay, Uruguay and southern Brazil. For explanation of 'ringed lizard' see Family Amphisbaenidae above.

Spotted Worm Lizard   *A. fuliginosus*
fuligo (L) genitive *fuliginis* soot; *-osus* (L) suffix meaning full of, or augmented; 'very sooty'; a reference to the blackish brown and white spots on the body. Ranging from Panama to Argentina and from the Andes to the Atlantic coast.

Small-headed Worm Lizard   *Leposternon microcephalum*
lepos (Gr) a scale; *sternon* (Gr) the breast, chest; a reference to the type of scales on the throat and chest; *mikros* (Gr) small; *kephalē* (Gr) the head. Inhabiting the northern part of South America.

African Single-shield Worm Lizard   *Monopeltis capensis*
monos (Gr) alone, single; *peltē* (Gr) a small shield; a reference to the snout-plate which is part of the digging structure; *-ensis* (L) suffix meaning belonging to; named from the Cape, it inhabits a large area of central South Africa ranging to northern Botswana.

Florida Worm Lizard   *Rhineura floridana*
rhis (Gr) genitive *rhinos* the nose; *eurus* (Gr) wide, broad; a reference to the wide shovel-shaped snout which is heavily armoured and adapted for digging. One authority gives: *rhinē* (Gr) rough, a rasp; *oura* (Gr) the tail; when taking refuge in its hole the rough scaly tail is sometimes used to block the

entrance; either derivation seems suitable. Inhabiting central Florida.

Family   TROGONOPHIDAE   8 species

trōgō (Gr) I gnaw, *trōgōn* (Gr) gnawing; *ophis* (Gr) a snake; it is doubtful whether they actually gnaw their way underground, but they certainly use the shovel-shaped head to burrow in the ground. They have short tapered tails and are known as the Sharp-tailed Worm Lizards.

Wiegmann's Worm Lizard   *Trogonophis wiegmanni*
Trogonophis see Family Troganophidae above; Named after A. F. A. Wiegmann (1802–1841), the author of several books on reptiles; he came to a tragic end, being killed by hostile natives when travelling in Sicily. This Worm Lizard inhabits north-west Africa.

(No recognised English name)   *Agamodon anguliceps*
agama + odōn (Ionic Gr) a tooth; the teeth are acrodont like the agamid lizard's teeth; that means they are set on the top of the jaw bone and not on the side, which is known as pleurodont; *angulus* (L) an angle, a corner; *ceps* (New L) from *caput* (L) the head; a reference to the chisel-shaped head; the worm-lizard heads show a variety of differently shaped snouts but all are adapted for digging their burrows. Inhabiting Somalia and south-eastern Ethiopia.

11 **Snakes** SQUAMATA: SERPENTES (or OPHIDIA)

Carpet Python
Morelia argus

The snakes, of the order Squamata, suborder Serpentes, with about 2,200 species, show a very great range of size. Some members of the family Typhlopidae are quite tiny at about 9 cm (3½ in) long, whereas the huge Anaconda *Eunectes murinus* reaches about 9.5 m (31 ft); some herpetologists claim that the Reticulated Python *Python reticulatus* is the largest reptile known and may grow to a length of 10 m (32 ft 6 in); however, measuring the length of a snake, either alive or dead, is a very difficult procedure.

There are a number of anatomical features peculiar to snakes which distinguish them without doubt from legless lizards, which may look very much like snakes e.g. the small Slow-worm *Anguis fragilis* and the much larger Pallas's Glass Lizard *Ophisaurus apodus*.

The anatomical details are not within the scope of this book, but it should be noted that they have no eyelids, they are deaf and they cannot shed their tails (autotomy) like most lizards. The eye is covered with a transparent scale that is part of the skin, so when the snake sheds its skin it will have a new eye covering with the new skin. The external ear and eardrum are missing, so it is virtually deaf, but it is very sensitive to vibrations in the ground, so that a person walking nearby will cause alarm, even if unseen. Hence the idea that a snake-charmer can influence the snake with his music is nonsense; the snake may sway from side to side in time with the music, but it is only following the movements of the snake-charmer. For further information about autotomy see Chapter 2, p. 20).

Family TYPHLOPIDAE   about 400 species

tuphlos (Gr) blind; *ops* (Gr) the eye, face; although known as Blind Snakes they are not blind, but the eyes are small and inefficient and can probably only distinguish light from dark.

Common Blind Snake   *Typhlops braminus*
Typhlops see Family Typhlopidae above; Brama, or more correctly Brahma, is the supreme being of the Hindu pantheon; *-inus* (L) suffix meaning belonging to; intended to mean 'of India'. It also inhabits Madagascar, Sri Lanka and parts of south-east Asia, including Malaysia and Indonesia.

Vermiform Blind Snake   *T. vermicularis*
vermis (L) a worm; diminutive *vermiculus*, a little worm; *-aris* (L) suffix meaning pertaining to. A worm-like creature about 35 cm (1 ft 2 in) in length, inhabiting Asia Minor and ranging east to the southern Caspian Sea area.

Family LEPTOTYPHLOPIDAE   more than 40 species

leptos (Gr) slender, see Family Typhlopidae above. Known as the Slender Blind Snakes.

Western Blind Snake   *Leptotyphlops humilis*
Leptotyphlops see Family Leptotyphlopidae above; *humilis* (L) small, dwarfish; one of the smallest in this family, about 10 cm (4 in) long. Inhabiting Mexico, Central America and part of southern North America including Florida.

Large-scaled Blind Snake   *L. macrolepis*
makros (Gr) long; *lepis* (Gr) a scale; considered to be the largest snake in this genus, about 30 cm (1 ft) in length. Inhabiting Venezuela.

Family   ANILIIDAE   about 9 species
anileōs (Gr) unmerciful, cruel (see also below). Known in England as the Pipe Snakes.

False Coral Snake   *Anilius scytale*
Anilius from *anileōs* (Gr) cruel; it bears a distinct likeness to the poisonous Coral Snake, with rings of black and brilliant red, which evidently gave rise to the name 'cruel'; however it lacks the yellow rings of the true Coral Snake, and is not poisonous, hence 'false'; *skutalē* (Gr) a staff; can mean a serpent of uniform roundness and thickness; this also accounts for the English name Pipe Snakes. Inhabiting the northern half of South America though not Peru and Ecuador.

Asian Pipe Snake   *Cylindrophis rufus*
kulindros (Gr) a cylinder; *ophis* (Gr) a snake; (see notes under False Coral Snake about the shape of Pipe Snakes); *rufus* (L) red; an allusion to the underside of the tail which is a bright red; it has a habit of lifting the tail when alarmed which probably serves as a warning: 'keep off'! Inhabiting southern India, Sri Lanka, the Burma-Vietnam area and Malaysia.

Family   UROPELTIDAE   43 species, possibly more
oura (Gr) the tail; *peltē* (Gr) a small shield; a reference to a much enlarged scale on the tail which takes various forms; the function of this special scale is not known with any certainty; it may be used for burrowing; known as the Shield-tailed Snakes.

Shield-tailed Snake   *Rhinophis oxyrhynchus*
rhinē (Gr) rough, a rasp; *ophis* (Gr) a snake; an allusion to the large roughened scale that protects about half the tail; some

herpetologists suggest that it may be used to block the entrance to their hole when they hide; *oxus* (Gr) sharp, pointed; *rhunkhos* (Gr) the beak, snout; it has an unusually pointed head and snout. Inhabiting Sri Lanka.

Nilgiri Shield-tailed Snake *Uropeltis ocellatus*
Uropeltis see Family Uropeltidae above; *ocellus* (L) a little eye; *-atus* (L) suffix meaning provided with; a reference to the eye-like spots on the body. Inhabiting southern India; it has been seen at high altitudes on the Nilgiri Hills.

Family XENOPELTIDAE 1 species

xenos (Gr) a stranger, foreigner; can mean strange, unusual; *peltē* (Gr) a small shield, in this case taken to mean the scales, skin, as in the English word 'pelt'; an allusion to the highly iridescent scales; this also gives them the English name Sunbeam Snakes.

Sunbeam Snake *Xenopeltis unicolor*
Xenopeltis see Family Xenopeltidae above; *unicolor* (L) of one colour, uniform in colour; it is a uniform brown in colour which makes the iridescence more remarkable. Inhabiting the Burma-Vietnam area and the Malay Peninsula.

Family ACROCHORDIDAE 2 species

akros (Gr) at the end, top; *khordē* (Gr) gut, string; *akrokhordōn* (Gr) a wart with a thin neck, as though on the top of a stalk; the scales of these snakes do not overlap, they abut against each other and have a sharp ridge, which gives the skin a rough warty appearance. Usually called Wart Snakes, they are sometimes known as Water Snakes, as they live in marshy areas and river estuaries and the anatomy is adapted for an aquatic life.

Javan Wart Snake *Acrochordus javanicus*
Acrochordus see Family Acrochordidae above; *-icus* (L) suffix

meaning belonging to; although named from Java, it inhabits a large coastal area in Vietnam, Malaysia, the Philippines, New Guinea and many islands in that area.

Indian Wart Snake   *Chersydrus granulatus*
khersudros (Gr) an amphibious serpent; *granulum* (L) a small grain; *-atus* (L) suffix meaning provided with; a reference to the rough skin (see Family Acrochordidae above). This snake is also frequently seen in the water; the range is more western, and includes the east coast of India and Sri Lanka, but does not extend east to New Guinea.

Family   BOIDAE   about 100 species

boa (L) a kind of snake.   The Boas and Pythons make up this family.

Subfamily   LOXOCEMINAE   1 species

Loxocemus could be translated from the greek to mean 'slanting muzzle', but the reason for this is obscure and not explained by Cope who wrote the original description.

Mexican Python   *Loxocemus bicolor*
Loxocemus see Subfamily Loxoceminae above; *bicolor* (L) of two colours; it has a brown body and pale belly; little is known about this snake and its correct place in classification is still being discussed by herpetologists. Inhabiting western and southern Mexico and the neighbouring part of Central America.

Subfamily   PYTHONINAE

puthōn (Gr) the serpent slain by Apollo, thence surnamed the Pythian.

Angola Python   *Python anchietae*
Python see Subfamily Pythoninae above. Named after a com-

paratively unknown Portuguese naturalist, J. de Anchieta, who died in 1897; presumably a woman because of the feminine genitive ending -*ae*; at one time she was resident in Angola. It ranges to neighbouring parts of south-west Africa.

Short-tailed or Blood Python   *P. curtus*
curtus (L) shortened; not only does it have a short tail, but the total length is less than most pythons at about 3 m (9 ft 9 in). Known as a Blood Python because the body is mainly dark red; it inhabits Malaysia, Sumatra and Borneo.

Indian or Asiatic Python   *P. molurus molurus*
molouros (Gr) a kind of serpent. It ranges from West Pakistan through India, including Sri Lanka, to southern China, and southwards to Malaysia. This is the nominate subspecies (see Chapter 2, p. 19).

Two-banded or Burmese Python   *P. m. bivittatus*
bi- (L) prefix meaning two; *vitta* (L) a ribbon, a band; -*atus* (L) suffix meaning provided with. This subspecies inhabits Burma and the Indo-Australian region.

Royal or Ball Python   *P. regius*
rex (L) genitive *regis* a king; *regius* (L) royal. A small python up to 2 m (6 ft 6 in) in length, it has an interesting habit, when frightened, of rolling itself up to form an almost perfect ball shape, with its head well protected in the centre. It is also interesting, but disappointing, that in captivity it soon becomes tame, is no longer frightened, and will not demonstrate this trick! It inhabits the tropical part of West Africa.

Reticulated Python   *P. reticulatus*
reticulum (L) a little net; *reticulatus* (L) net-like, reticulated; a reference to the marking on the body. Possibly the longest snake known; a length of 10 m (32 ft 6 in) has been recorded. Ranging from Thailand to the Malay Archipelago and the Philippine Islands.

African Python   *P. sebae*
Named after Albert Seba (1665–1736) a Dutch zoologist; in

1734, he published a four-volume work describing animals, which Linnaeus found was a valuable source of reference (Schmidt & Inger); in 1735, he founded the Seba Museum in Amsterdam. This python is quite widespread in Africa south of the Sahara.

Timor Python   *P. timorensis*
A small python, less than 3 m (about 9 ft) in length, and considered rare; it lives on the islands of Timor and Flores, Indonesia.

Amethystine Rock Python   *Liasis amethistinus*
Lias is a type of blue limestone, so both generic and specific names indicate blue; the snake is predominantly brown, but Schneider's original description says: 'colore corporis ex cinereo caerulescente', and although this suggests 'bluish', it probably refers to the well-defined surface iridescence. Amethyst is a type of blue quartz, possibly seen in the rocks where this python was first found; it is supposed to ward off intoxication, from *a-* (Gr) a prefix meaning not, or there is not, and *methustikos* (Gr) drunken. Inhabiting the Philippines, Indonesia, New Guinea and northern Australia.

Green Python   *Chondropython viridis*
khondros (Gr) like groats, coarse-grained; a reference to the type of scales; *viridis* (L) green; sometimes known as the Papuan Tree Python, it is emerald green with some white marking. Inhabiting Papua New Guinea, the Solomon Islands and the Aru Islands.

Subfamily   BOINAE

boa (L) a type of snake.

Black-tailed Boa   *Tropidophis melanura*
tropis (Gr) genitive *tropios* which later became *tropidos* a ship's keel; *ophis* (Gr) a snake; a reference to the keeled scales on the belly, which probably help it to crawl up rocks or walls; *melas* (Gr) genitive *melanos* black; *oura* (Gr) the tail. One of the quite

small boas, less than 1 m (about 3 ft) in length, it is widespread on the larger islands in the West Indies including the Bahamas, Cuba and Hispaniola.

Rainbow Boa   *Epicrates cenchris*

epikratēs (Gr) one who gets possession of a thing; a reference to the snake's ability to hold an animal in its coils and constrict it until it dies from suffocation; it does not crush it to death; *kenkhros* (Gr) millet, small grains, = *kenkhris* (Gr) a kind of serpent with bead-like scales; in spite of the English name, the colouring of this snake is not particularly bright, but the scales have an iridescent sheen in certain light conditions, especially noticeable after it has shed its skin. Inhabiting Central and South America.

Bahama Boa   *E. striatus*

stria (L) a furrow; *striatus* (New L) striped.

Rosy Boa   *Lichanura roseofusca*

likhanos (Gr) the forefinger; *oura* (Gr) the tail; 'a finger-tail'; a reference to the short tapered tail having the appearance of a finger; *roseus* (L) rose-coloured; *fuscus* (L) dark-coloured. Inhabiting the south-western part of the USA and Mexico.

Three-striped Boa   *L. trivirgata*

tria (L) three; *virgatus* (L) striped; sometimes known as the Rosy Boa, it has three reddish stripes extending the length of the body, one on the back and one on each flank. Inhabiting Lower California.

Javelin Sand Boa   *Eryx jaculus*

Eryx is the name of a mountain in Sicily famous for its Temple of Venus; according to fable it was named after the Sicilian King Eryx. There is probably no special reason for this generic name; it was given by Linnaeus in 1758 and he used a number of names from classical mythology without thought of any physical significance; other zoologists have sometimes followed his example; *jaculum* (L) a dart, a javelin; *jaculus* (L) a serpent that darts from a hole to seize its prey. Inhabiting sandy or desert areas in south-east Europe, Asia Minor and northern Africa.

Boa Constrictor   *Boa constrictor*

Boa see Subfamily Boinae, p. 127; *constringo* (L) I bind together, confine. The snakes in the family Boidae are not venomous; they seize their prey with powerful teeth and then coil round the animal's body; they do not crush it to death, but constrict it enough to stop it breathing; thus it dies from suffocation. They are big snakes, though the Boa Constrictor is not the longest at about 4 m (13 ft). Inhabiting Mexico, Central America and ranging south to central Argentina.

Anaconda   *Eunectes murinus*

eu- (Gr) prefix meaning well, nicely; *nēktēs* (Gr) a swimmer; 'a good swimmer'; *murinus* (L) mouse-like, mouse-coloured; although not so brightly coloured as some of the other big snakes the name does not seem really suitable. This big snake is largely aquatic and is usually seen in the water. There is some competition between the Anaconda and the Reticulated Python as to which can be called the longest reptile in the world; some herpetologists give the length of the Anaconda as up to 9.5m (31 ft) and say it is the longest, while others say that the Reticulated Python has been recorded as measuring 10 m (32 ft 6 in). The Anaconda inhabits South America east of the Andes from Colombia ranging south to Paraguay.

Subfamily   BOLYERIINAE

The genus was created by Gray, who gave no reason for the name. It seems likely that it refers to a little known collector named Bolyer.

Round Island Snake   *Bolyeria multicarinatus*

Bolyeria see Subfamily Bolyeriinae above; *multus* (L) many; *carina* (L) a keel; *carinatus* (L) keeled; a reference to spines on the underside of the vertebrae. Originally living on Mauritius but now found only on the nearby Round Islet; it is considered to be an endangered species and may soon be extinct.

Family   COLUBRIDAE   about 1,200 species

coluber (L) fem. *colubra* a serpent. Known as the Typical Snakes.

Subfamily   XENODERMINAE

xenos (Gr) a stranger; can mean strange, unusual; *derma* (Gr) skin, hide; there are several unusual features concerning the scales, for example the lip scales have upturned edges and the body scales are similar to those in some lizards, which do not overlap.

Javan Xenodermin   *Xenodermis javanicus*
Xenodermis see Subfamily Xenoderminae above; *-icus* (L) suffix meaning belonging to; inhabiting Java.

Stoliczka's Xenodermin   *Stoliczkaia khasiensis*
Named after Dr F. Stoliczka (1838–1874) who carried out a geological survey in India; *-ensis* (L) suffix meaning belonging to, usually a place; it was probably first found by Dr Stoliczka on the Khasi Hills in north-eastern India where this snake inhabits a large area of mountainous terrain.

Subfamily   SIBYNOPHINAE

sibynē (Gr) a spear; *ophis* (Gr) a snake; a reference to the shape of the head, hence the English name Spear Snakes.

Asian Spear Snake   *Sibynophis chinensis*
Sibynophis see Subfamily Sibynophinae above; *-ensis* (L) suffix meaning belonging to, usually a place; inhabiting eastern Asia including China and surrounding areas.

(No recognised English name)   *Scaphiodontophis annulatus*
skaphē (Gr) anything dug out, or scooped out; *odous* (Gr) genitive *odontos* a tooth, fang; *ophis* (Gr) a snake; a reference to the widened tips of the maxillary teeth; *annulatus* (L) having a ring; this is a reference to the striking pattern of alternating rings of red, black and yellow on the body. Inhabiting Central America and the north-western part of South America.

Subfamily   XENODONTINAE

xenos (Gr) a stranger; can also mean strange, unusual; *odōn* (Ionic Gr) a tooth, fang; a reference to the unusual dentition; they have enlarged *solid* and *ungrooved* fangs at the back of the upper jaw and yet some are known to produce venom. Into this subfamily come the American Hog-nosed Snakes.

Eastern Hog-nosed Snake   *Heterodon platyrhinos*
heteros (Gr) different; can mean other than usual; *odōn* (Ionic Gr) a tooth, fang, see Subfamily Xenodontinae above; *platus* (Gr) flat, wide; *rhis* (Gr) genitive *rhinos* the nose; a reference to the peculiar shape of the nose which is adapted for burrowing. Inhabiting eastern North America.

Subfamily   NATRICINAE

nato (L) I swim; *natrix* (L) genitive *natricis* a water-snake; the name Water Snakes may be misleading as they spend a lot of their time on land and hence are sometimes known as Grass Snakes; they are good swimmers, living near water, and frogs form an important part of their diet.

Ringed Snake or Grass Snake   *Natrix natrix natrix*
Natrix see Subfamily Natricinae above. This is the nominate subspecies; for an explanation of the generic name *Natrix* appearing three times, see TAUTONYMS, p. 13. The colour of this snake varies in different individuals from greyish brown to green, and the name Ringed Snake arises from two half-moon yellow patches just behind the head which almost completes a

ring; in some countries it is called 'The Snake with the Golden Crown' and is supposed to bring good luck. Inhabiting Europe, though not Ireland, northern Scotland, Iceland, or northern Scandinavia, but ranging east to part of Russia and south to northern Africa.

Collarless Grass Snake   *N. n. astreptophora*
a- (Gr) prefix meaning not, or there is not; *streptos* (Gr) a kind of collar; *phora* (Gr) a carrying, bearing; 'not having a collar'; the marking on the neck is not so distinct as with the Ringed Snake, see above. Inhabiting the Iberian Peninsula and north-west Africa.

Sicilian Grass Snake   *N. n. sicula*
siculus (L) a Sicilian. This subspecies is found only in Sicily and other subspecies are found on other islands such as Corsica and Sardinia.

Red-bellied Water Snake   *N. erythrogaster*
eruthros (Gr) red; *gaster* (Gr) the belly; inhabiting a large area in the south-eastern part of the USA, but not as far south as Florida.

Viperine Snake   *N. maura*
mauros (Gr) dark; the pattern of marking on the back is similar to a viper, but it is a harmless snake; people have been known to get bitten by picking up a viper thinking it was the harmless *N. maura*. Inhabiting the Iberian Peninsula and the south-eastern part of France.

Fishing Snake   *N. piscator*
piscator (L) a fisherman; an aquatic snake, it spends a lot of time in the water and the diet is mostly fish and frogs. Inhabiting a wide area in southern and south-east Asia.

Common Water Snake   *N. sipedon*
sēpedōn (Gr) rottenness, decay; can mean a snake whose bite causes putrefaction, but it is not venomous. Inhabiting a large area on the eastern side of North America ranging from the Great Lakes south to Florida.

Diced Snake   *N. tessellata*
tessellatus (L) set with small cubes; the body is marked with a diced pattern in rows. Inhabiting a range from West Germany across eastern Europe to central Asia.

(No recognised English name)   *N. trianguligera*
triangulum (L) a triangle; *gero* (L) I carry; a reference to the red triangular marks on the side of the body. Inhabiting Malaysia and Indonesia.

Kirtland's Water Snake   *Clonophis kirtlandi*
klonos (Gr) violent confused motion, persons fleeing in confusion; presumably a reference to persons who flee when seeing this snake; Liddell and Scott's *Lexicon* adds: 'and, comically, a turmoil in the bowels'; doubtless similar to our modern saying 'butterflies in the stomach'! *ophis* (Gr) a snake; it is a harmless snake. Named after Dr J. P. Kirtland (1793–1877), an American physician and zoologist, who founded the Cleveland Medical College, Ohio. This snake ranges from Ohio through Pennsylvania to New Jersey.

Common Garter Snake   *Thamnophis sirtalis*
thamnos (Gr) a bush, shrub; *ophis* (Gr) a serpent; it may be found in marshy areas, or where there are shrubs, and even near houses but it is not known to live mostly in shrubs; the colourful striped pattern is supposed to resemble a woman's garter. Ranging from southern Canada to the eastern part of the USA.

Subfamily   COLUBRINAE

coluber (L) fem. *colubra* a serpent.

Smooth Snake   *Coronella austriaca*
corona (L) a crown; *-ellus* (L) diminutive suffix; no reason is given for naming this snake 'a little crown'; *auster* (L) genitive *austri* the south wind; *-acus* (L) suffix meaning relating to, hence *austriaca*, of the south, southern. It is widespread in Europe including southern Scandinavia and southern Britain; the

scales are smooth, in contrast with most other snakes that have keeled scales.

Milk Snake   *Lampropeltis getulus*
lampros (Gr) shining, beautiful; *peltē* (Gr) a small shield; in this case taken to mean the scales, skin, as in the English pelt, 'shining pelt', a reference to the bright colouring specially noticeable after shedding the skin – 'Among the most beautiful colubrid snakes in the world' (Grzimek); *getulus* (L) belonging to the Getulians, a people inhabiting north-west Africa in ancient times; this is a mistake, as the snake was thought to have been found in the Morocco area, whereas it lives in the southern part of the USA and Mexico. It acquired the name Milk Snake because of a story that it sucked milk from cows while they slept; it is unlikely to be true.

Smooth Green Snake   *Opheodrys vernalis*
opheo (New L) from *ophis* (Gr) a snake; *drus* (Gr) a tree, usually an oak tree; *verno* (L) I flourish, I become green; *-alis* (L) suffix meaning relating to; it inhabits shrubs and grassland. The scales are smooth like those of *Coronella austriaca* p. 143; it is widespread in the USA.

Balkan Racer   *Coluber gemonensis*
coluber (L) a snake; *-ensis* (L) suffix meaning belonging to; named from Gemona, in north-east Italy, it also inhabits the adjacent part of Yugoslavia, the Balkan coast and some Greek islands such as Crete. Known as a 'racer' from its rapid movement when disturbed – 'On a wooded rocky slope they can disappear with lightning speed' (Schmidt & Inger).

Horseshoe Racer   *C. hippocrepis*
hippos (Gr) a horse; *krepis* (Gr) a shoe; a reference to a horseshoe-like mark on the head. Inhabiting the Iberian Peninsula, north-western Africa and the small island of Pantellaria off the north-east coast of Tunisia.

Slender Racer   *C. najadum*
naja (New L) from *nāga* (Sanskrit) a serpent; a very slender

snake, little more than 7 mm ($\frac{5}{16}$ in) in diameter, with a length of about 1.3 m (4 ft 2 in). Inhabiting the Balkan countries and ranging east through Asia Minor.

Coachwhip Snake *Masticophis flagellum*
mastic (New L) from *mastix* (Gr) a whip; *ophis* (Gr) a snake; *flagellum* (L) a small whip; there is a legend which says that this snake can coil round a man and bind him to a tree, and then whip him to death with its tail; it is, of course, no more than a legend, as the snake is not a constrictor and is far too small to perform such a feat. It is widespread along the southern part of the USA including Florida and ranging south to Mexico.

Aesculapian Snake *Elaphe longissima*
elaphos (Gr) a deer; the reason for the name is obscure. Linnaeus gave no explanation when he created it in 1758. *longus* (L) long; *longissima* (L) very long; in spite of the name, it is not the longest snake in this genus, at about 2 m (6 ft 6 in). Inhabiting south-eastern Europe and Asia Minor. In ancient times, this snake was associated with the snake symbol of Aesculapius, the god of medicine, and hence enjoyed protection; it is depicted on the badge of the Royal Army Medical Corps.

Sharp-headed Rat Snake *E. oxycephala*
oxus (Gr) sharp, pointed; *kephalē* (Gr) the head. An arborial snake inhabiting Burma, Malaysia, Indonesia and the Philippines. Some species are known as Rat Snakes because rats form a large part of the diet.

Four-lined Rat Snake *E. quattuorlineata*
quattuor (L) four; *linea* (L) a line; *lineata* (L) lined; the four lines along the body are usually seen best in the older snakes. The single 't' in the specific name is not correct Latin. Inhabiting Italy, Sicily, the Balkan Peninsula and ranging east through Turkey to the Caucasus.

Ladder Snake *E. scalaris*
scala (L) a ladder; *-aris* (L) suffix meaning pertaining to; there

are two lines forming a ladder-like pattern along the back
which is most noticeable in older animals. Inhabiting the
Iberian Peninsula.

Oriental Rat Snake   *Ptyas mucosus*
ptuas (Gr) the spitter, a kind of serpent which is supposed to spit
poison into the eye of an aggressor; *mucus* (L) the mucous
matter of the nose; *mucosus* (L) full of mucous; it is not recorded
that this snake actually spits, like the Spitting Cobra, but it
makes a peculiar coughing noise, as though it has a sore throat,
in addition to the usual hissing. Inhabiting a large area of
south-east Asia, including India and China, but not Kampuchea, Vietnam, and Malaysia, although it is found further
south in Java. Known as a Rat Snake because rats form an
important part of the diet.

Keeled Rat Snake   *P. carinatus*
carina (L) the keel of a ship; *-atus* (L) suffix meaning provided
with; a reference to the keeled scales of the under-belly.
Inhabiting eastern Indian, Malaysia and Indonesia.

Black Tree Snake   *Thrasops flavigularis*
thrasos (Gr) boldness, impudence; *opsis* (Gr) appearance, aspect; a harmless snake, but on account of its colour and size it is
easily mistaken for the Black Mamba, a very dangerous
venomous snake; *flavus* (L) yellow; *gula* (L) the throat, gullet;
-aris (L) suffix meaning relating to. Inhabiting western Africa.

Subfamily   CALAMARINAE

calamus (L) a reed; *-aria* (L) suffix meaning like or connected
with; the reason for the name Reed Snakes is not explained in
zoological records, but they are often found in marshy places.

Linné's Reed Snake   *Calamaria linnaei*
Calamaria see Subfamily Calamarinae above. Named after Carl
von Linné (1707–1778), the famous Swedish botanist; he
studied and named many animals, in addition to plants.
Inhabiting Java.

Subfamily LYCODONTINAE

lukos (Gr) a wolf; *odous* (Gr) genitive *odontos* a tooth; a reference to the long upper jaw teeth, the Wolf Snakes.

Indian Wolf Snake *Lycodon aulicus*
Lycodon see Subfamily Lycodontinae above; *aluix* (L) genitive *aulicis* a furrow; probably an anatomical mistake as the long upper teeth are not grooved. Widespread in south-eastern Asia, including India, Sri Lanka, the Burma-Vietnam area, Malaysia, Indonesia and the Philippines.

Cape Wolf Snake *Lycophidion capense*
ophis (Gr) a snake; diminutive *ophidion* a small snake; the length is about 45 cm (1 ft 5½ in) and should be compared with that of the Indian Wolf Snake (above) which is usually more than 50 cm (about 1 ft 8 in); *-ensis* (L) suffix meaning belonging to; 'of The Cape'. Inhabiting the southern part of South Africa.

African Striped Ground Snake *Bothrophthalmus lineatus*
bothros (Gr) a hole, a trench; *ophthalmus* (Gr) the eye; a reference to the small heat-sensitive pits in front of the eyes (see Pit Vipers, p.oo) *linea* (L) a line; *lineatus* (L) lined; a reference to the black and red stripes. Inhabiting the Ivory Coast and ranging south through Nigeria, to Congo and Zaire.

Subfamily DIPSADINAE

dipsa (Gr) thirst; *dipsas* (Gr) genitive *dipsados* a venomous snake whose bite causes intense thirst, hence the English name Thirst Snakes.

Indonesian Thirst Snake *Dipsas carinatus*
Dipsas see Subfamily Dipsadinae above; *carina* (L) a ship's keel; *carinatus* (L) keeled; a reference to the keeled scales of the belly. Inhabiting Sumatra and Java.

(No recognised English name) *Sibynomorphous ventrimaculatus*
sibynē (Gr) a spear; *morphē* (Gr) form, shape; an allusion to the shape of the head; *venter* (L) genitive *ventris* the belly; *macula* (L)

a mark, a spot; *maculatus* (L) spotted; a reference to the marks on the under-belly. Inhabiting North America.

Subfamily DASYPELTINAE

dasus (Gr) hairy, shaggy, can mean rough; *peltē* (Gr) a small shield, in this case taken to mean the scales, skin, as in the English word 'pelt'; a reference to the rough scales of the body. Known as the Egg-eating Snakes.

African Egg-eating Snake   *Dasypeltis scabra*
Dasypeltis see Subfamily Dasypeltinae above; *scaber* (L) fem *scabra*, rough. The egg-eating snakes have only small weak teeth which are of little use except perhaps to assist in swallowing the egg; however the vertebrae of the neck have sharp processes which penetrate the oesophagus, almost like a saw, and this cuts open the egg as it is being swallowed; the shell is regurgitated. This species is widespread in Africa.

Indian Egg-eating Snake   *Elachistodon westermanni*
elakhus (Gr) small, short; *elakhistos* (Gr) smallest, shortest; *odōn* (Ionic Gr) fang, tooth; it has only small, weak teeth (see *D. scabra* above). Dr G. F. Westerman (1807–1890) was for many years the Director of the Zoological Gardens in Amsterdam. This snake inhabits a small area in north-eastern India.

Subfamily HOMALOPSINAE

homalos (Gr) smooth, even; *opsis* (Gr) aspect, appearance; an allusion to the smooth scales. The species in this subfamily are largely aquatic, inhabiting both inland waters and the sea.

(No recognised English name)   *Homalopsis buccata*
Homalopsis see Subfamily Homalopsinae above; *bucca* (L) the cheek, especially when puffed out as in eating; *-atus* (L) suffix meaning provided with. Inhabiting south-east Asia, including Burma, Sumatra, Java and neighbouring small islands, parti-

cularly along the coast; the anatomy is adapted for an aquatic life.

Tentacled Snake *Erpeton tentaculatum*
herpeton (Gr) a reptile; *tentaculum* (L) a feeler; *-atus* (L) suffix meaning provided with; it has a peculiar feeler-like process on each side of the snout. Inhabiting Kampuchea and the southern part of Vietnam.

Subfamily BOIGINAE

The name Boiga was coined, with no explanation given, by the Austrian zoologist Dr Fitzinger (1803–1862).

Mangrove Tree Snake *Boiga dendrophila*
Boiga see Subfamily Boiginae above; *dendron* (Gr) a tree; *philos* (Gr) loved, dear. Inhabiting southern Asia, Indonesia and the Philippines.

Hooded Snake *Macroprotodon cucullatus*
makros (Gr) long; *prōtos* (Gr) first; *odōn* (Gr) a tooth; an allusion to the elongated upper front teeth; *cucullus* (L) a hood; *cucullatus* (L) hooded; this name is an allusion to the dark marking on the head and neck. Inhabiting southern Spain, the Balearic Islands and the north-western part of northern Africa.

(No recognised English name) *Dendrophis mycterizans*
dendron (Gr) a tree; *ophis* (Gr) a snake; this is a leaf-green arboreal snake; *mycterizō* (Gr) I turn up the nose; *mycterizans* (New L) a turning up of the nose; this refers to the tip of the snout, which is a moveable trunklike structure. Inhabiting the southern half of the Malay Peninsula.

Grey Vine Snake *Oxybelis acuminatus*
oxus (Gr) sharp, pointed; *belos* (Gr) a dart, a sting; *acumen* (L) genitive *acuminis* a point to sting with; *-atus* (L) suffix meaning provided with; although mildly venomous the poison is not usually strong enough to be dangerous to human beings. The

slender body has a long drawn-out snout; it is an arboreal snake and often lives in vine trees. Inhabiting South America.

Green Vine Snake   *O. fulgidus*
fulgidus (L) shining; a reference to the bright green and yellow colour, which is particularly noticeable after the snake has shed its skin. Inhabiting Central America and the adjacent part of South America.

Flying Snake   *Chrysopelea ornata*
khrusos (Gr) gold; *pēlēx* (Gr) a crown, helmet; *ornatus* (L) ornament, decoration; a reference to the bright green and yellow colours of the head and body. Certainly this snake cannot fly, although an ancient legend says that it turns into a bird when moving from tree to tree! Even its ability to glide a short distance has rarely been seen and recorded, although some unusual anatomical features suggest that short glides are possible. It inhabits Sri Lanka and ranges through south-eastern Asia including China, the Thailand–Vietnam area, Malaya, Indonesia and the Philippines.

Boomslang   *Dispholidus typus*
dus- (Gr) prefix meaning bad, ill, not normal, as in e.g. dyspepsia; *pholis* (Gr) genitive *pholidos* a horny scale, especially of reptiles; a reference to the unusual formation of the dorsal scales on the forepart of the body, being spread out on the sides and strongly overlapping near the spine; *tupos* (Gr) a blow, mark, or impression; can mean a type, and may be a reference to this unusual scale formation. Widespread in the southern half of Africa.

Family   ELAPIDAE   150 species

elops (Gr) mute, the mute one, the name of a type of fish; also a type of serpent, giving rise to *elaps* (New L) a serpent. The Cobras.

King Cobra   *Ophiophagus hannah*
ophis (Gr) a snake; *phagein* (Gr) to eat; it feeds almost entirely on

snakes, even poisonous snakes; the name *hannah* is taken to mean a giant; it is the largest venomous snake known and may reach a length of 5 m (16 ft 3 in) though 4 m (13 ft 1 in) is more usual. Inhabiting India, though not Sri Lanka, and ranging to southern China, Malaysia, the Sunda Islands, the Andaman Islands and the Philippines.

Indian Cobra   *Naja naja*
naja (New L) derived from Sanskrit *nāga*, a snake; in Hindu mythology this is a divine snake. The snake in one of Kipling's stories was called Nag. Inhabiting India and ranging through Asia to southern China, the Sunda Islands and the Philippines.

Egyptian Cobra   *N. haje*
hajj or *hadj* means a Mohammedan pilgrimage to Mecca. Widespread in northern Africa but not the Sahara and ranging south on the eastern side to Mozambique.

Spitting or Black-necked Cobra   *N. nigricollis*
niger (L) black; *collum* (L) the neck. It only spits the venom at intruders, not at its prey; it does this by means of the fangs, which have a hole in the front, so that pressure from the poison gland forces the venom directly forward; it does not spit in the ordinary sense of the word. Inhabiting Africa south of the Sahara, but not Madagascar or the most southern part of Africa.

Shield-nosed Snake   *Aspidelaps scutatus*
aspis (Gr) genitive *aspidos* a shield; *elaps*, see Family Elapidae, p. 140; *scutum* (L) an oblong shield, *scutatus* (L) armed with a shield; an allusion to the broad snout plate which is used for digging. This burrowing snake lives in the sandy deserts of South Africa.

(No recognised English name)   *Elaps corallinus*
elaps (New L) see Family Elapidae, p. 140; *corallinus* (L) coral-red; a reference to the bright red bands round the body that alternate with black bands; 'One of the prettiest in this genus' (H. Gadow). Inhabiting the forests of tropical South America

and the Lesser Antilles. Dr H. F. Gadow FRS (1855–1927) was
at one time Curator at the Cambridge University Museum.

Water Cobra   *Boulengerina annulata*
Named after George Albert Boulenger, the famous Belgian-
British herpetologist, who did a lot of work for the British
Museum (Natural History) relating to reptiles during the years
1882 to 1896; *annulus* (L) (or *anulus*) a ring; *annulata* (L) having
a ring; a reference to the black and brown bands on the body.
It inhabits shallow water and lives in the large lakes of Central
Africa.

Black Mamba   *Dendroaspis polylepis*
dendron (Gr) a tree; *aspis* (Gr) a type of snake, an asp; *polus* (Gr)
many; *lepis* (Gr) a scale; a reference to the smooth scales
arranged in diagonal rows. Although the generic name indicates a tree-snake it lives mostly on the ground but sometimes
climbs into bushes; it is not really black but dark brown or dark
grey. Inhabiting Africa south of the Sahara but not the extreme
southern part.

Banded Krait   *Bungarus fasciatus*
bungarus (New L) from *bangāru* (Telugu); this is the language of
a large group of people living in Andhra Pradesh State, India;
fascia (L) a band or girdle; *fasciatus* (L) banded; it has yellow
bands on a shining black body. Inhabiting south-eastern Asia
and the Sunda Islands.

Many-banded Krait   *B. multicinctus*
multus (L) many; *cinctus* (L) a girdle; it has closely set narrow
black and white rings. Inhabiting southern China.

South American Coral Snake   *Micrurus corallinus*
mikros (Gr) small; *oura* (Gr) the tail; *korallion* (Gr) red coral;
corallinus (New L) coral red; an allusion to the broad red bands;
these are interspersed with narrow bands of yellow and black.
The Coral Snakes are brightly coloured and venomous, some
having a very dangerous poisonous bite. Inhabiting Brazil.

Eastern or Common Coral Snake *M. fulvius*
fulvus (L) tawny, yellowish brown; 'tawny' does not do justice to this brightly coloured snake; it has wide red and black bands interspersed with narrow yellow bands. Inhabiting south-eastern USA and Mexico.

Arizona Coral Snake *Micruroides euryxanthus*
-oides (New L) from *eidos* (Gr) apparent shape, form; this snake, similar to *Micrurus*, has to be placed in a separate genus because the dentition is different; *eurus* (Gr) wide; *xanthos* (Gr) yellow; an allusion to the yellow bands which are wider than those of *Micrurus*. Inhabiting the western part of Arizona.

Taipan *Oxyuranus scutulatus*
oxus (Gr) sharp, pointed; *oura* (Gr) the tail; *-anus* (L) suffix meaning belonging to; it has a long pointed tail; *scutulata* (L) a chequered garment; an allusion to the type of scales on the body. Inhabiting coastal areas of northern Queensland and New Guinea, and islands in the Torres Strait, which is the name of the sea between New Guinea and Australia.

Death Adder *Acanthophis antarcticus*
akantha (Gr) a thorn, a prickle; *ophis* (Gr) a serpent; it has a short tail with a spiked tip and some people believed it could sting with its tail; *-icus* (L) suffix meaning belonging to; it does not live in the Antarctic; it inhabits central and western Australia and New Guinea. The name Death Adder is justified; it has a head similar to an adder and the poison is deadly, more deadly than the cobra; records suggest that about 50% of people bitten by this snake will die.

Family HYDROPHIIDAE 50 species
hudōr (Gr) water; *ophis* (Gr) a serpent. The Sea Snakes.

Subfamily LATICAUDINAE
latus (L) broad, wide; *cauda* (L) the tail.

Black-banded Sea Krait   *Laticauda laticauda*

Laticauda see Subfamily Laticaudinae above; the tail is rudder-shaped and flattened vertically, an adaptation for swimming; however Sea Kraits are often seen ashore sunning themselves and the females come ashore to lay and bury their eggs. Inhabiting coastal waters from the Bay of Bengal and Japan and ranging south-east to northern Australia and New Guinea. For an explanation of the specific name being the same as the generic name see TAUTONYMS, p. 13).

(No recognised English name)   *L. semifasciata*

semi- (L) half; *fascia* (L) a band or girdle; *fasciatus* (L) banded; the young have distinct dark bands on the body but adult animals gradually lose the bands, so at one stage they are 'half-banded'. Inhabiting the coastal waters of the Ryukyu Islands, the Philippines and the Moluccas.

Subfamily   HYDROPHIINAE

hudōr (Gr) water; *ophis* (Gr) a serpent; the Sea Snakes.

Blue-banded Sea Snake   *Hydrophis cyanocinctus*

Hydrophis see Subfamily Hydrophiinae above; *kuaneos* (Gr) dark blue; *cinctus* (L) a girdle; these are genuine water animals, much more so than those in the subfamily Laticaudinae. Their anatomy is adapted for a marine life, they bear live young and never come ashore; indeed if washed ashore by accident, or caught and put on land, they are almost helpless, and the rib cage, normally supported by the buoyancy of the water, is liable to collapse under its own weight and they suffocate; in fact just like a stranded whale. This snake is about 2m (6 ft 6 in) long and is found in a large area of eastern Asia, inhabiting coastal waters from the Persian Gulf to Japan and New Guinea.

Luzon Sea Snake   *H. semperi*

Dr C. G. Semper (1832–1893) explored the Philippines in 1858 and published his book *Travels in the Philippine Archipelago* in 1886. This snake inhabits Lake Taal, in Luzon; perhaps not

strictly a 'sea snake', it is the only species known in the Family
Hydrophiidae to live in fresh water.

Hardwicke's Sea Snake   *Lapemis hardwickii*

Lapemis has no real meaning, it is simply an anagram of *Pelamis*
(below). Major General T. Hardwicke FRS (1756–1835) was a
naturalist who served in the Indian Army. This snake inhabits
coastal waters in south-east Asia ranging from the Burma-
Vietnam area through Malaysia, Indonesia and the Philippines, to New Guinea and northern Australia.

Yellow-bellied Sea Snake   *Pelamis platurus*

pēlamus (Gr) the tunny fish; this seems a peculiar name for a
snake, but with its flattened, quite bulky body, it does resemble
a typical fish; *platus* (Gr) wide, flat; *oura* (Gr) the tail; the
rudder-like tail is flattened vertically and thus has a similar
action to the tail of a fish. Inhabiting the warm areas of the
Indian and Pacific Oceans; it does not always remain near the
coast.

Small-headed Sea Snake   *Microcephalophis gracilis*

mikros (Gr) small; *kephalē* (Gr) the head; *ophis* (Gr) a serpent;
with a thick body possibly 4 cm ($1\frac{1}{2}$ in) in diameter the
unusually thin neck and head are very noticeable; *gracilis* (L)
thin, slender. Inhabiting the warm areas of the Indian and
Pacific Oceans.

Family   VIPERIDAE   80 species

vipera (L) a viper, possibly derived from *vivus* (L) alive, and
pario (L) I bear, bring forth, i.e. 'born alive'; most vipers do not
lay eggs, they bear live young and are referred to as being
viviparous.

Sand Viper   *Vipera ammodytes*

Vipera, see Family Viperidae above; *ammos* (Gr) sand; *dutes* (Gr)
a burrower, a diver; it is not well named as on the whole it
prefers a rocky or stony habitat. Inhabiting south-eastern
Europe including Yugoslavia and Greece and ranging east to

Asia Minor. It is the most dangerous European snake, although the venom is not so powerful as the cobra, or the rattlesnake.

Asp Viper   *V. aspis*
aspis (Gr) a viper, an asp. Inhabiting the Mediterranean area and ranging north to Switzerland and the southern part of Germany.

Adder or Common European Viper   *V. berus*
berus (New L) a name that was at one time used for a snake, possibly *Natrix natrix*, the Water Snake. Inhabiting Europe as far north as the Arctic Circle and ranging east to the Caucasus Mountains.

Caucasus Viper   *V. kaznakovi*
A. N. Kaznakow was a Russian naturalist but little is known about him; the name was given by Nikolski in 1909. Inhabiting the western Caucasus and Turkey. Prof A. M. Nikolski (1858–1942) was a herpetologist at Kharkov University.

Radde's Viper   *V. raddei*
Prof G. F. R. Radde (1831–1903) was a Director at the Caucasian Museum, Tbilisi, in Georgia. This viper inhabits Turkey, Iran and Georgia.

Russell's Viper   *V. russelli*
Named after Dr Patrick Russell (died 1805), a British physician and naturalist. This viper possesses a powerful venom and has a reputation for causing many deaths; it inhabits India and Burma, southern China and some Australasian islands.

Mountain Viper   *V. xanthina*
xanthos (Gr) various shades of yellow; *-inus* (L) suffix meaning like; it is a reddish brown colour. Inhabiting the mountains of northern Asia Minor.

Saw-scaled Viper   *Echis carinatus*
ekhis (Gr) a viper, an adder; *carina* (L) a keel; *-atus* (L) suffix meaning provided with; an allusion to the scales which have prominent keels along the back and transverse keels on the

sides. It has a wide distribution ranging from desert areas in North Africa, through Saudi Arabia, Iraq and Iran to India and Sri Lanka.

Horned Viper   *Cerastes cerastes*
cerastes (L) a horned serpent; it has a sharp horn-like scale above each eye. Inhabiting sandy desert areas in North Africa and Arabia. For an explanation of the specific name being the same as the generic name, see TAUTONYMS, p. 13.

Common Sand Viper   *C. vipera*
In this case the generic name is misleading as it does not have the two horns. Inhabiting a similar area to *C. cerastes*, see above.

Puff Adder   *Bitis arietans*
bitis (New L) to bite; *arieto* (L) I butt like a ram; can mean I strike violently; *arietans* (L) striking violently; 'to bite striking violently'. Widespread in Africa from southern Morocco to the Cape of Good Hope and ranging east into Arabia. The name Puff Adder arises from its unusual form of hissing, which is alarmingly loud and caused by forcing air in and out of the lungs; i.e. 'puffing'.

Gabon Viper   *B. gabonica*
Named from Gabon, in western Africa, it inhabits a large area in western and central Africa and is especially common in Cameroon. Because of its widespread distribution it is reputed to have caused more deaths than any other African snake. In addition to 'puffing', some adders are able to inflate themselves to much more than normal size when in the mood for aggression, which may also account for the name 'puff adder'.

Rhinoceros Viper or River Jack   *B. nasicornis*
nasus (L) the nose; *cornu* (L) the horn of animals; the scales form two horn-like projections on the snout. The name River Jack arises from its liking for marshy areas near rivers; it is a good swimmer and widespread in Africa south of the Sahara.

Dwarf Puff Adder   *B. peringueyi*
Dr L. A. Peringuey (1855–1924) was a French naturalist and a

Director of the Natural History Museum in Cape Town from 1906 until his death in 1924. This small puff adder is only about 30 cm (1 ft) long; it inhabits the south-western part of Africa.

Burrowing or Mole Viper   *Atractaspis microlepidota*
atraktos (Gr) a thin shaft, an arrow; *aspis* (Gr) a viper, an asp; an allusion to the body being more slender than other vipers; the head is not triangular and not really distinct from the body, which assists in the burrowing activity; *mikros* (Gr) small; *lepis* (Gr) genitive *lepidos* a scale; *-ota* (New L) a suffix used to mean provided with; it has small granular scales. Inhabiting Africa south of the Sahara.

Family   CROTALIDAE   about 130 species
krotalon (Gr) a little bell, a rattle; this family owes its scientific name to the rattlesnakes, known as Pit Vipers, see *Bothrops atrox*, p. 160.

Eastern Diamond-back Rattlesnake   *Crotalus adamanteus*
adamanteus (L) hard as steel; comparing the hardness of steel to that of a diamond; this snake has a pattern of diamond-shaped marks along the back. Inhabiting the south-eastern part of the USA including Florida.

Western Diamond-back Rattlesnake   *C. atrox*
atrox (L) fearful, cruel; all rattlesnakes are venomous, with a powerful venom, and much feared by Man. Inhabiting a large area of desert and arid country in the south western USA.

Santa Catalina Rattlesnake   *C. catalinensis*
-ensis (L) suffix meaning belonging to; it lives on Santa Catalina Island, in the gulf of California and is unique in having no rattle; this is because the terminal scales which in normal rattlesnakes form the rattle are shed when it moults.

Sidewinder or Horned Rattlesnake   *C. cerastes*
cerastes (L) a horned serpent; the scales on the head form a small horn above each eye. A small rattlesnake, length only about 65

cm (2 ft 2 in) it inhabits desert areas in the south-western USA. The name 'sidewinder' refers to its peculiar method of movement on loose sand by throwing loops of its body forward, which gives the impression that it faces sideways to the actual direction of travel.

Timber Rattlesnake   *C. horridus*
horridus (L) rough, prickly; a reference to the type of scales. Inhabiting the south-eastern part of the USA, but not Florida.

Prairie Rattlesnake   *C. viridis viridis*
viridis (L) various shades of green; it is an olive colour with rows of brown spots. Inhabiting the north-western part of the USA and the neighbouring part of Canada. There are a number of subspecies in the western part of North America; this is the nominate subspecies and three other subspecies are given below. For an explanation of subspecies see Chapter 2, pp. 18–19.

Arizona Black Rattlesnake   *C. v. cerberus*
Named after Cerberus, the mythical 3-headed monster. It is not really black but a very dark brown or blue colour. Inhabiting Arizona.

Utah Rattlesnake   *C. v. concolor*
concolor (L) of one colour; it is a pale brown colour without the conspicuous markings of the other rattlesnakes. Inhabiting eastern Utah.

Oregon Rattlesnake   *C. v. oreganus*
-anus (L) suffix meaning belonging to; named from Oregon, it inhabits California, Oregon and British Columbia.

Massasauga   *Sistrurus catenatus*
sistrum (L) derived from *seistron* (Gr) a small rattle used in religious worship, a child's rattle; *oura* (Gr) the tail; 'a tail-rattle'; *catena* (L) a chain, a shackle; can mean a gold or silver chain worn by women as an ornament; a reference to the chain-like coloured pattern along the back. It inhabits Illinois, Indiana, Wisconsin and Michigan and ranges east to the southern part of Ontario and central New York State. Known

as the Massasauga Rattler, the name is a corruption of Missisauga, the name of a river in Ontario.

Mexican Pygmy Rattlesnake *S. ravus*
ravus (L) tawny; although known as a 'pygmy rattlesnake', measuring about 86 cm (2 ft 10 in) it is not so small as the Sidewinder (p. 158). Inhabiting Mexico.

Fer de Lance *Bothrops atrox*
bothros (Gr) a hole, a pit; *ops* (Gr) the eye, face; *atrox* (L) hideous, savage; the generic name refers to a small depression or pit on the head between the eye and the nostril. A membrane in the pit is richly supplied with nerve endings which are so sensitive to heat that the warmth of a small warm-blooded animal can be detected 30 cm (1 ft) or more away; obviously of great benefit in the dark. Inhabiting Central America and the northern half of South America. Fer de Lance is French for a spearhead.

Island Viper *B. insularis*
insula (L) an island; *aris* (L) suffix meaning pertaining to; it is confined to the small island of Queimada Grande, off the coast of Brazil. The island has no human inhabitants and the number of these snakes has been steadily decreasing during the past few decades, possibly because the continued interbreeding causes a deterioration in health; they may be in danger of extinction.

Jararaca *B. jararaca*
Jararaca is a native name for this snake derived from the tupi language; the Tupi are a group of people living in the Amazon valley. This viper inhabits southern Brazil and Argentina.

White-lipped Tree Viper *Trimeresurus albolabris*
The meaning of *Trimeresurus*, which suggests 'three-part tail', is obscure. (Further research was intended by the Author.) *albus* (L) white; *labrum* (L) a lip. Inhabiting India.

Okinawa Habu or Yellow-spotted Lance-head *T. flavoviridis*
flavus (L) yellow; *viridis* (L) green. Inhabiting Okinawa and the

Ryukyu Islands; Habu is a native name for this snake in these islands.

Large-scaled Lance-head   *T. macrolepis*
makros (Gr) long; *lepis* (Gr) a scale; a reference to the unusually large scales on the head. Inhabiting Sri Lanka.

Chinese Mountain Viper   *T. montecola*
mons (L) genitive *montis* a mountain; *colo* (L) I inhabit, I dwell in. Inhabiting the mountains of southern China.

Chinese Habu   *T. mucrosquamata*
mucro (L) a sharp point; *squamatus* (L) scaly; a reference to the sharply keeled scales. Inhabiting southern China. Habu see Okinawa Habu above.

Bushmaster   *Lachesis mutus*
Lachesis was one of the three Fates in Greek mythology and was supposed to assign to Man his term of life, a macabre thought by Linnaeus who named this snake as it can certainly do this; it is the largest venomous snake in North or South America. *mutus* (L) dumb, mute; related to the rattlesnakes and similar in appearance, it shakes its tail vigorously when alarmed, but has no rattle so it was named *mutus*. However, when in the undergrowth, the tail actually makes quite a loud rustling noise. Inhabiting almost the entire north-eastern half of South America and the southern part of Central America.

Chinese Copperhead   *Agkistrodon acutus* (= *Ancistrodon*)
agkistron (Gr) a fish-hook; *odon* (Gr) a tooth; a reference to the curved fangs, though of course they are not barbed; *acutus* (L) sharp, pointed. Inhabiting the Shanghai area of China and ranging west to North Vietnam.

Tropical Moccasin   *A. bilineatus*
bis (L) two, double; *linea* (L) a line; *-atus* (L) suffix meaning provided with; an allusion to the yellow lines, one above the eye and one below, along the side of the head. Inhabiting Mexico.

Copperhead   *A. contortrix*
contorqueo (L) I twist, turn; *contortus* (L) full of turns, twisted; *contortrix* (L) one that is able to turn, twist. Inhabiting the south-eastern part of the USA including Florida. The top of the head is reddish brown.

Halys Viper   *A. halys*
halusis (Gr) a chain; a reference to the chain-like pattern along the back; this species, together with several subspecies, is very widespread in south-east Asia.

Himalayan Viper   *A. himalayanus*
Inhabiting the mountains of the Himalayan region and sometimes found at an altitude of about 2,800 m (9,100 ft).

Cottonmouth or Water Moccasin   *A. piscivorus*
piscis (L) a fish; *voro* (L) I eat, devour; although known as the Water Moccasin it is not really an aquatic animal, but is a good swimmer; the diet is mostly fish and frogs. Inhabiting the south-eastern part of the USA but not Florida. The name Cottonmouth refers to the interior of the mouth, which is white and very conspicuous when the mouth is opened wide in a defensive or attacking posture.

PART THREE
The Bird Orders

12 Birds in General—The Class Aves

This group of animals—the class Aves—have spread to almost every part of the globe; ptarmigan, with feet protected by feathers, and and ability to burrow in the snow, can live in Arctic areas; penguins in Antarctic areas live entirely on the ice and never touch land. Mountain birds have been seen at an altitude of 8,000 metres (over 26,000 feet) on Mount Everest, and some sea birds can dive to a depth of 60 metres (200 feet).

There are about 8,600 species of birds, and about 4,000 species of mammals; together they are the only warm-blooded animals. Nowadays zoologists prefer not to use the term 'cold-blooded' when referring to fish, reptiles, and their kin, because the temperature of the blood varies with the surroundings; it is known as poikilothermic, from the Greek *poikilos*, variable, and *thermē*, heat; because the temperature of the blood varies according to the surroundings. In some cases the blood can become so hot in the sun that a lizard, for instance, must seek the shade or it would perish. It is now known that the body temperature of some 'cold-blooded' animals may at times be higher than the temperature of the surroundings.

Unlike mammals, birds do not have sweat glands; the moisture produced would make the feathers useless for flight and reduce their ability to preserve warmth; they are 'warm-blooded'. This condition, in birds and mammals, is known as homeothermic (or homoiothermic) from the Greek *homoios*, like or similar; i.e. the temperature of the blood remains about the same. It is higher in birds than mammals, about 41°C (106°F), less during sleep, and higher at times of great activity. Some regulation of blood flow to the legs and feet can control temperature, being less in cold conditions and more in hot; heat loss is also helped by rapid panting, and heat preservation by 'fluffing out' the feathers.

It is generally recognised that the birds developed from the reptiles, some 200 million years ago, the feathers being a form of reptilian scales, which latter can still be seen on the legs and feet; embryonic feathers are similar to the scales of a young reptile.

The well-known fossil of a reptile-like bird, *Archaeopteryx* (from the Greek meaning 'ancient wing'), which was discovered during the year 1860 in Germany, shows characteristics of both reptiles and birds. One peculiar feature was that it had claws on the leading edge of the wings which it could use for climbing, so it was really a quadruped; a living example of this is the Hoatzin, a remarkable bird that lives in Central and northern South America. When first hatched the young hoatzin has similar claws on the fore limbs. Thus it starts life as a quadruped, using the claws for hanging on to branches and clambering about in the tree where its nest has been built.

In this respect it resembles the prehistoric semi-reptilian birds known to us only as the fossils *Archaeopteryx* (mentioned above) and *Archaeornis* ('ancient bird'). The nest is usually built over water, and when alarmed the nestling drops out of the nest into the water, swims ashore, a remarkable feat in any case, and *climbs* back to its nest. After about a week, the claws begin to disappear as the normal wing feathers develop. Thus in a matter of weeks it goes through an evolutionary change that took the 'lizards to birds' millions of years to bring about.

Chapter 13, which follows the explanatory chart overleaf, sets out the Orders, Families and Subfamilies for ease of reference and in the following chapters, we look at each of the orders and the families which make up the class Aves.

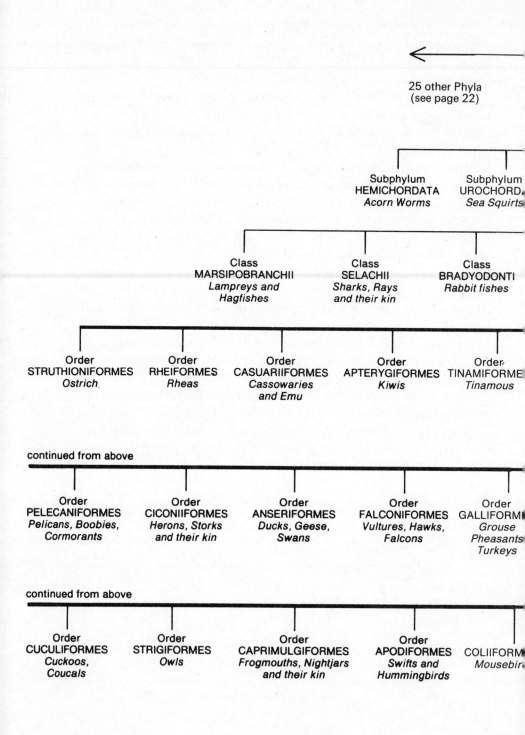

25 other Phyla
(see page 22)

Subphylum
HEMICHORDATA
Acorn Worms

Subphylum
UROCHORD
Sea Squirts

Class
MARSIPOBRANCHII
*Lampreys and
Hagfishes*

Class
SELACHII
*Sharks, Rays
and their kin*

Class
BRADYODONTI
Rabbit fishes

Order
STRUTHIONIFORMES
Ostrich

Order
RHEIFORMES
Rheas

Order
CASUARIIFORMES
*Cassowaries
and Emu*

Order
APTERYGIFORMES
Kiwis

Order
TINAMIFORME
Tinamous

continued from above

Order
PELECANIFORMES
*Pelicans, Boobies,
Cormorants*

Order
CICONIIFORMES
*Herons, Storks
and their kin*

Order
ANSERIFORMES
*Ducks, Geese,
Swans*

Order
FALCONIFORMES
*Vultures, Hawks,
Falcons*

Order
GALLIFORM
*Grouse
Pheasants
Turkeys*

continued from above

Order
CUCULIFORMES
*Cuckoos,
Coucals*

Order
STRIGIFORMES
Owls

Order
CAPRIMULGIFORMES
*Frogmouths, Nightjars
and their kin*

Order
APODIFORMES
*Swifts and
Hummingbirds*

COLIIFORM
Mousebir

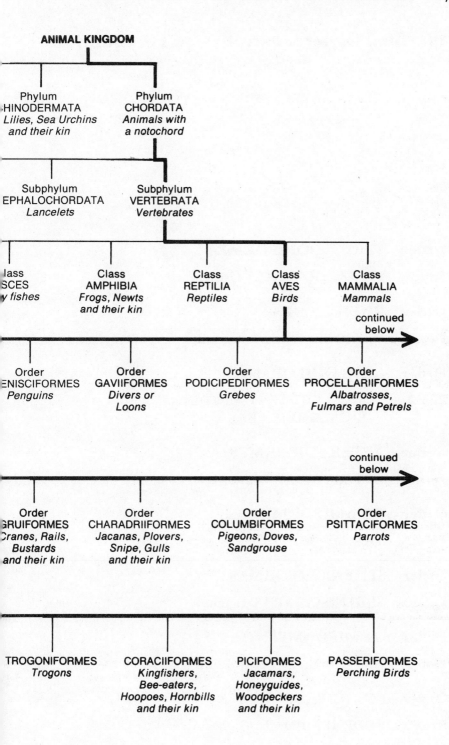

13 Bird Orders, Families and Subfamilies

Order STRUTHIONIFORMES

Family **STRUTHIONIDAE** Ostrich

Order RHEIFORMES

Family **RHEIDAE** Rheas

Order CASUARIIFORMES

Families **CASUARIIDAE** Cassowaries
  DROMAIIDAE Emu

Order APTERYGIFORMES

Family **APTERYGIDAE** Kiwis

Order TINAMIFORMES

Family **TINAMIDAE** Tinamous

Order SPHENISCIFORMES

Family **SPHENISCIDAE** Penguins

Order GAVIIFORMES

Family **GAVIIDAE** Divers, or Loons

Order PODICIPEDIFORMES

Family **PODICIPEDIDAE** Grebes

Order   PROCELLARIIFORMES

Families   **DIOMEDEIDAE**   Albatrosses
PROCELLARIIDAE   Shearwaters, Fulmars, Petrels
HYDROBATIDAE   Storm-Petrels
PELECANOIDIDAE   Diving-Petrels

Order   PELECANIFORMES

Families   **PHAETHONTIDAE**   Tropicbirds
PELECANIDAE   Pelicans
SULIDAE   Gannet and Boobies
PHALACROCORACIDAE   Cormorants
ANHINGIDAE   Darters
FREGATIDAE   Frigatebirds

Order   CICONIIFORMES

Families   **ARDEIDAE**   Herons, Bitterns
BALAENICIPITIDAE   Whale-headed Stork
SCOPIDAE   Hammerhead Stork
CICONIIDAE   Storks
THRESKIORNITHIDAE   Ibises, Spoonbills
PHOENICOPTERIDAE   Flamingos

Order   ANSERIFORMES

Families   **ANHIMIDAE**   Screamers
ANATIDAE   Ducks, Geese, Swans

Order   FALCONIFORMES

Families   **CATHARTIDAE**   Vultures
SAGITTARIIDAE   Secretary Bird
ACCIPITRIDAE   Hawks, Harriers
PANDIONIDAE   Osprey
FALCONIDAE   Falcons

Order   GALLIFORMES

Families   **MEGAPODIIDAE**   Scrub Hens
CRACIDAE   Curassows, Guans
TETRAONIDAE   Grouse

PHASIANIDAE Quails, Pheasants, Peacocks
NUMIDIDAE Guineafowls
MELEAGRIDIDAE Turkeys
OPISTHOCOMIDAE Hoatzin

Order GRUIFORMES

Families **MESITORNITHIDAE** Mesites
TURNICIDAE Button-Quails
PEDIONOMIDAE Plains Wanderer
GRUIDAE Cranes
ARAMIDAE Limpkin
PSOPHIIDAE Trumpeters
RALLIDAE Rails, Coots
HELIORNITHIDAE Sungrebe, Finfoots
RHYNOCHETIDAE Kagu
EURYPYGIDAE Sunbittern
CARIAMIDAE Seriemas
OTIDIDAE Bustards

Order CHARADRIIFORMES

Families **JACANIDAE** Jacanas
ROSTRATULIDAE Painted Snipe
HAEMATOPODIDAE Oystercatchers
CHARADRIIDAE Plovers
SCOLOPACIDAE Snipe, Woodcock
RECURVIROSTRIDAE Avocets, Stilts
PHALAROPODIDAE Phalaropes
DROMADIDAE Crab Plover
BURHINIDAE Thick-knees
GLAREOLIDAE Pratincoles, Coursers
THINOCORIDAE Seedsnipe
CHIONIDIDAE Sheathbills
STERCORARIIDAE Skuas
LARIDAE Gulls, Terns
RYNCHOPIDAE Skimmers
ALCIDAE Auklets, Murres, Guillemots

Order COLUMBIFORMES

Families **PTEROCLIDIDAE** Sandgrouse
COLUMBIDAE Pigeons, Doves

Order PSITTACIFORMES

Family PSITTACIDAE Parrots

Order CUCULIFORMES

Families MUSOPHAGIDAE Touracos
CUCULIDAE Cuckoos, Coucals

Order STRIGIFORMES

Families TYTONIDAE Barn Owls
STRIGIDAE Typical Owls

Order CAPRIMULGIFORMES

Families STEATORNITHIDAE Oilbird
PODARGIDAE Frogmouths
NYCTIBIIDAE Potoos
AEGOTHELIDAE Owlet-Nightjars
CAPRIMULGIDAE Nightjars

Order APODIFORMES

Families HEMIPROCNIDAE Tree Swifts
APODIDAE Swifts
TROCHILIDAE Hummingbirds

Order COLIIFORMES

Family COLIIDAE Colies or Mousebirds

Order TROGONIFORMES

Family TROGONIDAE Trogons

Order CORACIIFORMES

Families ALCEDINIDAE Kingfishers
TODIDAE Todies
MOMOTIDAE Motmots
MEROPIDAE Bee-Eaters
LEPTOSOMATIDAE Courol
CORACIIDAE Rollers

UPUPIDAE   Hoopoe
PHOENICULIDAE   Wood-Hoopoes
BUCEROTIDAE   Hornbills

Order · **PICIFORMES**

Families   **GALBULIDAE**   Jacamars
BUCCONIDAE   Puffbirds
CAPITONIDAE   Barbets
INDICATORIDAE   Honeyguides
RAMPHASTIDAE   Toucans
PICIDAE   Woodpeckers

Order   **PASSERIFORMES**

Families   **EURYLAIMIDAE**   Broadbills
DENDROCOLAPTIDAE   Woodcreepers
FURNARIIDAE   Ovenbirds
FORMICARIIDAE   Antbirds
CONOPOPHAGIDAE   Gnateaters
RHINOCRYPTIDAE   Tapaculos
PITTIDAE   Pittas
PHILEPITTIDAE   Asities, False Sunbirds
ACANTHISITTIDAE   New Zealand Wrens
TYRANNIDAE   Tyrant Flycatchers
OXYRUNCIDAE   Sharpbill
PIPRIDAE   Manakins
COTINGIDAE   Contingas, Becards
PHYTOTOMIDAE   Plantcutters
MENURIDAE   Lyrebirds
ATRICHORNITHIDAE   Scrub-Birds
ALAUDIDAE   Larks
HIRUNDINIDAE   Swallows, Martins
MOTACILLIDAE   Wagtails, Pipits
CAMPEPHAGIDAE   Cuckoo-Shrikes
PYCNONOTIDAE   Bulbuls
IRENIDAE   Leafbirds
PRIONOPIDAE   Helmet Shrikes
LANIIDAE   Shrikes
VANGIDAE   Vangas or Vanga Shrikes
PTILOGONATIDAE   Waxwings

DULIDAE Palm Chat
CINCLIDAE Dippers
TROGLODYTIDAE Wrens
MIMIDAE Mockingbirds, Thrashers
PRUNELLIDAE Accentors
MUSCICAPIDAE 11 subfamilies
Subfamilies **TURDINAE** Thrushes, Robins
ORTHONYCHINAE Logrunners
TIMALIINAE Babblers, Wren-Tits
PARADOXORNITHINAE Parrotbills
SYLVIINAE Old World Warblers
MALURINAE Wren-Warblers
ACANTHIZINAE Australian Warblers
MUSCICAPINAE Flycatchers
RHIPIDURINAE Fantail Flycatchers
MONARCHINAE Monarch Flycatchers
PACHYCEPHALINAE Whistlers, Shrike-Thrushes

Families **AEGITHALIDAE** Long-tailed Tits
REMIZIDAE Penduline Tits
PARIDAE Tits, Titmice
SITTIDAE Nuthatches
CERTHIIDAE Treecreepers
RHABDORNITHIDAE Philippine Creepers
CLIMACTERIDAE Australian Treecreepers
DICAEIDAE Flowerpeckers
NECTARINIIDAE Sunbirds
ZOSTEROPIDAE White-Eyes
EPTHIANURIDAE Australian Chats
MELIPHAGIDAE Honeyeaters
EMBERIZIDAE 5 subfamilies
Subfamilies **EMBERIZINAE** Buntings, American Sparrows
CATAMBLYRHYNCHINAE Plush-Capped Finch
CARDINALINAE Cardinal Grosbeaks
THRAUPINAE Tanagers, Honeycreepers
TERSININAE Swallow-Tanager

Families **PARULIDAE** Wood Warblers

DREPANIDIDAE   Hawaiian Honeycreepers
VIREONIDAE   Vireos
ICTERIDAE   Blackbirds, Orioles
FRINGILLIDAE   2 subfamilies
Subfamilies   **FRINGILLINAE**   Brambling, Chaffinch
CARDUELINAE   Goldfinches and their kin

Families   **ESTRILDIDAE**   Waxbills
PLOCEIDAE   2 subfamilies
Subfamilies   **VIDUINAE**   Whydahs
PLOCEINAE   Weavers, Sparrows

Families   **STURNIDAE**   Starlings
ORIOLIDAE   Orioles
DICRURIDAE   Drongos
CALLAEIDAE   Wattlebirds
GRALLINIDAE   Magpie Larks
ARTAMIDAE   Wood-Swallows
CRACTICIDAE   Butcherbirds, Currawongs
PTILONORHYNCHIDAE   Bowerbirds
PARADISAEIDAE   Birds of Paradise
CORVIDAE   Crows, Jays, and their kin

14 **Ostriches** STRUTHIONIFORMES

Ostrich
Struthio camelus

This order contains only one family, and in that family only one species of ostrich is alive today. However, in the past there were quite a number ranging over Syria, Arabia, and Africa. Now, as far as we know, only the African species remains, though there are some sub-species. It is well known that they cannot fly, but with their long and powerful legs they can run very fast, and speeds of 40 mph have been recorded.

It is interesting to note that the male incubates the eggs during the night and the female during the day; the young are able to leave the nest about 24 hours after hatching. Contrary to popular belief there is no reliable evidence that ostriches hide their heads in the sand when in danger, but they might crouch down behind a tuft or hillock in which case the body could show while the head and neck could be hidden.

This group of flightless birds, including the rheas, cassowaries, emus and kiwis, are known as ratites, from the Latin *ratis*, a raft or flat-bottomed boat, a boat without a keel; the breastbone of these birds lacks the ridge or keel which serves as an attachment for the powerful flight muscles of other birds that can fly; they are known as carinates, from the Latin *carina*, a keel.

Order STRUTHIONIFORMES

struthio (L) genitive *struthionis*, an ostrich *forma* (L) form, shape; can mean a species, a sort, a particular kind, so not necessarily referring to the actual shape.

Family STRUTHIONIDAE 1 species

Ostrich *Struthio camelus*
struthio (L) an ostrich *camelus* (L) a camel; 'camel-like ostrich'; it seems a suitable name for this fast-running bird. Inhabiting Africa, it has been introduced to Australia where it lives wild in the south.

Common Rhea
Rhea americana

Order **RHEIFORMES**

named after Rhea, the wife of Saturn and mother of Zeus.

Family RHEIDAE 2 species

Greater Rhea *Rhea americana*
Sometimes known as the Common Rhea, it ranges from north-eastern Brazil to the central part of Argentina, South America.

Lesser or Darwin's Rhea *Pterocnemia pennata*
pteron (Gr) feathers *knēmē* (Gr) the leg between the knee and ankle; 'feather-legged'; an allusion to the feathers that cover the top part of the leg *pennatus* (L) winged; they have wings but they are small and useless for flight; they can run very fast. Named after Charles R. Darwin (1809–1882), the English zoologist and author of the theory of organic evolution. Inhabiting South Africa.

16 Cassowaries CASUARIIFORMES

Cassowary
Casuarius casuarius

Another flightless bird order, the cassowaries are found in northern parts of Australia, in New Guinea, and adjacent islands. Like the ostrich and the rheas they have powerful legs and they are said to be able to kill a man with one blow. There are records of natives being killed by these birds, and their method is to leap into the air and strike down with their legs; the inner toe is armed with a long murderous claw, which can inflict a terrible wound. These birds are known as ratites (see notes under Ostrich on p. 176).

Order CASUARIIFORMES

from *kesuari* a Malayan word for the Cassowary.

Family CASUARIDAE 3 species

Bennett's Cassowary *Casuarius bennetti*
Dr G. Bennett (1804-1893) was an Australian surgeon, botanist, and zoologist, of New South Wales. Inhabiting New Guinea and some adjacent islands in the Pacific.

Australian Cassowary *C. casuarius*
Like all cassowaries it cannot fly, but can run very fast and is a good swimmer. Inhabiting north-eastern Queensland and New Guinea.

One-wattled Cassowary   *C. unappendiculatus*
unus (L) one   *appendix* (L) genitive *appendicis,* an appendage   *-culus*
(L) diminutive suffix   *-atus* (L) suffix meaning provided with.
Inhabiting New Guinea.

Family   DROMAIIDAE   1 species
dromaios (Gr) running at full speed, swift.

Emu   *Dromaius novaehollandiae*
Cannot fly, but a very fast runner and a good swimmer; New Holland
is an old name for part of Australia, hence the Latinised form. Unlike
the cassowaries the emu is a comparatively friendly bird, and does not
possess the dagger-like claw on the inner toe. This bird is known as a
ratite (see notes under Ostrich on p. 176). Inhabiting dry forest and
open plains in Australia.

Common Kiwi
Apteryx australis

The Kiwi is the national emblem of New Zealand; the name kiwi is a Maori word, and supposed to be an imitation of their cry. Unlike other birds they have a keen sense of smell; they belong to the group known as ratites (see notes under Ostrich on p. 176).

Order   APTERYGIFORMES

a- (Gr) not, without   *pteron* (Gr) feathers, can mean wings   *pterugion* (Gr) a little wing; the wings of the kiwi are rudimentary and not visible on the outside of the body.

Family   APTERYGIDAE   3 species

Common or Brown Kiwi   *Apteryx australis*
a- (Gr) not, without   *pterux* (Gr) a wing   *australis* does not mean Australia but derives from *auster* (L) the south wind, the south   *-alis* (L) relating to; hence *australis* (L) southern. They have a shrill cry— 'kiwi' and are confined to New Zealand.

Great Spotted Kiwi   *A. haasti*
Sir Julius von Haast (1824-1887) was a New Zealand explorer and geologist. He was a Director of the Canterbury Museum in 1866 and 1867.

Little Spotted Kiwi   *A. oweni*
Named after Sir Richard Owen (1804-1892), an English research anatomist and explorer and author. This kiwi, and *A. haasti*, above, inhabit South Island, New Zealand.

Martineta Tinamou
Eudromia elegans

The Tinamous look very like guinea fowl, partridges, and other game birds, but this is only external appearance although they make good eating. The anatomy shows relationship to the rheas, and in other ways they are similar to the group known as ratites (see notes under Ostrich on p. 176). However, they can fly, though only for short distances. They inhabit Central and South America.

Order **TINAMIFORMES**

tinamu is a native name from the Galibi, a tribe in French Guiana.

Family TINAMIDAE 46 species

White-throated Tinamou *Tinamus guttatus*
gutta (L) a drop of fluid, a spot *guttatus* (L) spotted, speckled; a reference to the inner flight feathers which are spotted pale buff; the

throat is not pure white but could be described as very pale fawn. Inhabiting southern Colombia and Venezuela, the Amazon basin, and south to northern Bolivia.

Slaty-breasted Tinamou   *Crypturellus boucardi*
kruptos (Gr) covered, hidden   *oura* (Gr) the tail   *-ellus* (L) diminutive suffix 'a little hidden tail'; the feathers on the rump usually hide the very short tail feathers. Named after A. Boucard (1839-1904), a French ornithologist and author who was in Mexico and the Amazon area in 1877. This tinamou inhabits Mexico, Central America, and northern South America.

Kalinowski's Tinamou   *Nothoprocta kalinowskii*
nothos (Gr) spurious, counterfeit   *prōktos* (Gr) hindpart, tail. Sclater and Salvin, in their *Nomenclator avium neotropicalium* (London, 1873) do not explain the reason for this name, but it could refer to the small tail being covered by the body feathers, thus giving a deceptive appearance. Dr P. L. Sclater, FRS (1829-1913) was Secretary of the Zoological Society of London from 1859 to 1903; O. Salvin, FRS (1855-1898) was Curator of Ornithology at Cambridge University from 1874 to 1882. They named the bird after J. Kalinowski (1860-c. 1942), a Polish naturalist who was living in Peru from 1889 probably until his death about 1942. This tinamou inhabits the northern part of South America.

Elegant Crested Tinamou or Martineta Tinamou   *Eudromia elegans*
eu- (Gr) a prefix meaning well, nicely   *dromos* (Gr) a running, escape; they usually rely on running, rather than flying, as an escape from predators   *elegans* (L) neat, elegant   *martinete* (Sp) a night heron; this tinamou has a long slender crest similar to that of the night heron. Inhabiting the southern area of South America.

19 **Penguins** SPHENISCIFORMES

Rockhopper Penguin
Eudyptes crestatus

Penguins are usually associated with snow and ice, yet there are none living in the Arctic regions of the north; mostly they inhabit the Antarctic and neighbouring islands, and the southern coasts of South America, Africa, and Australia. They are not restricted to icy regions; one species, the Galapagos Penguin, lives on the shores of the Galapagos Islands, near the equator.

They are flightless, but as divers and swimmers they are probably the most accomplished of all birds; the wings are short, and useless for flight, but play the largest part in swimming.

Order **SPHENISCIFORMES**

sphēn (Gr) a wedge   *-iscus* (New L) a diminutive suffix, derived from the Greek *-iskos*; 'a little wedge'; a reference to the shape and shortness of the wings.

Family **SPHENISCIDAE**   16 species

Adélie Penguin   *Pygoscelis adeliae*
pugē (Gr) the rump, buttocks   *skelos* (Gr) the leg; a reference to the position of the legs which are set well back on the body, on the rump in fact; thus they have to adopt an upright posture: the French sector of the Antarctic where these penguins were first found was named Terre Adélie by Admiral J. S. C. Dumont d'Urville (1790-1842) after his wife Adélie.

Emperor Penguin   *Aptenodytes forsteri*
a- (Gr) not, without   *ptēnos* (Gr) feathered, winged   *dutes* (Gr) a diver; 'a wingless diver'. Named after Johann R. Forster (1729–1798), a German traveller and ornithologist; he was on *H.M.S. Resolution* from 1772 to 1775. This is the largest penguin, more than 1 m (about four feet) tall and it lives on the coast of Antarctica.

King Penguin   *A. patagonica*
Patagonia is the region at the southern end of South America; this penguin also inhabits some islands such as South Georgia and Kerguelen, and the coasts of Tasmania and New Zealand. It is nearly as big as the Emperor Penguin.

Macaroni Penguin   *Eudyptes chrysolophus*
eu- (Gr) a prefix meaning well, nicely   *duptēs* (Gr) a diver   *khrusos* (Gr) gold   *lophos* (Gr) a crest or tuft on the head of birds; the name macaroni derives from the unusual structure of the yellow feathers that form the crest. Inhabiting the coasts of the southern parts of the Atlantic and Indian Oceans.

Rockhopper Penguin   *E. crestatus*
cristatus (L) crested; a misspelling of *crista* (L) a crest; it has an untidy crest of feathers on the forehead. Inhabiting southern islands and southern coasts of South America, South Africa, Australia, and New Zealand, in a rocky habitat.

Yellow-eyed Penguin   *Megadyptes antipodes*
megas (Gr) great   *duptēs* (Gr) a diver; it is not one of the big penguins, being about 76 cm (30 in) tall   *antipodes*, strictly speaking, means two points on opposite sides of the earth, though it is sometimes used to mean Australia or New Zealand. Inhabiting South Island, New Zealand, and Stewart, Auckland and Campbell Islands.

Jackass Penguin   *Spheniscus demersus*
Spheniscus, see notes under Order, above   *demergo* (L) I sink, I submerge   *demersus* (L) means depressed. The English name is derived from its donkey-like call. Inhabiting southern African coasts.

Galapagos Penguin   *S. mendiculus*
mendicus (L) beggarly, needy   *-culus* (L) dim. suffix; the opposite of *regillus* (L) royal, magnificent; this is taken to mean small, as it is one of the smallest penguins, only about 50 cm (20 in) tall. Inhabiting Galapagos and neighbouring islands.

Red-throated Diver
Gavia stellata

Loons live almost entirely on the water, and are expert divers and swimmers like the penguins; they swim fast and deep, and it is claimed that they can go to depths of about 47 m (150 ft) when chasing fish, their main diet. They can remain submerged for long periods, some observers say as long as 15 minutes, which is remarkable if true.

The webbed feet use a sculling action something like the action of a ship's propeller, and the legs being set far back on the body the whole bird appears to be a sort of living torpedo. They inhabit northern areas and arctic seas, and are known to visit Scotland and the east coast of England.

Order **GAVIIFORMES**

gavia (L) a bird, perhaps the seamew.

Family **GAVIIDAE** 4 species

Yellow-billed Loon *Gavia adamsii*
Named after E. Adams (1824–1856), a surgeon on *H.M.S. Enterprise* from 1849 to 1855, exploring the American arctic area. Inhabiting the northern areas of Canada, Alaska and eastern Asia. The name 'loon' is of Scandinavian origin and usually used in America, meaning

an awkward person; it is an allusion to their awkward gait when on land due to the legs being set so far back on the body.

Black-throated Diver or Arctic Loon   *G. arctica*
Breeding in arctic areas, it comes south in winter and may be seen on the shores of China, the Mediterranean area, and Mexico; the English name derives from a black patch on the throat.

Red-throated Diver or Loon   *G. stellata*
stella (L) a star   *stellatus* (L) set with stars, starry; a reference to the dark spots on the lower neck and flanks; it has a dark red throat patch. The range is similar to *G. arctica*, above.

Great Crested Grebe
Podiceps cristatus

Grebes are good swimmers and divers but not so expert as the divers of the family Gaviidae, and not able to remain under water for so long; their normal time does not exceed 2 to 3 minutes. They are the only birds known to dive with their young clinging to their back, as a means of escape from danger for the whole family.

They live almost entirely on the water, sleeping, feeding, courting and mating, and are well known for their elaborate courtship display; they have a wide distribution over many parts of the world.

Order PODICIPEDIFORMES

podex (L), genitive *podicis*, the rump   *pes* (L), genitive *pedis*, the foot; 'rump-foot', an allusion to the legs being set well back on the body, on the rump in fact. In some places the local name is 'arsefoot'!

Family PODICIPEDIDAE   19 species

Horned Grebe   *Podiceps auritus*
Podiceps is considered to be a contracted form of *podicipes*, which could be taken to mean 'rump-foot' (see above), but *-ceps* (L) as a suffix means 'headed', so this would translate as 'rump-headed', so the contraction is unfortunate   *auris* (L) the ear   *auritus* (L) having ears; a

reference to the conspicuous ear tufts, and these give rise to the English name 'horned'. Inhabiting temperate and subarctic regions of North America and Eurasia and migrating south in winter to subtropical areas.

Great Crested Grebe   *P. cristatus*
cristatus (L) crested; it has a prominent crest, in fact most grebes have a crest: the English name 'great' has been given because it is the largest grebe, being about 48 cm (19 in) long. It inhabits most of Europe including the British Isles, but not northern Scandinavia; it is found in many parts of Asia, Africa, Australia, and New Zealand.

Madagascar Little Grebe   *P. pelzelnii*
Named after A. von Pelzeln (1825–1891), a German zoologist and author who was Custodian of the Royal Museum in Vienna from 1859 to 1883. This grebe is confined to Madagascar.

Little Grebe or Red-throated Dabchick   *P. ruficollis*
rufus (L) red   *collum* (L) the neck; a reference to the chestnut-coloured throat patch; it is the smallest grebe, about 25 cm (10 in) long. Inhabiting Europe, including the British Isles, and southern Sweden, most of Africa, and ranging across Asia as far as the East Indies and Japan.

Western Grebe   *Aechmophorus occidentalis*
aikhmē (Gr) the point of a spear, a spear   *phora* (Gr) a carrying, bearing; a reference to the tarsal bones of the foot which are narrow and shaped like a blade   *occidentalis* (L) western. Inhabiting western North America and ranging south to southern Mexico.

Flightless or Short-winged Grebe   *Centropelma micropterum*
kentron (Gr) a spur, spike   *pelma* (Gr) the sole of the foot, a reference to the bones of the foot (see above)   *mikros* (Gr) small   *pteron* (Gr) wings; the wings are very small and useless for flight. It is found only in South America on Lake Titicaca in the Andes.

Atitlán Grebe   *Podilymbus gigas*
Podilymbus is a combination of *Podiceps* (see under Order and Horned Grebe, above) and *kolumbis* (Gr) a diving bird; another coined word and a mixture of Latin and Greek which really has no meaning   *gigas* (Gr) a giant; it is one of the biggest grebes, about 46 cm (18 in) long. This is a very rare grebe, probably almost flightless, and confined to Lake Atitlán in Guatemala.

22 Albatrosses, Fulmars, Shearwaters and Petrels PROCELLARIIFORMES

Black-footed Albatross
Diomedea nigripes

The Order Procellariiformes includes the albatrosses, fulmars, shearwaters, and petrels; they are sea birds and masters of flight 'par excellence', ranging over vast areas of the oceans. The albatross spends its life at sea, sleeping on the surface, feeding at sea, and only returning to land to breed.

Taken as a whole birds of this order, particularly the albatrosses, like stormy weather, as the gusts and upcurrents caused by the wind and waves enable them to stay airborne and glide for long distances; they do this apparently without effort, and with hardly a movement of their wings. Our most experienced glider pilots, their cockpits packed with modern instruments, have yet to achieve the mastery of flight and navigation demonstrated by these birds.

Order PROCELLARIIFORMES

procella (L) a violent wind, a storm; they like stormy weather at sea (see above).

Family DIOMEDEIDAE 13 species
Diomedes, according to the Greek legend, was driven by a storm on

to the coast of southern Italy, where he remained and died an old man; his companions were turned into birds.

Grey-headed Albatross   *Diomedia chrysostoma*
khrusos (Gr) gold   *stoma* (Gr) the mouth; the upper surface of the beak is golden yellow and the head is grey merging into white at the neck. Inhabiting subantarctic islands and ranging north into temperate regions.

Wandering Albatross   *D. exulans*
exul = *exsul* (L) an exile, a wanderer; it inhabits southern oceans and is given to widespread wandering, travelling great distances; the enormous wingspan may be more than 3 m (11 ft) which enables it to remain airborne for long periods in almost effortless gliding flight (see introductory notes, above).

Sooty Albatross   *D. fusca*
fuscus (L) dark-coloured; the plumage has a grey-brown tinge which distinguishes it from the usual white albatross. Inhabiting southern oceans and ranging north as far as the southern parts of South America, South Africa, and Australia and New Zealand.

Black-browed Albatross   *D. melanophrus*
melas (Gr), genitive *melanos*, black   *ophrus* (Gr) the eyebrow; an allusion to the dark plumage round the eye. Inhabiting southern oceans but sometimes seen north of the equator.

Black-footed Albatross   *D. nigripes*
niger (L) black   *pes* (L) the foot. Inhabiting the Pacific Ocean from the southern part of the Bering Sea to the coasts of Asia and North America, the Hawaiian Islands, and ranging south to the equator.

Family   PROCELLARIIDAE   53 species
procella (L) a storm, a violent wind; they like stormy weather and are 'at home' at sea like the albatrosses.

Giant Petrel or Giant Fulmar   *Macronectes giganteus*
makros (Gr) long; in general can mean large   *nēktēs* (Gr) a swimmer *giganteus* (L) gigantic; the largest fulmar, and may be up to nearly 1 m (38 in) long. The name petrel is said to come from St Peter, and the gospel story of his walking on the water; fulmar is from Old Norse *full*, meaning foul, and *mar*, a gull; it derives from the bird's ability to spit

a foul-smelling oily liquid at intruders. Inhabiting the southern oceans and coasts of South America, South Africa, Australia and New Zealand.

Northern Fulmar   *Fulmarus glacialis*
Fulmarus, see above   *glacialis* (L) glacial, icy; it is widespread in the northern hemisphere inhabiting small arctic islands and ranging south to temperate regions including the coasts of the British Isles.

Dove or Antarctic Prion   *Pachyptila desolata*
pakhus (Gr) thick, stout   *ptilon* (Gr) a feather; a reference to the soft dense feathers, a characteristic of the dove prion   *desolatus* (L) forsaken, desolate; a reference to the habitat, the lonely uninhabited antarctic region: the name dove refers to its appearance, about the size of a dove at 28 cm (11 in) and having blue-grey upperparts and pale grey underparts. Prion is from the Greek *priōn*, a saw, and refers to the bill which has serrated saw-like edges. Inhabiting the antarctic sea and islands.

Parkinson's Petrel   *Procellaria parkinsoni*
procella (L) a storm   *-arius* (L) suffix meaning pertaining to; they are associated with stormy weather (see introductory notes about albatrosses). Named after S. Parkinson (c. 1745–1771); a draughtsman to Sir J. Banks FRS who was with Capt Cook on *H.M.S. Endeavour* in 1770. Inhabiting antarctic regions and the southern coasts of South America and Australia.

Greater Shearwater   *Puffinus gravis*
Puffinus, a Latinized form of the vernacular name puffin (see p. 144), originally *pophyn* (Middle English)   *gravis* (L) heavy, ponderous; a large bird, measuring about 46 cm (18 in). The English name derives from their manner of flight, very near the surface of the water and tilting the body from side to side, 'shearing' the water with their wings. Inhabiting the Tristan da Cunha group of islands in the South Atlantic, and the Falkland Islands off the southern coast of South America, they have a wide range wandering north possibly as far as the arctic circle.

Wedge-tailed Shearwater   *P. pacificus*
pacificus (L) of the Pacific; it inhabits small islands in the southern Pacific and Indian Oceans: the English name refers to the long wedge-shaped tail.

Short-tailed or Tasmanian Shearwater *P. tenuirostris*
tenuis (L) thin, slender *rostrum* (L) the beak; it has a more slender beak than other shearwaters and a shorter tail. It breeds in Tasmania but has a wide migratory range circumnavigating almost the whole of the Pacific Ocean.

Bermuda Petrel or Cahow *Pterodroma cahow*
dromos (Gr) a course, a race, usually of horses; can mean any quick movement, e.g. flight *pteron* (Gr) feathers, wings; 'fast on the wing'; the name cahow is an imitation of the peculiar mating cry. It is confined to Bermuda, in the North Atlantic, and the surrounding sea and small islands.

Dark-rumped Petrel *P. phaeopygia*
phaios (Gr) dusky, brown *pugē* (Gr) the rump, buttocks. This petrel is widespread in the southern Pacific, ranging to the west coasts of Central and South America and across to the small islands off the west coasts of New Guinea and Australia.

Solander's Petrel *P. solandri*
Named after Dr D. C. Solander FRS (1733-1782), a Swedish botanist on *H.M.S. Endeavour* from 1768 to 1771 and for a number of years he was working for the British Museum (Natural History). This petrel inhabits the southern coast of Australia and the Indian Antarctic Sea.

Snow Petrel *Pagodroma nivea*
pagos (Gr) ice *dromos* (Gr) a running course *nix* (L) genitive *nivis*, snow *niveus* (L) snowy; the name is not only suitable for its habitat but the plumage which is almost entirely white. Widespread in the oceans of the Antarctic and ranging north to about the southern tip of New Zealand and South America, and further north after breeding.

Family HYDROBATIDAE 20 species
hudor (Gr) water *batēr* (Gr) the threshold on which one treads.

Wilson's Storm Petrel *Oceanites oceanicus*
oceanus (L) the ocean, derived from Ōkeanos, in Greek mythology the god of the river which was believed to encircle the earth *-ites* (L) suffix meaning having to do with, belonging to *-icus* (L) suffix meaning belonging to; the Storm Petrel is the smallest of the ocean birds and spends most of its life at sea; sailors associate its appearance at sea with stormy weather (see introductory notes to the Albatrosses).

Named after Alexander Wilson (1766–1813), the Scottish-American ornithologist; he wrote the nine volume *American Ornithology* published in 1805. This petrel is widespread in the Antarctic, breeding on the many islands and migrating north into the Indian Ocean, the Pacific, and the Atlantic.

White-faced Storm-Petrel   *Pelagodroma marina*
pelagos (Gr) the sea, especially the open sea, the high sea   *dromos* (Gr) a running course   *marinus* (L) of the sea. Breeding on islands in the southern oceans and migrating northwards.

Storm-Petrel   *Hydrobates pelagicus*
hudor (Gr) water   *batēs* (Gr) one that treads or covers; a reference to the petrel's habit of apparently treading on the water with a 'pattering' of feet when hunting for food   *pelagos*, see above   *-icus* (L) suffix meaning belonging to. Inhabiting the North Atlantic, including the sea around the British Isles, and the Mediterranean.

Markham's Storm-Petrel   *Oceanodroma markhami*
oceanus (L) the ocean, derived from Ōkeanos, in Greek mythology the god of the river which was believed to encircle the earth   *dromos* (Gr) a running course. Named after Sir C. R. Markham FRS (1830–1916), a naturalist and explorer and author of the book *Travels in Peru and India*. This petrel inhabits the seas around the coasts of North America, Central, and South America.

Family   PELECANOIDIDAE   4 species
pelekan (Gr) a water-bird of the pelican kind   *-oides* (New L) from *eidos* (Gr) apparent shape, resemblance; it does not actually resemble the pelican in appearance, but like the pelican it has an extensible throat pouch used for holding food while hunting.

Peruvian Diving-Petrel   *Pelecanoides garnotii*
Named after Dr P. Garnot (1794–1838), a French author and ornithologist. It ranges along the western coast of South America as far north as northern Peru.

Subantarctic Diving-Petrel   *P. urinatrix*
urinator (L) a diver   *-atrix* (L) suffix denoting feminine; probably suggested by the smallness of the bird, only about 18 cm (7 in). Breeding on remote islands in the Subantarctic and migrating north to the open sea.

23 Tropicbirds, Pelicans, Gannets, Cormorants, Anhingas and Frigatebirds PELECANIFORMES

European White Pelican
Pelecanus onocrotalus

The pelicans, in spite of their rather ungainly appearance, are graceful fliers and good swimmers and divers; their enormous pouch, attached to the lower bill and throat, is too well known to need description. They hunt for fish in fresh water and sea water. Although all the birds in this group inhabit the sea or inland waters, the frigatebird cannot swim; it feeds mostly by frightening other birds into dropping their prey which it then catches in mid-air, hence 'frigatebird' because the pirates used frigates. It also picks up small marine creatures from the surface of the sea, and it is said that if it accidentally comes down on the water it cannot rise again, and may perish.

The anhingas, sometimes known as snakebirds because of their long neck, frequent inland waters. They catch fish by spearing them with their beak; the pronounced 's-bend' in the snake-like neck can be straightened, like a spring suddenly released, and thus they impale a fish.

The gannets, and also the brown pelican, have an unusual way of catching their prey: they dive from a height and strike the water with such force that it can stun a fish just below the surface.

Tropicbirds also dive from a height, but instead of hitting the water they dive cleanly, and plunge some distance below the surface. Cormorants, on the other hand, swim on the surface before diving, and then under water using both the wings and the webbed feet. In China, Japan, and India they are trained to catch fish for their owners, in some cases swimming quite a distance under water on a long leash.

Order   PELECANIFORMES

pelecanus (L) a pelican   *forma* (L) form, shape; can mean sort, kind.

Family   PHAETHONTIDAE   3 species
phaethōn (Gr) shining, radiant; Phaethōn was the Greek sun god *phaethontis* (Gr) a poetical form with a similar meaning; an allusion to the pure white plumage which reflects the fluorescent greens and blues of the water when the birds are hovering near the surface.

White-tailed Tropicbird   *Phaethon lepturus*
leptos (Gr) slender   *oura* (Gr) the tail; a reference to the tail which consists of two long white and slender tail feathers, which stream out behind to a length greater than that of the body. Worldwide distribution in tropical waters and breeding on islands in the Caribbean, Bermuda, and far and wide in the Atlantic and Pacific Oceans.

Red-tailed Tropicbird   *P. rubricauda*
ruber (L) red   *cauda* (L) the tail of an animal. Ranging over great distances and breeding on islands in the warm areas of the Pacific and Indian Oceans.

Family   PELECANIDAE   7 species
pelecanus (L) a pelican.

American White Pelican   *Pelecanus erythrorhynchos*
eruthros (Gr) red   *rhunkhos* (Gr) the beak; the large beak with its enormous pouch is red; this is one of the big pelicans with a wingspan of about 3 m (10 ft). Inhabiting inland waters and lakes in North and Central America.

Brown Pelican   *P. occidentalis*
occidentalis (L) western; inhabiting the south Atlantic coast of North America, the Gulf coast, the West Indies, the Pacific coast from Cali-

fornia south to Chile, and the Galapagos Islands. It is a sea bird but never ventures far from land and breeds on small off-shore islands.

European White Pelican   *P. onocrotalus*
onocrotalus (L) the pelican. An almost pure white pelican inhabiting rivers, lakes, and some coasts in south-east Eurasia, Africa, and to some extent India and Malaysia. The name European is too restrictive; it is sometimes known as the Old World White Pelican.

Family   SULIDAE   7 species
sula is an Icelandic name for the gannet or booby and means a foolish person, probably because of their stupid-looking and ungainly gait on land, but also because they are so indifferent to the approach of people that it is possible to walk up to one and seize it in the hand. Some authors give the genus as *Morus*, for instance the Common Gannet as *Morus bassanus*, and the Cape Gannet as *M. capensis*; this is incorrect; at one time the scientific name for the genus was *Morus*, until it was discovered that an earlier name *Sula* had been given. According to the international rules the first name given as the genus is the correct one, so the name is *Sula*   *Mōros* (Gr) means foolish, stupid, so that name seems to agree with the general idea that the birds looked foolish or behaved in a foolish manner.

Common Gannet   *Sula bassana*
-*anus* (L) suffix meaning belonging to   *bassana* is a coined name meaning belonging to the Bass Rock, in the Firth of Forth, Scotland. Inhabiting the northern Atlantic ranging from Newfoundland across to the British Isles and the Channel Isles; some authorities now consider *S. capensis* of South Africa and *S. serrator* of southern Australia, Tasmania, and New Zealand, to be conspecific; they breed on rocky islands and coastal cliffs. *Sula bassana* is sometimes known as the Solan Goose; gannets are about the size of a goose.

Brown Booby   *S. leucogaster*
leukos (Gr) white   *gastēr* (Gr) the belly; the plumage is dark brown above with a white breast and belly. Worldwide distribution breeding on tropical islands right round the globe; when not breeding gannets spend their life at sea covering vast distances.

Red-footed Booby   *S. sula*
It has red feet and the bare skin on the face is sometimes red; the only booby that perches and builds its nest in trees. Inhabiting southern

regions of the Atlantic, Pacific, and Indian Oceans, and breeding on islands where there are trees.

Family PHALACROCORACIDAE   30 species
phalakros (Gr) bald-headed   *korax* (Gr), genitive *korakos*, a raven, a crow; anything hooked like a raven's beak; cormorants have a sharply hooked bill and usually part of the face is naked. The name cormorant is derived from a combination of the two Latin words corvus marinus, a 'sea-crow'.

Galapagos Flightless Cormorant   *Nannopterum harrisi*
nannos = *nanos* (Gr) a dwarf; strictly speaking it means one whose limbs are too small for his body   *pteron* (Gr) wings; the small rounded wings are useless for flight: named after Charles M. Harris (fl 1890–1899), Chief Naturalist in the Galapagos Islands in 1897 and 1898; he collected for the Tring Museum, Hertfordshire, England. Confined to Galapagos and neighbouring small islands.

Long-tailed Cormorant   *Phalacrocorax africanus*
Phalacrocorax, see under Family, above   *africanus* (L) of Africa; it has a longer tail than other cormorants. Living on lakes and rivers in Egypt and south of the Sahara ranging west across Africa to Gambia; also Madagascar.

Blue-eyed Shag   *P. atriceps*
ater (L) black   *ceps* (New L) from *caput* (L) the head; the head, and the upperparts, are black, but the eyes are not blue, they are brown; the name is an allusion to the eyelids which are blue and the bare patch on the face is quite a bright blue. There is no basic difference between a shag and a cormorant, but the shag is slightly smaller, however in some countries the name of a particular bird may be a shag and in others a cormorant. Inhabiting southern oceans and inland waters in Chile, Argentina, and Uruguay.

Peruvian Cormorant   *P. bougainvillii*
Named after Adm H. Y. P. Baron de Bougainville (c. 1781–1846), the French naturalist and author. This cormorant is found only on the coasts of Peru and Chile and sometimes Colombia.

Great Cormorant   *P. carbo*
carbo (L) coal, charcoal; i.e. black, a reference to the glossy blue-black plumage. The biggest cormorant, about 1 m (38 in), with a very wide

distribution from Atlantic coasts in northern North America, Greenland, Iceland, Scandinavia, the British Isles, and east through Africa and India on coasts and some inland waters as far as Japan, Australia, and New Zealand.

Auckland Island Shag *P. colensoi*
Named after the Rev. W. Colenso FRS (1811–1899), a naturalist and missionary in New Zealand, it inhabits Auckland Island which lies to the south of South Island, New Zealand.

Little Pied Cormorant *P. melanoleucos*
melas (Gr), genitive *melanos* black   *leukos* (Gr) white. Inhabiting Indonesia, Papua New Guinea, some of the Pacific Islands, and Australia.

Spotted Shag *P. punctatus*
punctum (L) a hole, a prick   *punctillum* (L) a dot, a spot   *punctatus* spotted; the upper parts and wings are marked with little dark spots. Inhabiting the coasts of New Zealand.

Family ANHINGIDAE 4 species
Anhinga is a Tupi-Guarani word of the tribes living in the Amazon regions of South America, meaning 'water-turkey', and this English name is sometimes used in America; it has been suggested that the word is derived from *anguinus* (L) snaky, on account of the long snake-like neck.

Anhinga *Anhinga anhinga*
Inhabiting the southern part of the USA, the Galapagos Islands, Central and South America ranging south to Argentina.

Asian Darter *A. melanogaster*
melas (Gr), genitive *melanos* black   *gastēr* (Gr) the belly. Inhabiting India, the Burma-Vietnam area, and Malaysia and Indonesia.

Australian Darter *A. novaehollandiae*
New Holland was at one time a name for part of Australia (see p. 365). This anhinga inhabits Australia and Papua New Guinea. Darter is probably the name usually used in several countries but in the USA it is called the Anhinga.

African Darter *A. rufa*
rufus (L) red, ruddy; a reference to the chestnut-coloured neck. Inhabiting parts of the Middle East, Africa south of the Sahara, and Madagascar.

Family   FREGATIDAE   5 species

fregata (New L) from *fregata* (It) a frigate; they often feed by frightening other birds into dropping their prey which they then catch in mid-air, hence 'frigatebirds' because of their piratical behaviour and the pirates used frigates.

Christmas Island Frigatebird   *Fregata andrewsi*

Named after Dr C. W. Andrews FRS (1866–1924), a zoologist who was working at the British Museum (Natural History) in 1892 and visited Christmas Island in 1892 and 1897; it is in the Indian Ocean about 200 miles south of Java. This frigatebird breeds only on this island and some neighbouring small islands.

Ascension Island Frigatebird   *F. aquila*

aquila (L) an eagle; an allusion to the hooked beak and its generally ferocious attack to obtain food. It breeds on Ascension Island far out in the South Atlantic, mid-way between Africa and South America.

Magnificent Frigatebird   *F. magnificens*

magnificus (L) eminent, magnificent; an allusion to its great size, the largest frigatebird, it can measure over 1 m (45 in). Inhabiting the tropical part of the Atlantic on the coasts of North, Central, and South America and some small islands, and the same in the eastern Pacific Ocean.

Greater Frigatebird   *F. minor*

minor (L) lesser, smaller; it is smaller than *magnificens*, at about 1 m (40 in); the English name 'greater' means it is not 'the greatest'. Inhabiting the tropical parts of the coasts and islands of the Atlantic, Indian, and Pacific Oceans.

24 Herons, Storks, Flamingos and their kin CICONIIFORMES

White Spoonbill
Platalea leucorodia

There are six families of birds in the order Ciconiiformes, and they are all associated with inland waters. They have long legs for wading, and some have partially webbed feet, and they frequent rivers, lakes, and marshy areas in most countries throughout the world.

Herons, egrets and bitterns eat mainly fish, and other aquatic animals; they have powder downs and a special serrated claw which are connected with preening. The powder down is an area of small feathers that easily crumble into a powder, and the feathers are continuously replaced by new growth. The beak is used to crumble the feathers, and the powder is then spread on those feathers that have been soiled by the slime and oil of fish. The powder soaks up the oil and the feathers are then 'combed out' with the special claw. The whale-headed stork also has powder downs, but other storks and the remainder of the birds in this group do not possess this specialised equipment. It is quite rare among birds but there are others that have it, for example the toucans and parrots.

Order CICONIIFORMES

ciconia (L) a stork   *forma* (L) form, shape; can mean a sort, kind, so not necessarily referring to the actual shape.

Family ARDEIDAE   62 species
ardea (L) a heron.

Grey Heron   *Ardea cinerea*
cinereus (L) ash-coloured. Inhabiting the British Isles, part of Iceland, Scandinavia, and northern Africa, and ranging across Eurasia to Japan, India, and south-east Asia.

Goliath Heron   *A. goliath*
In the Old Testament, Goliath was the Philistine giant who was killed by the boy David with a stone from his sling. This is a big heron which may be as much as 1·5 m (5 ft) in height. Inhabiting Senegal, Egypt, Madagascar, South Africa and parts of the Red Sea region.

Great Blue Heron   *A. herodias*
erōdios (Gr) a heron; a big heron with pale blue wings and upperparts inhabiting the southern part of North America, and Central America.

Little Green Heron   *Butorides virescens*
butio (L) a bittern   *-ides* (L) suffix denoting a relationship; a reference to the herons being related to the bitterns   *viridis* (L) green   *-escens* (L) suffix meaning beginning to, somewhat; 'greenish'. Inhabiting the southern part of North America and ranging south through Central America to the northern part of South America.

Madagascar Squacco Heron   *Ardeola idae*
ardea (L) a heron   *-olus* (L) diminutive suffix; a small heron: named after Frau Ida R. Pfeiffer (1797-1858), a German traveller and authoress who was in Madagascar from 1856 to 1858. The name squacco is from Italian dialect *sguacco*.

Squacco Heron   *A. ralloides*
rale (Fr) a rail, a water rail *Rallus aquaticus*   *-oides* (New L) from *eidos* (Gr) likeness of form, can mean a type of; an allusion to this heron being similar to the water rail. Inhabiting southern Europe, Turkey, parts of East Africa, and Madagascar.

Cattle Egret   *Bubulcus ibis*
bubulcus (L) a ploughman that ploughs with oxen; in a general sense a herdsman. It is an allusion to the habit of this egret of following buffalo or herds of domestic cattle searching for the insects that are disturbed by their hooves, or perching on their backs to find the insects that plague the cattle; a mutual benefit   *ibis* (Gr) an Egyptian bird that feeds on aquatic animals and to which divine honours were paid; this is not an ibis but an egret that feeds on insects. Originally inhabiting only Africa and southern Asia, it has spread to parts of North,

Central, and South America, Indonesia and Australia; in Africa it is found to the north and south of the Sahara, ranging south to Madagascar and South Africa.

Great White Heron or Great Egret *Egretta alba*
aigrette (Fr) à tuft of feathers or plume, an egret; derived from the plume of feathers worn by the male bird during the breeding season *albus* (L) white; a large white heron measuring up to 89 cm (35 in) of which the body is probably less than 46 cm (18 in). Inhabiting North, Central, and South America, Africa south of the Sahara and Madagascar, large areas of central Asia, India, southern Asia, Malaysia, Indonesia, Papua New Guinea, and Australia.

Little Blue Heron *E. caerulea*
caeruleus (L) dark blue. Inhabiting the southern part of North America, Central America, and the northern part of South America.

Chinese Egret *E. eulophotes*
eu (Gr) prefix meaning nicely, well, often used to mean typical *lophos* (Gr) a crest *-otēs* (Gr) suffix meaning possession of; many thousands of egrets of all species have been slaughtered in the past to obtain the crests for decoration, mostly for women's hats. Inhabiting China, south-eastern Asia, and ranging to Australia.

Little Egret *E. garzetta*
garza (It) derived from *garza* (Sp) an egret *-etto* (It) diminutive suffix; 'little egret'. An almost world-wide distribution but not known in Central and South America; it ranges across southern Europe and Asia, Africa, south-east Asia to Papua New Guinea and Australia.

Boat-billed Heron *Cochlearius cochlearius*
coclearum (L) a spoon; the bill is more like an upturned boat than a spoon; this heron must not be confused with the spoonbills in the family Threskiornithidae. Inhabiting the mangrove swamps of central Mexico and ranging south through Central America and Brazil to northern Argentina.

White-backed Night Heron *Nycticorax leuconotus*
nux (Gr), genitive *nuktos*, night *korax* (Gr) a raven, a crow; usually only seen during the hours of twilight, it has a harsh crow-like call *leukos* (Gr) white *notos* (Gr) the back. Inhabiting Africa.

Black-crowned Night Heron *N. nycticorax*
It has a glossy black head and neck. Inhabiting the temperate regions

of North America and ranging south to Argentina, the southern part of Eurasia and part of northern Africa, India, the Burma-Vietnam area, Japan, Hawaii, Malaysia and the Philippines.

Banded Tiger Heron   *Tigrisoma lineatum*
tigris (L) a tiger   also *tigris* (Gr) a tiger   *soma* (Gr) the body   *linea* (L) a string, a line   *-atus* (L) suffix meaning provided with; the body has dark bands or stripes on a brownish background giving very good protective colouring among the reeds. Inhabiting Central America and the northern part of South America.

Least Bittern   *Ixobrychus exilis*
ixos (Gr) mistletoe   *brukhō* (Gr) I roar, I howl; *ixos* probably means a reed, so the name would mean a 'reed-howler', and is a reference to the booming call for which bitterns are famous. The name *Ixobrychus* was given by G. J. Billberg in 1828, but he never gave a proper explanation. It used to be thought that the booming sound was made by the bittern blowing into the hollow stem of a reed, as with a reed instrument like the bassoon; several different explanations have been suggested but the most likely one is that the male's throat is specially modified to produce this sound; it is quite remarkable and has been compared to a distant foghorn, and in fact can be heard over a distance of nearly 5 km (3 miles)   *exilis* (L) thin, slender; it is one of the small bitterns at about 31 cm (12 in), even smaller than *minutus*, below. Inhabiting the tropical parts of America and ranging south to Brazil.

Little Bittern   *I. minutus*
minutus (L) small, minute; it is about 35 cm (14 in) long, very small compared to the Eurasian Bittern at 76 cm (30 in). It is occasionally seen in south-east England, and ranges across Eurasia, Africa, India, the Burma-Vietnam area, and Australia.

American Bittern   *Botaurus lentiginosus*
boo (L) I cry aloud, I roar   *taurus* (L) a bull; a reference to the booming call for which bitterns are famous; it is the large bitterns like this one, measuring about 66 cm (26 in), that are the real exponents of booming; the small bitterns really only make a deep croak   *lentigo* (L) genitive *lentiginis*, a lentil-shaped spot, a freckle   *-osus* (L) suffix meaning full of, very spotted; a reference to the spotted brown plumage. Inhabiting the southern part of North America and ranging south through Central America to the northern part of South America.

Eurasian Bittern   *B. stellaris*
stella (L) a star   *stellaris* (L) starry; the plumage is more streaked than
starred though it does have brown spots. Inhabiting the south-eastern
part of England, the temperate and tropical areas of Eurasia, Africa,
India, the Burma-Vietnam area, Malaysia, and Indonesia.

Family BALAENICIPITIDAE   1 species
balaena (L) a whale   *caput* (L), genitive *capitis*, the head; the head and
huge beak are said to resemble the head of a whale.

Whale-headed Stork or Shoebill   *Balaeniceps rex*
ceps (New L) from *caput* (L) the head   *rex* (L) a king; the bill is rather
like a shoe but is not much like that of a whale except that it is unusually large. For anatomical reasons it has been difficult to place this
bird except in a family on its own. Inhabiting Sudan, Uganda, and
Zaire, and probably other parts of Africa where it has not yet been
recorded.

Family SCOPIDAE   1 species
scopae (L) a besom, a broom made of twigs; the tufts of feathers which
project from the back of the head together with the large beak in
front give the appearance of a hammer.

Hammerhead Stork or Anvilhead   *Scopus umbretta*
umbra (L) a shade   *-etta* (New L) from *-etto* (It) diminutive suffix; a
small shade; a reference to the crest and beak together giving the
appearance of a sunshade or umbrella. Anatomically it is neither a
stork nor a heron though it bears a close resemblance to both, so it is
placed in a family on its own. Widespread in Africa from southern
Sudan to South Africa, Madagascar, and also parts of Arabia.

Family CICONIIDAE   17 species
ciconia (L) a stork.

Yellow-billed Stork   *Mycteria ibis*
ibis (Gr) an Egyptian bird that feeds on worms and aquatic animals,
to which divine honours were paid; it has also been translated as a
stork; this is a misleading name as it is not an ibis, of the family
Threskiornithidae. A white bird with a red face and a yellow bill, it
lives in Africa and Madagascar.

Painted Stork   *M. leucocephala*
leukos (Gr) white   *kephalē* (Gr) the head; mainly a white bird display-

ing various colours such as red legs, an orange bill, and yellow and red skin on the bare face. Inhabiting India and ranging through the Burma-Vietnam area and south-western China.

Asian Open-billed Stork   *Anastomus oscitans*
anastomōsis (Gr) an opening   *oscitans* (L) listless, sluggish, from *oscito* (L) I gape, I yawn; these names refer to the peculiar bill which never closes except at the tip, and it has been suggested that this assists the stork when picking up snails and mussels, which form part of its diet. Inhabiting the Indian subcontinent, Sri Lanka, the Burma-Vietnam area, and south-east China.

Abdim's Stork   *Ciconia abdimii*   (formerly *Sphenorhynchus*)
ciconia (L) a stork; named after Bey Al-Arnaut Abdim (1780–1827), at one time Governor of the Wadi Halfa area in northern Sudan. Inhabiting Sudan and other parts of tropical Africa.

White Stork   *C. ciconia*
This is the well-known stork that sometimes builds its nest on the roofs of houses, is supposed to bring good luck, and is strictly protected throughout Europe and Africa. Widespread in Eurasia from Denmark and Holland south to southern Iberia, north-western Africa, east through Asia Minor, central Asia, India and Japan.

Saddle-bill Stork   *Ephippiorhynchus senegalensis*
ephippios (Gr) for putting on a horse, a saddle   *rhunkhos* (Gr) the beak; a reference to the black band round the middle of the large orange-red bill   *-ensis* (L) suffix meaning belonging to; named from Senegal in West Africa, it is widespread in Africa except the extreme south.

Marabou   *Leptoptilos crumeniferus*
leptos (Gr) slender   *ptilon* (Gr) a wing   *crumena* (L) a small money purse or bag, usually hanging from the neck   *fero* (L) I bear; an allusion to the naked pink throat-pouch hanging from the neck about 46 cm (18 in) long; a rather ugly appendage which has some connection with the breathing: *marabout* (Fr) a priest; can mean an ugly and mis-shapen man; this stork is ugly and even grotesque when seen on the ground but is graceful in the air. The range is similar to the saddle-bill stork, above.

Family   THRESKIORNITHIDAE   31 species
thrēskos (Gr) religious   *ornis* (Gr), genitive *ornithos*, a bird. The most famous species of ibis, the Sacred Ibis *Threskiornis aethiopica*, was

formerly worshipped by the Egyptians and was supposed to preserve
the country from plagues and serpents. It was zealously preserved in
temples, and numerous mummified remains of ibises wrapped in
linen in the ordinary way have been found at Memphis near Cairo,
and Thebes in Upper Egypt.

Sacred Ibis   *Threskiornis aethiopica*
See above *-icus* (L) suffix meaning belonging to; of Ethiopia.
Although once common in Egypt this ibis is no longer found there;
it inhabits Ethiopia, from where it takes its name, and is widespread
in the wetter regions of Africa including Madagascar; it also inhabits
parts of south-east Asia, Malaysia, Indonesia, and some Pacific islands.

Indian White Ibis   *T. melanocephalus*
melas (Gr), genitive *melanos*, black, dusky   *kephalē* (Gr) the head; the
body feathers are white but the naked areas of the head are bluish
black. In addition to India it inhabits parts of Malaysia and Indonesia.

Australian White Ibis   *T. molucca*
The Moluccas are a group of islands in the Banda Sea, about 960 km
(600 miles) to the north of Australia; it also inhabits Papua New
Guinea and parts of Australia, and occasionally Tasmania.

Bald Ibis   *Geronticus calvus*
gerōn (Gr) genitive *gerontos*, an old man   *-icus* (L) suffix meaning belonging to or pertaining to   *calvus* (L) bald. Inhabiting South Africa.

Hermit Ibis   *G. eremita*
erēmitēs (Gr) a hermit. Sometimes known as the Waldrapp, it inhabits
Africa and parts of southern Asia; known to have bred in central and
southern Europe until about 1800.

Japanese Ibis   *Nipponia nippon*
Nippon is the Japanese name for Japan. This ibis is now very rare.

White Ibis   *Eudocimus albus*
eudokimos (Gr) to be of good repute, famous   *albus* (L) white. Inhabiting southern parts of North America and the north of South America,
and Central America.

Scarlet Ibis   *E. ruber*
ruber (L) red; the bright red plumage of this ibis is quite remarkable.
Inhabiting the northern part of South America.

Glossy Ibis   *Plegadis falcinellus*
plēgas (Gr), genitive *plēgados*, a sickle   *falx* (L), genitive *falcis*, a sickle, shaped like a sickle; a reference to the curved beak   *-ellus* (L) diminutive suffix, 'a little sickle'. It has a wide range including southern Europe, Africa, Asia, the East Indies and Australia; an occasional visitor to the south of England in the autumn.

Roseate Spoonbill   *Platalea ajaja*   (formerly *Ajaia*)
platalea (L) a water bird, the spoonbill   *ajaja* is a South American native name for the spoonbill. Unlike the other spoonbills, which are almost pure white, this bird is tinged with pink and darkening to red in some parts. Inhabiting the southern part of North America, Central America, and the northern half of South America.

African Spoonbill   *P. alba*
albus (L) white. Inhabiting Africa and Madagascar.

Yellow-billed Spoonbill   *P. flavipes*
flavus (L) yellow   *pes* (L) the foot; it has yellow legs and feet as well as the yellow bill. Inhabiting Australia.

White Spoonbill   *P. leucorodia*
leukos (Gr) white   *rodeos* (Gr) of roses; *leucorodia* could mean rose-pink but the name is misleading as the plumage is white, however an orange patch can be seen on the throat during summer. Inhabiting southern Europe, Asia, north-eastern Africa, and may visit southeastern England in the winter.

Lesser Spoonbill   *P. minor*
minor (L) smaller. Inhabiting eastern China and Japan.

Royal Spoonbill   *P. regia*
regius (L) royal. Inhabiting New Guinea and Australia.

Family   PHOENICOPTERIDAE   4 species
phoinix (Gr), genitive *phoinikos*, deep purple or crimson; the word derives from the fact that the discovery of this colour was ascribed to the Phoenicians   *pteron* (Gr) feathers, can mean a wing; 'red-feathered': hence *phoinikopteros* (Gr) the flamingo.

Flamingos, in spite of their long legs and neck which give them a heron-like appearance, are related to the geese, ducks and swans, of the order Anseriformes. The evidence is not only anatomical, but in

their habits, and the behaviour of the newly hatched chicks. Sometimes they have been placed in a separate order, the Phoenicopteri-
formes, but now they are usually placed in the order Ciconiiformes,
with the herons and storks.

They are gregarious birds, flying in large formations and feeding
and breeding together in their thousands. They like shallow water,
scooping and dredging the mud with their specially shaped beaks they
sieve out the small invertebrates and vegetable matter on which they
feed. Their red plumage gives a wonderful colour to huge areas of the
lakesides where they congregate.

Lesser Flamingo   *Phoeniconaias minor*
phoinix (Gr), genitive *phoinikos*, deep purple or crimson   *naias* (L) a
water nymph   *minor* (L) smaller. Inhabiting the southern part of
Africa and some parts of India.

Greater Flamingo   *Phoenicopterus ruber*
ruber (L) red, ruddy. This bird is an example of the confusion that can
occur when using the English names of animals. In books on ornithology it is variously referred to as the Greater, the Common, the
American, and the European Flamingo; the only way to establish for
certain the species referred to is to use the Latin name. It has been
said that the name 'flamingo' is derived from *flamenco* (Sp) an Anda-
lusian folk-song, and another possible source is *flamma* (L) a flame, a
fire. This flamingo is widespread and inhabits the Bahama Islands,
parts of Mexico, the Galapagos Islands, the southern part of South
America, southern Europe, Africa, India, and the Burma-Vietnam
area.

Andean Flamingo   *Phoenicoparrus andinus*
parra (L) a bird of ill omen, and variously translated as 'perhaps the
owl', and 'perhaps the wheatear'; any connection is obscure. Inhabiting the Andes in southern Peru, western Bolivian plateaus, and the
Andes in northern Chile and north-western Argentina. It is the largest
species.

James's Flamingo   *P. jamesi*
Named after H. B. James (1846-1892), a naturalist and a business
man living in Chile. Inhabiting high Andean lakes, possibly up to
about 11,000 feet (3,500 m), and now a very rare species.

25 Screamers, Geese, Swans and Ducks
ANSERIFORMES

Black Swan
Cygnus atratus

The three species of screamers have been allotted a family on their own as there seems no obvious position for them with the other birds in the order Anseriformes. They are a peculiar mixture of game birds, soaring birds of prey, and waterfowl, the latter being the reason why they are placed in this order; they possess various peculiar anatomical features including a pair of sharp spurs on the shoulder of each wing.

The English name 'screamer' arises from their harsh screaming calls which carry a great distance, probably up to 3 km (2 miles); when many of them assemble in groups they all take up the cries and the resulting noise is almost unbearable.

The geese, swans, and ducks is a large group of waterfowl with an almost world-wide distribution, though mainly in the Northern Hemisphere. Ancient literature tells us that as long ago as the year 500 B.C. these birds were known about and recorded—no doubt because, as we know today, they make good eating.

Geese and swans have a close relationship among individuals, sometimes the 'pair bond' lasting for life, which in the case of swans could be 40 to 50 years or more. They communicate by a posture and voice language, for instance when the gander has driven off an intruder he utters a special cry, and the female repeats the cry as if to show her approval, and stretches out her neck parallel to the ground; even the baby goslings copy their parents by making this gesture. There are birds other than geese and swans that show a similar behaviour.

Many species, notably the Greylag Goose *Anser anser*, fly in 'V' and other special formations now adopted by man in aeroplanes; it is tiring for the leaders and sometimes they are changed during flight.

Order ANSERIFORMES

anser (L) a goose.

Family ANHIMIDAE 3 species
anhima (New L) derived from a Portuguese word, derived from *anhuma*, a Tupi native name for the bird.

Horned Screamer *Anhima cornuta*
cornutus (L) horned; a reference to the remarkable horn about 10 cm (4 in) long curving forward from the top of the head. Inhabiting Venezuela, Bolivia, and the southern part of Brazil.

Northern Screamer *Chauna chavaria*
khaunos (Gr) gaping; metaphorically can mean foolish, silly *chavaria* is a coined name probably from *charivari* (Fr) rough music, cacophonous noise, a reference to their harsh cries. Inhabiting Colombia and Venezuela.

Southern Screamer *C. torquata*
torquatus (L) wearing a collar; a reference to the band of dark feathers round the neck. Inhabiting Bolivia, Brazil, and Argentina.

Family ANATIDAE 146 species
anas (L), genitive *anatis*, a duck.

Magpie Goose *Anseranas semipalmata*
Anseranas, see above *semi-* (L) prefix meaning half *palma* (L) the palm of the hand; this is a reference to the feet, which are only partly webbed. Any ornithologist could be forgiven for not knowing where to place this bird in the scheme of classification; it has certain characteristics typical of swans, geese, and ducks, and it has been suggested that it should be placed in an order on its own; among other peculiarities it perches in trees, like the Tree Ducks (below). The name

Magpie derives from the colour of the plumage; it has a black head and neck, a white body, and a black rump and tail; it inhabits southern Papua New Guinea and the northern part of Australia.

Black-billed Tree Duck   *Dendrocygna arborea*
dendron (Gr) a tree   *kuknos* (Gr) a swan   *cycnus = cygnus* (L) a swan; the tree ducks and the swans are alike in some of their habits   *arbor* (L) a tree; some species of tree ducks may be seen perching in trees. Inhabiting North and Central America.

Black-bellied Tree Duck   *D. autumnalis*
autumnalis (L) relating to autumn; probably a reference to the brownish colour of the back and breast. They have a distinctive whistling call and are sometimes known as Whistling Ducks. Inhabiting North, Central, and South America.

Fulvous Tree Duck   *D. bicolor*
bicolor (L) of two colours; it is light brown with a dark brown back and pale stripes on the flanks   *fulvous* means dull yellow, tawny, from *fulvus* (L) yellowish brown, tawny. Ranging from California in North America to Argentina in South America, and also inhabiting Africa and India.

Eyton's Tree Duck   *D. eytoni*
Named after T. C. Eyton (1809–1880), a naturalist and author; he founded the Eyton Hall Museum. Inhabiting Australia and New Guinea.

Spotted Tree Duck   *D. guttata*
gutta (L) a drop of fluid, a spot   *guttatus* (L) spotted, speckled. Inhabiting islands in the south-west Pacific.

Indian Tree Duck   *D. javanica*
Sometimes known as the Lesser Whistling Duck, it is not restricted to Java and inhabits India, southern China, and other areas of south-east Asia.

Coscoroba Swan   *Coscoroba coscoroba*
coscoroba (Sp) probably derived from *cosaroba*, a Tupi native name for 'a swan-like diving bird'. This swan has a shorter neck than other swans, and really looks more like a duck; for several reasons it is considered to be intermediate between the ducks and the swans. Inhabiting southern South America, ranging from southern Brazil to Tierra del Fuego and including the Falkland Islands.

Black Swan *Cygnus atratus*
cygnus (L) a swan *ater* (L) black *atratus*, clothed in black as for mourning. Inhabiting Australia.

Whooper Swan *C. cygnus*
It has a 'whooping' type of call and is sometimes known as the Whistling Swan. Inhabiting Scandinavia and Iceland and may visit the British Isles in winter.

Black-necked Swan *C. melanocoryphus*
melas (Gr), genitive *melanos*, black *koruphē* (Gr) the head, the top part. Inhabiting South America, the range is similar to the Coscoroba Swan (above).

Mute Swan *C. olor*
olor (L) a swan. This is the common swan that originally bred in Europe and Asia and has now been introduced to many parts of the world by man; it is the one usually seen on lakes and rivers in the British Isles. It is not entirely mute as it can make grunting and hissing noises.

White-fronted Goose *Anser albifrons*
anser (L) a goose *albus* (L) white *frons* (L) forehead, brow. Inhabiting western North America, Central America, Africa, parts of Eurasia including the British Isles, India, and the Burma-Vietnam area.

Greylag Goose *A. anser*
The plumage is predominantly grey to greyish-brown, and it is suggested that 'lag' came about because it lags behind when other kinds of geese have departed for their breeding grounds. It breeds in northern parts such as Iceland and Scandinavia, also northern Scotland and ranging east to the Balkans and the Black Sea area; it winters in southern Europe, northern Africa, and an eastern race ranges across southern Asia to China. It can sometimes be seen migrating across England in huge formation flights; I have seen over 100 flying south in three large 'V' formations of more than 30 birds each.

Pink-footed Goose *A. brachyrhynchus*
brakhus (Gr) short *rhunkhos* (Gr) the beak; this is a close relative of the Bean and Greylag but with a shorter beak, and is sometimes considered to be a subspecies of the Bean Goose *A. fabalis* (below). Breeding in Greenland, Iceland, and the Norwegian archipelago of Spitzbergen, and migrating south to Scotland, England, Wales, and coastal areas in northern France.

Snow Goose   *A. caerulescens*
caeruleus (L) dark blue   *-escens* (L) suffix meaning beginning to, somewhat; the plumage is almost white, hence the English name, but there
is a grey variety usually known as the Blue Goose in which the main
body colour is blue-grey. Inhabiting North America and an occasional
visitor to western Europe, and also known in southern parts of Asia.

Lesser White-fronted Goose   *A. erythropus*
eruthros (Gr) red   *pous* (Gr) the foot; it is similar to the White-fronted
Goose (above) only smaller, and there is a larger patch of white on the
forehead extending onto the crown. Breeding in northern Scandinavia
and Eurasia and migrating south to Romania, Bulgaria, Asia Minor,
and other areas of southern Asia including India and the Burma-
Vietnam area.

Bean Goose   *A. fabalis*
faba (L) the broad bean   *-alis* (L) suffix meaning pertaining to; the
name was given by Dr Thomas Pennant FRS (1726-1798) and refers
to its habit of feeding on the beans and other crops left in the fields
after the harvest. Breeding in northern areas from Greenland to
eastern Siberia and migrating to southern Europe (and occasionally
Britain), to North America and to parts of southern Asia.

Brent (or Brant) Goose   *Branta bernicla*
Branta is a name coined from the Anglo Saxon *brennan*, to burn, and
refers to the reddish-brown colour   *bernicla* (New L) from *bernicle* (Fr)
a barnacle (see Barnacle Goose, below). Inhabiting North and
Central America, northern Europe including the British Isles and
Scandinavia, and ranging south-east to the Burma-Vietnam area.

Canada Goose   *B. canadensis*
-ensis (L) suffix meaning belonging to; inhabiting Canada and
northern areas of North America and migrating south to the Gulf of
Mexico and other southern areas. It is now a familiar sight in Britain
with its graceful black neck and white chin; it was introduced to
Europe in the seventeenth century.

Barnacle Goose   *B. leucopsis*
leukos (Gr) white   *opsis* (Gr) aspect, appearance; 'white-faced'; it has
a white forehead and face. The name arises from an ancient belief that
this goose was an offspring of the barnacle, and even more bizarre the
belief that these barnacles grew on trees. In fact, when the shell of a

barnacle is broken open the body shows a distinct resemblance to a goose. It is a mutual exchange of names, as there is the Goose Barnacle *Lepas anatifera   lepas* (Gr) a limpet   *anas* (L), genitive *anatis*, a duck *fero* (L) I bear, I bring forth. The Barnacle Goose inhabits Greenland, Iceland, Spitzbergen, and Norway and migrates south in winter to Ireland and parts of northern France.

Red-breasted Goose   *B. ruficollis*
rufus (L) red   *collus = collum* (L) the neck of man and animals; not only the neck but also the breast is quite a striking red in marked contrast to the brown and white body. Inhabiting northern areas of western Asia and migrating south to Asia Minor, India, and other parts of southern Eurasia.

Ne-ne or Hawaiian Goose   *B. sandvicensis*
-ensis (L) suffix meaning belonging to. The Hawaiian Islands were formerly known as the Sandwich Islands, a territory of the USA; they were named by Capt James Cook (1728-1779) the explorer and navigator, after John Montagu Sandwich, the 4th Earl of Sandwich (1718-1792). Ne-ne is the Hawaiian name for this goose; it is found only in Hawaii except that some pairs were brought to the Wildfowl Trust in England for breeding, to rescue them from extinction; still very rare.

Andean Goose   *Chloephaga melanoptera*
khloë (Gr) the tender shoot of plants   *phagein* (Gr) to eat; in addition to various grasses they are fond of seaweed   *melas* (Gr), genitive *melanos*, black   *pteron* (Gr) feathers, or can mean wings. Inhabiting rocky shores along the west coast of South America.

Upland Goose   *C. picta*
pictus (L) painted; a reference to the colour pattern of the plumage which is white above with narrow black lines across and a yellow to buff-coloured breast. Named Upland Goose because it usually lives on higher ground and not near the sea; it inhabits Chile and southern Argentina and ranges south to Tierra del Fuego.

Ruddy-headed Goose   *C. rubidiceps*
rubidus (L) red, reddish   *ceps* (New L) from *caput* (L) the head. Inhabiting South America.

Blue-winged Goose   *Cyanochen cyanoptera*
kyaneos (Gr) dark blue, glossy blue   *khēn* (Gr) a goose   *pteron* (Gr)

feathers, can mean wings. Inhabiting Africa; really a sheldrake but superficially rather like a goose.

Ruddy Shelduck   *Tadorna ferruginea*
tadorna (New L) derived from *tadorne* (Fr) a sheldrake   *ferrugineus* (L) rusty, 'rust-coloured'   *sheld* (Old E) variegated, particoloured (see below). Inhabiting southern parts of Eurasia, northern Africa, and ranging east to China.

Shelduck   *T. tadorna*
Sometimes known as the Sheld Duck which is probably a better name, from *sheld* (Old E) variegated, particoloured; the plumage includes the colours black, white, chestnut, and dark green, and is more colourful than *ferruginea* (above). Widespread in Eurasia including the British Isles and Scandinavia, breeding coastal areas; also Mediterranean, northern Africa, and ranging east to Arabia, India, China, and Japan.

Paradise Duck   *T. variegata*
varius (L) variegated   *varie* (L) with diverse colours   *variegatus* (L) provided with diverse colours. The name Paradise Duck arises from the colourful plumage, like the Bird of Paradise. Inhabiting Australia.

Patagonian Crested Duck   *Lophonetta specularioides*
lophos (Gr) the crest of birds, and other crests   *nētta* (Attic Gr) a duck, derived from *neō* (Gr) I swim   *speculum* (L) a mirror; the English word speculum can mean the bright patch on a bird's wing, specially a duck's wing   *-oides* (New L) derived from *eidos* (Gr) apparent shape, resemblance; on a mostly brown and dark brown plumage a pink patch, or speculum, shows up on the wings. Inhabiting western Argentina, southern Chile, Patagonia, and ranging south to Tierra del Fuego and the Falkland Islands.

Falkland Island Flightless Steamer Duck   *Tachyeres brachypterus*
takhus (Gr) swiftness, speed   *eressō* (Gr) I row   *takhuērēs* (Gr) fast-rowing   *brakhus* (Gr) short   *pteron* (Gr) feathers, can mean wings. This flightless duck with only small wings can achieve quite a speed on the surface of the water by beating the water with its wings as though rowing, and this causes much flying spray and general commotion which has given rise to the name 'steamer duck'; an allusion to the paddle steamers in general use when the name was given. It is confined to the Falkland Islands.

Flying Steamer Duck   *T. patachonicus*
-icus (L) suffix meaning belonging to   *patachonicus*, of Patagonia (the southern part of Argentina). Only this one of the three Steamer Ducks is able to fly, having larger wings, but even so often prefers to escape from intruders by 'steaming' along the surface of the water. Inhabiting southern Chile and Argentina and ranging south to Tierra del Fuego, and the Falkland Islands.

Magellan Flightless Steamer Duck   *T. pteneres*
ptēna (Gr) fowls, birds   *pteneres* is a coined word intended to mean 'rowing birds' (see *T. brachypterus* above). Inhabiting coastal areas along the Magellan Straits and ranging north to some extent, and the islands around Tierra del Fuego.

Pintail   *Anas acuta*
anas (L) a duck   *acutus* (L) sharp, pointed; the central tail feathers are elongated and sharply pointed, forming the 'pintail'. It has a widespread distribution in the Northern Hemisphere.

American Wigeon   *A. americana*
-anus (L) suffix meaning belonging to; inhabiting North, Central, and South America.

Bahama Pintail   *A. bahamensis*
-ensis (L) suffix meaning belonging to, usually referring to a place. Not entirely confined to the Bahamas, and known in North and South America.

Madagascan Teal   *A. bernieri*
Named after Dr J. A. Bernier, a French naval surgeon and botanist; he spent three years in Madagascar from 1831 to 1834.

Chestnut Teal   *A. castanea*
kastana (Gr) the chestnut tree; an allusion to the chestnut-coloured plumage. Inhabiting Australia.

Northern Shoveler   *A. clypeata*
clypeus = *clipeus* (L) a shield   *-atus* (L) suffix meaning provided with; probably a reference to the white breast which stands out in contrast to the darker plumage of the body; it has a shovel-like broad-fronted bill. Widespread throughout the northern part of the world including North, Central and South America, Europe, including south-east England, parts of Scandinavia, Africa, Asia, including India and the Burma-Vietnam area.

South American or Speckled Teal   *A. flavirostris*
flavus (L) yellow   *rostrum* (L) the beak. Inhabiting South America.

Baikal Teal   *A. formosa*
formosus (L) finely formed, beautiful; the English name is from Lake Baikal, a very large fresh-water lake in southern Siberia. It also lives in parts of North America, south-east Asia, and is a very rare visitor to Europe.

South Georgia Pintail   *A. georgica*
South Georgia is a British island about 1,287 km (800 miles) to the east of the Falkland Islands in the South Atlantic; it is a very remote place, but although named from this island other subspecies of this pintail duck inhabit parts of South America and Africa.

Meller's Duck   *A. melleri*
Named after Dr C. J. Meller (c. 1836–1869), who was Superintendent of the Botanical Gardens in Mauritius in 1865; Mauritius is one of the Mascarene Islands in the Indian Ocean about 1,126 km (700 miles) to the east of Madagascar. This duck inhabits Africa.

Eurasian Wigeon   *A. penelope*
penelops (Gr) a kind of duck; Penelope was the wife of Odysseus, and there is a Greek legend about her being thrown into the sea as a baby and rescued by sea-birds, but whether the bird was named after her or she was named after the bird is not known. This duck is widespread in areas of North and Central America, Europe, Asia, Africa, India, the Burma-Vietnam area, Malaysia, Indonesia, and New Guinea.

Mallard   *A. platyrhynchus*
platus (Gr) wide or flat   *rhunkhos* (Gr) the beak; a reference to the heavy and rather flat beak. Widespread distribution similar to *A. penelope*, above, and including Australia.

Garganey   *A. querquedula*
querquedula (L) a kind of duck: garganey is the Italian name used in the area of the lakes of northern Italy. Inhabiting Europe but only the south-eastern part of England; Africa, Asia, India, the Burma-Vietnam area, Malaysia, Indonesia, New Guinea and Australia.

Gadwall   *A. strepera*
strepo (L) I make a noise, I rattle; a reference to the call which has also been described as reedy. Inhabiting North and Central America, Europe including the British Isles but little known in Scandinavia;

Africa, Asia, India, and the Burma-Vietnam area. The origin of the name Gadwall is obscure, but was first used in the seventeenth century.

Torrent Duck   *Merganetta armata*
mergus (L) a diver, a water fowl   *nētta* (Attic Gr) = *nēssa* (Gr) a duck, from *nēo* (Gr) I swim   *armatus* (L) armed; a reference to the sharp spurs on its wings. The name Torrent Duck has been given because it frequents rapid mountain streams and likes the rough tempestuous water because it finds the essential small invertebrate food there; these streams are in the Andes Mountains and the birds range along the whole of the western part of South America.

Freckled Duck   *Stictonetta naevosa*
stiktos (Gr) dotted, dappled   *nētta* (Gr) see above   *naevus* (L) a mole or wart on the body; can mean a spot, a blemish   *-osus* (L) suffix meaning full of; 'very spotted'. Inhabiting Australia.

Steller's Eider   *Polysticta stelleri*
polus (Gr) many, much   *stiktos* (Gr) dotted, dappled; not a very apt name as the drake does not appear to be spotted though the duck has a mottled brown plumage, however the drake takes on the brown colour in winter. George W. Steller (1709-1769) was a German zoologist and traveller; several other animals have been named after him. This eider inhabits arctic Canada and Siberia, and ranges south as far as the northern coasts of the Atlantic and Pacific Oceans.

Common Eider   *Somateria mollissima*
sōma (Gr), genitive *sōmatos*, the body   *erion* (Gr) wool   *mollis* (L) soft, *mollissima*, very soft; 'very soft body-wool'. Noted for the very soft down, hence the name eiderdown, or quilt. Inhabiting the colder parts of the Northern Hemisphere including the British Isles.

Redhead   *Aythya americana*
aythya (New L) from *aithuia* (Gr) a sea-gull or diving bird; an important American game bird akin to the Canvasback and European Pochard but with a brighter rufous head. Inhabiting North and Central America.

Australian Pochard   *A. australis*
auster (L), genitive *austri*, the south   *-alis* (L) suffix meaning relating to   *australis* (L) southern; it does not necessarily mean Australia. This pochard inhabits Indonesia, New Guinea, some of the small Pacific islands, and Australia.

Baer's Pochard   *A. baeri*
Named after Prof K. E. V. Baer (1792–1876), a German Professor of
Zoology, of Königsberg. This pochard inhabits North America,
Eurasia, India, and the Burma-Vietnam area.

Common Pochard   *A. ferina*
ferina (L) the flesh of wild animals, game. Widespread distribution
including North America, Eurasia, Africa, India, the Burma-Vietnam
area, Malaysia, and Indonesia. It is considered an important game
bird and makes good eating.

Tufted Duck   *A. fuligula*
fuligo (L) soot   *-ulus* (L) dim. suffix; can also mean somewhat, a
tendency; 'somewhat sooty'; it is mostly black except for the white
flanks, and has a trailing crest at the back of the head, hence the
English name. The distribution is similar to *A. ferina* (see above),
ranging from North America, and including Iceland, across Eurasia
to Kamchatka, the peninsula at the eastern end of the USSR.

New Zealand Scaup   *A. novaeseelandiae*
Scaup is a Scottish word for scalp, the mussel-scalp or mussel-scaup;
this pochard has been given the name because it feeds on mussel-
scaups.

Canvasback   *A. valisineria*
Indirectly named after Prof Antonio Vallisnieri (1661–1730), an
Italian naturalist who was a Professor of Medicine in Padua; its
favourite food is the waterweed *Vallisnera spiralis*, named after him.
The name canvasback derives from the colour and the marking of grey
and white on the back. Inhabiting North and Central America, it is
an important game bird, and considered one of the best ducks for the
table.

African or Red-eyed Pochard   *Netta erythrophthalma*
nētta (Attic Gr) = *nēssa* (Gr) a duck, from *nēo*, I swim   *eruthros* (Gr)
red   *ophthalmos* (Gr) the eye; the eye is quite a bright red in the drake,
but brown in the duck. Inhabiting Africa and South America.

Red-crested Pochard   *N. rufina*
rufus (L) red   *-inus* (L) suffic meaning like, pertaining to; a reference
to the crest. It is known in East Anglia, and found in various regions
of Eurasia, but local.

Maned Goose   *Chenonetta jubata*
khēn (Gr), genitive *khēnos*, a goose   *nētta* (Attic Gr) see under African
Pochard, above   *juba* = *iuba* (L) the mane of an animal   *iubatus* (L)
maned; a reference to the crest. Inhabiting Australia.

Brazilian Teal   *Amazonetta brasiliensis*
nētta (Attic Gr) see above; 'Amazon duck'   *-ensis* (L) suffix meaning
belonging to; it is not confined to Brazil and ranges from the northern
part of South America south to Buenos Aires.

Mandarin Duck   *Aix galericulata*
aix (Gr) a water bird of the goose kind; the appearance is more that
of a typical duck rather than a goose   *galerum* (L) a skull cap
galericulum (L) a small cap for the head   *galericulata*, provided with a
small cap; an allusion to the colouring of the head and a crest which
give the impression of a cap. The name Mandarin has been given
because of the brilliant and decorative plumage, and for this reason
it has been introduced to many countries and domesticated and can
often be seen on ponds in parks and gardens. Originally inhabiting
China and other far-eastern countries.

Wood Duck   *A. sponsa*
sponsa (L) a betrothed woman, a bride; the name is taken from the
Mandarin Duck, above; in the old days in China a pair was given as
a traditional wedding present to symbolise marital fidelity. Although
living in quite different parts of the world these two ducks are very
closely related and the females appear to be almost identical. Inhabiting the temperate regions of North America.

Knob-billed Goose   *Sarkidiornis melanotos*
sarx (Gr), genitive *sarkos*, flesh   *idios* (Gr) separate, distinct; hence
strange   *ornis* (Gr) a bird; a reference to the peculiar knob on the bill
of the male   *melas* (Gr), genitive *melanos*, black   *nōton* (Gr) the back;
it has a black back in contrast to the white breast and grey flanks.

Muscovy Duck   *Cairina moschata*
-inus (L) suffix meaning belonging to; it is unlikely that this duck is
named from Cairo, Egypt, as it inhabits Central and South America;
it could be from Cairu, Brazil   *moschatus* (New L) musky; originally
known as the 'musk-duck', it has a strong musky smell; 'Muscovy'
does not suggest any connection with Moscow.

Spur-winged Goose   *Plectropterus gambensis*
plēktron (Gr) a cock's spur   *pteron* (Gr) feathers, can mean wings; a
reference to the spur on the wing shoulder   *-ensis* (L) suffix meaning
belonging to; it is not confined to Gambia although named from
there; it ranges across Africa from Sudan and Ethiopia south to the
Zambesi.

Barrow's Goldeneye   *Bucephala islandica*
bu = bous (Gr), genitive *boos*, a bull   *kephalē* (Gr) the head; 'bull-
headed' is a reference to the large head on a rather short neck   *-icus*
(L) suffix meaning belonging to; named from Iceland it is not confined to that island; it breeds there and also in Greenland and other
parts of northern North America but in winter it ranges further south.
It is named in honour of Sir John Barrow (1764–1848), an English
explorer and Admiralty official; the small town of Barrow in northern
Alaska, and Barrow Point, are named after him.

Smew   *Mergus albellus*
mergus (L) a sea bird, a diver, from *mergo* (L) I dip, I plunge in   *albus*
(L) white   *-ellus* (L) diminutive suffix, used here to suggest 'whitish';
the male is mostly white with grey flanks and black marking. The
name smew derives from the Old German word *smiehe*. Breeding in
northern Scandinavia and northern Asia and ranging south in winter
to southern Europe, northern India, China, and Japan; an occasional
visitor to southern England.

Auckland Island Merganser   *M. australis*
auster (L), genitive *austri*, the south   *australis*, southern; it does not
necessarily mean Australia; Auckland Island is about 483 km (300
miles) south of New Zealand. The name merganser is derived from
mergus and *anser* (L) a goose; this merganser is very rare and may even
now be extinct.

Hooded Merganser   *M. cucullatus*
cucullus (L) a hood   *-atus* (L) suffix meaning provided with; a reference to the fan-shaped crest which is white with a black border and
has the appearance of a hood. Inhabiting North America from the
southern part of Canada ranging south to Central America.

Goosander or Common Merganser   *M. merganser*
merganser, see *M. australis*, above. The name goosander is probably
from goose combined with *and* (Old Norse), plural *ander*, a duck;
'goose-duck' would suggest a large duck and this is the largest of all

the ducks. Inhabiting North America, Europe including Iceland, Scandinavia, and the British Isles, though not Ireland, and ranging east across Asia to Kamchatka, China, and Japan.

Red-breasted Merganser   *M. serrator*
serratus (L) toothed like a saw; a reference to the bill which has backward-pointing serrations which enable it to hold the slippery fish on which it feeds; all the mergansers have the serrated bill. *M. serrator* breeds in the northern parts of North America and Eurasia including Iceland, Scandinavia, and the northern parts of the British Isles, and ranges south throughout the whole of this area in the winter.

Blue-billed Duck   *Oxyura australis*
oxus (Gr) sharp, pointed   *oura* (Gr) the tail; a reference to the tail consisting of stiff feathers and this group of ducks are sometimes known as 'stifftails'   *auster* (L), genitive *austri*, the south   *-alis* (L) suffix meaning relating to; southern; it does not necessarily mean Australia although in fact this duck does live in the southern part of Australia. The male has a pale blue bill.

Ruddy Duck   *O. jamaicensis*
-ensis (L) suffix meaning belonging to; named from Jamaica it also inhabits Canada, the USA, Central America, and parts of South America.

White-headed Duck   *O. leucocephala*
leukos (Gr) white   *kephalē* (Gr) the head. Inhabiting certain isolated areas in southern Europe, most of the northern part of North Africa, and some areas in Asia including India.

Lake Duck   *O. vittata*
vitta (L) a chaplet, a ribbon   *vittatus*, bound with a ribbon, can mean striped. Inhabiting South America.

White-backed Duck   *Thalassornis leuconotus*
thalassa (Gr) the sea   *ornis* (Gr) a bird; it seems a strange name for a bird that usually frequents inland waters rather than the sea   *leukos* (Gr) white   *nōton* (Gr) the back. Inhabiting Africa.

Musk Duck   *Biziura lobata*
Bizi, obscure   *oura* (Gr) the tail; the origin of 'bizi' has not been traced but is probably an obscure coinage by Shaw; the stiff tail has an unusual shape and is normally held upright   *lobatus* (New L) lobed, from *lobos* (Gr) a lobe; a reference to the strange lobe under the

bill. In the Official Checklist of the Birds of Australia, H. Wolsten-holme, BA, of Wahroonga, Sydney, referred to the MS of Dr Herbert Langton (deceased), an expert on the derivation of bird names; he says: 'Like myself he could not find the origin of certain obscure names, such as *Biziura*, *Epthianura*, and *Aplonis*.' Dr G. Shaw, FRS (1751–1813) was a professor at the British Museum (Natural History) from 1807 until his death in 1813. This duck has a musk gland which produces an unpleasant odour; it inhabits southern Australia and Tasmania.

Black-headed Duck   *Heteronetta atricapilla*
heteros (Gr) other, different   *nētta* (Attic Gr) = *nēssa* (Gr) a duck, from *nēo*, I swim   *ater* (L) fem. *atra*, black   *capillus* (L) the hair of the head, usually of men, can mean the hair of animals: 'a different duck'; it is not closely related to other stifftails and is unique in being the only truly parasitic member of the duck family, laying its eggs in other birds' nests. Inhabiting the southern part of the Andes, South America.

26 Vultures, Hawks, Eagles, Falcons and their kin FALCONIFORMES

Goshawk
Accipiter gentilis

This group consists of the big vultures, the scavengers, and the diurnal birds of prey. The word diurnal comes from the Latin *diurnus*, meaning belonging to a day, as these birds hunt in the daytime, unlike the owls for instance, who hunt at night. It includes the Osprey and the Secretary Bird.

They range in size from the Falconets, about 80 cm (7 in) long, to the huge Andean Condor, *Vultur gryphus*, with a wingspan of about 3·5 m (12 ft); it is the largest flying bird in the world still alive today. There is no doubt that the vultures, although they look rather repulsive, do a good job; I know of someone who says they clean up his ranch, and when they visit his ranch-house to see what there is to scavenge, he refers to them as 'my refuse disposal unit'. They rank among the finest soaring birds in the world as demonstrated by their effortless circling in thermals.

Order FALCONIFORMES

phalkōn (Gr) = *falco* (L), genitive *falconis*, a falcon, derived from *falx* (L), genitive *falcis*, a sickle; said to be on account of the curved talons.

225

Family CATHARTIDAE   7 species
kathartēs (Gr) a cleanser, a purifier; being scavengers they clear up
rotting carcases and other refuse.

Lesser Yellow-headed Vulture   *Cathartes burrovianus*
-anus (L) suffix meaning belonging to; a curious alteration of the spelling, it is named after a Dr Burrough but further details about him are
not known; the name was given in 1845. This vulture inhabits South
America.

Black Vulture   *Coragyps atratus*
korax (Gr) a raven, a crow   *gups* (Gr) a vulture; 'a raven-vulture';
the black plumage and general appearance is that of a raven   *ater* (L)
black   *atratus* (L) clothed in black as for mourning. Inhabiting the
middle areas of North America and ranging south through Central
America to Patagonia.

California Condor   *Vultur californianus*
vultur (L) a vulture   *-anus* (L) suffix meaning belonging to. Inhabiting the western part of North America and ranging south to Central
America, but now almost extinct.

Andean Condor   *V. gryphus*
grupos (Gr) hook-nosed, with an aquiline nose. The largest flying bird
in the world with a wing-span of about 3·5 m (12 ft), it ranges along
the entire length of the Andes from sea level up to about 3,047 m
(10,000 ft).

Family SAGITTARIIDAE   1 species
sagitta (L) an arrow   *sagittarius* (L) an archer, a bowman; '. . . from
the way their upright carriage and dignified stride reminded people
of an archer about to loose an arrow' (Purnell's *Encyclopedia of Animal
Life*). In appearance and anatomy it does not have any obvious affinities that would place it in any particular group so it has been given a
family on its own; its proper place in classification remains in doubt.

Secretary Bird   *Sagittarius serpentarius*
-arius (L) suffix meaning pertaining to; it is associated with snakes
because it is adept at killing them and they form part of its diet: the
name 'secretary' is an allusion to the spray of quills projecting from
the back of the head, reminiscent of the old-fashioned quill pens that
were tucked behind the ears of secretaries. Widespread throughout
most of Africa.

Family ACCIPITRIDAE 211 species
accipiter (L), genitive *accipitris*, a hawk.

Honey Buzzard *Pernis apivorus*
pernis (New L), said to be a corruption of *pternis* (Gr) a bird of prey
apis (L) a bee *voro* (L) I devour; it eats bees and wasps, nipping off
the tail before eating, and the honey, and also eats frogs and lizards
and small mammals. Inhabiting most of Europe and occasionally
breeding in southern England, ranging east across Russia to the
eastern side of the Ural Mountains.

Barred Honey Buzzard *P. celebensis*
-ensis (L) suffix meaning belonging to; inhabiting Celebes, now known
as Sulawesi, one of the islands of Indonesia; it also inhabits the Philippines. The name 'barred' is a reference to the dark brown markings
in the plumage.

Swainson's Kite *Gampsonyx swainsonii*
gampsos (Gr) bowed, curved *onux* (Gr) a claw; a reference to the
strongly hooked talons: named after W. Swainson, FRS (1789-1855),
a zoologist, artist, and taxidermist; he spent a number of years in
Brazil, Sicily, Africa, and New Zealand. This kite inhabits Central
and South America.

Swallow-tailed Kite *Elanoides forficatus*
elanus (New L) derived from *elanos* (Late Gr) a kite *-oides* (New L)
from *eidos* (Gr) apparent shape, resemblance *forficatus* (New L)
forked, derived from *forfex* (L), genitive *forficis*, scissors; a very beautiful bird with a deeply forked tail; the whole appearance in silhouette
is that of a large swallow. Inhabiting the southern part of the USA
and ranging south to southern South America.

White-tailed Kite *Elanus leucurus*
Elanus, see above *leukos* (Gr) white *oura* (Gr) the tail. Inhabiting
the southern borders of the USA and ranging south through Central
and much of South America.

Snail Kite *Rostrhamus sociabilis*
rostrum (L) the beak *hamus* (L) a hook; the upper mandible is slender
and strongly hooked, a real 'winkle-picker', which enables it to extract
the snails on which it feeds, and this is its only food *sociabilis* (L)
easily united, compatible; a reference to its habit of nesting in colonies
up to 100 strong, the nests being in close proximity; an unusual breeding behaviour for kites. Ranging from Mexico to Argentina; an

isolated group, sometimes known as Everglade Kites, live in the Everglades of Florida, but these are now very rare.

Mississippi Kite   *Ictinia mississippiensis*
iktinos (Gr) a kite or hen-harrier   *-ensis* (L) suffix meaning belonging to; named from the River Mississippi, it is found in North, Central, and South America.

Square-tailed Kite   *Lophoictinia isura*
lophos (Gr) the crest of birds   *iktinos* (Gr) a kite or hen-harrier; 'a crested kite'   *isos* (Gr) equal   *oura* (Gr) the tail; a reference to the tail feathers being of equal length, thus forming the square tail. Inhabiting Australia.

Black-breasted Buzzard   *Hamirostra melanosternon*
hamus (L) a fish-hook; can mean anything hooked   *rostrum* (L) the beak   *melas* (Gr), genitive *melanos*, black   *sternon* (Gr) the chest or breast. Inhabiting the Australian region.

Black Kite   *Milvus migrans*
milvus (L) a bird of prey, a kite   *migro* (L) I move from place to place; migrans, wandering; '. . . encountered almost everywhere throughout the warmer parts of the Old World' (*Living Birds of the World*, Gilliard); northern populations migrate south in winter. The plumage is very dark brown above and rufous brown below, rather than black.

Red Kite   *M. milvus*
It is not red but the plumage is more rufous than the Black Kite (above); common, and the range is similar.

Brahminy Kite   *Haliastur indus*
hals (Gr) the sea   *astur* (L) a species of hawk: named from the River Indus, it has a wide distribution from India to the Solomon islands and the Philippines. Sometimes known as a Sea Eagle, it often frequents rivers and sea coasts. Brahman or Brahmin means having to do with prayer; this kite is held sacred by the Hindus.

White-tailed Eagle   *Haliaeetus albicilla*
hals (Gr), genitive *halos*, the sea   *aietos* = *aetos* (Gr) an eagle; the birds in this genus are sometimes known as sea-eagles as they usually inhabit sea coasts, rivers, and lakes, to obtain the fish which they snatch from the water with their claws; mammals and birds also form part of their diet   *albus* (L) white   *-illus* (L) diminutive suffix; this is supposed to mean 'white-tailed' but it does not; the mistake may

originate from the name of the wagtail, *Motacilla* (for explanation see Motacillidae p. 236). Ranging from Greenland, Iceland, and Norway through Eurasia to Japan and usually migrating south in winter to the Oriental region.

Bald Eagle *H. leucocephalus*

leukos (Gr) white *kephalē* (Gr) the head; it is not bald, but the white head and neck give it the appearance of being bald. It is the national emblem of the USA and appears on the dollar bill; at one time it was widespread in North America but owing to shooting and destruction of habitat it is now very rare, and only seen in any quantity in Alaska and Florida.

White-bellied Sea Eagle *H. leucogaster*

leukos (Gr) white *gastēr* (Gr) the belly. Ranging from India to southern China, Papua New Guinea and nearby islands, Australia and Tasmania.

Steller's Sea Eagle *H. pelagicus*

pelagos (Gr) the sea *-icus* (L) suffix meaning belonging to: named after George W. Steller (1709-1769), a well-known German zoologist and traveller; a number of other animals have been named in his honour. Inhabiting the Pacific coast of Asia.

African Fish Eagle *H. vocifer*

vociferor (L) to cry aloud, shout; a very noisy bird calling with a clear, yelping voice. Inhabiting rivers, lakes, and swamps as high as 1,219 m (4,000 ft), in Africa.

Grey-headed Fishing Eagle *Ichthyophaga ichthyaetus*

ikhthus (Gr) a fish *phagein* (Gr) to eat *aetos* (Gr) an eagle; 'a fish-eating fish eagle'. Inhabiting India and ranging south-east to Indonesia.

Palm-nut Vulture *Gypohierax angolensis*

gups (Gr) a vulture *hierax* (Gr) a hawk or falcon *-ensis* (L) suffix meaning belonging to; named from Angola it is found in other parts of Africa where there are oil palms; although the husks of oil palm nuts appear to be its favourite food it also eats other food such as fruit, molluscs, and locusts.

Bearded Vulture or Lammergeier *Gypaetus barbatus*

gups (Gr), genitive *gupos*, a vulture *aetos* (Gr) an eagle *barbatus* (L) bearded; a reference to the black feathers under the beak that form the beard. The name lammergeier comes from the German *lammer*, lambs, and *geier*, a vulture; they are reputed to carry off live lambs for food but this has never been recorded by reliable observers. Inhabiting southern parts of Eurasia and northern Africa ranging from the Pyrenees to China in mountainous areas.

Indian White-backed Vulture   *Gyps bengalensis*
gups (Gr) a vulture   *-ensis* (L) suffix meaning belonging to; it ranges from the Indian sub-continent to Burma, Laos, Cambodia, and Vietnam. The plumage is almost black with a conspicuous white patch on the lower back.

Cape Vulture   *G. coprotheres*
kopros (Gr) dung; can mean dirt in general   *thēreuō* (Gr) I hunt; a reference to its habit of feeding on decaying carcases. Inhabiting Africa.

Indian Black Vulture   *Sarcogyps calvus*
sarx (Gr), genitive *sarkos*, flesh   *gups* (Gr) a vulture   *calvus* (L) bald; it has a bare head and neck of a reddish orange colour and the body is black with some white markings. Inhabiting India and the Burma-Vietnam area.

White-headed Vulture   *Trigonoceps occipitalis*
trigōnos (Gr) three-cornered, triangular   *ceps* (New L) from *caput* (L) the head; a reference to the white down on the head which gives it a triangular outline   *occiput* (L), genitive *occipitis*, the back part of the head, can mean the head as a whole   *-alis* (L) suffix meaning relating to. Ranging from Eritrea in East Africa to the Orange River in South West Africa.

Snake Eagle   *Circaetus cinereus*
kirkos (Gr) a kind of hawk or falcon that flies in circles   *aetos* (Gr) an eagle   *cinereus* (L) ash-coloured; the plumage is a dark slaty-brown: a fearless hunter of snakes, it has been seen to kill and eat a large poisonous snake about 1 m (3 ft) in length. Inhabiting Africa.

Philippine Serpent Eagle   *Spilornis holospilus*
spilos (Gr) a blemish, a spot   *ornis* (Gr) a bird   *holos* (Gr) whole, entire; a brown plumage almost entirely spotted with white: it eats small mammals, lizards, and other reptiles, chiefly snakes. Inhabiting the Philippines.

Marsh Harrier *Circus aeruginosus*

kirkos (Gr) a kind of hawk or falcon that flies in circles; in fact this harrier is seldom seen soaring and flying in circles; it usually flies low over marshy ground searching for frogs, small mammals, small water-fowl, and fish *aerugo* (L) rust of copper *aeruginosus* (L) full of copper-rust, rusty; can mean the red rust of iron, being a reference to the upper plumage which is a reddish brown. It has a wide distribution in Europe and Asia, including the south of England, Africa including Madagascar, India, ranging south-east through Burma, Malaysia, the Philippines, New Guinea, and Australia.

Pallid Harrier *C. macrourus*

makros (Gr) long *oura* (Gr) the tail; actually the tail is not much longer than other closely related harriers: the plumage is distinctly pale with grey above and a white breast. It has a wide distribution rather similar to the Marsh Harrier (above), but not extending so far west in Europe and not to the south-east in Australasia.

Montagu's Harrier *C. pygargus*

pugē (Gr) the rump *argos* (Gr) shining, bright *pugargos* (Gr) has been used to mean a type of eagle, one with a white rump, probably the White-tailed Eagle *Haliaeetus albicilla*; this harrier, *C. pygargus*, has a well-defined white rump. Inhabiting Eurasia including the southern part of England, Africa, India, and the Burma-Vietnam area. Named after Col George Montagu (1751-1815), a writer on natural history and an early member of the Linnaean Society; he had a fine collection of animals including birds which was eventually purchased by the British Museum (Natural History).

Pale Chanting Goshawk *Melierax canorus*

melos (Gr) a song *ierax* (Gr) a hawk or vulture *cano* (L) I sing *canorus* (L) melodious; it does not sing like an ordinary song-bird but has a melodious whistle and calls from a tree-top; the plumage above is pale grey. Inhabiting the eastern and southern part of Africa.

Cooper's Hawk *Accipiter cooperii*

accipiter (L) the common hawk: named after W. Cooper (1798-1864), an American zoologist; he founded the New York Lyceum of Natural History. In most species the female is larger than the male and in this hawk the female is about one-third larger than the male; it inhabits North and Central America.

Red-thighed Sparrowhawk   *A. erythropus*
eruthros (Gr) red   *pous* (Gr) the foot; in this case taken to mean the
leg. Inhabiting Africa.

Goshawk   *A. gentilis*
gentilis (New L) noble; it has been given this name due to its prowess
at hunting in the days of falconry; it has been described as 'the most
noble of all falcons' (R. D. Macleod). Inhabiting North and Central
America, Europe including occasionally the British Isles, Asia to India
and the Burma-Vietnam area.

Black-mantled Goshawk   *A. melanochlamys*
melas (Gr), genitive *melanos*, black   *khlamus* (Gr) a short cloak or
mantle; a reference to the black plumage that forms a short 'cape'
round the head and neck. Inhabiting New Guinea.

Sparrowhawk   *A. nisus*
According to the Greek legend Nisus, king of the Megara, was changed
into a sparrowhawk. Widespread and common in Europe including
the British Isles, Africa, ranging east to Japan, including India, the
Burma-Vietnam area, and Indonesia.

African Long-tailed Hawk   *Urotriorchis macrourus*
oura (Gr) the tail   *triorkhos* (Gr) a kind of falcon or kite   *makros* (Gr)
long   *oura* (Gr) the tail. Inhabiting Africa.

Rufous-winged Buzzard-Eagle   *Butastur liventer*
buteo (L) a kind of falcon or hawk   *astur* (L) a kind of hawk   *livens*
(L), genitive *liventis*, bluish, the colour of lead. Inhabiting Indonesia.

White Hawk   *Leucopternis albicollis*
leukos (Gr) white   *pternis* (Gr) a bird of prey   *albus* (L) white   *collum*
(L) the neck. Inhabiting Central and South America.

Slate-coloured Hawk   *L. schistacea*
skhistos (Gr) divided, cleft   *schist* (New L) slate   *-aceus* (L) suffix
meaning made of, like; slate-coloured. Inhabiting South America.

Black Hawk   *Buteogallus anthracinus*
buteo (L) a kind of falcon or hawk   *gallus* (L) a cock   *anthrax* (Gr),
genitive *anthrakos*, coal or charcoal   *-inus* (L) suffix meaning like;
coal-coloured. Inhabiting North, Central, and South America.

Solitary Eagle   *Harpyhaliaetus solitarius*
harpē (Gr) a bird of prey   *haliaetos* (Gr) a sea-eagle   *solitarius* (L)

solitary. Inhabiting Central and South America.

White-tailed Hawk   *Buteo albicaudatus*
buteo (L) a kind of falcon or hawk   *albus* (L) white   *cauda* (L) the tail of an animal   *-atus* (L) suffix meaning provided with; the underparts, rump, and tail are white. Ranging from Texas, through Central America to South America.

Common Buzzard   *B. buteo*
Although most of the birds in this genus are known as hawks, those that live in Europe or Africa are often known as buzzards; this one is common and widespread in Africa and Europe including the British Isles though not Ireland; it ranges through southern Asia to India, the Burma-Vietnam area, Malaysia, and Indonesia.

Galapagos Hawk   *B. galapagoensis*
-ensis (L) suffix meaning belonging to, usually referring to a place; it is confined to Galapagos and the surrounding small islands.

Red-tailed Hawk   *B. jamaicensis*
-ensis (L) see above; it has a rather short reddish-brown tail. It ranges from Alaska south to Panama and the West Indies.

Rough-legged Hawk   *B. lagopus*
lagōs (Gr) a hare   *pous* (Gr) the foot; 'hare-footed'; a reference to the legs being feathered more extensively than others in the genus. Inhabiting Europe including the eastern part of the British Isles but not Spain, and ranging east to the Burma-Vietnam area.

African Mountain Buzzard   *B. oreophilus*
oros (Gr), genitive *oreos*, a mountain   *philos* (Gr) loved, pleasing; a 'mountain-lover'. Inhabiting Africa.

Harpy Eagle   *Harpia harpyia*
Arpuiai (Gr) the Snatchers, derived from *arpazō* (Gr) I snatch away, carry off; there are various interpretations but in late mythology they appear as hideous winged monsters; a very large eagle and able to carry off various mammals such as monkeys, opossums, and agoutis. It ranges from southern Mexico to southern Brazil and northern Argentina.

Monkey-eating Eagle   *Pithecophaga jefferyi*
pithekos (Gr) an ape   *phagein* (Gr) to eat: named after Jeffrey Whitehead (fl 1840–1909), an English naturalist. A very large eagle feeding

on various mammals in addition to monkeys and also birds. Very rare and probably now found only on Mindanao in the Philippines.

Wedge-tailed Eagle *Aquila audax*
aquila (L) an eagle *audax* (L) bold, audacious. Inhabiting New Guinea, Australia, and Tasmania.

Golden Eagle *A. chrysaetos*
khrusos (Gr) gold *aetos* (Gr) an eagle; the colour is mostly dark tawny to brown with a golden crown and nape. A widespread species but becoming rare in places; inhabiting North America and ranging south to Mexico, Eurasia including Scandinavia and Scotland and ranging south to northern Africa, and east to Arabia, India, and the Burma-Vietnam area.

Gurney's Eagle *A. gurneyi*
Named in honour of J. H. Gurney Jnr (1848–1922), an ornithologist and author, it lives in New Guinea.

Tawny Eagle *A. rapax*
rapax (L), genitive *rapacis*, rapacious, grasping: the plumage is rufous brown, or tawny. Inhabiting most of Africa and ranging east to Arabia, the southern part of India, and the Burma-Vietnam area.

Bonelli's Eagle *Hieraaetus fasciatus*
hierax (Gr) a hawk or falcon *aetos* (Gr) an eagle *fascia* (L) a band or girdle; *fasciatus*, banded; it has bars across the long tail terminating with a wide black band. Named after Prof F. A. Bonelli (1784–1830), an Italian Professor of Natural History. It inhabits southern parts of Europe and northern Africa and ranges east to southern India and the Burma-Vietnam area.

Long-crested Eagle *Lophaetus occipitalis*
lophos (Gr) the crest of birds, and other crests *aetos* (Gr) an eagle *occiput* (L), genitive *occipitis*, the back part of the head *-alis* (L) suffix meaning relating to; the plumage is very dark brown or black with a prominent crest on the head. Widespread in Africa south of the Sahara.

Hodgson's Hawk-Eagle *Spizaetus nipalensis*
spizias (Gr) the sparrow-hawk *aetos* (Gr) an eagle *-ensis* (L) suffix meaning belonging to; named from Nepal it inhabits southern Asia, India, and the Burma-Vietnam area: the English name is in honour

of B. H. Hodgson, FRS (1800-1894) who was Resident in Nepal from 1833 to 1843.

Ornate Hawk-Eagle   *S. ornatus*
ornatus (L) dress, ornament; a reference to the very colourful plumage and striking crest. Not a well-known hawk-eagle although fairly widespread from Mexico to Argentina.

Crowned Eagle   *Stephanoaetus coronatus*
stephanos (Gr) a crown, garland   *aetos* (Gr) an eagle   *corona* (L) a crown   -*atus* (L) suffix meaning provided with; it has a prominent black and white double crest. Inhabiting Africa.

Family PANDIONIDAE   1 species
Named after Pandion, a legendary king of Athens.

Osprey   *Pandion haliaetus*
haliaeetos (L) the osprey or sea-eagle. Widespread in almost the whole of the northern hemisphere and migrating south in winter; it breeds in America, south to Brazil, Africa and Australasia.

Family FALCONIDAE   61 species
falx (L), genitive *falcis*, a sickle, a reference to the bird's curved talons.

Red-throated Caracara   *Daptrius americanus*
daptēs (Gr) fem *daptria*, an eater, bloodsucker   -*anus* (L) suffix meaning belonging to; it inhabits Central and South America from southern Mexico to Peru and southern Brazil. It has a naked red area on the face and throat   *caracara* is an Argentinian word for this bird of prey and said to be an imitation of its call.

White-throated Caracara   *Phalcoboenus albogularis*
phalkōn (Gr) a falcon (=*falco* (L), genitive *falconis*, a falcon)   *boenus*, derived from *bainō* (Gr) I walk, step; 'a walking falcon'; with their long legs they are quite at home on the ground and frequently search for their prey on foot; 'characteristic head-bobbing walk' (*World Atlas of Birds*, Mitchell Beazley); 'caracaras run rapidly on the ground' (*Living Birds of the World*, Gilliard)   *albus* (L) white   *gula* (L) throat -*aris* (L) suffix meaning pertaining to. Ranging from Florida and southern Texas through South America to Tierra del Fuego.

Common Caracara   *Polyborus plancus*
poluborus (Gr) much-devouring; *plancus* (L) flat-footed. Ranging from

the southern part of North America, through Central America including Cuba, south to Tierra del Fuego including the Falkland Islands.

Chimango Caracara   *Milvago chimango*
milvus (L) a bird of prey, a kite   *-ago* (New L) suffix meaning resemblance   *chimango* (Argentine/Spanish) a beetle-eater. Inhabiting
South America.

Laughing Falcon   *Herpetotheres cachinnans*
herpeton (Gr) a creeping thing, a reptile   *thēraō* (Gr) I hunt, chase,
wild beasts; can mean I catch, take, wild beasts; an allusion to its
main prey which is snakes and lizards   *cachinno* (L) one who laughs
violently, a laugher; a reference to the two-note call which is usually
repeated many times. Inhabiting Mexico and ranging south to
Bolivia, Paraguay, Brazil, and northern Argentina.

Barred Forest-Falcon   *Micrastur ruficollis*
mikros (Gr) small   *astur* (L) a species of hawk; at about 36 cm (14 in)
it is one of the smaller falcons   *rufus* (L) red   *collum* (L) the neck; the
upperparts, throat, and breast are a rufous brown; the underparts
have black and white bands and there are white bands across the black
tail. Inhabiting the southern part of Mexico and ranging south
through Panama to Brazil and northern Argentina.

African Pygmy Falcon   *Polihierax semitorquatus*
polios (Gr) hoary, grey, usually referring to the hair   *hierax* (Gr) a
hawk or falcon; a reference to the blue-grey upperparts   *semi-* (L)
prefix meaning half   *torquatus* (L) wearing a collar; a reference to the
white collar. A very small falcon measuring about 20 cm (8 in); it
inhabits Somalia and Sudan ranging south to southern Africa but not
the extreme south.

Pied Falconet   *Microhierax melanoleucos*
mikros (Gr) small   *hierax* (Gr) a hawk or falcon   *melas* (Gr), genitive
melanos, black   *leukos* (Gr) white; a reference to the black and white
plumage. The falcons in this genus, known as falconets, are quite
small being only about 18 to 23 cm (7 to 9 in) long; they range through
southern Asia and the oriental region from India to the Philippines,
Malaysia, and Indonesia.

Fox Kestrel   *Falco alopex*
falx (L), genitive *falcis*, a sickle; a reference to the birds curved talons

alōpēx (Gr) a fox; a reference to the plumage which is 'fox-red'. Inhabiting western and northern Central Africa.

Merlin *F. columbarius*
columba (L) a pigeon, a dove *-arius* (L) suffix meaning pertaining to; referring to its occasional capture of a pigeon: Merlin is from Old French *esmerillon*. Widespread in the Northern Hemisphere including the British Isles, and moving south in winter.

African Hobby *F. cuvieri*
Named after the French zoologist and author M. F. Cuvier (1775–1838), brother of the more well-known Baron Georges Cuvier. The name hobby, through various changes, is said to come from Old French *hober*, to move. Inhabiting Africa.

Dickinson's Kestrel *F. dickinsoni*
Named after Dr J. Dickinson (1832–1863), at one time a missionary and doctor in Nyasaland, now known as Malawi. Inhabiting Africa.

Grey Falcon *F. hypoleucus*
hupo (Gr) under, less than usual *leukos* (Gr) white; i.e. less than white, 'off-white'. Inhabiting Australia.

Prairie Falcon *F. mexicanus*
This falcon usually favours bare and arid regions; it inhabits Mexico and parts of Central America.

Peregrine Falcon *F. peregrinus*
peregrinus (L) that comes from foreign parts, strange, foreign; here taken to mean a wanderer; it is known throughout the world except in the polar regions.

Gyrfalcon *F. rusticolus*
rusticola (L) an inhabitant of the country. The derivation of gyrfalcon is obscure but could be from the German *gier*, greedy; the German name for the bird is *gierfalke*. Inhabiting the Arctic round the world and sometimes migrating south in winter.

American Kestrel or American Sparrowhawk *F. sparverius*
sparver (Middle E) from *esprevier* (Middle F) the sparrowhawk *sparverius* (New L) pertaining to sparrows; the sparrow is one of the birds that form its prey. Inhabiting North, Central, and South America, ranging from Alaska to Tierra del Fuego.

Common Kestrel   *F. tinnunculus*

tinnio (L) I ring, tinkle   *tinnulus* (L) ringing, tinkling   *tinnunculus* (L) a kind of hawk, the kestrel; the name means 'little bell-ringer' and refers to the birds repeated high-pitched call. Inhabiting North America, Eurasia including the British Isles, Africa, and ranging east to India, the Burma-Vietnam area, Malaysia, and Indonesia.

27 Grouse, Pheasants, Peacocks, Guineafowls, Turkeys and their kin

GALLIFORMES

Bobwhite
Colinus virginianus

The order Galliformes includes the well-known game birds and also the domestic hen. Then there are the curassows which also make good table birds, but living in forests in uninhabited territory, perhaps 2,400 m (8,000 ft) above sea level, they are not a popular source of food. In addition there are the little-known megapodes (from the Greek 'big-feet'), a strange assortment consisting of the brush turkeys, mallee fowls, incubator birds and others that incubate their eggs by heat, but not heat from their own bodies. In some cases it is the heat generated by a mound of rotting vegetation, which they build for the purpose; or it can be the heat of the sun on sand, or, most surprising of all, the heat from underground volcanic action.

Finally there is the hoatzin, a most remarkable bird, in some respects resembling the others in this order, but with certain marked differences. For example when first hatched the young hoatzin is practically naked, but has claws on the fore limb. Thus it starts life as a quadruped, using the claws for hanging on to branches or clambering about in the trees where its nest has been built. In this respect it resembles certain prehistoric semi-reptilian birds, known to us only as fossils; equally remarkable the young bird can swim. After about a week the claws begin to disappear as the normal wing feathers develop, thus in a few days it goes through an evolutionary change that took the lizard-like birds millions of years to bring about.

Order   GALLIFORMES

gallus (L) a cock (pertaining to poultry)   *forma* (L) form, shape; can mean sort, kind, so not necessarily referring to the actual shape.

Family   MEGAPODIIDAE   12 species
megas (Gr) large   *pous* (Gr), genitive *podos*, a foot; an allusion to their unusually large and strong legs and feet.

Marianas Scrub Hen   *Megapodius laperouse*
Capt J. F. Comte de la Pérouse (1741-1788) was a French naturalist: Marianas is the name of the group of islands in the Pacific Ocean to the north of New Guinea where these birds live.

Mallee Fowl   *Leipoa ocellata*
leipō (Gr) I leave, I abandon   *ōon* (Gr) an egg, plural *ōa*; 'a deserter of eggs'; a reference to their habit of leaving the eggs to be hatched by the heat of the sun or of rotting vegetation   *ocellatus* (L) spotted with eyes; a reference to the spotted plumage   *mallee* is a native Australian name for scrubby kinds of eucalyptus, and probably refers to their habitat. They live in Australia.

Brush Turkey   *Alectura lathami*
alektōr (Gr) a cock   *oura* (Gr) the tail: named after Dr John Latham, FRS (1740-1837) an English zoologist. Inhabiting Australia.

Red-billed Brush Turkey   *Talegalla cuvieri*
tale, a New Guinea native name for a water-hen   *gallus* (L) a cock: named after Baron Georges Cuvier (1769-1832) the famous French anatomist and zoologist; he was at one time Professor of Natural History in the Collège de France. Inhabiting New Guinea.

Bruijn's Brush Turkey   *Aepypodius bruijnii*
aipus (Gr) high   *pous* (Gr), genitive *podos*, a foot; another way of saying 'big feet': named after Dr J. Bruijn (1811-1895) a Dutch zoologist and botanist. Inhabiting New Guinea.

Family   CRACIDAE   44 species
krazō (Gr) I croak, scream   *crax* (New L), genitive *cracis*, a screamer;

a reference to their noisy call which is magnified by a special formation of the unusually long trachea.

Plain Chachalaca   *Ortalis vetula*
ortalis (Gr) a young bird   *vetula* (L) a little old woman. It inhabits North America, Central America, and other lands bordering the Caribbean Sea   *Chachalaca* is Spanish from Nahuatl, meaning the twittering of a bird.

Crested Guan   *Penelope purpurascens*
Pēnelopē (Gr) a feminine name, actually a weaver   *purpurascens* (L) purplish; this bird has a variety of colours, with a dark greenish olive back with bronze reflections, and the naked part of the face a slaty blue. It is wide-ranging from Mexico to Argentina.   *Guan* is an American-Spanish name for the bird.

Horned Guan   *Oreophasis derbianus*
oros (Gr), genitive *oreos*, a mountain   *phasianos* (Gr) the Phasian bird or pheasant; the name comes from the River Phasis, in Colchis (see *Phasianus colchicus* p. 248). Named after the 13th Earl of Derby, formerly the Hon. E. S. Stanley (1775-1851). He was President of the Zoological Society of London from 1831 until his death in 1851. This guan inhabits the mountain forests from southern Mexico to Guatemala.

Crested or Black Curassow   *Crax alector*
crax (New L) see above, under Family   *alektōr* (Gr) a cock. Inhabiting South America and particularly widespread in Amazonia. The name *curassow* is said to derive from the island of Curacao, in the West Indies, but it is not known that any of the curassows ever lived there.

Red-wattled Curassow   *C. globulosa*
globus (L) a ball, globe   *-osus* (L) suffix meaning full of, prone to; the wattle takes the form of red globules round the base of the beak. Inhabiting South America.

Helmeted Curassow   *C. pauxi*
Paoxi is a native bird name in the northern part of Venezuela. This curassow lives in Venezuela and a large area in Bolivia, South America.

Family   TETRAONIDAE   16 species
tetraōn (Gr) a bird of the grouse kind.

Siberian or Black-billed Capercaillie   *Tetrao parvirostris*
parvus (L) small   *rostrum* (L) the beak. Inhabiting eastern Siberia.

Black Grouse   *Lyrurus tetrix*
lura (Gr) a lyre   *oura* (Gr) a tail; it has lyre-shaped tail feathers:
tetrix = *tetrax* (Gr) a kind of grouse. Inhabiting Europe and parts of
Asia.

Willow Ptarmigan   *Lagopus lagopus*
lagōpous (Gr) the ptarmigan; literally 'hare-foot', from *lagōs* (Gr) a
hare, and *pous* (Gr) a foot; this is a reference to the lower leg and foot,
which in winter is covered with feathers, like a gaiter. Inhabiting
Arctic areas in both the New and Old World.

White Tailed Ptarmigan   *L. leucurus*
leukos (Gr) white   *oura* (Gr) the tail. Inhabiting Alaska, Canada, and
ranging south to the Rockies. The name ptarmigan is derived from
the Gaelic *tarmachan*.

Rock Ptarmigan   *L. mutus*
mutus (L) dumb, silent; it is not dumb but the call is rough and un-
musical. Widespread in the far north around the world and the
mountains of Eurasia.

Spruce Grouse   *Dendragapus canadensis*
dendron (Gr) a tree   *agapaō* (Gr) I love, I welcome   *-ensis* (L) suffix
meaning belonging to; 'of Canada'. It is widespread in North
America.

Blue Grouse   *D. obscurus*
obscurus (L) obscure, secret, reserved; it lives high in the Rocky
Mountains of North America, and is seldom seen, unlike the Spruce
Grouse (above) which is often seen in bush camps and is comparatively
tame.

Ruffed Grouse   *Bonasa umbellus*   (replaces *Tetrastes*)
bonasus (L) a species of bison; probably a reference to the low hooting
noises and drumming sounds made by the male   *umbella* (L) a
parasol or umbrella; during courtship display the male raises the tail
feathers to make a fan-like shape. Inhabiting the northern part of
North America.

Prairie Chicken   *Tympanuchus cupido*   (includes *T. pallidicinctus*)
tumpanon (Gr) a drum + *-ochos* (New L) from *ekhō* (Gr) I hold = drum-

holding; a reference to the drumming sounds made by the male during courtship display *cupido* (L) physical desire, love. Inhabiting the central part of the USA.

Sharp-tailed Grouse   *T. phasianellus*   (replaces *Pediocetes*)
phasianus (L) the pheasant (see p. 248)   *-ellus* (L) dim. suffix, 'a small pheasant'; grouse are among the smallest in the family Tetraonidae. Inhabiting North America.

Sage Grouse   *Centrocercus urophasianus*
kentron (Gr) a spike, a spur   *kerkos* (Gr) the tail   *oura* (Gr) the tail *phasianos* (Gr) the pheasant (see p. 248). Inhabiting North America.

Family   PHASIANIDAE   183 species
phasianus (L) a pheasant   *phasianos* (Gr) the Phasian bird or pheasant (see p. 248 under *Phasianus colchicus*).

Bearded Wood Partridge   *Dendrortyx barbatus*
dendron (Gr) a tree   *ortux* (Gr) the quail   *barbatus* (L) bearded. Inhabiting Mexico, Central America, and parts of the Greater and Lesser Antilles.

Mountain Quail   *Oreortyx pictus*
oros (Gr), genitive *oreos*, a mountain   *ortux* (Gr) the quail   *pictus* (L) painted; a reference to its bright colours. Inhabiting the mountains of California and Nevada.

California Quail   *Lophortyx californica*
lophos (Gr) a crest   *ortux* (Gr) the quail   *-icus* (L) suffix meaning belonging to; 'of California'. From the south-western part of the USA it ranges south to Mexico.

Bobwhite   *Colinus virginianus*
colín (Sp) derived from the Nahuatl zolin, a partridge; in Mexico colín means a quail   *-anus* (L) suffix meaning belonging to; 'of Virginia'. The male has black and white stripes on the head and a white throat patch; inhabiting the eastern part of the USA.

Marbled Wood Quail   *Odontophorus gujanensis*
odous (Gr), genitive *odontos*, a tooth   *phora* (Gr) carrying, bearing; it has tooth-like markings on the plumage   *-ensis* (L) suffix meaning belonging to, usually referring to a place; 'of Guiana'. Inhabiting Central and South America.

Stripe-faced Wood Quail   *O. balliviani*
Named in honour of Gen. J. Ballivian (1804-1852), President of the
Bolivian Republic from 1842 to 1847. Range as above.

Ocellated Quail   *Cyrtonyx ocellatus*
kurtos (Gr) curved   *onux* (Gr) talon, claw   *oculus* (L) the eye; dim.
ocellus, a little eye   *-atus* (L) suffix meaning provided with; a reference
to the 'eye-like' markings. Inhabiting south-western USA and ranging
south to Mexico.

Himalayan Snowcock   *Tetraogallus himalayensis*
tetraōn (Gr) a bird of the grouse kind; also *tetrao* (L) a heath-cock
gallus (L) a cock   *-ensis* (L) suffix meaning belonging to, usually referring to a place. Inhabiting the Himalayan area and Tibet.

Szecheny's Pheasant Grouse   *Tetraophasis szechenyii*
tetraōn (Gr) a bird of the grouse kind   *phasianos* (Gr) the Phasian bird,
or pheasant: named after Count Bela Szecheny (1837-1918), a
naturalist and author who was in Central Asia from 1877 to 1880.
Ranging through the Himalayan area from eastern Afghanistan to
Bhutan.

Red-legged Partridge   *Alectoris rufa*
alektōr (Gr) a cock   *rufus* (L) red. Inhabiting western Europe, Corsica,
and the Canary Islands.

Black Francolin   *Francolinus francolinus*
francolin (Fr) from *francolino* (Old It) a partridge. Ranging from
Cyprus through western Asia to Assam.

Hartlaub's Francolin   *F. hartlaubii*
Named after Dr K. J. G. Hartlaub (1814-1900), at one time Prof. of
Zoology at Bremen. This francolin inhabits South Africa.

Somali Greywing Francolin   *F. lorti*   (or *africanus*)
Named after E. E. Lort Phillips (1857-1944), a big game hunter and
collector in East Africa; he spent many years attempting to establish
the identity of this bird and finally succeeded during a special
expedition to Somaliland about the year 1895. Inhabiting East
Africa.

Swainson's Francolin   *Pternistis swainsoni*   (or *Francolinus*)
pterna (Gr) the heel   *pternistēs* (Gr) one who strikes with the heel; the
legs have sharp spurs on them used in fighting. Named after W. Swain-

son, FRS (1789-1855) a much travelled English zoologist and artist. This francolin inhabits South Africa and is particularly common in the Kruger National Park.

Common or Grey Partridge   *Perdix perdix*
perdix (L) a partridge. Inhabiting Europe including the British Isles, southern Scandinavia, northern Spain, and ranging across Europe to Asia Minor and Central Asia.

Mrs Hodgson's Partridge   *P. hodgsoniae*
Named in honour of Mrs Hodgson in 1857, the wife of B. H. Hodgson, FRS (1800-1894), Resident in Nepal from 1833 to 1843. This partridge lives in Tibet.

Madagascar Partridge   *Margaroperdix madagascarensis*
margarōdēs (Gr) pearl-like   *perdix* (L) = *perdix* (Gr) a partridge; a reference to the grey plumage   *-ensis* (L) suffix meaning belonging to, usually referring to a place. Inhabiting Madagascar.

Common Quail   *Coturnix coturnix*
coturnix (L) a quail. Inhabiting Europe including the British Isles, Asia, and Africa.

Swamp Quail   *C. ypsilophorus*
Upsilon (Gr) the letter u   *phoreō* I bear; a reference to the markings on the back which resemble the Greek capital letter U; it is rather like our capital letter Y. Inhabiting Australia and Papua New Guinea.

Stubble Quail   *C. novaezealandiae*
Although originally inhabiting New Zealand and Australia it has not been seen in New Zealand for the past 100 years.

Manipur Bush Quail   *Perdicula manipurensis*
perdix (L) a partridge   *-cula* (L) diminutive suffix; 'a little partridge' *-ensis* (L) suffix meaning belonging to, usually referring to a place; named after Manipur in southern Assam, this quail inhabits southern Assam and northern Burma.

Tree Partridge   *Arborophila torqueola*
arbor (L), genitive *arboris*, a tree   *philos* (Gr) dear, pleasing   *torques* (L) a twisted necklace or collar   *-olus* (L) diminutive suffix, 'a little collar'; a reference to the dark ring round the neck. Tree partridges live in the forests of eastern India ranging to southern China and Malaya.

Bamboo Partridge   *Bambusicola fytchii*
bambu (Malayan) probably from Tulu; the generic name of the plant
is *Bambusa   colo* (L) I inhabit: named after Maj Gen A. Fytche
(1820-1892), Commander in Chief of the British Army in Burma from
1867 to 1871. This partridge inhabits Assam and possibly northern
Burma.

Ceylon Spur-fowl   *Galloperdix bicalcarata*
gallus (L) a cock   *perdix* (L) a partridge   *bi-* (L) prefix meaning two
calcar (L) a spur   *-atus* (L) suffix meaning provided with; it has two
spurs on the legs. Inhabiting Ceylon, now known as Sri Lanka.

Western Tragopan   *Tragopan melanocephalus*
tragos (Gr) a he-goat: Pan is a Greek god of pastures and woods, and
is usually represented with horns and goats legs; the name tragopan
is a reference to the two fleshy erectile horns on the head of the male
bird   *melas* (Gr), genitive *melanos*, black   *kephalē* (Gr) the head; the
head is not entirely black but has black markings. At one time
tragopans were widespread from western Pakistan to Burma and
China, but many are now rare and only found in certain areas; the
Western Tragopan probably still inhabits parts of Pakistan and
Kashmir.

Satyr Tragopan   *T. satyra*
The Satyr is a mythical Greek god, half man and half beast, with a
goat's horns, ears and tail; the name is a reference to the two fleshy
erectile horns on the head of the male, but in fact all male tragopans
have the two horns. Inhabiting Nepal and Bhutan.

Temminck's Tragopan   *T. temminckii*
Named after Prof C. J. Temminck (1778-1858), a Dutch zoologist at
one time Director at the Natural History Museum, Leyden. Inhabiting eastern India and Burma and ranging to central China.

Cabot's Tragopan   *T. caboti*
Named after Dr S. Cabot (1815-1885) an American zoologist and
Curator of the Boston Natural History Society in 1845. Inhabiting
eastern China.

Himalayan Monal or Impeyan Pheasant   *Lophophorus impejanus*
lophos (Gr) the crest of birds   *phoros* (Gr) a bearing, carrying: named
in honour of Sir Elijah Impey (1732-1809) and Lady Impey who
introduced the bird into England; Sir Elijah was Chief Justice in

Bengal from 1774 to 1789   *Monāl* is a Nepalese name for the bird.
Ranging from Afghanistan to Assam.

Chinese Monal   *L. lhuysii*
Named after E. Drouyn de L'Huys (1805-1881) a French zoologist
Monāl is a Nepalese name for this pheasant; it inhabits China.

Brown Eared Pheasant   *Crossoptilon mantchuricum*
krossoi (Gr) a fringe, tassels   *ptilon* (Gr) feathers, a wing; a reference
to the two brown 'earlike' tufts of feathers on the head   *-icus* (L) suffix
meaning belonging to; 'of Manchuria'. Inhabiting the mountains of
north-eastern China and Manchuria.

Kalij Pheasant   *Lophura leucomelana*
lophos (Gr) the crest of birds   *oura* (Gr) the tail; 'tail-crest'   *leukos*
(Gr) white   *melas* (Gr), genitive *melanos*, black; the males are white
above and black underneath   *Kalij* or *Kaleege* (Nepali and Pahari) is
the native name for this pheasant inhabiting India.

Crested Fireback   *L. ignita*
ignitus (L) fiery, glowing; the male has a glowing crimson back and
rump. Inhabiting Borneo and Sumatra.

Red Jungle Fowl   *Gallus gallus*
gallus (L) a cock; this is the ancestor of domestic poultry; a very
colourful bird with bright red plumage and iridescent green. It is
widespread in northern India ranging to southern China and
Malaysia but not Borneo.

Domestic Hen   *G. domesticus*
Most zoologists give it the name *G. domesticus*. Widespread throughout
the world, it is descended from *G. gallus*, above.

Ceylon Jungle Fowl   *G. lafayetii*
Named in honour of M. J. P. La Fayette, Marquis (1757-1834) the
French politician. It is found only in Ceylon, now Sri Lanka.

Grey Jungle Fowl   *G. sonneratii*
Named after P. Sonnerat (1745-1814), French Navy and naturalist;
he was in the Philippines in 1771 and India in 1775. Inhabiting
Afghanistan, West Pakistan, and India.

Koklass Pheasant   *Pucrasia macrolopha*
Pucrasia is derived from *pukras* (Nepalese) a pheasant   *makros* (Gr)
long   *lophos* (Gr) a crest   *koklass* is a native name in Nepal for this

pheasant, inhabiting northern India and the Himalayas.

Common or Ring-neck Pheasant   *Phasianus colchicus*
phasianos (Gr) the Phasian bird or pheasant; the name comes from the
River Phasis in Colchis, which flows into the Black Sea; also *phasianus*
(L) the pheasant: Colchis is an ancient territory in Asia situated
between the Black Sea and the Caspian Sea; it does not appear on
modern maps and only exists as a Greek legend, and once famous as
a home of sorcery   *-icus* (L) suffix meaning belonging to. Known as
'ring-necks' because there is usually a prominent white collar, these
game birds are now widespread in Eurasia and America but originally
came from Asia.

Reeves's Long-tailed Pheasant   *Syrmaticus reevesii*
surma (Gr), genitive *surmatos*, a robe that trails   *-icus* (New L) a suffix
sometimes used to emphasize a certain characteristic; 'a long robe
that trails': named after J. Reeves (1774-1856), an English naturalist
who was in China from 1812 to 1831. This pheasant inhabits the
mountainous regions of eastern Asia.

Mrs Hume's Long-tailed Pheasant   *S. humiae*
Named in honour of Mrs Mary Hume (1880-1890), the wife of A. O.
Hume (1829-1912) naturalist and author; they lived in India. This
pheasant inhabits eastern India and Burma and Thailand.

Elliot's Long-tailed Pheasant   *S. ellioti*
Named after Dr D. G. Elliot (1835-1915), an American zoologist,
who was Curator of the Chicago Field Museum from 1894 to 1908.
A rare pheasant inhabiting a small area in south-western China.

Golden Pheasant   *Chrysolophys pictus*
khrusos (Gr) gold   *lophos* (Gr) a crest   *pictus* (L) painted; a reference
to the bright colours of this bird with its fine golden crest. Inhabiting
the mountainous regions of eastern Asia.

Lady Amherst's Pheasant   *C. amherstiae*
Named in honour of Lady Sarah Amherst, Countess (1762-1838) the
wife of William Pitt Amherst, 1st Earl (1773-1857) British diplomat.
He was in China in 1816 and later became Governor-General of India
from 1823 to 1828. The range is similar to *C. pictus*, above.

Bronze-tailed Peacock Pheasant   *Polyplectron chalcurus*
polus (Gr) many   *plēktron* (Gr) anything to strike with, a cock's spur;
the male has two and often three spurs on each leg   *khalkos* (Gr)

copper *oura* (Gr) the tail. Inhabiting Malaysia, probably mainly the mountains of Sumatra.

Crested Argus Pheasant *Rheinardia ocellata*
Capt Rheinard, a French Army officer, was in Annam from 1880 to 1883; it is now usually known as Vietnam *ocellata* (L) provided with little eyes; a reference to the 'eyelike' markings. Ranging from Vietnam to the Malay Peninsula.

Argus Pheasant *Argusianus argus*
Named after Argus, the mythical son of Arestor, who had 100 eyes, and Hera transplanted his eyes to the tail of the peacock. The inner wing feathers of the Argus Pheasant show rows of colourful 'eyelike' markings *-anus* (L) suffix meaning belonging to. Inhabiting Malaya and Borneo.

Common or Blue Peafowl *Pavo cristatus*
pavo (L) a peacock *cristatus* (L) crested. Usually known as the Peacock, and Peahen; inhabiting India and Ceylon, now known as Sri Lanka.

Congo Peafowl *Afropavo congensis*
pavo (L) a peacock; 'African peacock' *-ensis* (L) belonging to; it inhabits the forests of Congo, now known as Zaire.

Family NUMIDIDAE 8 species
Numidea is an old Roman name for a district in north-west Africa, now the Morocco area; a flock of *Numida meleagris* still live there.

Black Guineafowl *Agelastes niger*
agelastikos (Gr) disposed to herd together; flocks of these birds can number several hundred *niger* (L) black. Inhabiting Africa south of the Sahara.

Tufted Guineafowl *Numida meleagris meleagris*
The nominate subspecies. *Numida*, see above under Family. Meleagros was the son of Oeneus, king in Calydon; according to the legend his sisters were changed to guineafowls on his death. It has a tuft on the head. Inhabiting Africa south of the Sahara and an isolated flock in the Morocco area; the English name derives from Guinea in West Africa.

Helmeted Guineafowl *N. m. mitrata*
mitratus (L) wearing a turban or high head-dress; a reference to the

bony casque on the head; this is a subspecies that some authors classify as a separate species. Inhabiting Africa south of the Sahara.

Kenya Crested Guineafowl   *Guttera pucherani*
gutta (L) a drop of fluid   *guttatus* (L) spotted, speckled; a reference to the speckled plumage; Dr J. Pucheran (1817–1894) was a French zoologist who was working at the Natural History Museum in Paris in 1843. This guineafowl inhabits East Africa.

Vulturine Guineafowl   *Acryllium vulturinus*
akron (Gr) highest, topmost; can mean a point, a peak   *-illion* (Gr) diminutive suffix, 'a small point'; a reference to the pointed tail *vulturinum* (L) vulture-like; a reference to its appearance, having the head and upper part of the neck nearly naked. Inhabiting Africa.

Family   MELEAGRIDIDAE   2 species
meleagris (Gr), genitive *meleagridos*, a sort of guineafowl; Meleagros was the son of Oeneus, king in Calydon; according to the legend his sisters were changed into guineafowls on his death.

Turkey   *Meleagris gallopavo*
meleagris (Gr) a sort of guineafowl   *gallus* (L) a cock (of poultry) *pavo* (L) a peacock; a misleading mixture of names; the turkey, although related, is not any of these birds. Inhabiting North America and Mexico and parts of Central America.

Ocellated Turkey   *Agriocharis ocellata*
agrios (Gr) wild, living in the fields   *kharis* (Gr) grace, beauty *ocellus* (L) dim. of *oculus*, a little eye   *-atus* (L) provided with; it has eye-like markings on the tail feathers. Inhabiting the Yucatan Peninsula, Guatemala, and Honduras, Central America.

Family   OPISTHOCOMIDAE   only 1 species
opisthokomos (Gr) wearing the hair long at the back; this strange bird, the Hoatzin, has a long straggly crest of feathers on the top and at the back of the head.

Hoatzin   *Opisthocomus hoatzin*
opisthokomos (Gr) see above   *uatzin* (Nahuatl) a pheasant; the Nahuatl are a people of southern Mexico and Central America. The hoatzin inhabits the northern part of South America. For more information about this unusual and interesting bird see page 239.

Spotted Crake
Porzana porzana

Order GRUIFORMES

grus (L) a crane.

Family MESITORNITHIDAE (or MESOENATIDAE)
3 species
Dr A. L. Rand says that 'mesite' is a 'classical word of irrelevant meaning'. *ornis* (Gr), genitive *ornithos*, a bird. Dr Rand is a U.S. ornithologist and author of *The New Dictionary of Birds*.

Brown Mesite *Mesitornis unicolor*
unicolor (L) of one colour. Inhabiting eastern Madagascar.

White-breasted Mesite *M. variegata*
vario (L) I variegate; it is reddish brown above, a grey collar, white on the chest spotted with black. Inhabiting Madagascar.

Bensch's Monias *Monias benschi*
monias (Gr) solitary; although they sometimes gather in flocks of 30 or more, the male works alone to construct the nest; J. H. E. Bensch

(born 1868) was a Colonial Administrator in Madagascar in 1903. The three mesites are found only in Madagascar.

Family TURNICIDAE   15 species
from *coturnix* (L), genitive *coturnicis*, a quail   *turnix* (New L) a shortened form, said to indicate that they lack the hind toe; the hind toe is missing.

Black-breasted Bustard-Quail   *Turnix melanogaster*
turnix (New L) see above   *melas* (Gr) black   *gaster* (Gr) the belly. Inhabiting Australia.

Bustard-Quail   *T. suscitator*
suscito (L) I arouse   *suscitator* (L) an awakener; the female makes a booming call at dawn. Inhabiting the Philippines, also India, Malaysia, China and Japan.

Little Button-Quail   *T. sylvatica*
silva (L) a wood   *-icus* (L) suffix meaning belonging to. Inhabiting southern Spain, Africa, and ranging across Asia to the Philippines.

Quail Plover or Lark Quail   *Ortyxelos meiffrenii*
ortux (Gr) the quail   *elos* (Gr) low ground by rivers, marsh-meadows; it is sometimes known as the Marsh Quail. Baron Meiffren-Laugier de Chartreuse (fl. 1829) was a French zoologist and author of several books on ornithology. This lark-like quail inhabits Senegal to Uganda.

Family PEDIONOMIDAE   1 species
pedion (Gr) a plain   *nomos* (Gr) an abode, a feeding place.

Plains Wanderer   *Pedionomus torquatus*
torquatus (L) having a necklace; it has a white collar spotted with black. Inhabiting the south-eastern part of Australia.

Family GRUIDAE   14 species
grus (L) a crane.

Whooping Crane   *Grus americana*
This crane, inhabiting North America and parts of Central America, is now very rare, and possibly only about 50 are still in existence. It is named for its loud call; the trachea, or windpipe, is about 1·5 m (5 ft) long and partly coiled within the breast bone, which helps to produce

the loud 'trombone-like' noise. All cranes have an unusually long trachea.

Common Crane   *G. grus*
grus (L) a crane. Widespread in Eurasia and ranging south as far as South Africa in winter.

Japanese Crane   *G. japonensis*
Inhabiting eastern Asia and Japan, it is becoming rare.

Brolga   *G. rubicunda*
rubicundus (L) red, ruddy   *brolga* is a native Australian name. Inhabiting New Guinea and Australia.

Crowned Crane   *Balearica pavonina*
The generic name is based on hearsay, and is a mistake; Howard Saunders writes: 'I am not aware of the existence of a single authentic specimen of *Balearica pavonina* either on the mainland of Southern Spain or in the Balearic Islands.' *pavoninus* (L) of a peacock; can mean coloured like a peacock's tail; a reference to the colourful plumage and crown, or crest. It ranges all across Africa north of the equator from Senegal on the west coast to Ethiopia and Somalia in the east.

Family   ARAMIDAE   1 species
aramus   a bird name of unknown origin.

Limpkin   *Aramus guarauna*
The specific name probably refers to *Guarauno*, American Spanish for the Warrau, a people of Venezuela and Guyana; the English name *limpkin* is said to refer to the peculiar gait which gives the impression of limping. Inhabiting the northern part of South America, the Caribbean Islands, Central America, and the southern part of North America.

Family   PSOPHIIDAE   3 species
psophos (Gr) a noise; they make loud trumpeting noises.

Common Trumpeter   *Psophia crepitans*
crepito (L) I rattle, I clatter   *crepitans*, clattering; the call is more a deep booming sound rather than a rattle. Inhabiting the northern part of South America.

White-winged Trumpeter *P. leucoptera*
leukos (Gr) white *pteron* (Gr) feathers; can mean wings. Inhabiting
South America, particularly the Amazon region.

Green-winged Trumpeter *P. viridis*
viridis (L) green. Inhabiting Brazil.

Family RALLIDAE 124 species
rallus (New L) from *rallus* (Medieval L) a rail, for example the Water
Rail or the Clapper Rail.

Water Rail *Rallus aquaticus*
aquaticus (L) living in water, belonging to water; usually seen among
reeds in swampy areas or ponds. Inhabiting Europe, including Great
Britain, and ranging across Asia to Japan.

Cape Rail *R. caerulescens*
caeruleus (L) blue *-escens* (L) suffix meaning beginning to, slightly;
'bluish'. Found in the southern part of Africa.

Virginia Rail *R. limicola*
limus (L) mud *colo* (L) I dwell in, I inhabit. Found in the southern
part of Canada and ranging south to South America.

Clapper Rail *R. longirostris*
longus (L) long *rostrum* (L) the beak. The English name is a reference
to its rattling call; it inhabits both the east and west coasts of North
America, ranging south through Central and South America.

Slate-breasted Rail *R. pectoralis*
pectus (L) the breast *pectoralis*, relating to the breast. Inhabiting New
Guinea and Australia.

White-striped Rail *Rallicula leucospila*
rallus, see above *-culus* (L) dim. suffix; 'a small rail' *leukos* (Gr)
white *spilos* (Gr) a mark, a blemish. Inhabiting New Guinea.

Brown Wood-Rail *Aramides wolfi*
aramus is a bird name of unknown origin: Dr T. Wolf (born 1841) a
geologist in Ecuador in 1883, named this bird in 1884. It inhabits
South America.

Mountain-Rail *Edithornis sylvestris*
Edith, a female name + *ornis* (Gr) a bird; named in honour of Mrs

Edith Baker in 1937, the wife of G. F. Baker (1878–1937) a New York banker and Trustee of the American Museum of Natural History *silvestris* (L) belonging to woods. This mountain-rail inhabits Malaysia.

Weka   *Gallirallus australis*
gallus (L) a cock (of poultry)   *rallus*, see above   *australis* does not necessarily mean Australia   *auster* (L) the south   *-alis* (L) suffix meaning relating to; hence southern *Weka* is the Maori name for the flightless rail; although the wings seem well developed it does not fly. Inhabiting New Zealand.

Corncrake   *Crex crex*
crex (Gr) a bird with a sharp knotched bill and long legs; Liddell & Scott's Greek Lexicon says—'This description does not suit the *crex rallus* Linn., our rail, though its cry is well expressed by the name (which is onomatopoeic)'. The main food is insects and other invertebrates though it does eat corn and other seeds when available. Inhabiting northern parts of Europe including the British Isles and ranging south in winter to Africa, Madagascar, and parts of Arabia.

African Black Crake   *Limnocorax flavirostra*
limnē (Gr) a marsh, a pond   *korax* (Gr) a raven or crow   *flavus* (L) yellow   *rostrum* (L) the beak; it is a black moorhen-like bird that lives in swamps and on lakesides; the beak is greenish-yellow. Common in Africa south of the Sahara, specially in the area of the lakes of Central Africa.

Carolina Crake   *Porzana carolina*
porzana (New L) from Italian, a crake; not confined to Carolina, it inhabits northern parts of North America and ranges south to northern South America.

Spotted Crake   *P. porzana*
Inhabiting North America, Europe, Africa, western Asia, India, and the Burma-Vietnam area. It has brown upper parts and a grey white-speckled breast.

Baillon's Crake   *P. pusilla*
pusillus (L) very small; a small crake, about the size of a starling; named after L. A. F. Baillon (1778–1851), a French zoologist from Abbeville. Inhabiting Eurasia, Africa, the Oriental Region, New Guinea, and Australia.

Galápagos Rail   *Laterallus spilonotus*
latus (L) side, flank   *rallus* (New L) from *rallus* (Medieval L) a rail;
a reference to the white bars on the back and sides   *spilos* (Gr) a mark,
blemish   *nōtos* (Gr) the back. Inhabiting South America.

Moorhen   *Gallinula chloropus*
gallina (L) a hen (of poultry)   *-ulus* (L) diminutive suffix   *khlōros*
(Gr) greenish-yellow   *pous* (Gr) the foot; it has greenish-yellow legs
and feet. The distribution is almost world-wide but not the Australian
region.

Purple Gallinule   *Porphyrula martinica*
porphurula, dim. of *porphuriōn* (Gr) a red-coloured water bird   *por-
phureōs* (Gr) purple; the plumage is dark purple rather than red:
named from the island Martinique, in the West Indies, it is also found
in North America, Central, and South America.

Purple Gallinule or Swamp Hen   *Porphyrio porphyrio*
Porphyrio, see above. Distribution is widespread, ranging through
southern Europe, Asia, India and Malaysia, Australia and New
Zealand.

Takahe   *Notornis mantelli*   (or *Porphyrio*)
notos (Gr) the south   *ornis* (Gr) a bird: named after W. D. B. Mantell
(1820-1895) a New Zealand naturalist   *takahe* is a Maori name for
this flightless bird, now almost extinct. Inhabiting New Zealand.

Coot   *Fulica atra*
fulica (L) a coot   *atra* (L) black. Distribution is widespread in Europe
and Asia, parts of Africa north of the Sahara, and ranging east to
Japan.

Horned Coot   *F. cornuta*
cornutus (L) horned. Inhabiting the Andes area in Bolivia, Chile and
northern Argentina.

Crested or Knob-billed Coot   *F. cristata*
cristatus (L) having a crest. Inhabiting the northern part of Africa, and
Spain.

Giant Coot   *F. gigantea*
gigas (L) a giant   *gigantea* (L) belonging to the giants. Inhabiting the
Andes area of South America, mainly in Peru.

Family HELIORNITHIDAE 3 species
hēlios (Gr) the sun *ornis* (Gr), genitive *ornithos*, a bird; from the original vernacular name 'sunbird', but records do not explain the name; it is now called 'finfoot', or 'sungrebe'.

African Finfoot *Podica senegalensis*
pous (Gr), genitive *podos*, a foot; hence *podicus* (New L) belonging to the foot; a reference to the long lobed toes, the lobes acting as paddles instead of webbed feet which gives it greater agility when running on land; this also gives it the name 'finfoot' *-ensis* (L) suffix meaning belonging to; it is not confined to Senegal and is widespread in Africa south of the Sahara.

Asian or Masked Finfoot *Heliopais personata*
hēlios (Gr) the sun *pais* (Gr) a child; 'child of the sun' *personatus* (L) wearing a mask; it has a black front to the face. Inhabiting Bengal and Malaya.

Sungrebe *Heliornis fulica*
Heliornis, see above under Family *fulica* (L) a coot. Inhabiting Mexico, Central America, and ranging south through South America.

Family RHYNOCHETIDAE 1 species
rhis (Gr), genitive *rhinos*, the nose *khaitē* (Gr) flowing hair or mane; it has a long shaggy crest, on the head rather than on the nose, which is raised when the bird is excited.

Kagu *Rhynochetos jubatus*
jubatus = *iubatus* (L) having a mane, crested. Found only on the island New Caledonia, off the east coast of Australia, it is very rare. Kagu is a local native name.

Family EURYPYGIDAE 1 species
eurus (Gr) wide *pugē* (Gr) the rump; 'wide-rumped'.

Sunbittern *Eurypyga helias*
hēlios (Gr) the sun *hēlias*, daughter of the sun. Inhabiting Central America and the Amazon area; it is very rare.

Family CARIAMIDAE 2 species
cariama and *seriema* are Portuguese forms of a native word (prob. *Tupi*) for this bird.

Crested Seriema   *Cariama cristata*
cristata (L) having a crest. Inhabiting the southern half of South America including part of Brazil.

Burmeister's Seriema   *Chunga burmeisteri*
Chunga is a native name in Argentina for a crane-like bird: Dr K. H. K. Burmeister (1807–1892) was Director of the Zoological Museum at Halle University from 1842 to 1848. This seriema inhabits the southern half of South America.

Family   OTIDIDAE   21 species
ōtis (Gr), genitive *ōtidos*, a bustard with long ear feathers.

Little Bustard   *Otis tetrax*
tetrax (Gr) name of two kinds of birds, one probably of the grouse kind (*tetrao* L). Inhabiting Europe, northern Africa, and ranging to the south-west of Asia.

Denham's Bustard   *Neotis denhami*
neos (Gr) new, recent   *ōtis* (Gr) a bustard: Lt Col D. Denham (1786–1826) explored north-west Africa from 1822 to 1825; he named this bustard in 1826. Inhabiting Africa.

Australian Bustard   *Ardeotis australis*   (formerly *Choriotis*)
ardea (L) a heron   *auster* (L) the south   *-alis* (L) suffix meaning belonging to; 'southern'. Found only in Australia.

Crested Bustard   *Eupodotis ruficrista*   (formerly *Lophotis*)
eu- (Gr) prefix meaning well, nicely; often used to indicate typical *pous* (Gr), genitive *podos*, the foot, claws, + *ōtis*; 'a bustard having strong or typical claws'; they are fast runners: *ruficrista* (coined from L) red-crested. Inhabiting Africa.

29 Snipe, Oystercatchers, Plovers, Curlews, Skuas, Gulls, Auks and their kin CHARADRIIFORMES

Avocet
Recurvirostra avosetta

The order Charadriiformes includes many familiar birds, and they are mostly to be found by the sea or near fresh-water lakes, rivers, and marshes. Almost without exception they eat various crustaceans, and the diet includes worms and insects; in some cases, for example the plovers and lapwings, they also eat seeds.

The Jacana, or Lily-trotter, is interesting for its enormously long toes and nails, which enable it to walk about on floating vegetation, hence the name 'lily-trotter'. Some species in this order change the usual role of male and female, for instance in the case of the American Painted Snipe *Nycticryphes semicollaris* the male incubates the eggs and feeds and cares for the young. This is also the case with the Dotterel *Eudromias morinellus*, and even more surprising the female Phalarope *Phalaropus lobatus* also initiates the courtship display. She is larger and more colourful than her different mate, who might almost be called 'hen-pecked'.

Order CHARADRIIFORMES

kharadra (Gr) a mountain stream, a ravine   *kharadrios* (Gr) a bird dwelling in clefts, perhaps the lapwing or the curlew; a misleading name as they do not live in clefts or frequent mountain streams, and usually the nest is a ground-scrape or an open cup.

Family JACANIDAE 7 species
jaçanã (Port) a name used for these birds, probably by the South-American Indians; in English the soft 'c' is usually replaced by a hard 'k' sound.

African Jacana *Actophilornis africanus*
aktē (Gr) a headland on the coast; can mean the rugged banks of rivers *philos* (Gr) loved, pleasing   *ornis* (Gr) a bird; although sometimes seen on mossy rocks and river banks, the jacana is usually found walking about on floating vegetation, made possible by its enormously long toes and nails. Inhabiting Africa south of the Sahara.

Australian Jacana *Irediparra gallinacea*
Named after Tom Iredale in 1911, an English naturalist from Cumberland   *parra* (L) a bird of ill omen; can mean the lapwing   *gallus* (L) a cock   *gallinaceus* (L) of poultry. Inhabiting Indonesia, New Guinea, and Australia.

Pheasant-tailed Jacana *Hydrophasianus chirurgus*
hudor (Gr) water   *phasianos* (Gr) from the River Phasis, the Phasian bird (see p. 248) 'a water-pheasant'; jacanas always live on or near water   *kheirourgos* (Gr) working or doing by hand, a surgeon; the reason for the specific name is obscure. From the Indo-Malayan region eastwards to Taiwan and southwards to the Philippines.

Bronze-winged Jacana *Metopidius indicus*
metopidios (Gr) on the forehead; the jacana has a frontal shield on the forehead   *indicus* (L) of India. Inhabiting India, the Thailand and Vietnam area and Malaysia.

American Jacana *Jacana spinosa*
jacana (see above, under Family)   *spinosus* (L) thorny, prickly; jacanas have a spike on the front edge of the wing, said to be used in fighting. Ranging from southern Texas to Argentina.

Family ROSTRATULIDAE 2 species
rostrum (L) the beak   *-atus* (L) suffix meaning provided with   *-ulus*

(L) dim. suffix; 'provided with a small beak'; the beak could not be described as small, but it is shorter than that of the true snipe (Family Scolopacidae).

Painted Snipe *Rostratula benghalensis*
-*ensis* (L) suffix meaning belonging to; in addition to the Bengal area of India it is found in the southern part of Asia ranging from Asia Minor and Africa, including Madagascar, to Japan, Malaysia, the Philippines, Indonesia, and Australia, but rarely in Sarawak and Borneo.

South American Painted Snipe *Nycticryphes semicollaris*
nux (Gr), genitive *nuktos*, night *kruphios* (Gr) hidden, secret; active only at dawn and dusk, they are secretive and solitary *semi-* (L) half *collare* (L) a chain for the neck *collaris*, pertaining to the neck; it has a white collar which does not completely encircle the neck. Inhabiting southern South America.

Family HAEMATOPODIDAE 6 species
haima (Gr), genitive *haimatos*, blood *pous* (Gr), genitive *podos*, the foot; 'blood-coloured feet'; oystercatchers have pink legs and feet.

Black Oystercatcher *Haematopus ater*
ater (L) black. Inhabiting southern South America.

European Oystercatcher *H. ostralegus*
ostreon (Gr) oyster *legō* (Gr) I gather, I pick up; oystercatcher is a misnomer as they do not eat oysters, but they do open and eat various shell-fish, probably mostly mussels and cockles. This oystercatcher has a wide range apart from Europe, including the Canaries, North and South America, South Africa, Asia, Australia, and New Zealand.

American Oystercatcher *H. palliatus*
palliatus (L) clad in a Greek mantle. Inhabiting New Jersey and California ranging to northern South America.

Family CHARADRIIDAE 62 species
(see notes under Order, opposite).

Southern Lapwing *Vanellus chilensis*
vanellus (New L) from *vanneau* (Fr) the lapwing, from *van* (Fr) a winnowing fan, a reference to the slow flapping of the wings in flight

-ensis (L) suffix meaning belonging to; named from Chile as it inhabits South America.

Spur-winged Lapwing   *V. spinosus*
spinosus (L) thorny, prickly; there is a spur on the front edge of the wing. Inhabiting southern Europe, the Middle East, and northern and central Africa.

Common Lapwing   *V. vanellus*
Wide distribution including Europe, Africa, southern Asia including the Thailand and Vietnam area.

Golden Plover   *Pluvialis apricaria*
pluvialis (L) relating to rain, bringing rain; several explanations of this name have been given by various authors, all having some connection with rain. The Oxford English Dictionary suggests that it is because the birds foretell rain by their restlessness; the German name for the plover means 'rain-piper' because they are known to sing in the rain   *apricus* (L) places open to the sun; can mean loving the sun *-arius* (L) suffix meaning belonging to; probably a reference to the bird's golden colour. Inhabiting Eurasia and India.

American Golden Plover   *P. dominica*
-icus (L) suffix meaning belonging to; a reference to the Dominican Republic, where the typical specimen was found probably during migration as they do not live there. The Golden Plovers are widespread, breeding in the Arctic around the world and migrating south in winter to tropical countries including South America, Africa, and Australia.

Grey or Black-bellied Plover   *P. squatarola*
squatarola (New L) from Italian dialect, the black-bellied plover. Inhabiting northern coastal areas of Siberia, Alaska, Canada and some islands, migrating south and wintering all round the world from South America to Australia.

Little Ringed Plover   *Charadrius dubius*
kharadrios (Gr) a bird living in clefts (see under Order p. 260)   *dubius* (L) uncertain, doubtful; J. A. Scopoli (1723–1788) a professor of botany in Padua, who named the typical form, was doubtful whether it was a distinct species. Widespread in Europe including the southeastern part of England, northern Africa, Asia and ranging to Malaysia and New Guinea. Given its English name because of the

white and black bands round the neck.

Ringed Plover   *C. hiaticula*
hiatus (L) a cleft, an opening (cf. *karadrios,* under Order p. 260) *cula,*
probably a misspelling of *colo* (L) I inhabit, I dwell (see p. 260).
Widespread, breeding in the north around the world and migrating
south in winter to South America, Africa, and southern Asia.

Mountain Plover   *C. montana*
montanus (L) of a mountain. Inhabiting North and Central America.

Killdeer   *C. vociferus*
vociferor (L) I cry aloud, I shout; they are noisy birds, the call sounding
like 'kill-dee', hence the vernacular name Killdeer. Ranging from
Canada to Mexico and the West Indies, and coasts of Peru and Chile.

Dotterel   *Eudromias morinellus*
eu- (Gr) prefix meaning well, nicely   *dromeus* (Gr) a runner; when
hunting they run in short bursts, halt and freeze, pick up an insect,
then dash off in a new direction   *morinellus* (New L) according to Dr
Gesner (1516-1565), and Webster's International Dictionary, it is a
coined word, a diminutive of *mōros* (Gr) stupid, foolish, and in part
refers to the Morini, an ancient name for the people of Flanders where
the bird used to be very common; it has a reputation for stupidity (see
Australian Dotterel, p. 271). Inhabiting scattered areas of North
America, northern Europe especially Scandinavia, Scotland, the
Netherlands, and migrating south in winter to southern Italy, Greece,
North Africa, and ranging south-east to Thailand and Vietnam.

Wrybill   *Anarhynchus frontalis*
ana- (Gr) prefix meaning up, upon, back   *rhunkhos* (Gr) snout, beak;
there is a most peculiar shape to the beak, the outer or distal part being
bent to the right; it is thought that in some way this helps the bird to
obtain insects that hide under stones   *frons* (L), genitive *frontis,* the
forehead   *-alis* (L) suffix meaning relating to; it has a white forehead.
Inhabiting New Zealand.

Family SCOLOPACIDAE   81 species
skolopax (Gr), genitive *skolopakos,* a bird of the snipe kind, perhaps a
woodcock.

Upland Sandpiper   *Bartramia longicauda*
Named after W. Bartram (1739-1823) in 1851, an American natu-

ralist *longus* (L) long *cauda* (L) the tail of an animal. Inhabiting North, Central, and South America.

Common Curlew *Numenius arquata*
noumēnius (Gr) derived from *neo-mēnios*, meaning used at the new moon; can also mean the curlew on account of the bird's crescent-shaped bill *arcus* (L) a bow *-atus* (L) provided with, hence *arquata* (New L) provided with a bow, another reference to the bird's crescent-shaped bill. Inhabiting the northern parts of Eurasia including the British Isles and ranging south in winter to Africa and southern Asia as far as Malaysia and Borneo.

Long-billed Curlew *N. americanus*
-anus (L) suffix meaning belonging to. Breeding in the western part of North America and migrating south in winter to southern Mexico.

Eskimo Curlew *N. borealis*
boreus (L) pertaining to the north wind *borealis* (L) northern. This curlew is very rare; it breeds in Alaska and northern North America and migrates south in winter to southern South America.

Far Eastern Curlew *N. madagascariensis*
-ensis (L) suffix meaning belonging to; the name is a mistake, it does not belong to Madagascar and has never lived there; it breeds in eastern Siberia and ranges south in winter to Malaysia, New Guinea and many islands in that part of the Pacific Ocean, and Australia.

Whimbrel *N. phaeopus*
phaios (Gr) dusky, grey *pous* (Gr) the foot; the legs are greenish grey. Breeds in parts of Alaska, Greenland, Iceland, Scandinavia, and northern USSR and migrates south in winter to South America, Africa and Australia.

Bristle-thighed Curlew *N. tahitiensis*
-ensis (L) suffix meaning belonging to. Breeding in the mountains of Alaska it migrates across the Pacific Ocean to winter in Tahiti and other islands of the Polynesian group as far south as New Caledonia.

Bar-tailed Godwit *Limosa lapponica*
limus (L) mud *-osus* (L) suffix meaning full of, prone to; they often inhabit mud flats and marshy places *-icus* (L) suffix meaning belonging to; 'of Lapland'. Breeding in northern areas of Europe, and occasionally Britain, it migrates south in winter sometimes as far as South Africa, Australia and New Zealand. The tail is white with black

bars. The name godwit is obscure, but is said to come from Old English meaning 'good creature', a reference to it being good to eat.

Black-tailed Godwit *L. limosa*
Breeds in northern Eurasia and migrates south in winter to south-eastern Asia, Africa and Australia.

Wood Sandpiper *Tringa glareola*
trungas (Gr) a white-rumped water bird, the sandpiper *glarea* (L) gravel *-olus* (L) dim. suffix, 'little gravel', to indicate sand, though they tend to inhabit marshy places rather than sandy places. Breeding in the northern part of Eurasia and migrating south in winter to south-east Asia, Africa, and Australia.

Yellowshank *T. melanoleuca*
melas (Gr) black *leukos* (Gr) white; it has a white upper rump and black tips to the wings. Inhabiting North America and migrating south in winter to Central and South America.

Greenshank *T. nebularia*
nebula (L) mist *-arius* (L) suffix meaning belonging to; a reference to the bird's habitat on marshy ground where it is often misty. Inhabiting Europe including the British Isles and northern Asia, and migrating south in winter to southern Europe, Africa, south-east Asia, Malaysia, New Guinea, and Australia.

Green Sandpiper *T. ochropus*
ōkhros (Gr) pale, yellow *pous* (Gr) foot, talons or claws; actually the feet and legs are a greenish colour. Wide distribution in the northern part of Eurasia and migrating south in winter to Africa, south-east Asia and Malaysia.

Solitary Sandpiper *T. solitaria*
Inhabiting North America and migrating in winter to Central and South America. Usually seen singly or at the most in pairs.

Common Snipe *Gallinago gallinago*
gallinago, a coined word derived from *gallina* (L) a hen (poultry). Almost world-wide distribution migrating south in winter.

Japanese Snipe *G. hardwickii*
Named after Maj. Gen. T. Hardwick, FRS (1756-1835) a naturalist in India. This snipe inhabits Japan, Vietnam, Thailand, New Guinea and Australia.

Swinhoe's Snipe   *G. megala*
megalē (Gr) great: R. Swinhoe, FRS (1836–1877) was in the Consular Service in China. Inhabiting eastern Siberia and migrating south to India, the Vietnam-Thailand area, Malaysia, New Guinea and neighbouring small islands, and Australia.

Pintail Snipe   *G. stenura*
stenos (Gr) narrow   *oura* (Gr) the tail. Breeds in Siberia and migrates to India and south-east Asia.

Ruddy Turnstone   *Arenaria interpres*
arena (L) sand   *arenarius* (L) relating to sand, sandy; it is found on rocky and pebbly shores and also on sandy flats   *interpres* (L) a negotiator, a messenger; an allusion to the alarm call which warns other birds of the approach of danger. Breeds on the coasts of Alaska, northern Canada, Greenland, Scandinavia and Siberia, and migrates south to an almost world-wide distribution including South America, South Africa, Australia and New Zealand.

Black Turnstone   *A. melanocephala*
melas (Gr), genitive *melanos*, black   *kephalē* (Gr) the head; the head and breast are mainly black. Breeds in Alaska and migrates south in winter to southern California and Central America.

Short-billed Dowitcher   *Limnodromus griseus*
limnē (Gr) a marsh, a pond   *dromeus* (Gr) a runner; 'marsh-runner' *griseus* (New L) grey. The name *dowitcher* is of Iroquoian origin; the Iroquois are an American-Indian people. This snipe breeds in North America and winters in Central and South America.

American Woodcock   *Scolopax minor*   (formerly *Philohela*)
skolopax (Gr) a woodcock   *minor* (L) lesser. Inhabiting the south-eastern part of North America.

Eurasian Woodcock   *S. rusticola*
rusticola (L) an inhabitant of the country; a reference to the bird's habitat, woodland rather than marshes. Breeds in Europe though not in the far north or south, and ranging across Asia to Japan, also some islands such as the Canaries, Maderia, and Azores and parts of Asia Minor and the Himalayas; it does not migrate far.

Jack Snipe   *Lymnocryptes minimus*
limnē (Gr) a marsh, a pond   *kruptos* (Gr) secret, hidden; an allusion

to the bird's habit of hiding among reeds   *minimus* (L) smallest; it is
the smallest snipe. Breeds in north-east Europe and northern Asia
migrating south in winter to Africa, India, Vietnam and Thailand,
and Malaysia.

Surfbird   *Aphriza virgata*
aphros (Gr) froth, sea foam   *zaō* (Gr) I live; they hunt for prey very
close to the surf as they seem to prefer rocks and pebbles that have just
had a wave break over them   *virga* (L) a twig; *virgatus*, made of twigs,
can mean striped; the bird is usually mottled black and white rather
than striped. Breeds in Alaska and migrates south in winter to Central
and South America as far as Chile.

Sanderling   *Calidris alba*   (formerly *Crocethia*)
Calidris is a coined word derived from *skalidris* (Gr) a spotted bird
albus (L) white; a reference to the birds plumage in winter. Sanderling
is derived from an Icelandic name for the sandpiper. Almost world-
wide distribution, breeding in the Northern Hemisphere and migrating south in winter to all southern countries.

Baird's Sandpiper   *C. bairdii*
Named after Spencer Fullerton Baird (1823-1887) the American
ornithologist. Breeds in the far north of North America and migrates
south to South America.

Purple Sandpiper   *C. maritima*
maritimus (L) of the sea; apart from the breeding season it is always
found along the shore or nearby, and prefers rocky coasts. Breeds in
Arctic areas on coasts and islands of the Arctic Sea and migrates south
in winter.

Temminck's Stint   *C. temminckii*
Named after Prof C. J. Temminck (1778-1858) the Dutch zoologist
at one time Director of the Natural History Museum, Leyden. The
origin of the name stint is obscure. Breeds subarctic and arctic areas
from Scandinavia to the Bering Straits, and occasionally in Britain
and migrates south in winter mainly to Africa and south-east Asia as
far as Malaysia.

Spoonbilled Sandpiper   *Eurynorhynchus pygmeus*
eurunō (Gr) I make wide or broad   *rhunkhos* (Gr) the beak   *pugmaios*
(Gr) a foot tall, dwarfish. Breeding in northern North America and

Eurasia and migrating south in winter, the Siberian species to India, and the Thailand-Vietnam area.

Ruff   *Philomachus pugnax*
philos (Gr) loving   *makhē* (Gr) battle, combat   *pugnax* (L) fond of fighting, quarrelsome; there is ritual fighting among the males as part of the courtship ceremony but it is really only pretence and bravado. Breeding in the far north of Eurasia and migrating south in winter to South Africa and the Oriental region.

Family   RECURVIROSTRIDAE   7 species
recurvo (L) I curve backwards   *rostrum* (L) a beak; only the avocets have the upward-curved bill from which the name is derived.

Ibisbill   *Ibidorhyncha struthersii*
ibis (L), genitive *ibidis*, a sacred Egyptian bird that lived on water animals, the ibis   *rhunkhos* (Gr) a beak, bill; the beak is curved downward like that of the ibis, quite different from the avocet's beak which is curved upward: named after Dr John Struthers (fl 1855), a Scottish zoologist who was at Glasgow University in 1831. Inhabiting the mountains of central Asia, and ranging to Kashmir and Burma.

Black-winged Stilt   *Himantopus himantopus*
himas (Gr), genitive *himantos*, a leather thong   *pous* (Gr) the foot; 'strap-leg'; a reference to the slender lower leg. Almost world-wide distribution with slight variations in different countries which would be classified as sub-species or races. The name 'stilt' is a reference to the long thin legs.

Banded Stilt   *Cladorhynchus leucocephalus*
klados (Gr) a shoot, a twig   *rhunkhos* (Gr) the beak; a reference to the long slender beak which resembles a long thin shoot or twig   *leukos* (Gr) white, pale   *kephalē* (Gr) the head. Inhabiting Australia. It has a chestnut-brown band across the breast.

American Avocet   *Recurvirostra americana*   (includes *H. mexicanus*)
Recurvirostra, see notes under Family, above. Inhabiting North and Central America. Avocet is from the Italian *avosetta*.

Avocet   *R. avosetta*
Inhabiting Eurasia, Africa, India, and the Vietnam-Thailand area.

Family   PHALAROPODIDAE   3 species
phalaris (Gr) a coot, so called from its bald white head (*phalakros* (Gr)

bald) *pous* (Gr), genitive *podos*, the foot, 'coot-footed'; an allusion to the lateral lobes on the feet which resemble those of coots and form paddles that assist when swimming.

Red or Grey Phalarope *Phalaropus fulicarius*
fulica (L) a coot *-arius* (L) suffix meaning belonging to (see above). The plumage is red in summer and grey in winter. Breeds in arctic areas almost throughout the northern hemisphere, and migrates south in winter ranging to South America, Africa, the oriental region, and Australia.

Northern Phalarope *P. lobatus* (formerly *Lobipes*)
lobatus (New L) lobed (see above). The range is similar to the Red Phalarope, above.

Wilson's Phalarope *P. tricolor* (formerly *Steganopus*)
In summer the female has a white rump, dark grey wings, and reddish-brown patterns on the side of the neck and wing coverts. Named after Alexander Wilson (1766-1813) an American ornithologist. Breeds inland in central Canada and USA and migrates to South America. All three phalaropes are occasionally seen in Britain.

Family DROMADIDAE 1 species
dromas (Gr), genitive *dromados*, running; spends a lot of time walking and running looking for crabs, its favourite food. Can fly and also swims well.

Crab Plover *Dromas ardeola*
ardeola (L) a little heron; it is rather like a small heron. Inhabiting Africa, parts of southern Asia including India.

Family BURHINIDAE 9 species
bous (Gr) an ox *rhis* (Gr), genitive *rhinos*, the nose; 'bull-nosed'; the head is relatively large and the bill short and thick.

Double-striped Thick-knee *Burhinus bistriatus*
stria (L) a furrow, a channel *striatus* (New L) striped; *bi-striatus*, having two stripes; there are two stripes on the head. Thick-knee is a misleading name; it refers to the bulging joint between the tibia and tarsus which is in fact the ankle. Inhabiting Mexico and Central America and the northern part of South America.

Southern Stone-Curlew *B. magnirostris*
magnus (L) great *rostrum* (L) the beak; the beak is short and thick.

The name 'stone-curlew' refers to their liking for stony and pebbly ground. Inhabiting Australia.

Stone-Curlew *B. oedicnemus*
oidos (Gr) a swelling *knēmē* (Gr) the knee, lower part of the leg; 'a thick knee'. Ranging from south-east England southern Eurasia and India to Africa.

Peruvian Thick-knee *B. superciliaris*
supercilium (L) the eyebrow *-arius* (or *aris*) (L) suffix meaning belonging to; there is a white streak above the eye. Ranging from Ecuador to Peru.

Beach Stone-Curlew *Esacus magnirostris* (or *Orthorhamphus*)
Esacus, obscure; evidently a coined name and not explained in the ornithological records in the British Museum (Natural History) *magnus* (L) great *rostrum* (L) the beak; the beak is short and thick; note that the Southern Stone-Curlew has the same specific name. Inhabiting the Indo-Australasian region.

Family GLAREOLIDAE 17 species
glarea (L) gravel *-olus* (L) diminutive suffix *glareola* (New L) a coined word to indicate dry sandy places; 'small gravel'; the family frequent dry stony places, deserts, ploughed land, etc.

Egyptian Plover *Pluvianus aegyptius*
pluvius (L) rainy, bringing rain *-anus* suffix meaning belonging to; a coined word, 'associated with rain' (see Golden Plover p. 262) *Aegyptus* (L) Egypt *-ius* (L) suffix meaning belonging to. Inhabiting Egypt and ranging south across Africa from Senegal to Ethiopia, and south to Uganda.

Indian Courser *Cursorius coromandelicus*
cursor (L) a runner *-ius* (L) suffix meaning belonging to; they are good runners, usually running rather than flying when escaping from danger; most coursers can run faster than a man: the Coromandel Coast is the east coast of India, forming part of the Bay of Bengal *-icus* (L) suffix meaning belonging to. Inhabiting India.

Cream Coloured Courser *C. cursor*
cursor (L) a runner (see above). Mainly found in Africa, but sometimes ranging to southern Europe and south-west Asia.

Temminck's Courser   *C. temminckii*
Named after Prof C. J. Temminck (1778-1858) a Dutch zoologist.
The distribution is widespread in Africa except the Congo forests, the
north-east area, and the south-west.

Australian Dotterel   *Peltohyas australis*
peltē (Gr) a small shield; Sharpe gives no explanation of *Peltohyas* but
it may refer to a shield-shaped mark at the base of the beak   *auster*
(L) the south   *-alis* (L) suffix meaning relating to. Inhabiting
Australia. Dotterel is derived from *dote*, and the diminutive suffix *-rel*;
'a little fool'; they have a reputation for stupidity, being so tame and
easily caught. Here the name is incorrect, as it is not in the same family
as the lapwings, plovers, and dotterels, the Charadriidae. Dr R. B.
Sharpe (1847-1909) was an ornithologist in the British Museum
(Natural History) from 1872 to 1909.

Black-winged Pratincole   *Glareola nordmanni*
Glareola, see under Family (p. 270): named after Alexander v. Nord-
mann (1803-1866), a Russian zoologist who travelled in southern
Russia and Crimea in 1842. Inhabiting southern Eurasia and Africa.
Pratincole is a name derived from Latin *pratum*, a meadow, and *incola*
(L) a dweller, alluding to a common habitat.

White-collared Pratincole   *G. nuchalis*
nuchalis (New L) of the neck; a reference to the white line round the
neck. Inhabiting Africa.

Pratincole   *G. pratincola*
pratincola, see above under Black-winged Pratincole. Inhabiting
southern Eurasia, Africa and India.

Family   THINOCORIDAE   4 species
this (Gr), genitive *thinos*, the beach, shore   *korax* (Gr) a raven; they
inhabit rocky shores but can also be found high in the mountains.

Gay's Seedsnipe   *Attagis gayi*
attagas (Gr) a long-billed bird fond of the water, perhaps the godwit:
named after Dr Claude Gay (1800-1873), a French botanist who
lived in Chile from 1824 to 1841. Inhabiting western South America
along the Andes, ranging south from Ecuador. The English name
seedsnipe is a reference to its diet which is mainly seeds and occasionally some vegetable matter and insects.

D'Orbigny's Seedsnipe   *Thinocorus orbignyianus*
Thinocorus, see above under Family: named after Prof A. D. d'Orbigny
(1803-1857), a French zoologist who was in South America from 1826
to 1833 and was Professor at the Museum of Natural History, Paris, in
1854. Inhabiting Peru and ranging south to Tierra del Fuego.

Pygmy or Patagonian Seedsnipe   *T. rumicivorus*
rumex (L), genitive *rumicis*, sorrel   *voro* (L) I devour; a reference to
their liking for the sorrel plant. Breeding in the southern part of South
America it migrates north to Argentina and some valleys in the high
Andes. The smallest seedsnipe, about 18 cm (7 in) long; Patagonia is
the southern part of Argentina.

Family   CHIONIDIDAE   2 species
khiōn (Gr) snow; a reference to the completely white plumage and to
their habitat in lands of snow, ice and rock.

Snowy Sheathbill   *Chionis alba*
albus (L) white. Inhabiting sub-antarctic islands such as South
Georgia, South Orkneys, and other small islands off the tip of southern
South America, it migrates north as far as Uruguay. The name sheathbill derives from a horny casing that covers the base of the bill.

Lesser Sheathbill   *C. minor*
minor (L) smaller, less; actually very little smaller than *C. alba*. Inhabiting sub-antarctic islands such as Prince Edward, Crozet, Kerguelen and Heard; these lie between the southern part of the Indian
Ocean and the Antarctic.

Family   STERCORARIIDAE   5 species
stercus (L), genitive *stercoris*, dung   *stercorarius* (L) having to do with
dung; with reference to their fondness of offal and carrion.

Long-tailed Jaeger   *Stercorarius longicaudus*
longus (L) long   *cauda* (L) the tail; the central tail feathers are much
lengthened. Breeds in northern arctic areas of North America and
Eurasia and migrates to tropical areas in the southern Hemisphere
where it spends a lot of its time at sea. The name jaeger is from *jäger*
(G) a hunter, and derives from the fierce attacks it makes on other sea
birds, causing them to drop or even disgorge their prey; however
apparently *S. longicaudus* is less ferocious than its relatives.

Pomarine Jaeger   *S. pomarinus*
pōma (Gr) a cover, a lid   *rhis* (Gr), genitive *rhinos*, the nose; an allusion to the rim over the base of the bill which appears in summer. The range is similar to *S. longicaudus*, above.

MacCormick's Skua   *S. maccormicki*
Dr R. MacCormick (1800–1890) was surgeon with the British Antarctic Expedition in 1839–1843. This skua breeds in the Antarctic and migrates to Australia, New Zealand and surrounding islands, but like other skuas spends much of its time at sea. The name skua is derived from *skufr*, the Icelandic name for the bird.

Great Skua   *S. skua*   (or *Catharacta*)
skugvur (Old Norse) = *skufr*, a predatory gull. The range differs from other skuas in that it breeds in both the Arctic and Antarctic; in the north, Iceland, the Faeroes and Scotland, and in the south Antarctic and Sub-antarctic islands, the southern tip of South America and the Falkland Islands and the extreme south of New Zealand; both kinds migrate to warmer areas in winter and spend much of their time at sea.

Family   LARIDAE   86 species
laros (Gr) a sea bird, probably the gull or cormorant.

Herring Gull   *Larus argentatus*
argentatus (L) ornamented with silver; a reference to the silvery-grey plumage. Widespread on the sea coasts of the Northern Hemisphere.

Red-legged Kittiwake   *L. brevirostris*
brevis (L) short   *rostrum* (L) the beak. Breeds in the Bering Straits area and North Pacific and migrates south in winter. The name kittiwake is imitative of the bird's call.

Common Gull   *L. canus*
canus (L) hoary, grey; a reference to the predominant colour; like several other gulls that have grey wings with black wing-tips, these gulls are difficult to distinguish one from another. The range is similar to *L. argentatus*, above, and often seen inland on lakes and rivers and scavenging on rubbish dumps.

Lesser Black-backed Gull   *L. fuscus*
fuscus (L) dark-coloured. Inhabiting mainly the eastern Atlantic shores from Scandinavia ranging south to Africa.

Ross's Gull   *L. rosea*   (or *Rhodostethia*)
roseus (L) red; a white gull strongly tinged with pink. Discovered by
and named after Rear Admiral Sir James C. Ross (1800–1862), the
Arctic and Antarctic explorer. This gull is rarely seen south of the
Arctic area; it breeds in the Kolyma River area in north-eastern
Siberia and migrations in winter do not reach very far south.

Black-headed Gull   *L. ridibundus*
rideo (L) I laugh   *ridibundus* (L) laughing; an allusion to its supposedly
laughing call. Breeds in Britain, Ireland and western Europe ranging
to eastern Asia, migrating in winter to northern Africa, and ranging
to Asia, Persian Gulf, India, Philippines, China and Japan.

Black-legged Kittiwake   *L. tridactylus*
tria (Gr) three   *daktulos* (Gr) a finger, a toe; the three front toes are of
normal size but the hind toe is small and almost rudimentary. Breeding on North Atlantic coasts and migrating south in winter. The name
kittiwake is an imitation of their call.

White-winged Black Tern   *Chlidonias leucoptera*
Chlidonias is actually a misspelling; it should be *khelidonios* (Gr) of the
swallow, like the swallow, a reference to the tern's swallow-like wings
and forked tail; there is a rare Greek word *khelidonias*, meaning the
spring wind because the swallows come with it   *leukos* (Gr) white
pteron (Gr) feathers, can mean wings. Widespread distribution including sea coasts of Eurasia, Africa, the Oriental Region, New Guinea
and Australia. The name tern is of Scandinavian origin, akin to Old
Norse *therna*, a tern.

Black Tern   *C. nigra*
niger (L) black. Breeds in south-western Europe ranging to western
Asia, and winters on coasts and inland waters of Africa ranging south
to Angola and Tanzania.

Bridled Tern   *Sterna anaethetus*
sterna (New L) a tern, of Scandinavian origin, cf. *therna* (Old Norse) a
tern; *anaethetus* is a misspelling of *anaisthetos* (Gr) without feeling, without sense of a thing; can mean without common sense, stupid; an
allusion to its stupidity in allowing people to approach so near;
English sailors called it a 'noddy', Linneus named a related bird
Sterna solida, 'the dense tern'. Very widespread distribution and known
in almost every country in the world though rare in some parts. The

name bridled refers to a bridle-like black line from the base of the bill past the eyes.

Greater Crested Tern   *S. bergii*
Named after Dr F. G. Berg (1843-1902), a zoologist and Director of the Buenos Aires Museum from 1892 to 1901. Range is similar to *S. anaethetus*, above, but not recorded in the Americas.

Caspian Tern   *S. caspia*
Named from the Caspian Sea, the distribution is widespread, similar to above and including the Americas.

Roseate Tern   *S. dougallii*
Named after Dr Patrick Macdougall (c. 1770-c. 1817) a Scottish zoologist; Glasgow University in 1808   *roseus* (L) rose-coloured. Worldwide distribution, like *S. anaethetus*, above.

Forster's Tern   *S. forsteri*
Thomas Nuttall (1786-1859), the English ornithologist and botanist, described this tern and it was named after Dr Johann Reinhold Forster (1729-1798), the German naturalist who accompanied Capt Cook on his voyage to the Pacific on *HMS Resolution* in 1772. Inhabiting North America and South America.

Sooty Tern   *S. fuscata*
fuscus (L) dark, dusky; it has very dark wings and back. Inhabiting southern seas around the world.

Common Tern   *S. hirundo*
hirundo (L) a swallow; a reference to the swallow-like wings and the deeply-forked tail. It has a world-wide distribution.

Gull-billed Tern   *S. nilotica*
niloticus (L) of the Nile; the bird takes the name from the River Nile but it is known in most parts of the world, although rare in the New Guinea-Australia-Pacific Islands area. The bill is noticeably heavier than in other terns.

Trudeau's Tern   *S. trudeaui*
Named after Dr J. de B. Trudeau (1817-1887) a New York doctor, naturalist, and author of books on zoology. Inhabiting South America.

Inca Tern   *Larosterna inca*
'A gull-tern': the Incas were an American-Indian tribe of Peru. This

interesting tern, with its red beak and feet and long white trailing moustache, inhabits the coasts of Ecuador, Peru and Chile.

Brown Noddy   *Anous stolidus*
anous (Gr) without understanding, silly   *stolidus* (L) dull, slow of mind; authors of Latin names seem determined to condemn some birds as stupid just because they are not afraid of man! (cf *S. anaethetus*, p. 274). Inhabiting the Southern Hemisphere around the world.

Black or Lesser Noddy   *A. tenuirostris*
tenuis (L) thin, slender   *rostrum* (L) the beak; noddies have a slender pointed beak. The Black Noddy, although smaller than the Brown Noddy, is not one of the smallest. Found in most tropical and sub-tropical areas.

White Tern   *Gygis alba*
gugēs (Gr) a water-bird   *albus* (L) white. Found in most tropical and sub-tropical seas.

Family   RYNCHOPIDAE   3 species
rhunkhos (Gr) snout, beak   *ops* (Gr) eye, face, appearance; 'snout-like'; they have a compressed and blade-like bill, the lower mandible being longer than the upper to assist in skimming their prey from the surface of the water.

Indian Skimmer   *Rynchops albicollis*
albus (L) white   *collum* (L) the neck; it has a white neck and breast and upper parts dark to black. Inhabiting the coasts and inland waters of India, Burma, Vietnam, Thailand and Cambodia.

African Skimmer   *R. flavirostris*
flavus (L) yellow   *rostrum* (L) snout, beak; the beak is yellow and red. Inhabiting coasts and inland waters of Africa south of the Sahara.

Black Skimmer   *R. nigra*
niger (L) black; it is not entirely black, having a white breast and neck. It is found only in the New World, ranging from New Jersey in North America to the Amazon Basin and Argentina.

Family   ALCIDAE   21 species
alka (Icelandic) the auk.

Dovekie   *Alle alle*
allex (L) the great toe; probably a reference to the lack of the hind toe,

the 'x' being deleted to show that the toe is missing. Sometimes known as the Little Auk, dovekie is the diminutive of dove; the Great Auk is now extinct. Breeds in the Arctic on coasts of Greenland and groups of islands such as Spitzbergen and Franz Joseph Land, migrating south in winter to coasts of the British Isles and France, and from Greenland to the northern parts of the east coast of North America.

Razorbill *Alca torda*
alka (Icelandic) the auk *torda* is a local name for this auk, the general Swedish name being *tordmule*. Breeds on coasts of the northern Atlantic such as New England, western Greenland, Iceland, Scandinavia and the British Isles, migrating south in winter when much of the time is spent far out to sea. The name razorbill is a reference to the compressed unusually sharp bill.

Common Murre or Guillemot *Uria aalge*
ouria (Gr) a water bird *aalge* (Dan) the murre, a type of guillemot; the origin of the word murre is obscure. Inhabiting northern seas and ranging south to North Korea and San Francisco in the Pacific, and Portugal in the Atlantic; it is known on the coasts of Scandinavia, Iceland, the British Isles, northern France, and Newfoundland.

Brünnich's Guillemot *U. lomvia*
lomvia (D dialect) a bird. Named after M. T. Brünnich (1737–1827), a Danish zoologist and Director of the Natural History Museum, Copenhagen, in 1772. Sometimes known as the Thick-billed Murre, it breeds in islands in the Arctic seas such as Baffin Island, Spitzbergen and Franz Joseph Land, migrating south in winter.

Black Guillemot *Cepphus grylle*
kepphos (Gr) a sea-bird of the petrel kind *gryllus* (L) a cricket, a grasshopper; can mean a kind of comic figure; on land, standing upright like a small penguin and its long webbed feet sticking out in front, it is a comical sight; it sits down on the tarso-metatarsus bone giving it a flat-footed appearance. The range is similar to *U. aalge* and *U. lomvia*, above.

Kittlitz's Murrelet *Brachyramphus brevirostris*
brakhus (Gr) short *rhamphos* (Gr) beak, bill *brevis* (L) short *rostrum* (L) beak, snout; the name in Greek and repeated in Latin for the species, 'short-billed'. Named after F. H. Kittlitz (1779–1874) the Russian zoologist on the expedition to Kamchatka from 1826 to 1829. From arctic eastern Sibera ranging to Alaska and wintering in Kam-

chatka, the Aleutian Islands, and ranging as far south as Japan. Murrelet, dim. of murre; see *Uria aalge*, above.

Cassin's Auklet   *Ptychoramphus aleuticus*

ptux (Gr), genitive *ptukhos*, the folds of a garment, anything that appears to be in folds   *rhamphos* (Gr) the crooked beak of birds; generally a beak, bill; a reference to the compressed beak which appears to be in folds   -*icus* (L) suffix meaning belonging to; it ranges from the Aleutian Islands along the Alaska Peninsula and the Pacific shores of British Columbia and migrating south in winter to Central America. It is named after John Cassin (1813-1869), Curator of Birds, Philadelphia Academy of Sciences, who described many species of birds.

Crested Auklet   *Aethia cristatella*

aithuia (Gr) a seagull or diver   *crista* (L) the crest of an animal -*ellus* (L) diminutive suffix, 'little crest'; it has a small crest just above the base of the beak. Inhabiting the Bering Straits area, the Aleutian Islands, the Kamchatka Peninsula, and the Kuril Islands, and migrating south in winter.

Puffin   *Fratercula arctica*

fraterculus (New L) a friar; considered to be an allusion to the birds habit of clasping its feet together as though in prayer when rising from the sea   -*icus* (L) suffix meaning belonging to; 'of the Arctic'. Range similar to *Alca torda* (p. 277) but extending to Spitzbergen. The origin of the name puffin is obscure, but Francesca Greenoak in her book, *All the Birds of the Air*, says that she has read about a puffin, kept for a time in captivity, that would cry—'pupin, pupin'!

Horned Puffin   *F. corniculata*

cornu (L) the horn of an animal   -*culus* (L) diminutive suffix   -*atus* (L) suffix meaning provided with; 'having a little horn'; it has small fleshy horns above the eyes. Inhabiting coasts of the northern Pacific in the area of the Bering Straits and the Bering Sea and moving south in winter although some puffins remain in the breeding areas all the year round.

30 Sandgrouse, Pigeons and Doves
COLUMBIFORMES

Pallas's Sandgrouse
Syrrhaptes paradoxus

Pigeons and doves, in the order Columbiformes, are all basically the same, and all have similar characteristics; but the sandgrouse are a problem for taxonomists; they are not grouse. Although they resemble them in general shape and colour, and they live in sandy places which gives them their English name, they are related to the pigeons and doves. Some authors place them in an order on their own, the Pteroclidiformes. They frequent mostly desert and grassy areas whereas the pigeons and doves are to be found in almost all land habitats and have an almost world-wide distribution; the diet is mostly seeds and berries, though insects are also taken by sandgrouse, and insects and worms by the doves and pigeons.

The various kinds of domestic pigeons kept by pigeon fanciers and those seen in European city streets are descended from the rock-pigeon, or rock-dove, *Columba livia*; this process of selective breeding has probably been going on for some 5,000 years.

Order COLUMBIFORMES

columba (L) a pigeon or dove.

Family PTEROCLIDIDAE   16 species
pteron (Gr) feathers, can mean wings; Temminck says 'The generic name *Pterocles* that I propose for this genus indicates that there is something special about the wings; that is the unusual length of the feathers, the primary being the longest'. He gives no explanation for the second element, *cles*.

Pallas's Sandgrouse   *Syrrhaptes paradoxus*
syr- = *syn-* (Gr) prefix meaning together   *rhapto* (Gr) I sew; a reference to the vestigial web joining the toes   *paradoxos* (Gr) contrary to opinion, strange; a reference to the web and other unusual abnormal anatomical features. Named after Prof P. S. Pallas (1741-1811), a German zoologist and explorer and a professor at St Petersburg University; he made contributions to most of the natural sciences. This sandgrouse inhabits a large area in Asia to the east of the Caspian Sea and migrates south in winter as far as northern India and China.

Tibetan Sandgrouse   *S. tibetanus*
-anus (L) suffix meaning belonging to. It breeds high in the mountains of Central Asia and migrates south in winter to India, Burma, and the Thailand-Vietnam area.

Variegated Sandgrouse   *Pterocles burchelli*
Pterocles, see under Family, above: named after Dr W. J. Burchell (1782-1863), an English zoologist who made an expedition to Africa in 1811. Inhabiting Africa including parts of the Sahara and the Kalahari Desert.

Namaqua Sandgrouse   *P. namaqua*
Namaqualand is on the western side of South Africa and is partly a desert area. Towards the end of the nineteenth century the naturalist E. G. B. Meade-Waldo asserted that desert-living sandgrouse deliberately soak their breast feathers in water and in this manner carry it many miles to water their young, though other naturalists at the time considered this to be nonsense. The idea ceased to be seriously considered until quite recently ornithologists observing the Namaqua and other desert-living sandgrouse have actually seen this water-carrying system being used by the bird, including the deliberate soak-

ing of the feathers at the water hole and the chicks taking the water from the feathers at the nest.

Madagascar or Masked Sandgrouse *P. personatus*
persona (L) a mask *personatus* (L) masked; it has a black border round the beak. Inhabiting Madagascar.

Family COLUMBIDAE 280 species
columba (L) a pigeon or dove.

Nilgiri Wood Pigeon *Columba elphinstonii*
Named after the Hon M. Elphinstone (1779-1859), in Afganistan in 1808 and Governor of Bombay from 1819 to 1827, and author of books on natural history. This pigeon inhabits India and is named from the Nilgiri District in the south-west.

Rock Dove *C. livia*
Livius is a Roman name; in this case it is coined from *lividus* (L) bluish-grey. Originally from Eurasia, and the ancestor of the domestic pigeon, the distribution is now almost world-wide.

Wood Pigeon *C. palumbus*
palumbes (L) a wood pigeon, a ring dove. Inhabiting Europe including the British Isles but not northern Scandinavia; the Mediterranean islands and ranging east into Russia and part of Asia and south to India and Cambodia.

Collared Turtle Dove *Streptopelia decaocto*
streptos (Gr) a necklace *peleia* (Gr) a dove or wood-pigeon *dekaoktō* (Gr) 18; a Greek legend says that a servant girl was once employed for a miserable 18 pieces a year, and she prayed to the gods to let the world know of this meanness; they took pity on her and created a dove that calls 'dekaokto, dekaokto'; it is not much like the dove's call which is more like 'coo-COO-cook', but the Latin name is derived from the legend. Originally inhabiting Asia it is now fairly widespread in Europe.

Laughing Dove *S. senegalensis*
-ensis (L) suffix meaning belonging to, usually in relation to a place, 'of Senegal'. Though named from Senegal it inhabits much of Africa except desert areas and is not known in Madagascar; found in Turkey and Arabia and ranging to eastern India, it has been introduced to

western Australia. The call usually consists of five 'coos' which sounds as though the bird is laughing.

Turtle Dove   *S. turtur*
turtur (L) a turtle dove. Inhabiting Europe including the British Isles and ranging east into Asia beyond the Caspian Sea, south to northern Africa, the Middle East, India and the Burma area.

Barred or Zebra Ground-Dove   *Geopelia striata*
gē (Gr) the earth, ground   *peleia* (Gr) a dove or wood-pigeon   *stria* (L) a furrow, a groove   *striatus* (New L) striped. An eastern bird, inhabiting the Burma area, Indonesia, Malaysia, the Philippines, Australia, New Guinea, and the Pacific islands. Known as a ground-dove because it spends a lot of its time on the ground, but it is capable of flight.

Namaqua Dove   *Oena capensis*
oinas (Gr) the vine; can mean a kind of wild pigeon the colour of ripening grapes   *-ensis* (L) suffix meaning belonging to, usually in relation to a place, 'of The Cape'. Though named from South Africa it inhabits a large part of Africa including the Red Sea coast and south-western Arabia; Namaqualand is a little-known coastal region in south-west Africa.

Brush Bronzewing   *Phaps elegans*
phaps (Gr) a kind of wild dove or pigeon   *elegans* (L) neat, elegant. Inhabiting the southern part of Australia.

Crested Pigeon   *Ocyphaps lophotes*
ōkus (Gr) swift, fast   *phaps* (Gr) a kind of wild dove or pigeon; it is reputed to be able to outpace the Little Falcon *Falco longipennis* *lophos* (Gr) a crest   *lophōtos* (Gr) crested. Inhabiting Australia, mostly western area.

Blue or Common Crowned Pigeon   *Goura cristata*
Goura is a native name in New Guinea for the crowned pigeons *cristata* (L) crested. Inhabiting New Guinea.

African Green Pigeon   *Treron australis*
trērōn (Gr) timid, shy   *australis* (L) southern; this Latin word does not indicate Australia; it is derived from *auster* (L) the south wind. Inhabiting Africa.

Golden Dove   *Ptilinopus luteovirens*
ptilon (Gr) feathers   *pous* (Gr) the foot; 'feather-footed'; a reference
to the upper half of the leg (tarsus) being covered with feathers
luteus (L) golden-yellow, saffron-yellow   *virens* (L) becoming green.
It is found only on Viti Levu, one of the Fiji Islands.

Comoro Blue Pigeon   *Alectroenas sganzini*
alektoris (Gr) a hen (poultry), there is a rare form *alektruonis*; or the
final element could be from *oinas* (Gr) a kind of wild pigeon: named
after Capt Victor Sganzin (fl 1830 died 1841), a German zoologist
who was in Madagascar in 1831-1832. The Comoro Islands lie
between the northern tip of Madagascar and the mainland of Africa.

Baker's Imperial Pigeon   *Ducula bakeri*
dux (L), genitive *ducis*, a leader, cf English duke (imperial)   *ducula*
(New L) a little duke: named after Dr J. R. Baker, FRS (born 1900)
an Oxford zoologist and at one time President of the Microscopic
Society. This pigeon inhabits the New Hebrides.

Topknot Pigeon   *Lopholaimus antarcticus*
lophos (Gr) the crest or tuft on the head of birds   *laimos* (Gr) the
throat, can also mean the neck; there is a crest on the head and on the
back of the neck   *-icus* (L) suffix meaning belonging to, 'of the Antarctic'. Inhabiting the east coast of Australia from Cape York to
Victoria.

Masked Lovebird
Agapornis personata

In the order Psittaciformes and the family Psittacidae are all the parrot-like birds, including the well known budgerigar; they live in tropical and sub-tropical forests so are not known in the wild in Europe. Almost without exception they are brightly coloured and the sexes are usually alike, and they range in size from the little pygmy parrots of New Guinea to the huge macaws of South America.

Distinctive features are the feet with two toes turned forward and two backward, the powerful hooked beak which is also used for climbing, and the habit of holding up food in one foot to eat it. They are mostly vegetarian, and the diet consists of fruit, berries, seeds, nuts, etc., though some eat insects and occasionally carrion.

Some have become well known as mimics, the two best 'talkers' being the African Grey and the Budgerigar; one parrot kept as a pet earned a reputation by giving a good imitation of the flushing of a lavatory!

Order PSITTACIFORMES

psittacus (L) a parrot.

Family PSITTACIDAE 328 species

Black Lory *Chalcopsitta atra*
khalkos (Gr) copper, can mean bronze coloured *psitta*, abbreviated
form of *psittacus* (L) a parrot *ater* (L) black. Inhabiting New Guinea.

Duivenbode's or Brown Lory *C. duivenbodei*
Named after C. W. R. v. Duivenbode, a well-known merchant of
Ternate; sometimes known as 'King of Ternate', the dates of his birth
and death are obscure, but his son L. D. H. A. v. Duivenbode was born
in 1831 which gives some idea of his age. Ternate is an island in the
Moluccas, Indonesia. Inhabiting New Guinea, where most lories are
found.

Scaly-breasted Lorikeet *Trichoglossus chlorolepidotus*
thrix (Gr), genitive *trikhos*, hair *glossa* (Gr) the tongue; this lorikeet
likes nectar and it forms a large part of its diet; it has a 'brush-tipped'
tongue for seeking and lapping up the nectar *khlōros* (Gr) pale green,
can mean yellow *lepidōtos* (Gr) scaly, covered with scales; a reference
to the breast feathers, which have the appearance of overlapping
scales. Inhabiting Australia.

Yellow and Green Lory *T. flavoviridis*
flavus (L) yellow *viridis* (L) green. Inhabiting the Indonesian area.

Stresemann's Lory *Lorius amabilis*
luri = lori (Malay) a lory or parrot *amabilis* (L) worthy of love,
lovely: named after Prof Dr Erwin Stresemann (born 1889), a
German zoologist, Director of the Ornithological Dept Zoological
Museum Berlin in 1921. This lory inhabits New Britain.

Purple-crowned Lorikeet *Glossopsitta porphyrocephala*
glossa (Gr) the tongue *psitta*, abbreviated form of *psittacus* (L) a
parrot; like the Scaly-breasted Lorikeet, above, it has a brush-tipped
tongue for collecting nectar *porphureos* (Gr) purple *cephalē* (Gr) the
head. Ranging across southern Australia.

Little Lorikeet *G. pusilla*
pusillus (L) small. Inhabiting the Malay Archipelago, Polynesia, and
Australia.

Black or Great Palm Cockatoo *Probosciger aterrimus*
pro- (Gr) a prefix with many meanings, e.g. forward, for *boskō* (Gr)
I feed *gero* (L) I bear, carry; hence *proboscis + ger*, 'carried for feed-

ing'; a reference to the thick prehensile tongue *ater* (L) black
aterrimus (L) very black; the name 'cockatoo' is probably derived
from the call. Inhabiting Australia and New Guinea.

Galah *Eolophus rosiecapillus*

ēōs (Gr) dawn, the red colour at dawn *lophos* (Gr) the crest of birds
roseus (L) red *capillus* (L) hair; in this case taken to mean feathers;
it has pink breast feathers and a white crest with a pink tinge; the
crest is erectile. This cockatoo lives in Australia and Galah is the native
name.

Sulphur-crested Cockatoo *Cacatua galerita* (formerly *Kakatoe*)

Cacatua is from a Malayan name kakatoe or kokatua, derived from the
birds call, hence the English cockatoo *galeritus* (L) wearing a skull-
cap or hood; it has a bright yellow crest. Inhabiting Australia, New
Guinea, and nearby islands.

Little Corella *C. sanguinea*

sanguis (L) blood *sanguinea* (L) blood red; sometimes known as the
Blood-stained Cockatoo on account of the red plumage; Corella is a
native name in Australia. Inhabiting Australia and New Guinea.

Kaka *Nestor meridionalis*

Nestor was a legendary Greek hero known for long life and wisdom,
and the name is often used to mean a wise counsellor, a leader; there
is probably no special reason for the name. In 1758 Linnaeus used
several names from classical mythology for the generic or specific
names of animals and other authors have done the same; it is probable
that they were given without thought of any special significance
meridies (L) mid-day, noon *-alis* (L) suffix meaning relating to;
meridionalis (L) southern; it is found only in New Zealand, South
Island, and lives in the southern part, and the Stewart Islands. Kaka
is the Maori name.

Geelvink Pygmy Parrot *Micropsitta geelvinkiana*

mikros (Gr) little, small *psitta*, abbreviated form of *psittacus* (L) a
parrot; these are the smallest birds in the parrot family, only about
9 cm ($3\frac{1}{2}$ in) long: the Geelvink Channel separates the Houtman
Abrolhos group of islands from the west coast of Australia, where these
miniature parrots live.

Meek's Pygmy Parrot *M. meeki*

Named after Albert G. Meek (1871-1943), a London-born Australian naturalist and author; he lived in the New Guinea area from 1894 to 1913 and collected for Lord Rothschild. This Pygmy Parrot lives in New Guinea and there are other species living in the Pacific islands.

Crimson Rosella *Platycercus elegans*
platus (Gr) flat, can mean broad *kerkos* (Gr) the tail; it has a broad tail *elegans* (L) neat, elegant: the name rosella may be an allusion to the bright red plumage, but it is also suggested that it derives from Rosehill, a district near Sydney in south-east Australia; certainly the Crimson Rosella is known in that area, and it ranges from the north-east part of Queensland to near Adelaide.

Bourke's Parrot *Neophema bourkii*
neos (Gr) new, recent *phēmē* (Gr) a voice, a word; indicating a newly discovered parrot and named after Maj Gen Sir R. Bourke (1777–1855), Governor of New South Wales, Australia, 1831 to 1837. This parrot inhabits a large area in the southern part of Australia including part of New South Wales.

Turquoise Parrot *N. pulchella*
pulcher (L) beautiful, handsome *-ellus* (L) diminutive suffix, hence small and beautiful, 'a little beauty'; certainly an apt name for this small colourful parrot. Inhabiting a large area in the west of Australia from the North West Cape south to Perth, and ranging east to the Mount Heuglin area.

Budgerigar *Melopsittacus undulatus*
melos (L) a tune, a song *psittacus* (L) a parrot *undulatus* (L) waved, undulated; a reference to the dark wavy lines on the back of the head and neck and the adjacent part of the wings. Fairly widespread in Australia but only coastal in the south.

Night Parakeet *Geopsittacus occidentalis*
ge (Gr) earth, ground *psittakos* (Gr) a parrot; sometimes known as the ground parakeet because it is an almost flightless bird, and spends much of its time on the ground *occidentalis* (L) western. The name Night Parakeet refers to its nocturnal habits as it usually hunts at night; it is very rare and possibly extinct. Inhabiting Australia.

Kakapo *Strigops habroptilus*
strix (Gr), genitive *strigos*, a night bird, the owl *ops* (Gr) the eye, face, 'owl-faced'; the head and face have the appearance of an owl *habros* (Gr) soft, delicate *ptilon* (Gr) feathers; the feathers are unusually

soft, like those of an owl which enable it to fly almost silently and so useful when catching prey; however the kakapo is now a flightless bird and eats fruit, leaves and roots of grasses. It is rare and found only in one small area in the south-west of South Island, New Zealand; *kakapo* is a Maori name for this bird.

African Grey Parrot   *Psittacus erithacus*
psittacus (L) a parrot   *erithakos* (Gr) a solitary bird that could be taught to speak; sometimes it is taken to mean the robin-redbreast. Very well known as a domestic pet, in the wild it lives in West and Central Africa.

Madagascar or Grey-headed Lovebird   *Agapornis cana*
agapē (Gr) love   *ornis* (Gr) a bird   *canus* (L) hoary, grey-haired. Known as lovebirds because of their sociable habits; pairs will spend hours sitting together preening each other or just sitting bill-to-bill. Inhabiting Madagascar.

Fischer's Lovebird   *A. fischeri*
Named after Dr Gustav A. Fischer (1848-1886), a German zoologist who explored East Africa during the years 1876 to 1886. Inhabiting Africa south of the Sahara.

Masked Lovebird   *A. personata*
personatus (L) masked; it has a black face, the remainder of the plumage being bright yellow and green. Inhabiting Africa, mainly in the Tanzania area.

Philippine Hanging Parrot   *Loriculus philippensis*
lori (Malay) a lory or parrot   *-culus* (L) diminutive suffix, 'a little parrot'; the parrots in the genus *Loriculus* are very small, some being no bigger than a sparrow, and some have developed a habit of hanging upside down like bats and even sleeping in this position   *-ensis* (L) suffix meaning belonging to. It inhabits islands in the Philippine group.

Blossom-headed Parakeet   *Psittacula cyanocephala*
psittacus (L) a parrot   *-culus* (L) diminutive suffix   *kuaneos* (Gr) blue *kephalē* (Gr) the head. Inhabiting India and Sri Lanka.

Ring-necked Parakeet   *P. krameri*
Named after W. H. Kramer (fl. 1740-died 1765), an Austrian zoologist and author. This parakeet has a rosy half-collar; it inhabits Africa, India, and the Burma to Vietnam area.

Hyacinthine Macaw *Anodorhynchus hyacinthinus*
anodous (Gr) without teeth *rhunkhos* (Gr) the beak; the original description is in Latin: '. . . *rostrum totum nigrum, quam maxime crassum, compressum, altum, superius apice uncinatum, edentatum . . .*' which briefly translated means 'beak black, thick, compressed, upper hooked, toothless'. When Spix gave this name in 1824 it was probably not generally recognised that recent birds do not have teeth, although ancient birds such as *Archaeornis* did *hyacinthinus* (L) belonging to the hyacinth, can mean blue; the entire plumage is bright blue, a very beautiful macaw, inhabiting the tropical forests of eastern Brazil. Dr J. B. v. Spix (1781-1826) was a German zoologist and Professor of Zoology at the Munich Museum; he was in Brazil from 1817 to 1820.

Red-and-Green Macaw *Ara chloroptera*
ara (New L) probably an abbreviation of *arara* (Tupi) a macaw *khlōros* (Gr) green *pteron* (Gr) wings; the plumage is red with green wings. Inhabiting South America ranging from Panama to Brazil.

Coulon's or Blue-headed Macaw *A. couloni*
Named after Dr L. Coulon (1804-1894), a French zoologist who was for many years President of the Society for Natural Sciences in Neuchatel. The Blue-headed Macaw inhabits South America.

Red-masked Parakeet *Aratinga erythrogenys*
ara (New L) probably an abbreviation of *arara* (Tupi) a macaw *tinga* (Tupi) bright, ornamented *eruthros* (Gr) red *genus* (Gr) cheek, jaw. A macaw-like parrot inhabiting South America.

Thick-billed Parrot *Rhynchopsitta pachyrhyncha*
rhunkhos (Gr) the beak *psittakos* (Gr) a parrot *pakhus* (Gr) thick. Inhabiting Central America, Mexico, and ranging north to Arizona.

Hoffmann's or Sulphur-winged Parakeet *Pyrrhura hoffmanni*
purros = purrhos (Gr) red, flame-coloured *oura* (Gr) the tail; named after Dr Carl Hoffmann (1823-1859), a German zoologist who was in Costa Rica with Dr A. von Frantzius from 1854 to 1859. This parakeet is found in Central America.

Slender-billed Parakeet *Enicognathus leptorhynchus*
henikos (Gr) single *gnathos* (Gr) jaw; Gray does not explain the peculiar name 'single jaw' *leptos* (Gr) small, slender *rhunkhos* (Gr) the snout, beak. Inhabiting Chile. G. M. Gray, FRS (1808-1872) was an ornithologist at the British Museum (Natural History) from

1831 until his death in 1872; he published *The Genera of Birds* in 1840.

Monk or Grey-breasted Parakeet   *Myiopsitta monachus*
muia (Gr) a fly   *psitta* is an abbreviated form of *psittakos* (Gr) a parrot; 'a fly-parrot'; it is not established that this parakeet eats flies though many parrots eat insects of various kinds   *monakhos* (Gr) a monk, solitary; the name seems inappropriate as these parakeets are known to breed and move about in large flocks and are considered to be gregarious. Inhabiting Bolivia, Paraguay, Uruguay, and Argentina.

Golden-winged Parakeet   *Brotogeris chrysopterus*
brotogērus (Gr) with human voice; a word probably only used in connection with parrots   *khrusos* (Gr) gold   *pteron* (Gr) wings. Inhabiting the northern part of South America.

White-fronted Parrot or Amazon   *Amazona albifrons*
albus (L) white   *frons* (L) forehead, brow. Inhabiting southern Mexico and Central America.

St Vincent Parrot   *A. guildingii*
Named after The Rev L. Guilding (1798–1831), a naturalist and chaplain on St Vincent Island in the West Indies group; this parrot inhabits this island.

Yellow-headed Parrot   *A. ochrocephala*
ōkhros (Gr) pale, pale yellow   *kephalē* (Gr) the head; one of the largest Amazons, about 38 cm (15 in) long, a green bird with a pale yellow head. Inhabiting Mexico and ranging south to Brazil.

32 Turacos, Cuckoos and their kin

CUCULIFORMES

Common Cuckoo
Cuculus canorus

The turacos are cuckoo-like birds, and although usually placed in the order Cuculiformes there is some question as to whether this is correct, and at least one zoologist who has studied their anatomy gives them a separate order, Musophagiformes. Included in this could be the Hoatzin *Opisthocomus hoazin*, of the order Galliformes; the young of both these birds have an unusual claw on the wing which assists them to scramble about in the trees before they can fly.

Turacos are specially noted for their bright red and green plumage; the red pigment turacin gets its name from these birds, and the green pigment is called turacoverdin. It is the only true green pigment found in birds, as other birds with green colouring get it from the actual structure of the feathers, or from a mixture of other pigments.

Cuckoos are familiar to us through their unusual song, which gives them their common name in many countries, being an imitation of this song. The parasitic habit of using another bird's nest is well known, but by no means all the birds in the family Cuculidae use this rather unpleasant anti-social method of rearing their young. When a cuckoo lays its egg in another birds nest it means certain death to the host bird's own chicks, as they are thrown out by the fledgling cuckoo, possibly in the egg before it has time to hatch.

Some cuckoos exhibit most remarkable forms of migration, for example the Bronzed Cuckoo *Chrysococcyx lucidus*, from the islands off the coast of New Zealand. The parent birds, having laid their eggs in the host-bird's nest, depart to the Solomon Islands some 3,220 km (2,000 miles) away. Then, by some quite incredible instinct of navigation, the young birds follow them soon after they have learnt to fly, accomplishing this dog-leg journey via Australia without parental guidance.

The anis are unlike the true cuckoos, in that they build their nests, sometimes many birds sharing one large nest and all taking turns with the incubation. The coucals and couas also build their own nests; the couas live only on the island of Madagascar, while the coucals have a wide range, including Africa, Asia and Australia.

Order   CUCULIFORMES
cuculus (L) the cuckoo   *forma* (L) form, shape; can mean sort, kind, so not necessarily referring to the actual shape.

Family   MUSOPHAGIDAE   22 species
musa (New L) derived from the Arabic *môza*, a banana, a plantain *phagēma* (Gr) food; turacos are fruit-eating birds but they do not confine their diet entirely to bananas or plantains; some will eat small animals such as insects, snails, and other invertebrates.

Knysna Turaco   *Tauraco corythaix*
touraco (Fr) an imitation of the bird's cry   *koruthaix* (Gr) helmet-shaking, i.e. with waving plume; the prominent crest on these birds gives the head the appearance of wearing a military-type helmet. Knysna is a small town on the south coast of South Africa famous for its forests of hardwood trees, used for making furniture; the turacos live in these forests.

Hartlaub's Turaco   *T. hartlaubi*
Named after Dr K. J. G. Hartlaub (1814–1900) a German zoologist of Bremen who explored and collected in Africa in 1857. Inhabiting the mountain forests of Kenya and northern Tanzania.

White-cheeked Turaco   *T. leucotis*
leukos (Gr) white   *ous* (Gr), genitive *ōtos*, the ear; it has white patches on the sides of the head. Inhabiting north-east Africa.

Livingstone's Turaco *T. livingstonii*
Named after the famous missionary and explorer Dr David Livingstone (1813-1873), he travelled extensively in Central Africa during the years 1840 to 1873. It inhabits the forests of Tanzania, Mozambique and Zimbabwe.

Prince Ruspoli's Turaco *T. ruspolii*
Named after Prince E. Ruspoli (1866-1893), an Italian zoologist who was in north-eastern Africa from 1891 to 1893. This turaco is confined to a small area of Ethiopia (formerly Abyssinia).

Black-billed Turaco *T. schuettii*
Named after O. Schutt (1843-c.1888), a railway engineer who worked in various parts of the world, and in Angola from 1877 to 1879. Inhabiting Zaire and southern Uganda.

Lady Ross's Turaco *Musophaga rossae*
Musophaga, see notes under Family, above. Named after Lady J. C. Anne Ross (c. 1810-1857), wife of Rear Admiral Sir James C. Ross (1800-1862) the Arctic and Antarctic explorer and navigator. Inhabiting high mountain forests in Zaire, Africa.

Great Blue Turaco *Corythaeola cristata*
korus (Gr) a helmet *koruthaiolos* (Gr) moving the helmet quickly, waving *cristata* (L) crested; a reference to the large black crest; the plumage is mostly blue. Inhabiting equatorial Africa from the west coast to western Kenya.

Eastern Grey Plantain-eater *Crinifer zonurus*
crinis (L) hair *fero* (L) I bear *zonē* (Gr) a girdle, a band *oura* (Gr) the tail; references to the prominent crest and a white band on the tail. Inhabiting East Africa particularly Uganda and surrounding areas.

White-bellied Go-away Bird *Corythaixoides leucogaster*
koruthaix (Gr) helmet shaking *-oides* (New L) from *eidos* (Gr) shape, form; can mean resemblance; a reference to the crest being like the top of a military-type helmet *leukos* (Gr) white *gastēr* (Gr) the belly; with a pale grey back, head, and chest, it has a pure white belly. It has earned the name Go-away Bird because of its loud nasal cry 'g'way gwaay'; inhabiting Ethiopia and southern Sudan and ranging south to northern Tanzania.

Family   CUCULIDAE   127 species
cuculus (L) the cuckoo.

Red-winged Indian Cuckoo   *Clamator coromandus*
clamator (L) a shouter, a noisy speaker: *coromandus* refers to the Coromandel Coast of south-eastern India. Inhabiting India and ranging east through Burma, and south to the Thailand-Vietnam area and Malaysia.

Common Cuckoo   *Cuculus canorus*
canorus (L) melodius, sweet-sounding. Ranging from western Europe to Asia and migrating south in winter.

Pallid Cuckoo   *C. pallidus*
pallidus (L) pale; it is a light brownish-grey with pale underparts. It inhabits New Guinea, Australia and Tasmania.

Long-tailed Cuckoo   *Cercococcyx mechowi*
kerkos (Gr) the tail   *kokkux* (Gr) the cuckoo: a reference to the long tail: named after Maj A. von Mechow (born c. 1840), an Austrian zoologist who was in Angola from 1878 to 1881. Inhabiting Africa south of the Sahara.

Emerald Cuckoo   *Chrysococcyx cupreus*
khrusos (Gr) gold   *kokkux* (Gr) the cuckoo   *cuprum* (L) copper *cupreus* (L) coppery. Inhabiting Africa south of the Sahara.

Golden Bronze Cuckoo   *C. lucidus*   (formerly *Chalcites*)
lucidus (L) shining. Inhabiting New Guinea, some small Pacific islands, Australia, Tasmania and New Zealand.

Black-billed Cuckoo   *Coccyzus erythropthalmus*
kokkux (Gr) the cuckoo   *kokkuzō* (Gr) I cry cuckoo   *eruthros* (Gr) red *ophthalmos* (Gr) the eye; it has a black beak and a red ring round the eye; it breeds in North America and migrates to South America in winter.

Pearly-breasted Cuckoo   *C. euleri*
Named after Carlos Euler (1834–1901), a Swiss naturalist; he was in Brazil in 1853 and later became Vice Consul in Rio de Janeiro from 1867 to 1901. Inhabiting South America.

Koel   *Eudynamis scolopacea*
eudunamos (Gr) powerful   *skolopax* (Gr) the woodcock   *-aceus* (L) suffix forming an adjective from a noun, 'woodcock-like'. Inhabiting

India, Burma and the Thailand-Vietnam area, Malaysia, New Guinea, and some Pacific islands. *Koel* is a Hindustani name for the cuckoo.

Great Lizard Cuckoo *Saurothera merlini*
sauros (Gr) a lizard *thēraō* (Gr) I hunt; lizards form part of the diet: named after Comtesse M. de la Mercedes de Jaruco Merlin (1788–1852), a Cuban naturalist who married General Count Merlin in 1811. Inhabiting Central America, and the Greater Antilles.

Leschenault's Cuckoo *Phaenicophaeus leschenaultii*
phaeinos (Gr) shining, especially of burnished metal *phaios* (Gr) dusky, brown; a reference to the shining bronze effect of the bird's plumage: named after J. B. Leschenault de la Tour (1773–1826), a French naturalist who was in India from 1816 to 1822. Inhabiting India.

Smooth-billed Ani *Crotophaga ani*
krotōn (Gr) a louse, a tick *phagein* (Gr) to eat *-phagos* as a suffix means '-eating'; tick-eating: *ani* is a Spanish-American name for this cuckoo in Mexico and Brazil; insects of various kinds form the main diet. Known as smooth-billed as opposed to the groove-billed (below) this ani inhabits the southern part of North America, Mexico, Central America, Cuba and the West Indies, ranging south to Argentina.

Groove-billed Ani *C. sulcirostris*
sulcus (L) a furrow *rostrum* (L) the beak. Inhabiting the southern part of North America and ranging south through Central America to Argentina.

Greater Roadrunner *Geococcyx californianus*
gē (Gr) the earth *kokkux* (Gr) the cuckoo *-anus* (L) suffix meaning belonging to, 'of California': not confined to California it inhabits the southern part of North America and ranges south to Central America. Can run, assisted by the wings, at speeds up to about 20 mph, and is reluctant to fly.

Lesser Roadrunner *G. velox*
velox (L) rapid, swift; see notes above re *G. californianus*. Inhabiting Mexico and Central America.

Crested Coua *Coua cristata*
Coua is a native Malagasy name and is probably an imitation of the call *crista* (L) the crest of an animal *-atus* (L) suffix to form an

adjective from a noun   *cristatus* (L) crested. Found only on the island of Madagascar.

Red-capped Coua   *C. ruficeps*
rufus (L) red   *ceps* (New L) derived from *caput* (L) the head. All couas, ten species, are confined to the island of Madagascar.

Pheasant Coucal   *Centropus phasianinus*
kentron (Gr) a spur   *pous* (Gr) the foot or claw; a reference to the murderous-looking long hind claw   *phasianos* (Gr) the Phasian bird or pheasant (see p. 114 under *Phasianus colchicus*)   *phasianus* (L) a pheasant   *-inus* (L) suffix forming an adjective from a noun; 'pheasant-like'; a reference to the long tail, unusually long even for a cuckoo. Inhabiting New Guinea and Australia.

Common Coucal   *C. sinensis*
Sinai (Gr) the Chinese   *-ensis* (L) suffix meaning belonging to; 'of the Chinese'. Inhabiting south-east Asia, including part of China and ranging south to India, the Burma-Thailand-Vietnam area, and Malaysia.

Steere's Coucal   *C. steerei*
Named after Prof J. B. Steere (1842–1940), an American naturalist and author who was in the Philippines area from 1874 to 1875, and 1887 and 1888. Inhabiting the Philippine Islands, Malaysia and Indonesia.

White-browed Coucal   *C. superciliosus*
supercilium (L) the eyebrows   *-osus* (L) suffix meaning prone to, full of; it has distinct white eyebrows. Inhabiting the southern part of Arabia and most of Africa. The name coucal is French and is thought to derive from *coucou*, the French for a cuckoo, with the addition of the 'al' from *alouette*, a lark.

33 **Owls** STRIGIFORMES

Little Owl
Athene noctua

In the order Strigiformes there are two families, but only small differences separate the 10 Barn Owls in the family Tytonidae from the 123 species in the main family Strigidae. Nearly all have the typical flat round face, but in Tytonidae the face is more heart-shaped; one exception is the Hawk Owl *Surnia ulula* in which the face is more hawk-like. Some zoologists think that the flat round face has a part to play in the acute sound reception.

They are all carnivorous, with a varied diet including mammals, fish, crustaceans, lizards, insects, and birds and their eggs. In most species the plumage is unusually soft, resulting in almost noiseless flight, and this combined with an acute sense of hearing and keen eyesight makes them very efficient predators; they are mostly nocturnal.

Order **STRIGIFORMES**

strix (L), genitive *strigis*, a screech owl.

Family **TYTONIDAE** 10 species
tutō (Gr) a night-owl *tuton* (New L), genitive *tutonis*.

Barn Owl *Tyto alba*
albus (L) white; it is not entirely white, the upper parts are an orange

to buff colour; the face, front and belly are white. It has an almost world-wide distribution including Australia and Tasmania but not New Zealand or the Antarctic; it is not found in northern parts such as Alaska, northern Canada, northern Scandinavia, and northern Asia.

Sooty Owl   *T. tenebricosa*
tenebrae (L) darkness, especially of night   *tenebricosus* (L) full of darkness, shrouded in darkness; a reference to its nocturnal habits. Inhabiting New Guinea and Australia.

Tanzanian Bay Owl   *Phodilus prigoginei*
phaos = phōs (Gr) light, especially daylight   *deilos* (Gr) cowardly, craven, 'afraid of light'; a reference to its nocturnal habits. Named after Dr A. Prigogine (born 1913), a Belgian Doctor of Science and author. The name 'bay' refers to the chestnut colour of this owl which inhabits eastern Zaire.

Family STRIGIDAE   123 species
strix (L), genitive *strigis*, a screech owl.

White-throated Screech-Owl   *Otus albogularis*
otus (L) a horned owl, an owl with ear-tufts   *albus* (L) white   *gula* (L) the throat   *-aris* (L) a suffix meaning pertaining to: owls are well known for their fearful screeching and hooting noises. Inhabiting South America.

Flores Scops-Owl   *O. alfredi*
Named after Alfred H. Everett (1848–1898), a zoologist who was in Sarawak from 1869 to 1890; the owl takes its name from the island of Flores, Indonesia, and the name 'scops-owl' from *skōps* (Gr) a kind of owl, probably the little horned owl; the owls in the genus *Otus*, of which there are about 36, have ear-tufts of feathers that look like horns. Inhabiting parts of Indonesia.

Crested Owl   *Lophostrix cristata*
lophos (Gr) the crest of an animal   *strix* (Gr) an owl   *cristata* (L) crested. Inhabiting Central and South America.

Great Eagle-Owl   *Bubo bubo*
bubo (L) a horned owl; a rare Latin word probably incorrectly used; it really means to cry like a bittern; this owl has a deep hooting cry.

One of the largest owls, and known to take a roe deer fawn, it inhabits Eurasia, from Scandinavia and Spain in the west to the Pacific in the east, northern Africa, and India.

Great Horned Owl *B. virginianus*
Although named from Virginia it ranges from Alaska and northern Canada south to Tierra del Fuego in South America.

Pel's Fishing Owl *Scotopelia peli*
skotos (Gr) darkness *peleia* (Gr) a wood-pigeon: named after H. S. Pel (fl 1854) a Dutch naturalist; he was Governor of the Dutch Gold Coast from 1840 to 1850. This rare owl hunts for fish at night, catching them with its claws; it inhabits the rivers and lakes of western Africa.

Rufous Fishing Owl *S. ussheri*
Named after H. T. Ussher (1836–1880), an English naturalist who was in the Gold Coast from 1866 and Governor from 1879 to 1880. Inhabiting the forested waterways of western Africa.

Snowy Owl *Nyctea scandiaca*
Nyctea (New L) from *nukteros* (Gr) by night; a misleading name as this owl is known to be a daytime hunter of lemmings and other small mammals: named from northern Scandinavia but not confined to that area, it also inhabits northern Siberia, Spitzbergen, Iceland, northern Greenland, and the Hudson Bay area. It is almost pure white or creamy, sometimes with brown spots.

Hawk Owl *Surnia ulula*
Surnia is a coined name for this owl given in 1806 by Prof A. M. C. Duméril (1774–1860), the French zoologist *ulula* (L) a screech owl whose cry was an ill omen; it lacks the flat round face of the typical owls and resembles a hawk more than an owl. Inhabiting the forests of northern North America and Eurasia and not usually migrating south in winter.

Collared Pygmy-Owl *Glaucidium brodiei*
glaukos (Gr) glancing, gleaming, originally without notion of colour, later pale blue, grey *glaux* (Gr) an owl, so called from its glaring eyes; *glaukidion*, dim., suggesting a small owl: named after Sir B. C. Brodie, Bart. FRS (1783–1862), President of the Royal College of Surgeons in 1844. Inhabiting central parts of Eurasia, ranging south to India, the Burma-Vietnam-Thailand area, and Malaysia.

Papuan Hawk-Owl   *Uroglaux dimorpha*
oura (Gr) the tail   *glaukos* (Gr) pale blue, grey; it has a silvery-grey
tail   *dis-* or *di-* (Gr) two   *morphē* (Gr) form, shape; can mean sort,
kind; a reference to the two kinds of marking, horizontal bars on the
back and vertical streaks on the front. A rare nocturnal hunter inhabiting the forests of New Guinea, and having the appearance of a hawk
rather than an owl.

Solomon Islands Hawk-Owl   *Ninox jacquinoti*
Ninox (New L) a coined name for this genus of 16 hawk-like owls:
named after Vice-Admiral C. H. Jacquinot (1796–1879). Inhabiting
the Solomon Islands in the Pacific; they lie to the east of New Guinea.

Little Owl   *Athene noctua*
The name refers to Pallas Athene, the Goddess of Wisdom; 'As wise
as an owl'; *nox* (L), genitive *noctis*, night   *noctua* (L) a night-owl; in
spite of this name it is less nocturnal than many of the owls and may
often be seen perched on posts or telegraph poles, and hunting in daylight. Inhabiting Europe including Denmark and the southern part
of England where it was introduced about 1880; it ranges east to
central Russia and Arabia and south to Italy, Spain, and northern
Africa.

African Wood Owl   *Ciccaba woodfordii*
kikkabau (Gr) a cry in imitation of the screech-owl, hence *kikkabē* (Gr)
the screech-owl: named in honour of Col E. J. A. Woodford (c. 1761–
c. 1825) Adj 5th Foot Guards, 1888.

Tawny Owl   *Strix aluco*
strix (L) a screech-owl   *alucus = ulucus* (L) a screech-owl; a very rare
word and probably only used once by Vergil. Widespread in Europe
including the British Isles, but not Ireland; parts of Asia as far as the
eastern Himalayas and the northern part of Africa.

African Marsh Owl   *Asio capensis*
asio (L) a horned owl; it has small rather inconspicuous ear-tufts that
resemble horns   *-ensis* (L) suffix meaning belonging to; 'of The
Cape'; it is not confined to South Africa and has even been seen on
rare occasions in southern Spain, it is fairly widespread in Africa but
not in the east. Marsh Owl refers to the habitat, which is marshy
grassland and waterways.

Long-eared Owl   *A. otus*
ous (Gr), genitive *ōtus*, the ear; it has bigger ear-tufts than *A. capensis* (above). A wide distribution, including North and Central America, Eurasia from Ireland to Japan, part of China, and the northern part of Africa.

Fearful Owl   *Nesasio solomonensis*
nēsos (Gr) an island   *asio* (L) a horned owl   *-ensis* (L) suffix meaning belonging to; 'of the Solomon Islands'. The name fearful is probably taken from its screeching cry in the night; owls have long been known as birds of ill omen.

Buff-fronted Owl   *Aegolius harrisi*
aegolios = *aigōlios* (Gr) a night-bird of prey: named after Edward Harris (1799–1863); he was with Audubon on his Missouri River trip in 1843. A number of birds named by Audobon bear Harris's name, which he gave to commemorate their friendship. This buff-coloured owl lives in South America.

34 Frogmouths, Potoos, Nightjars and their kin CAPRIMULGIFORMES

Tawny Frogmouth
Podargus strigoides

The order Caprimulgiformes contains some unusual and rare birds such as the Oilbird, the Potoos, and the Owlet Nightjars. The place in classification of the unique Oilbird, with affinities to the owls, is still under discussion among ornithologists and is given a family on its own; the fat of the young birds can be made into a fine oil suitable for cooking and lighting, and which keeps well without turning rancid.

As suggested by the name nightjar all the birds in this order are nocturnal, though some in the family Caprimulgidae are also crepuscular (appearing at twilight). With the exception of the Oilbird, which eats fruit, they are all insect eaters, although the Frogmouth's diet is more varied and may include fruit, small mammals, and even reptiles.

Order CAPRIMULGIFORMES

caprimulgus (L) a milker of goats; can be used to mean a countryman. There was an ancient idea that the nightjars could suck milk from domestic goats, hence an alternative name for the birds is goatsucker; it has never been authenticated and today the idea is discounted, though it would be almost impossible to prove that it never happens

forma (L) form, shape; can mean sort, kind, so not necessarily referring to the actual shape.

Family STEATORNITHIDAE 1 species
stear (Gr) genitive *steatos*, fat, tallow *ornis* (Gr) genitive *ornithos*, a bird.

Oilbird *Steatornis caripensis*
In ancient times the natives penetrated the caves where the oilbirds live and slaughtered the chicks by the thousand to obtain the oil (see introductory notes, above) *-ensis* (L) belonging to; the name refers to the Caripe Caves; Caripe is a small town in the hills of north-eastern Venezuela, and thousands of the birds live in these caves. Oilbirds also inhabit Trinidad, Colombia, Ecuador and Peru. Also known by the Spanish native name *Guacharo*.

Family PODARGIDAE 12 species
podargos (Gr) swift-footed; however *argos* (Gr) can mean lazy, slow, so *podargos* could mean slow-footed; except when flying they are sluggish birds and seldom move in daylight.

Papuan Frogmouth *Podargus papuensis*
-ensis (L) suffix meaning belonging to; 'of Papua', New Guinea; it is also found in Australia.

Tawny Frogmouth *P. strigoides*
strix (Gr), genitive *strigos*, an owl *-oides* (New L) derived from *eidos* (Gr) form, shape, resemblance; they are like owls in size, plumage, and colouring. In habiting Australia and Tasmania.

Hodgson's Frogmouth *Batrachostomus hodgsoni*
batrakhos (Gr) a frog *stoma* (Gr) the mouth: named after B. H. Hodgson, FRS (1800–1894), a naturalist who was in Nepal from 1833 to 1843. This frogmouth inhabits India and the Burma to Vietnam area.

Javan Frogmouth *B. javensis*
-ensis (L) suffix meaning belonging to; in addition to Java it is found in the Burma to Thailand area and other parts of Malaysia.

Family NYCTIBIIDAE 5 species
nux (Gr), genitive *nuktos*, night *bios* (Gr) living, a means of living *nuktibios* = *nuktobios* (Gr) living at night, seeking one's food by night.

Great Potoo *Nyctibius grandis*
grandis (L) large, great. Ranging from Panama to Peru and Brazil.

Common Potoo *N. griseus*
griseus (New L) grey. The name potoo is derived from an imitation of
the birds call. Ranging from Mexico to Argentina, and Trinidad and
some islands in the West Indies.

Family AEGOTHELIDAE 7 species
aegos (Gr) a goat *thēlē* (Gr) a nipple; apparently a Greek form of
Caprimulgidae, suggesting that the birds use a goat's nipple to obtain
milk (see p. 302).

Mountain Owlet-Nightjar *Aegotheles albertisi*
Named after C. Luigi M. d'Albertis (1841–1901), a zoologist and
botanist who was in New Guinea from 1871 to 1877. Anatomically
these birds seem to be a mixture between the owls, the frogmouths,
and the nightjars, hence the name 'owlet nightjar', and sometimes the
name 'owlet frogmouth'; this one inhabits New Guinea.

Owlet Frogmouth *A. cristatus*
cristatus (L) crested; the crest consists of small plumes on the head,
and hardly noticeable. Inhabiting Papua New Guinea, Australia and
Tasmania. Frogmouth is a reference to the very wide gape.

New Caledonian Owlet-Nightjar *A. savesi*
Named after Théo Savés (1855–1918), a French zoologist; he was in
New Caledonia from 1873 to 1885. Inhabiting New Caledonia and
other islands in the Pacific.

Family CAPRIMULGIDAE 72 species
capra (L) a she-goat *mulgeo* (L) I milk *caprimulgus* (L) a milker of
goats; can be used to mean a countryman (see notes under Order
Caprimulgiformes, p. 302).

Archbold's Nightjar *Eurostopodus archboldi*
eurōstos (Gr) stout, strong *pous* (Gr), genitive *podos*, the foot: a reference to the feet which are bigger than usual for this family, which
have rather small feet. Named after R. Archbold (born 1907), an
American zoologist who was in the Mammals Dept of the American
Museum of Natural History in 1928, and in New Guinea in 1933 and
1934. Inhabiting New Guinea.

Spotted Nightjar   *E. guttatus*
gutta (L) a drop, a spot   *-atus* (L) suffix meaning provided with, i.e.
spotted. Inhabiting Australia, and Aru Island and New Ireland.

White-throated Nightjar   *E. mystacalis*
mustax (Gr), genitive *mustakos*, the upper lip, moustache   *-alis* (L)
suffix meaning pertaining to; it has bristles surrounding the mouth,
considered to be an aid to catching insects. Inhabiting Australia and
some of the Pacific islands.

Poor-will   *Phalaenoptilus nuttallii*
phallaina or *phalaina* (Gr) a sea monster, a whale; strangely, can mean
a butterfly or a moth, and used in this sense by Aristotle   *ptilon* (Gr)
feathers, or a wing; the flight of the poor-will is moth-like and waver-
ing as it chases flying insects: Thomas Nuttall (1786–1859) was an
English ornithologist, botanist and author who lived in the USA
from 1808 until his death in 1859; several other birds were named
after him. The poor-will ranges from south-west Canada through
North Dakota and central parts of the USA, west to the Pacific and
south to Mexico. The English name is an imitation of its cry.

Egyptian Nightjar   *Caprimulgus aegyptius*
caprimulgus (L) a milker of goats, can be used to mean a countryman
(see notes under Order Caprimulgiformes, p. 302). It is a rare visitor to
Egypt and the Mediterranean area in summer, from Africa, and
southern Asia as far east as the Burma area.

Common Nightjar   *C. europaeus*
It is widespread in Europe including the British Isles and southern
Scandinavia, and ranging east to Eurasia as far as India and the
Burma area, and south to Africa; it migrates in winter to the southern
areas, flying at night.

Long-tailed Nightjar   *C. macrurus*
makros (Gr) large, long   *oura* (Gr) the tail. Inhabiting India, the
Burma-Thailand-Vietnam area, Malaysia, New Guinea and Australia.

Rufous Nightjar   *C. rufus*
rufus (L) red, ruddy. Inhabiting Central and South America.

Whip-poor-will   *C. vociferus*
vocifero (L) I cry aloud, I vociferate; the whip-poor-will is well-known

for its continuous calling during the night, and the English name is an imitation of this call. It inhabits southern Canada and the USA and migrates south in winter to Mexico and parts of Central America.

Standard-winged Nightjar   *Macrodipteryx longipennis*
makros (Gr) large, long   *dis = di-* (Gr) two   *pterux* (Gr) a wing   *longus* (L) long   *penna* (L) a feather, a wing; these two names do not refer to the main wings but to the two much lengthened primaries of the male. During courtship flight these are trailed behind the main wing like two streamers, pennants or flags; i.e. the 'standard' in the English name; in order better to display these spectacular pennants the birds often fly in daylight which is most unusual for nightjars; after the mating period they are moulted and disappear. Inhabiting the southern part of Africa and migrating north in winter to the north of the equator.

Pennant-winged Nightjar   *M. vexillarius*
vexillum (L) a standard, a flag   *vexillarius* (L) a standard-bearer; see notes above with reference to the remarkable lengthened primary feathers. Inhabiting Africa, similar to *M. longipennis*.

Ladder-tailed Nightjar   *Hydropsalis climacocerca*
hudōr (Gr) water   *psalis* (Gr) a pair of shears or a single-bladed shearer like a razor; it opens and shuts the long rectrices when over water; these are the 'rudder' feathers of the tail   *klimax* (Gr), genitive *klimakos*, a ladder   *kerkos* (Gr) a tail; 'ladder-tailed'; the bars on the tail resemble the rungs of a ladder. Inhabiting South America.

35 Swifts and Hummingbirds
APODIFORMES

Long-tailed Sylph
Aglaiocercus kingi

The order Apodiformes is divided into three families, though there are only three recorded species of crested swifts, 71 species of swifts, and over 300 species of hummingbirds, their nearest relatives. They seem a strange assortment to be grouped together, but they are all related for various anatomical reasons; superficially the swifts are like the swallows but they are not closely related.

It has long been a popular belief that once a swift is on the ground it cannot take off again, on account of its very small and weak legs, but this is not true, although they rarely land on the ground and then probably by accident. The very small and inconspicuous legs have given rise to the scientific name *Apus*, meaning 'without feet'; when resting it would normally cling to the side of a tree, a cliff, or possibly a wall; it cannot perch in the normal way. They spend most of their life in the air, in which element they are completely at home and highly efficient, their manoeuvrability only exceeded by the humming birds which can fly forwards or backwards or remain completely stationary, like a helicopter. This is necessary for birds that obtain all their nourishment from the nectar in flowers or the insects they find there; to assist in the process the end of the tongue has developed many different shapes, including in one case a tubular tip.

Order APODIFORMES

a- (Gr) prefix meaning not, without   *pous* (Gr), genitive *podos*, a foot; 'without feet' (see notes above)   *forma* (L) form, shape; can mean sort, kind.

Family HEMIPROCNIDAE   3 species
hemi- (Gr) prefix meaning half; according to the Greek legend Proknē was transformed by the gods into a swallow and her sister Philomela into a nightingale, though some writers have Proknē the nightingale and Philomela the swallow. The prefix *hemi-* would not mean 'half a swallow' but 'like a swallow'; in any case this is misleading, as mentioned above, because although the swifts resemble swallows superficially, they are not swallows and are not even closely related.

Crested Tree Swift   *Hemiprocne longipennis*
longus (L) long   *penna* (L) a feather, a wing; it has long pointed wings and a long tail. The crest, a tuft of feathers on the head at the base of the beak, is erectile. Widespread in the oriental region from India to Malaysia where there are trees; unlike the swiftlets and swifts with their small weak legs it can perch in the trees, its favourite position.

Whiskered Tree Swift   *H. mystacia*
mustax (Gr), genitive *mustakos*, the upper lip, a moustache; it has long white whiskers at the side of the head. Inhabiting the Papua New Guinea area.

Family APODIDAE   71 species
a- (Gr) prefix meaning not, without   *pous* (Gr), genitive *podus*, a foot; 'without feet'; the legs are small, weak and inconspicuous, and cannot be used for perching in the normal way.

White-bellied Swiftlet   *Collocalia esculenta*
kolla (Gr) glue   *kalia* (Gr) a dwelling, a bird's nest   *esculentus* (L)

good to eat; the nest is made entirely of saliva, which solidifies and hardens, and is used to make the well-known bird's nest soup. Widespread in the oriental region from India to Papua New Guinea, the Pacific Islands, and Australia.

Giant Swiftlet   *C. gigas*
gigas (Gr) a giant; swiftlets, as the name suggests, are small birds with an average length of about 10 cm (4 in); this one may measure 15 cm (6 in) or more. Inhabiting Malaysia.

Spot-fronted Swift   *Cypseloides cherriei*
kupselos (Gr) the sand-martin   *-oides* (New L) from *eidos* (Gr) apparent shape, form; can mean resemblance; the sand-martin, the swallows, and the swifts all look much alike with their forked tails and long pointed wings: named after G. K. Cherrie (1865-1948), an American ornithologist and author; he was Curator of the Costa Rica Nature Museum from 1894 to 1897. It inhabits Central and South America.

Black Swift   *C. niger*
niger (L) black. Inhabiting North, Central, and South America.

André's Spine-tailed Swift   *Chaetura andrei*
chaeta (New L) a bristle from *khaitē* (Gr) flowing hair, a mane   *oura* (Gr) the tail; a reference to the stiff spine-like projecting shafts of the tail feathers: named after Eugene André (1861-1922), a French naturalist who was in Venezuela from 1897 to 1900. Inhabiting Central and South America.

Vaux's Spine-tailed Swift   *C. vauxi*
William S. Vaux (1811-1882), was an American zoologist and a member of the Philadelphia Academy of Natural Sciences; this swift was named in his honour in 1839 by his friend J. K. Townsend, an American ornithologist and author. It lives in south-west Canada and north-west USA and migrates south in winter, probably as far as Central America and northern South America. The swifts in the genus *Chaetura*, of which there are about 19, are often known as Spinetails without adding the name 'swift'.

Dark-backed Swift   *Apus acuticaudus*
a- (Gr) prefix meaning not, without   *pous* (Gr) a foot (see introductory notes on page 307)   *acutus* (L) sharp, pointed   *cauda* (L) the tail; a reference to the pointed tail feathers. Inhabiting India.

House Swift   *A. affinis*
affinis (L) adjacent, a neighbour. Inhabiting Eurasia, Africa, the Oriental Region from India to Malaysia.

Common Swift   *A. apus*
Widespread and common in Europe including the British Isles, ranging east as far as China and migrating south in winter to Africa including Madagascar.

Pallid Swift   *A. pallidus*
pallidus (L) pale; nearly all swifts have a dark or even black plumage; this species is pale brown with white on the throat. Inhabiting Mediterranean coasts, North Africa, and Asia Minor.

White-throated Swift   *Aeronautes saxatilis*
aēr (Gr), genitive *aeros*, the air   *nautēs* (Gr) a sailor; swifts spend a great deal of their lives in the air, and are famous for their skill and speed; this one is considered to be the fastest North American swift *saxatilis* (L) dwelling among rocks; a reference to its habitat, rocky places and cliffs, where it builds a nest composed of saliva, plant debris, and feathers, in a crevice on a rock face or in a cave. Inhabiting North and Central America.

Pygmy Swift   *Tachornis furcata*
takhus (Gr) fast, swift   *ornis* (Gr) a bird   *furca* (L) a fork   *-atus* (L) suffix meaning provided with; a reference to the forked tail. One of the smallest swifts, about 9 cm ($3\frac{1}{2}$ in) in length. Inhabiting north-western South America.

Palm Swift   *Cypsiurus parvus*
kupselos (Gr) a bird, the sand-martin   *oura* (Gr) the tail; 'sand-martin-like tail'; the sand-martins, swallows and swifts all have rather similar forked tails, but the swift, in the family Apodidae, is not closely related to them; they belong to the family Hirundinidae   *parvus* (L) small. Inhabiting Africa, and the Oriental Region from India to Malaysia.

Family   TROCHILIDAE   315 species
trokhilos (Gr) a small bird of the wagtail or sandpiper kind; has also been used to mean the wren; this can be misleading as the hummingbird is not closely related to any of these birds.

Pale-bellied Hermit   *Phaethornis anthophilus*
phaethōn (Gr) beaming, radiant   *ornis* (Gr) a bird; an allusion to the

brilliant plumage   *anthos* (Gr) a flower   *philos* (Gr) dear, pleasing;
hummingbirds spend much of their time hovering close to flowers to
obtain the nectar with their specially adapted beaks. The name hermit
is used for the hummingbirds in the genus *Phaethornis* on account of
their habit of frequenting dark forest recesses. Inhabiting northern
South America.

Broad-tipped Hermit   *P. gounellei*
Named after E. Gounelle (1850–1914) a French zoologist from Paris
who visited Brazil in 1887. Inhabiting South America.

Long-tailed Hermit   *P. superciliosus*
supercilium (L) the eye-brow   *-osus* (L) suffix meaning full of; it has
very noticeable white stripes above the eye. The proper meaning of
the Latin word *superciliosus* is haughty, disdainful. Inhabiting southern
Mexico, Central America, the north-west part of South America
including the Amazon Basin in Brazil.

White-tipped Sicklebill   *Eutoxeres aquila*
eu- (Gr) prefix meaning well, nicely   *toxērēs* (Gr) furnished with the
bow (the bow as in archery); a reference to the remarkable sharply
curved sickle-like bill   *aquila* (L) an eagle; an allusion to the eagle's
curved bill: white-tipped refers to the tips of the tail feathers. From
Costa Rica in Central America south to Colombia, Ecuador, and
north-eastern Peru in South America.

Lazuline Sabrewing   *Campylopterus falcatus*
kampulos (Gr) bent, curved   *pteron* (Gr) feathers or wings; a reference
to the thickened bent shafts of the primary feathers which resemble
sabres   *falx* (L), genitive *falcis*, a sickle   *falcatus* (L) sickle-shaped.
Inhabiting northern Venezuela, Colombia, and eastern Ecuador.
Lazulite is a blue mineral; the name refers to the iridescent dark violet
blue of the throat and breast plumage.

Glittering-bellied Emerald   *Chlorostilbon aureoventris*
khloros (Gr) green   *stilbōn* (Gr) the planet Mercury; in this case taken
to mean the shining or glittering one   *aurum* (L) gold   *aureus* (L)
golden   *venter* (L), genitive *ventris*, the belly. Inhabiting South
America.

Fork-tailed Emerald   *C. canivetii*
Named after E. Canivet De Carentan (fl 1876) a French amateur
collector. Inhabiting Mexico to Trinidad.

White-bellied Hummingbird   *Amazilia chionogaster*
amazilia (New L) probably derived from a Spanish-American word
and has some connection with the River Amazon; the name was given
to this genus of 30 hummingbirds by Lesson, the French zoologist.
khiōn (Gr) snow   *gastēr* (Gr) the belly. Inhabiting South America.

Honduras Emerald   *A. luciae*
Named in honour of Lucy Brewer, the daughter of Dr T. M. Brewer
(1814-1880) an American ornithologist and author. Inhabiting the
Honduras area in Central America.

Amethyst-throated Hummingbird   *Lampornis amethystinus*
lampē (Gr) a torch   *ornis* (Gr) a bird: amethyst, a bluish-violet quartz
-inus (L) suffix meaning like; 'amethyst-coloured'. Inhabiting Central
America.

Sword-billed Hummingbird   *Ensifera ensifera*
ensifera (L) sword-bearing; an allusion to the remarkably long bill,
longer than the head and body together, for obtaining nectar from
flowers with a long tube. Inhabiting Venezuela, Colombia, Ecuador
and Peru.

Chestnut-breasted Coronet   *Boissonneaua matthewsii*
Named after A. Boissonneau (fl 1839) a French ornithologist and
author, and Andrew Mathews (fl 1825-1841) a botanist and gardener
at Chiswick in 1825; he was in Peru and Chile from 1830 to 1841
where he died. This chestnut-breasted hummingbird inhabits Ecuador
and Peru.

Black-tailed Trainbearer   *Lesbia victoriae*
lesbias (L) a precious brilliant coloured stone found on the island of
Lesbos: named after Mde Victoire Mulsant, dates not recorded but
the name was given in 1846. The name 'trainbearer' refers to the
trailing central pair of tail feathers, about 15 cm (6 in) long. Inhabiting the Ecuador area in South America.

Long-tailed Sylph   *Aglaiocercus kingi*
aglaia (Gr) beauty, adornment   *kerkos* (Gr) the tail; a reference to
the long tail, about 13 cm (5 in), the upper parts being an iridescent
purple: named after Rear Admiral P. P. King, FRS (1791-1856), on
a government survey in command of HMS *Adventure* in South America
in 1825 to 1830. Inhabiting Venezuela, Colombia, Ecuador, eastern
Peru and Bolivia.

Racket-tailed Hummingbird   *Loddigesia mirabilis*
Named after G. Loddiges (1784-1846), a taxidermist who collected
about 1,000 hummingbirds in the Americas   *mirabilis* (L) wonderful;
the tail has only four feathers as against the usual ten in hummingbirds, the outer feathers being elongated; they take the shape of sickles
furnished with a purple flag at the tip and inverted so that the flags
are carried out to the sides. A rare species and only known from a
small valley 2,133 m (7,000 ft) up in the Andes in a wild part of Peru.

Amethyst Woodstar   *Calliphlox amethystina*
kallos (Gr) a beauty   *phlox* (Gr) a flame, reddish: amethyst, a bluish-
violet quartz   *-inus* (L) suffix meaning like; 'amethyst-coloured'.
Inhabiting South America.

Bee Hummingbird   *Mellisuga helenae*
mel (L) honey   *sugo* (L) I suck: named after Princess Hélène
d'Orleans (1814-1858) of Mecklenburg, East Germany; she married
the Duke of Orleans in 1837. This tiny bird is reputed to be the
smallest bird in the world, the male being only about 5 cm (2 in) long
including the bill and tail; the female is slightly larger. It is confined
to Cuba and the Isle of Pines, in the Greater Antilles.

Broad-tailed Hummingbird   *Selasphorus platycercus*
selas (Gr) light, brightness   *phoreō* (Gr) I bear, I carry; 'light-
bearing'; a reference to the bright iridescent plumage   *platus* (Gr)
broad, flat   *kerkos* (Gr) the tail. Inhabiting western North America
and ranging from Alaska south to Mexico.

Rufous Hummingbird   *S. rufus*
rufus (L) red, ruddy. Inhabiting western North America and Mexico.

36 **Mousebirds or Colies** COLIIFORMES

Red-faced Mousebird
Colius indicus

The order Coliiformes consists of only six birds, all in the family Coliidae. They have been classified as a separate order because zoologists have so far failed to find a satisfactory interrelation with any other group. Some of their anatomical features point to the parrots, and their behaviour in using the beak to help with climbing; on the other hand there are anatomical features that bear no relationship to parrots.

Order **COLIIFORMES**

colius (New L) derived from *kolios* (Gr) the green woodpecker, but there seems to be no good reason for the name.

Family **COLIIDAE** 6 species

Red-faced Mousebird *Colius indicus*
Indicus (L) of India; the name is misleading as it does not live in India;

the six mousebirds are found only in Africa; this one inhabits the
Congo River area, southern Tanzania, and ranging south. The name
mousebird has been given because they creep along branches of trees
with the body horizontal and close to the branch; in fact like a mouse.
'Red-faced' is an allusion to the bright red band surrounding the eye
and extending to the top of the bill.

White-headed Mousebird   *C. leucocephalus*
leukos (Gr) white   *kephalē* (Gr) the head. Inhabiting Africa south of
the Sahara.

Blue-naped Mousebird   *C. macrourus*
makros (Gr) long   *oura* (Gr) the tail; the mousebirds have long tails
and this species has a bright blue patch on the nape. Ranging from
Senegal in the west to Somalia in the east and south to Tanzania.

37 **Trogons** TROGONIFORMES

Collared Trogon
Trogon collaris

The order Trogoniformes consists of only one family and all the members are much alike in body form. Like the birds in the order Coliiformes they are classified as a separate group as they appear to have no near relatives. The Resplendent Quetzal *Pharomachrus mocino* is the national bird of Guatemala; a coin named the quetzal is the unit of currency and the bird is pictured on the coin, on the stamps, and on the flag. It is a colourful spectacular bird with a long train.

Order **TROGONIFORMES**

trōgō (Gr) I gnaw   *trōgōn* (Gr) gnawing   cf *trōglē* (Gr) a hole made by gnawing; trogons nest in old tree trunks in holes which they enlarge by gnawing; they also excavate holes in termite's nests. The trogon's beak is small and not particularly strong, so they have to choose dead and partly rotting trees.

Family **TROGONIDAE**   36 species

Resplendent Quetzal   *Pharomachrus mocino*
pharos (Gr) a mantle, a cloak   *makros* (Gr) long; a reference to the

long trailing tail: named after Dr J. M. Mocino (1757-1819), a Mexican naturalist. Quetzal is derived from *quetzalli*, an Aztec word for the bird's long tail feathers; it lives in the mountain forests of Central America, ranging from southern Mexico to Costa Rica and Panama. (See introductory notes, above.)

Eared Trogon   *Euptilotis neoxenus*
eu- (Gr) prefix meaning well, nicely   *ptilon* (Gr) feathers   *ous* (Gr), genitive *ōtis*, the ear; 'well-feathered ears'; it has tufts of feathers that look like ears   *neos* (Gr) new, recent   *xenos* (Gr) a stranger; probably so named because it was a newly discovered species in an unusual locality; more northern than other trogons. Inhabiting Central America.

Collared or Red-bellied Trogon   *Trogon collaris*
collum (L) the neck   *-aris* (L) suffix meaning belonging to; it has a white bar across the upper part of the breast, and a red belly. Inhabiting Central America and ranging south to northern Bolivia, Brazil, the western part of Colombia and Ecuador, and Trinidad and Tobago.

Mountain Trogon   *T. mexicanus*
-anus (L) suffix meaning belonging to; 'of Mexico'. Inhabiting the mountains of Mexico and Central America.

White-tailed Trogon   *T. viridis*
viridis (L) green; the upper parts are bronze green and the outer tail feathers are black with white tips. Inhabiting Panama and ranging south to north-west Bolivia and Brazil, the western part of Colombia and Ecuador, and Trinidad.

Narina's Trogon   *Apaloderma narina*
hapalos (Gr) soft, tender   *derma* (Gr) the skin; taxidermists have found this bird to be a problem as the delicate skin easily breaks and the feathers are liable to fall out; all trogons are alike in this respect: named after a Hottentot girl, Narina, who lived in the Knysna District of Cape Province about the years 1765 to 1782; little is known about her but she evidently died young. This trogon is widespread in Africa south of the Sahara.

Scarlet-rumped Trogon   *Harpactes duvauceli*
harpaktēs (Gr) a robber; it robs the nests of wasps and termites and then uses the nest as its own: named after A. Duvaucel (1796-1824)

a French naturalist who was at one time collecting in Sumatra. This trogon has a scarlet belly; it lives in the Burma to Vietnam area, and Malaysia including Sumatra and Borneo.

Red-headed Trogon   *H. erythrocephalus*
eruthros (Gr) red   *kephalē* (Gr) the head. Inhabiting India and ranging through Burma to Vietnam and Malaysia.

38 Kingfishers, Bee-eaters, Rollers, Hornbills and their kin CORACIIFORMES

Hoopoe
Upupa epops

The order Coraciiformes includes the kingfishers, todies, motmots, bee-eaters, rollers, hoopoes and hornbills. They all have hooked or slightly curved beaks except the kingfishers whose beaks are straight, large and pointed, and all are adapted for digging nesting holes in the banks of rivers or enlarging holes in trees. They are all carnivorous, and the diet may include small mammals and birds, fish, crustaceans, and insects; motmots, wood-hoopoes and hornbills also eat some fruit and berries.

There are strange legends about the kingfishers; the ancient Greeks thought they conceived at sea and built floating nests, and so at this time the gods favoured them and kept the sea calm. The Greek word for kingfisher, *alkuon*, is derived from *hals* (Gr) of the sea, and *kuō* (Gr) I conceive, hence 'halcyon days', calm days, kingfisher days.

Order CORACIIFORMES

korax (Gr) a crow or raven; also anything hooked like a raven's beak (see notes above).

Family   ALCEDINIDAE   90 species
alcedo (L) a kingfisher.

Pied Kingfisher   *Ceryle rudis*
kērulos (Gr) a sea bird, the halcyon; it is not really a sea bird (see notes above)   *rudis* (L) a foil; a reference to the long sharp beak. Inhabiting southern Europe, Africa south of the Sahara, southern Asia including the Oriental region. Known as the Pied Kingfisher for its entirely black and white plumage.

Amazon Kingfisher   *Chloroceryle amazona*
khloros (Gr) green   *ceryle*, see above: inhabiting Central America as well as the Amazon basin in South America.

Green Kingfisher   *C. americana*
Inhabiting Texas and ranging to Argentina.

Common Kingfisher   *Alcedo atthis*
alcedo (L) the kingfisher   *Atthis* (L) Athenian. Widespread in Europe including the British Isles, southern Asia, Africa, the Oriental region including India, the Burma-Cambodia area, Malaysia, New Guinea and the Pacific Islands.

Three-toed Kingfisher   *Ceÿx erithacus*
keux (Gr) a seagull; another reference to the kingfisher as though it is a sea bird, which it is not   *erithakos* (Gr) the name of a solitary bird which can be taught to speak; this translation is given in Liddell and Scott's Greek lexicon; it is true that the kingfisher is a solitary bird, or sometimes in pairs, but there is no known case of one being taught to speak. All kingfishers in the genus *Ceyx* are three-toed; this species inhabits the Oriental region ranging from India to Malaysia.

Kookaburra or Laughing Jackass   *Dacelo gigas*
(or *D. novaeguineae*)
Dacelo is simply an anagram of *alcedo* (L) a kingfisher   *gigas* (L) a giant; this is a large bird and may measure up to 46 cm (18 in) in length. Sometimes known as the Forest Kingfisher, it does not need to live near water and fish is probably not included in its normal diet of small reptiles, crabs, frogs and insects; originally found only in south-eastern Australia it has been introduced to western areas and Tasmania and New Zealand. *Kookaburra* is a native Australian name and people living in Australia are familiar with its terrible cackling laughter, said to be rather like a donkey braying.

Blue-winged Kookaburra *D. leachii*
Named after Dr W. E. Leach (1790-1836), zoologist at the British
Museum (Natural History). Inhabiting Australia and New Guinea.

Shovel-billed Kingfisher *Clytoceyx rex*
klutos (Gr) famous, glorious *keux* (Gr) a seagull; another reference
to the kingfisher as though it is a sea bird, which it is not *rex* (L) a
king; 'the king of seagulls'; it is certainly one of the largest kingfishers
and like the kookaburra may measure up to 46 cm (18 in) in length.
The large shovel-shaped bill is used for digging up earth-worms;
apparently its food consists largely of earth-worms, which is unusual
for a kingfisher. It is found only in the mountain forests of New Guinea.

White-backed Kingfisher *Halcyon albonotata*
alkuōn (Gr) the kingfisher; see introductory notes about derivation on
p. 319 *albus* (L) white *nōtos* (Gr) the back *-atus* (L) suffix meaning
provided with; 'having a white back'. Inhabiting New Britain.

Chestnut-bellied Kingfisher *H. farquhari*
Named after Admiral Sir A. M. Farquhar (1855-1937); the name was
given in 1899 after his voyage to the New Hebrides in 1894, where this
kingfisher lives. The New Hebrides is a group of islands lying to the
east of the Coral Sea, eastern Australia.

Grey-headed Kingfisher *H. leucocephala*
leukos (Gr) grey or white *kephalē* (Gr) the head. Inhabiting most of
Africa south of the Sahara and southern Arabia.

Forest Kingfisher *H. macleayii*
Named after Alexander Macleay (1767-1848), the Scottish zoologist
who was Secretary of the Linnaean Society of London from 1798 to
1825; he then went to Australia where he was Colonial Secretary of
New South Wales. This Forest Kingfisher does not need to live near
water; it inhabits New Guinea and Australia.

Brown-backed Paradise Kingfisher *Tanysiptera danae*
tanusipteros = *tanupteros* (Gr) long-winged *pteron* (Gr) wings, can
mean feathers; this refers to the long tail feathers which form a train:
the bird is named after Danae of Greek mythology; in the legend she
was the mother of Perseus; it inhabits New Guinea.

Family TODIDAE 5 species
todus (L) name of a small bird, a tody.

Narrow-billed Tody   *Todus angustirostris*
angustus (L) narrow   *rostrum* (L) the beak. Inhabiting the eastern part
of the Dominican Republic on the Island of Hispaniola, in the West
Indies.

Puerto Rican Tody   *T. mexicanus*
Although named from Mexico it is found only on the Island of Puerto
Rico, West Indies.

Jamaican Tody   *T. todus*
Found only on the Island of Jamaica.

Family   MOMOTIDAE   8 species
momot (New L) from American-Spanish *motmot*, supposed to be an
imitation of the birds call.

Tody Motmot   *Hylomanes momotula*
hulē (Gr) a wood, forest   *mania* (Gr) madness; as a suffix it is used to
mean 'mad on'; 'mad about forests'   *-ulus* (L) diminutive suffix; it is
the smallest motmot: it inhabits the forests of Mexico and Central
America.

Broad-billed Motmot   *Elektron platyrhynchum*
ēlektron (Gr) amber; sometimes used to mean bright, gleaming; in this
sense 'brightly coloured motmot'   *platus* (Gr) wide, broad   *rhunkhos*
(Gr) snout, beak. Inhabiting Central and South America.

Turquoise-browed Motmot   *Eumomota superciliosa*
eu- (Gr) prefix meaning well, nicely; in this case probably means
'nicely coloured motmot'   *supercilium* (L) eyebrow   *superciliosus* (L)
in this case meant to mean 'much eyebrow' though the Latin word
really means haughty, supercilious. Inhabiting Mexico and ranging
south to Costa Rica, Central America.

Rufous Motmot   *Baryphthengus ruficapillus*
barus (Gr) deep   *phthongos* (Gr) the voice, a sound   *baruphthongos* (Gr)
deep-voiced; they utter a 'ghostly hoot'   *rufus* (L) reddish   *capillus*
(L) hair. Inhabiting Central America from Nicaragua ranging south
to Columbia and on the east of the Andes to Paraguay and Argentina.

Blue-crowned Motmot   *Momotus momota*
Momotus, see above under Family: the crown and throat are light blue.
Inhabiting Central and northern South America on both sides of the
Andes; also Trinidad and Tobago.

Family   MEROPIDAE   23 species
merops (Gr) a bird, the bee-eater.

Celebes Bearded Bee-eater   *Meropogon forsteni*
merops (Gr) a bird, the bee-eater   *pōgōn* (Gr) the beard: named after
Dr E. A. Forsten (1811–1843), a naturalist from Java, in the Dutch
East Indes; he was Director of the Natural History Museum in
Batavia in 1839. It inhabits Celebes.

European or Common Bee-eater   *Merops apiaster*
apis (L) a bee   *-aster* (L) a suffix sometimes used to denote the diminutive; 'a small bee-eater'. It is widespread in Europe and sometimes
seen in the British Isles in summer; it also inhabits parts of Asia and
northern Africa, migrating south to Arabia and the whole of Africa.

Red-throated Bee-eater   *M. bulocki*
bulocki is a mis-spelling; the bird is named after William Bullock
(1775–c. 1840), an English traveller, naturalist, and mine owner; he
founded the Bullock Museum in 1799. Inhabiting western Africa to
Uganda.

Carmine Bee-eater   *M. nubicus*
-icus (L) suffix meaning belonging to; Nubia is a tract of country with
no precise limits in north-eastern Africa, to the south of Egypt; this
bee-eater inhabits a belt of country stretching from Senegal in the
west to Somalia in the east, and migrating south to Kenya, Zaire, and
Tanzania. A very beautiful bird, mostly carmine, with other parts
blue-green and pale blue.

Rainbow Bee-eater   *M. ornatus*
ornatus (L) dress, ornament; another beautiful and brightly-coloured
bird, mostly a brilliant light green with bands of black, orange and
blue on the throat and face and pale blue underparts. Inhabiting
Celebes, New Guinea, some Pacific islands, and Australia.

Family   LEPTOSOMATIDAE   1 species
Although in appearance and behaviour it is like the true rollers, it
differs anatomically in several respects which places it in a family and
genus on its own   *leptos* (Gr) slender, small   *sōma* (Gr) body, genitive
somatos.

Cuckoo Roller or Courol   *Leptosomus discolor*
discolor (L) of different colours; the male is a metallic green above with

grey to light grey underparts. The name roller refers to its undulating and spectacular aerial manoeuvres; it is not related to the cuckoos; *courol* is probably a native Malagasy name. It is found only in Madagascar and Comoro Islands.

Family   CORACIIDAE   16 species

korax (Gr), genitive *korakos,* a crow or raven; anything hooked like a raven's beak. (See introductory notes on p. 319).

Pitta-like Ground Roller   *Atelornis pittoides*

a- (Gr) not   *teleios* (Gr) complete   *ornis* (Gr) a bird; a reference to the unusually short primary feathers of the wing: *pitta* (Telugu or Tamil; these are Dravidian languages of India) a small bird   *-oides* (New L) from *eidos* (Gr) apparent shape, form, resemblance; the Pittidae form a complete family (see p. 353). Confined to the island of Madagascar.

Short-legged Ground Roller   *Brachypteracias leptosomus*

brakhus (Gr) short   *pteron* (Gr) feathers; can mean wings; it has unusually short wings, and short legs   *-acias* (New L) a suffix derived from Latin meaning tendency   *leptos* (Gr) slender, small   *sōma* (Gr) the body: like the Cuckoo Roller, above, the birds in this family are known as rollers because of the undulating and aerobatic flight behaviour, and the five species of ground rollers because they are usually seen on the ground and run for cover when alarmed. Like other ground rollers this species is confined to Madagascar.

Indian Roller   *Coracias benghalensis*

korakias (Gr) like a raven or crow; from *korax* (Gr), genitive *korakos,* a raven or crow; it is not really like a crow, being much smaller, about 30 cm (12 in) compared to the crow at about 46 cm (18 in); it is not black, but brightly coloured like all rollers   *-ensis* (L) suffix meaning belonging to; named from the Bengal area of India, it is also known in Iran and other parts of eastern Arabia, and in Burma and Thailand.

Blue-bellied Roller   *C. cyanogaster*

kuaneos (Gr) dark blue   *gaster* (Gr) the belly. Inhabiting Africa.

European Roller   *C. garrulus*

garrulus (L) talkative; the male and female have been seen to perch together calling to each other alternately and continuously in a sort of conversation. Inhabiting central Europe and ranging east to the Caspian area; most of the Mediterranean area including north-west Africa and ranging east to Turkestan; they nearly all migrate to East Africa in winter.

Broad-billed Roller or Dollar Bird   *Eurystomus orientalis*
eurus (Gr) wide   *stoma* (Gr) the mouth   *-alis* (L) suffix meaning relating to; 'of the Orient'; sometimes called Dollar Bird because of the white circular patches on the open wings. Ranging from northern India to Manchuria and Japan; the Solomon Islands; the Burma-Thailand-Vietnam area; Malaysia, New Guinea, and Australia.

Family   UPUPIDAE   1 species
upupa (L) a hoopoe; supposed to be an imitation of the bird's cry.

Hoopoe   *Upupa epops*
epops (Gr) a hoopoe. Closely related to the wood-hoopoes, although there are some slight anatomical differences; some authorities consider they belong in the same family. Inhabiting central and southern Europe, and has been seen in England and Iceland; Africa except central areas, Madagascar, central and southern Asia as far east as Japan and ranging south to India, Sri Lanka, and Malaysia.

Family   PHOENICULIDAE   8 species
phoinix (Gr), genitive *phoinikos*, purple-red, crimson   *-culus* (L) diminutive suffix; can mean somewhat, e.g. 'purplish'; the plumage of these birds is usually blackish with a metallic gloss of blue, green or purple.

Black Wood-Hoopoe   *Phoeniculus aterrimus*
ater (L) black   *aterrimus* (L) very black; it has a metallic gloss. Inhabiting tropical Africa.

White-headed Wood-Hoopoe   *P. bollei*
Named after Dr C. A. Bolle (1821-1909) a German zoologist; he was in the Canary Islands in 1852 and 1856. The head is really a very pale buff colour, and the body has a metallic gloss of blue, green and purple. Inhabiting the tropical forests of Africa.

Green Wood-Hoopoe   *P. purpureus*
purpureus (L) usually purple-coloured, but can mean shining, and

various colours such as reddish, violet, brownish, blackish; the general colour is blackish glossed with green and violet or blue, and a bright red bill. Inhabiting Africa south of the Sahara.

Family BUCEROTIDAE 44 species

boukerōs (Gr) horned like an ox; a reference to the large bill, and the horny casque which some species have on the base of the bill and the head.

Crowned Hornbill *Tockus alboterminatus*

Tockus is derived from a Portuguese imitation of the bird's cry 'toco-toco-tock' *albus* (L) white *terminus* (L) limit, boundary, end *-atus* (L) suffix meaning provided with, 'having a white end'; a reference to the white tips of the outer tail feathers. It does not have the horny casque on the bill but there is a small crest of feathers on the head. It inhabits the eastern part of Africa from Somalia south to Mozambique (formerly Portuguese) and ranging west to Zambia and south-east Congo.

Van der Decken's Hornbill *T. deckeni*

Named after Baron Van der K. C. Decken (1833–1865), a Dutch naturalist; he was exploring in East Africa from 1857 to 1865. Similar to *T. alboterminatus* but paler head and all white outer tail feathers. Inhabiting Ethiopia and southern Somalia and south through Kenya to northern Tanzania.

Malabar Grey Hornbill *T. griseus*

griseus (New L) grey. Named from the Malabar Coast which is the western coast from Goa to the southern tip; this hornbill inhabits that part of India.

White-throated Brown Hornbill *Ptilolaemus tickelli*

ptilon (Gr) feathers *laimos* (Gr) the throat. Named after Col S. R. Tickell (1811–1875) a British Army officer and naturalist who was stationed in India and Burma from 1834 to 1865. Inhabiting Assam and the Burma-Thailand area.

Rufous-necked Hornbill *Aceros nipalensis*

a- (Gr) prefix meaning not, or there is not *keras* (Gr) the horn of an animal *akerōs = akeratos* (Gr) hornless; this hornbill does not have the horny casque on the bill, though some in this genus do *-ensis* (L) suffix meaning belonging to; 'of Nepal'; Nepal is a small country on the north-east border of India bounded on the north by the Himalayas.

Blyth's Hornbill *A. plicatus*
plicatus (L) braided, folded; the male has 4 to 8 folds in the casque, the female rather less: named after E. Blyth (1810-1873), a zoologist who was Curator of the Natural History Museum in Calcutta from 1842 to 1864. Inhabiting the Burma-Thailand area, Malaysia, New Guinea, the Solomon Islands and other Pacific islands in that area.

Narcondam Hornbill *A. narcondami*
Some authorities include this species with *A. undulatus* (below). Narcondam is a small rocky island of the Andaman group of islands in the Bay of Bengal; it is too small to be named in ordinary maps, and this hornbill is considered to be confined to this island and not found anywhere else in the world. It is a typical fairly large hornbill with marked ridges across the casque like the Indian Pied Hornbill *Anthracoceros malabaricus* (below).

Wreathed Hornbill *A. undulatus*
undulatus (L) undulated, having waves; this, and the name 'wreathed', refers to the ridges across the casque. Inhabiting India, the Burma-Thailand area, and Malaysia.

Indian Pied Hornbill *Anthracoceros malabaricus*
anthrax (Gr), genitive *anthrakos*, coal or charcoal *keras* (Gr) the horn of an animal; this suggests a black horn, but the bill and the casque in this species is pale yellow; it is probably a reference to the plumage which is generally black with a green or dark blue sheen; there is some white on the wings and a white belly, giving the pied appearance when in flight: named from the Malabar Coast *-icus* (L) a suffix meaning belonging to; the Malabar Coast of India is the south-western coast from Goa to the southern tip. This hornbill also inhabits Burma and south-east China.

Trumpeter Hornbill *Bycanistes bucinator*
bukanē (Gr) a trumpet *-istēs* (Gr) suffix of agent, one who acts, hence *bukanistēs* (Gr) a trumpeter *bucina* (L) a curved trumpet *bucinator* (L) a trumpeter; most hornbills make loud and sometimes raucous noises. Inhabiting East Africa; although many different species live in Africa they are not known in Madagascar.

Rhinoceros Hornbill *Buceros rhinoceros*
boukerōs (Gr) big-horned, horned like an ox *rhis* (Gr), genitive *rhinos*, the nose *keras* (Gr) the horn of an animal *rhinokerōs* (Gr) a nose-horn; a reference to the prominent upcurved casque on the top

of the bill. A very large hornbill, about 127 cm (50 in), living in the Thailand area and Malaysia.

Abyssinian Ground Hornbill   *Bucorvus abyssinicus*
bou- (Gr) a prefix often used to mean something huge, as in *boukerōs* (above)   *corvus* (L) a raven or rook; 'a huge raven'; it is all black except for some white on the primary wing feathers, usually only noticeable during flight   *-icus* (L) a suffix meaning belonging to; from Ethiopia (formerly Abyssinia) it ranges south to Kenya and west across Africa to the Senegal area keeping north of Congo forests. Known as a Ground Hornbill because it rarely flies, and may walk a long distance searching for insects and small reptiles. There is only one other Ground Hornbill, *B. cafer*, which inhabits the southern half of Africa, and is very similar to *B. abyssinicus*.

39 Jacamars, Honeyguides, Toucans, Woodpeckers and their kin PICIFORMES

Spot-billed Toucanet
Selenidera maculirostris

This group of birds, of the order Piciformes, range in size from about 9 to 61 cm ($3\frac{1}{2}$ to 24 in), say from a sparrow to a large raven. They are not well known except for the toucans, of Central and South America, on account of the enormous and colourful bill, and the woodpeckers which have an almost world-wide distribution except for Madagascar and Australasia. They are all related for anatomical reasons, and all are insect eaters, though some, for instance the barbets and the woodpeckers, also eat seeds and fruit. The toucans also eat small reptiles, mammals, and nestlings, and the honeyguides eat honey, beexwax, grubs and the bees.

Not so well known are the 15 species of jacamars of tropical America, some of which are very colourful with a brilliant metallic sheen of bronze or green, their close relatives the 32 species of puffbirds, also from tropical America, so called because of the habit of 'puffing out' the loosely packed feathers round the head and neck, and the 78 species of barbets of tropical America, Africa, and the Oriental region. The puffbirds are not so brightly coloured as the jacamars, and some species could be described as dowdy, and their behaviour when not flying is lethargic and evidently appears foolish, which has earned them a local native name that translated means 'Silly John'.

Order   PICIFORMES

picus (L) a woodpecker; according to a legend Picus, son of Saturn, was changed by Circe into a woodpecker.

Family   GALBULIDAE   15 species
galbula (L) diminutive of *galbina* (L) a small bird, perhaps a species of thrush, the yellow thrush, or the yellow oriole   cf *galbinus* (L) greenish-yellow. The birds in this family and in the order Piciformes are not related to the thrushes; the name probably originates from the colour as several jacamars are a metallic green or metallic golden green.

White-eared Jacamar   *Galbalcyrhynchus leucotis*
galbula (L) a small bird   *alkuōn* (Gr) the kingfisher   *rhunkhos* (Gr) a beak, a bill; the heavy beak and short tail give it the appearance of a kingfisher   *leukos* (Gr) white   *ous* (Gr), genitive *ōtos*, the ear. Ranging from southern Mexico to southern Brazil but mainly in the Amazon region. The name jacamar is from a Tupi native name *jacama-ciri*.

Pale-headed Jacamar   *Brachygalba goeringi*
brakhus (Gr) short   *galba*, a shortened form of *galbina* (L) a small bird, probably meant to emphasize the smallness of the bird; it is about the size of a sparrow: named after Prof Anton Goering (1836–1905) a German naturalist and artist; he was in Venezuela from 1866 to 1872 collecting for the Zoological Society of London. It inhabits Panama and ranges south to Peru and Bolivia.

Three-toed Jacamar   *Jacamaralcyon tridactyla*
alkuōn (Gr) the kingfisher; jacamars are similar to kingfishers in colour, shape of beak, and behaviour   *tria* (Gr) three   *daktulos* (Gr) a finger or toe; it has two toes directed forward but only one backward; other jacamars have two directed forward and two backward. Inhabiting south-eastern Brazil.

Yellow-billed Jacamar   *Galbula albirostris*
galbula (L) diminutive of *galbina* (L) a small bird; see notes above

under Family   *albus* (L) white   *rostrum* (L) the snout, the beak; in spite of this name 'white-beaked' the beak is mainly yellow. Inhabiting the Guyanas, Venezuela, the eastern side of the Andes in Columbia, Ecuador, and Peru, and the Amazon area in Brazil.

Rufous-tailed Jacamar   *G. ruficauda*
rufus (L) red, ruddy   *cauda* (L) the tail; the underparts including the tail are red or bright chestnut. A very common jacamar in Central America and the northern part of South America including Colombia, Ecuador, Brazil and south to Paraguay and Argentina; also Trinidad and Tobago.

Great Jacamar   *Jacamerops aurea*
merops (L) a bee-eater 'a jacamar-bee-eater'; jacamars do eat bees and usually their diet consists entirely of insects   *aureus* (L) golden; it has a brilliant metallic golden green head and most of the body. This is the largest jacamar, measuring some 30 to 36 cm (12 to 14 in) in length; it inhabits Central and northern South America from Costa Rica to Brazil.

Family   BUCCONIDAE   32 species
bucco (L), genitive *bucconis*, a babbler, a fool, derived from *bucca* (L) the cheek, specially when puffed out in speaking, hence the English name 'puff-bird'; the plumage on the head and neck is easily puffed out giving an enlarged appearance.

Chestnut-capped Puffbird   *Bucco macrodactylus*
makros (Gr) large   *daktulos* (Gr) a finger or toe; it has relatively large feet for a small bird of about 13 cm (5 in) in length. The head is capped by chestnut-brown plumage. Inhabiting southern Venezuela and the Andes in Colombia, ranging south to eastern Peru, the western part of the Amazon basin in Brazil, and northern Bolivia.

Spot-backed Puffbird   *Nystalus maculatus*
nustaleos (Gr) drowsy; a reference to their lethargic behaviour when not flying (see introductory notes on p. 329)   *macula* (L) a mark, a spot   *-atus* (L) suffix meaning provided with. Inhabiting South America, mostly north-western area.

White-whiskered Puffbird or Softwing   *Malacoptila panamensis*
malakos (Gr) soft   *ptilon* (Gr) a feather; properly the soft feathers or down of a bird; the feathers are unusually soft, hence the English

name 'softwing'   -*ensis* (L) suffix meaning belonging to, particularly a locality, 'of Panama'. Inhabiting southern Mexico and Central America and ranging south to Colombia and western Ecuador. 'White-whiskered' refers to the tufts of white feathers on the cheeks.

Lance-billed Monklet   *Micromonacha lanceolata*

mikros (Gr) small   *monakhos* (Gr) a monk; 'a small monk', hence monklet; the name refers to the sombre dark brown colour; it is the smallest puffbird, about 13 cm (5 in) or less   *lanceolatus* (L) armed with a small lance or spike; a reference to the small pointed beak. Inhabiting Costa Rica in Central America and ranging south to the Amazon area in Brazil, South America.

Grey-cheeked or Red-capped Nunlet   *Nonnula ruficapilla*

nonna (Gr) a nun   -*ulus* (L) diminutive suffix, 'a nunlet'; it is one of the smallest puffbirds, about 15 cm (6 in) long, and the name refers to the rather drab brown colour   *rufus* (L) red   *capillus* (L) the hair of men, usually the hair of the head; can mean the hair of animals, in this case it means the plumage of the head, which is reddish brown; the cheeks are grey. Inhabiting Panama to northern Colombia, eastern Peru and western Brazil.

Yellow-billed Nunbird   *Monasa flavirostris*

Monasa is evidently a coined word from *monakhos* (Gr) a monk, and the name alludes to the mostly black colour; the beak is yellowish *flavus* (L) yellow   *rostrum* (L) the snout, the beak. Inhabiting Colombia and the eastern side of the Andes to eastern Peru and the western part of the Amazon basin in Brazil.

Swallow-Wing   *Chelidoptera tenebrosa*

khelidōn (Gr) a swallow   *pteron* (Gr) feathers, wings; usually in the plural; can mean a bird's wing, usually wings; it has long swallow-like wings, white underneath   *tenebrosus* (L) dark, gloomy; usually means darkness but in this case refers to the dark plumage which is almost black. Inhabiting the northern part of South America east of the Andes including Brazil and northern Bolivia.

Family   CAPITONIDAE   78 species

capito (L), genitive *capitonis*, one that has a large head, big-headed; these birds, the barbets, have a large head and rather thick heavy bill.

Black-spotted Barbet   *Capito niger*

niger (L) black; the male has black upperparts and tail; the female has the throat, breast and flanks spotted with dark brown. Inhabiting Venezuela south of the River Orinocco, Colombia, Peru and Bolivia east of the Andes, and the western part of the Amazon basin in Brazil.

Red-headed Barbet   *Eubucco bourcierii*
eu (Gr) a prefix meaning well, nicely   *bucco* (L) a babbler, derived from *bucca* (L) the cheek, specially when puffed out in speaking; a hybrid word of Greek and Latin, could means 'fat-cheeked', alluding to the size of the head: named after Jules Bourcier (1797–1873), a French naturalist who was the French Consul General in Ecuador from 1849 to 1850. The male has a scarlet head and throat, but the female has a yellow and black head with blue cheeks. The English name barbet, or 'little beard', is from *barbe* (Fr) and alludes to the hair-like bristles that surround the rather large head and throat and on the chin. Inhabiting Costa Rica and Panama and ranging south to Colombia, Ecuador, and north-eastern Peru.

Prongbilled Barbet   *Semnornis frantzii*
semnos (Gr) grand, fine   *ornis* (Gr) a bird: named after Dr A. von Frantzius (1821–1877), a German naturalist who was in Brazil from 1849 to 1853, and Costa Rica from 1853 to 1868. This barbet has two sharp prongs on the tip of the lower mandible and one on the tip of the upper mandible which fits between them when the bill is closed; a good weapon for securing wriggling insects; it inhabits Costa Rica and Panama in Central America.

Golden-whiskered Barbet   *Megalaima chrysopogon*
megas (Gr) big, wide   *laimos* (Gr) the throat; barbets are stout-bodied with big heads and thick necks   *khrusos* (Gr) gold   *pōgōn* (Gr) the beard: inhabiting Sumatra and Borneo.

Red-vented Barbet   *M. lagrandieri*
Named after Vice Admiral P. P. de Lagrandiere (1807–1876); in 1865 he was Governor of what was then known as French Cochin-China, now Vietnam. This barbet, which has red underparts, inhabits the Burma-Thailand-Vietnam area.

Naked-faced Barbet   *Gymnobucco calvus*
gumnos (Gr) naked   *bucco* (L), see under Red-headed Barbet above *calvus* (L) bald. Inhabiting Africa.

Golden-rumped Tinkerbird   *Pogoniulus bilineatus*
pōgōn (Gr) the beard   *-ulus* (L) diminutive suffix, 'the little bearded

one'; it is only about 10 cm (4 in) overall   *bilineatus* (L) having two
lines; it has a white line above and below the eye, the plumage is
glossy black above with a golden yellow rump. The Tinkerbirds, of
which there are 10 species in the genus *Pogoniulus*, get this name from
their monotonous call which is a continuous bell-like 'tink', sounding
like a metal hammer on an anvil and this can go on for hours. Inhabiting Uganda and Kenya and ranging south through East Africa to
Natal.

Pied Barbet   *Lybius leucomelas*   (formerly *Tricholaema*)
A mis-spelling of Libya, although this barbet does not live there;
named in 1783, in those days Libya had a very wide African con-
notation   *leukos* (Gr) white   *melas* (Gr) black; a black and white
barbet with a red forehead. Distribution is widespread in southern
Africa.

Black-breasted Barbet   *L. rolleti*
Named after Brun Rollet (1810–1855) a French elephant hunter in
the White Nile area in Sudan, and in Ethiopia, from 1831 to 1851.
Inhabiting Africa south of the Sahara.

D'Arnaud's Barbet   *Trachyphonus darnaudii*
trakhus (Gr) rough   *phōnē* (Gr) a sound, the voice of men or animals:
named after D'Arnaud (initials and dates not recorded) a French
explorer and elephant hunter in the Blue Nile area and Ethiopia from
1839 to 1843. Inhabiting Sudan and ranging across to the Horn of
Africa and south to Tanzania.

Levaillant's Barbet   *T. vaillantii*
Named after F. Levaillant (1753–1824), a French naturalist and explorer in South Africa from 1781 to 1784. This barbet is found in the
southern part of Africa from Angola across to Tanzania and ranging
south to eastern South Africa.

Family   INDICATORIDAE   14 species
indico (L) I show, I point out; indicator, one who points out; there are
many stories about these birds leading animals, and men, to bees'
nests, hence the English name 'honeyguide'. For many years naturalists dismissed them as 'tall stories' but it is now accepted that some
species, particularly the Greater Honeyguide *Indicator indicator*, do lead
people and animals to bees' nests by continually calling, a churring

sound, and then moving on until the man or animal follows; the ratel or honeybadger is known to follow in this way. The bird is not strong enough to break into the nest but the ratel, with its powerful claws, breaks it open and takes its fill, and the honeyguide then takes its share of the honey, the bees, their grubs, and even the wax; this latter is another puzzle for zoologists as wax is virtually indigestible by any animal except the honeyguides.

Wahlberg's Honeybird   *Prodotiscus regulus*

prodo (L) I disclose, I show; can mean I make known, I report   *-iscus* (L) diminutive suffix; 'a little reporter'   *rex* (L), genitive *regis*, a king diminutive *regulus* (L) a little king; named after J. A. Wahlberg (1810–1856) a Swedish traveller and naturalist. One of the smallest honeyguides, only about 10 cm (4 in) long, it inhabits Africa south of the Sahara.

Greater or Black-throated Honeyguide   *Indicator indicator*

indico (L) I show, I point out; see notes under Family, above. One of the largest honeyguides, about 20 cm (8 in) long, and best known for its habit of leading men and animals to bees' nests; the male has a black patch on the throat. Wide distribution in Africa but not seen in the West African rain forests, the Sahara, or the Kalahari Desert.

Lesser Honeyguide   *I. minor*

minor (L) lesser, smaller; it is about 14 cm ($5\frac{1}{2}$ in) long. This species is not known to be one of those that guide men and animals to bees' nests though it does eat bees' wax, grubs, and other insects. Distribution similar to *I. indicator*, above.

Indian or Orange-rumped Honeyguide   *I. xanthonotus*

xanthos (Gr) various shades of yellow including golden, orange, and even reddish   *nōtos* (Gr) the back; mostly yellow in colour with a large orange-yellow patch on the rump. Inhabiting India, the Himalayan forests, and the Burma-Thailand-Vietnam area.

Lyre-tailed Honeyguide   *Melichneutes robustus*

meli (Gr) honey   *ikhneutēs* (Gr) a tracker, a hunter   *robustus* (L) strong, robust. The central tail feathers are elongated and lyre-shaped; distribution is widespread in Africa and similar to *I. indicator*.

Family   RAMPHASTIDAE   38 species

rhamphos (Gr) a beak   *-astus* (Gr) augmentative suffix, 'a huge beak'.

Emerald Toucanet   *Aulacorhynchus prasinus*
aulax (Gr), genitive *aulakos*, a furrow   *rhunkhos* (Gr) the beak; it has
grooved mandibles   *prason* (Gr) a leek   *prasinos* (Gr) of a leek-green;
the plumage is almost entirely shades of green, some quite bright.
Inhabiting Mexico and ranging south through Central America to
northern Venezuela, Colombia, eastern Ecuador and eastern Peru,
in mountain forests from 900 to 3,000 m (3,000 to 10,000 ft).

Brown-mandibled Araçari   *Pteroglossus mariae*
pteron (Gr) feathers or wings   *glossa* (Gr) the tongue; the long tongue
is fringed, probably to assist in catching and dealing with insects:
named in honour of Max J. Beauharnais (1817–1852), a prince distinguished for his love and support of science, 3rd Prince of Eichstadt
and Duke of Leuchtenberg, and particularly in honour of his wife
The Duchess Marie Beauharnais (1819–1876), Duchess of Leuchtenberg. Inhabiting the upper Amazon basin, including the eastern parts
of Colombia, Ecuador, and Peru, and the western part of Brazil.
Araçari is a Portuguese name for the 11 toucans in this genus; it is
derived from the Tupi language.

Many-banded Araçari   *P. pluricinctus*
plus (L), genitive *pluris*, more   *pluri-* as a prefix usually means several,
but although the English name is 'many-banded' there are in fact
only two bands   *cinctus* (L) a girdle; there are two black bands across
the reddish-yellow underparts. Inhabiting Venezuela and nearby
north-western Brazil, eastern Colombia and ranging south to north-
eastern Peru.

Spot-billed Toucanet   *Selenidera maculirostris*
selēnē (Gr) the moon   *selēnion* (Gr) any small moon-shaped object
derē (Gr) the neck, throat; in allusion to the crescent-shaped golden-
yellow collar round the hind neck   *macula* (L) a mark, a spot
rostrum (L) the beak, snout. Inhabiting Brazil.

Black-billed Mountain Toucan   *Andigena nigrirostris*
andinus (New L) Andean, of the Andes   *genus* (L) a race, a kind
niger (L) black   *rostrum* (L) the beak, snout.

Plate-billed Mountain Toucan   *A. laminirostris*
andinus (New L) Andean, of the Andes   *genus* (L) a race, a kind
lamina (L) a plate or thin piece of metal or other material   *rostrum*
(L) the beak, the bill; it has plates on the sides of the bill. Inhabiting
the Andes Mountains in Ecuador and Peru.

Orange-billed Toucan   *Ramphastos aurantiirostris*
rhamphos (Gr) a beak or bill   *-astus* (Gr) augmentative suffix, 'a huge beak'   *aurantium* (New L) an orange, derived from *aurum* (L) gold *aureus* (L) golden   *rostrum* (L) the beak, bill. Inhabiting northern South America.

Chestnut-mandibled Toucan   *R. swainsonii*
Named after W. Swainson, FRS (1789-1855), a much travelled zoologist, artist and taxidermist; he was in South America from 1816 to 1818 and he also worked in Africa and New Zealand. Inhabiting the Panama area and particularly Barro Colorado Island; this island is in the Gatun Lake in the Panama Canal Zone.

Family   PICIDAE   206 species
picus (L) the woodpecker.

Red-breasted Wryneck   *Jynx ruficollis*
iugx (or *iunx*) (Gr) the wryneck   *rufus* (L) red, reddish   *collum* (L) the neck; it has a reddish-brown collar. Inhabiting Africa from Cameroun in the west across to Ethiopia in the east and ranging south to The Cape, but not known in South-West Africa. The wryneck is related to the woodpeckers in several aspects of its anatomy but it does not have the strong beak and is not able to excavate holes in trees; it has a habit of twisting its neck into peculiar contortions, particularly when disturbed, hence the name 'wryneck'.

Wryneck   *J. torquilla*
torqueo (L) I twist   *-illus* (L) diminutive suffix; 'little twister' (see above). Inhabiting Europe, including southern Scandinavia, Great Britain, where it is now seldom seen, Africa north of the equator, and ranging across Asia to Japan.

Olivaceous Piculet   *Picumnus olivaceus*
Picumnus and Pilumnus were two brother deities of the Romans; Picumnus was a personification of the woodpecker   *oliva* (L) an olive *-aceus* (L) suffix meaning similar to, like; olive coloured. Inhabiting Central and South America. A piculet is a small woodpecker, this one is tiny, no bigger than a blue tit; it has olive-green upperparts and upper breast.

White-bellied Piculet   *P. spilogaster*
spilos (Gr) a spot, a blemish   *gaster* (Gr) the belly; the belly is white

with black markings. Inhabiting Venezuela, Guyana, Surinam, and ranging south into Brazil.

Antillean Piculet   *Nesoctites micromegas*

nēsos (Gr) an island   *ktites* = *ktistes* (Gr) an inhabitant; Antilles is the name applied to the group of islands, including Cuba, that lie to the north of the Caribbean Sea   *mikros* (Gr) small   *megas* (Gr) big; 'a small big one'; the piculets are the small woodpeckers.

Ground Woodpecker   *Geocolaptes olivaceus*

gē (Gr) the earth   *kolaptō* (Gr) I cut, chisel   *kolaptēs* (Gr) a cutter, a chiseller   *oliva* (L) an olive   *-aceus* (L) suffix meaning like; in this case olive coloured. This woodpecker spends almost its entire life on the ground, only occasionally seen in trees, in which case it climbs there; the upperparts are olive brown. It is confined to south-east and southern South Africa.

Yellow-shafted Flicker   *Colaptes auratus*

Colaptes, see above   *auratus* (L) golden, adorned with gold; the upperparts are not golden but there are yellow shafts to the flight feathers and the underwings and undertail are bright yellow; this gives it the English name. Inhabiting Alaska, Canada, Newfoundland, and ranging south to Florida; it keeps to the eastern side of the Rocky Mountains.

Golden-green Woodpecker   *Piculus chrysochloros*

picus (L) the woodpecker   *-ulus* (L) diminutive suffix, 'a little woodpecker'; woodpeckers range in size from 9 to 61 cm ($3\frac{1}{2}$ to 24 in); this is one of the small ones, about 15 cm (6 in) long   *khrusos* (Gr) gold *khloros* (Gr) green. Inhabiting Central America.

Golden-tailed Woodpecker   *Campethera abingoni*

kampē (Gr) a caterpillar   *thēraō* (Gr) I hunt, chase, wild animals; can mean I take, catch, wild animals; it feeds on various insects and grubs including caterpillars and ants: named after M. Abingdon, the 5th Earl of Abingdon (1784-1854), an English naturalist; this is a misspelling of Abingdon, but as the name was established and universally recognised in 1836 it cannot now be changed although recently certain exceptions to this rule have been allowed in cases of mistakes in spelling. Inhabiting the Senegal area in West Africa and ranging across to Sudan, and from Central and East Africa and Somalia south to Natal.

Fine-spotted Woodpecker   *C. punctuligera*
punctus (L) a puncture   *punctatus* (L) spotted as with punctures
-ulus (L) diminutive suffix   *gero* (L) I bear, carry; 'bearing small
spots'; the upperparts are finely spotted with white, and the underparts pale with small dark spots. Inhabiting the Senegal area,
northern Zaire and southern Sudan.

Chestnut Woodpecker   *Celeus elegans*
keleos (Gr) the green woodpecker   *elegans* (L) neat, elegant; the
plumage is mainly a chestnut colour. Inhabiting Venezuela, Colombia
east of the Andes, ranging south to the Amazon basin and northern
Bolivia; also Trinidad.

Grey-headed Woodpecker   *Picus canus*
picus (L) the woodpecker; see notes under Order Piciformes on p. 196
canus (L) white, hoary; often used to mean the grey hair of the aged.
Inhabiting Eurasia from southern Scandinavia and France ranging
east to Japan and south-east to Sumatra and also the Himalayas.

Green Woodpecker   *P. viridis*
viridis (L) green. Inhabiting a similar area to *P. canus* above, but not
extending as far as Sumatra.

Lineated Woodpecker   *Dryocopus lineatus*
drus (Gr), genitive *druos*, a tree   *kopos* (Gr) striking, beating; 'tree
striking'   *lineatus* (L) lined. Inhabiting Central and South America.

Black and White Woodpecker   *Phloeoceastes melanoleucos*
phloios (Gr) peel or bark of trees   *keazō* (Gr) I cleave, I split wood
-istes (Gr) suffix meaning one who acts; 'a cleaver of bark'   *melas* (Gr),
genitive *melanos*, black   *leukos* (Gr) white; the colour is mostly black
with white patches and white lines down the neck, and underparts
dull white and black. Inhabiting Central America, Trinidad, South
America on the eastern side of the Andes ranging south to northern
Argentina.

Acorn Woodpecker   *Melanerpes formicivorus*   (formerly *Centurus*)
melas (Gr), genitive *melanos*, black   *herpēs* (Gr) a creeper, one that
moves slowly, 'a black creeper'; a reference to the way a woodpecker
creeps slowly up the trunk of a tree searching for insects and grubs;
this woodpecker has a glossy black back, wings and tail but it does not
follow that all species in this genus are black, though a number of
them are; the generic name is simply a label and they are grouped

together because of similarities in their anatomy and not because of the colour of the plumage   *formica* (L) an ant   *voro* (L) I devour; although ants form an important part of the diet, this woodpecker is well known as a lover of acorns; groups of five or six will live together and spend hours drilling thousands of holes in the trunk of a tree, and into each hole an acorn is inserted as a reserve food for times when insects are scarce. Inhabiting coastal areas on the western side of the USA from Oregon ranging south through Central America to Colombia.

Red-crowned Woodpecker   *M. rubricapillus*
rubor (L) redness   *capillus* (L) hair, usually the hair of the head, can mean the hair of animals; the crown and nape of the neck are crimson. Inhabiting Panama and Costa Rica and ranging south to Venezuela and Colombia; also Tobago.

Red-breasted Sapsucker   *Sphyrapicus ruber*
sphura (Gr) a hammer   *picus* (L) the woodpecker   *ruber* (L) red; it hammers on tree trunks and other objects, even metal, in a fast and then slow rhythm. The English name comes from its liking for sap; it drills many small holes usually in deciduous trees and drinks the sap, however its main food is insects and these are attracted by the sap oozing from the holes. Inhabiting the western part of Canada and the USA.

Arabian Woodpecker   *Picoides dorae*   (formerly *Dendrocopos*)
picus (L) the woodpecker   *-oides* (New L) from *eidos* (Gr) shape, form, resemblance: the specific name is in honour of Mrs H. Dora Philby (1888-1957), the wife of J. B. H. Philby, a naturalist and explorer who led the British political mission to central Arabia in 1917 and 1918 and later explored the southern province of Nejd in 1920. This woodpecker inhabits Arabia and Egypt.

Great Spotted Woodpecker   *P. major*
major = *maior* (L) greater; woodpeckers range in size from about 9 to 61 cm ($3\frac{1}{2}$ to 24 in); this is not one of the biggest, being about 23 cm (9 in), but the comparison is with the Middle and Lesser Spotted Woodpeckers; the plumage is predominately black with white spots and bars. Inhabiting Europe including Great Britain, but not Ireland, ranging south to north-west Africa, Asia and east to Japan.

Middle Spotted Woodpecker   *P. medius*
medius (L) middle. The range is similar to *P. major*, above, but not in

the British Isles and not ranging so far north or south in Asia.

Lesser Spotted Woodpecker   *P. minor*
minor (L) lesser. Inhabiting Europe including the southern part of Great Britain, north-west Africa, Turkey, and Asia ranging east to Japan. One of the smaller woodpeckers, about 15 cm (6 in) long.

Brown-backed Woodpecker   *P. stricklandi*
Named after H. E. Strickland, FRS (1811–1853) an English zoologist. Inhabiting Mexico.

Three-toed Woodpecker   *P. tridactylus*
treis (Gr) = *tri* (L) three   *daktulos* (Gr) a finger, or a toe; it has two toes pointing forwards and one pointing backwards, though most woodpeckers are zygodactyl, i.e. having two forwards and two backwards; there are other three-toed woodpeckers. Distribution is widespread and includes North America, Eurasia, ranging south to India and the Burma-Thailand-Vietnam area.

Cuban Green Woodpecker   *Xiphidiopicus percussus*
xiphos (Gr) a sword   *xiphidion* (Gr) a small sword   *picus* (L) the woodpecker   *percussus* (L) a knocking, striking; references to its sharp beak and habit of hammering on trees in search of insects. Inhabiting Cuba.

Lesser Bay Woodpecker   *Blythipicus rubiginosus*
Named after E. Blyth (1810-1873) a zoologist and Curator of the Calcutta Museum from 1842 to 1864   *picus* (L) the woodpecker   *ruber* (L) red   *rubigo* (L), genitive *rubiginis*, rust   *-osus* (L) suffix meaning full of   *rubiginosus*, rusty, rust-coloured. Inhabiting the Burma-Thailand-Vietnam area and Malaysia.

Crimson-backed Woodpecker   *Chrysocolaptes lucidus*
khrusos (Gr) gold   *kolapto* (Gr) I cut, chisel   *kolaptēs* (Gr) a cutter, a chiseller   *lucidus* (L) bright, shining; a reference to the bright crimson crest, crown and back. Inhabiting India including Ceylon, the Burma-Thailand-Vietnam area, and Malaysia.

Magellanic Woodpecker   *Campephilus magellanicus*
kampē (Gr) a caterpillar   *philos* (Gr) dear, pleasing   *-icus* (L) prefix meaning belonging to, 'of Magellan's Strait'. Inhabiting the southern tip of Chile, Patagonia, and the island Tierra del Fuego; Magellan's Strait separates this island from the mainland.

White-throated Seedeater
Sporophila albogularis

Order   PASSERIFORMES

passer (L) a sparrow, or any small bird   *forma* (L) form, shape; can
mean sort, kind, so not necessarily referring to the actual shape; the
birds in this very big order are by no means all 'sparrow-like' but they
do have certain anatomical features in common, e.g. regarding the
claws, there are three toes in front and one behind; they are sometimes referred to as 'perching birds'.

Family   EURYLAIMIDAE   14 species
eurus (Gr) wide   *laimos* (Gr) the throat; the birds in this family have
a broad flat head and a wide bill, hence the English name broadbill.

Rufous-sided Broadbill   *Smithornis rufolateralis*
Named after Sir Andrew Smith (1797–1872), an army surgeon who
was a pioneer in scientific research in southern Africa   *ornis* (Gr) a
bird   *rufus* (L) red, ruddy   *latus* (L), genitive *lateris*, side, flank
-alis (L) suffix meaning relating to. Inhabiting West and Central
Africa; it has prominent reddish-brown patches on the sides of the
chest.

Grauer's or African Green Broadbill   *Pseudocalyptomena
graueri*
pseudēs (Gr) false   *kaluptos* (Gr) covered, wrapped round; the bill is
covered at the base with a tuft of green feathers; this bird was first

discovered by a collector named Grauer in Tanzania (then known as Tanganyika) and later in 1909 was described by Lord Rothschild and given the name *Pseudocalyptomena graueri*, 'false' because at that time Green Broadbills of the genus *Calyptomena* were not considered to be African birds, and only known in the Malaysian area. The specific name refers to Rudolph Grauer (1871-1927), an Australian ornithologist who was collecting in Africa from 1904 to 1911. The bird is very rare, and seldom seen, and then only in two areas; one in eastern Zaire and the other in southern Uganda.

Dusky Broadbill   *Corydon sumatranus*
Corydon is taken from *Korudōn*, the name of a well-known Greek shepherd   *-anus* (L) suffix meaning belonging to; the name is taken from Sumatra but it also inhabits other parts of south-eastern Asia and Malaysia. The plumage is very dark, almost black, hence 'dusky'.

Black and Yellow Broadbill   *Eurylaimus ochromalus*
Eurylaimus, see notes under Family, above   *ōkhros* (Gr) pale yellow *malus* (L) derived from *melas* (Gr) black; it is a black bird with parts of the wings and back yellow. Inhabiting the Burma-Thailand-Vietnam area and Malaysia.

Long-tailed Broadbill   *Psarisomus dalhousiae*
psaros (Gr) speckled, like a starling   *soma* (Gr) the body: named after Countess C. Dalhousie (1786-1839), the wife of the Governor General of India in 1829. Inhabiting India and ranging east to Malaysia.

Lesser Green Broadbill   *Calyptomena viridis*
kaluptos (Gr) covered, wrapped round; the bill is covered at the base with a tuft of green feathers   *viridis* (L) green. This is the smallest of the four Green Broadbills, about 15 cm (6 in) long; the range is Burma, Malaysia, Indonesia including Borneo.

Family   DENDROCOLAPTIDAE   50 species
dendron (Gr) a tree   *kolaptēr* (Gr) a chisel   *kolaptēs* (Gr) a chiseller, of birds a pecker.

Plain Brown Woodcreeper   *Dendrocincla fuliginosa*
dendron (Gr) a tree   *kinklos* (Gr) a bird, a kind of wagtail   *fuligo* (L), genitive *fuliginis*, soot   *-osus* (L) suffix mean full of, prone to, 'very sooty'; a reference to the rather dull brown plumage. Inhabiting Central America and ranging south to Brazil.

Olivaceous Woodcreeper   *Sittasomus griseicapillus*
sittē (Gr) a kind of woodpecker   *soma* (Gr) the body   *griseus* (New L)
derived from *greis* (G) grey   *capillus* (L) the hair of men, usually the
hair of the head, can mean the hair of animals; the plumage is mainly
olive-coloured. Inhabiting Central America and ranging south to
southern Brazil.

Wedge-billed Woodcreeper   *Glyphorynchus spirurus*
gluphō (Gr) I carve, cut   *rhunkhos* (Gr) the bill   *speiroō* (Gr) I wrap
round   *oura* (Gr) the tail; probably a reference to the way the tail
spreads when used as a support for climbing vertical tree trunks. The
bill is short and the upper mandible flattened, hence 'wedge-billed';
woodcreepers do not actually cut into the tree like woodpeckers but
prize off the bark with their strong bills, searching for insects, grubs,
and spiders. Inhabiting Central America and South America on the
eastern side of the Andes, south as far as Bolivia and Brazil.

Cinnamon-throated Woodcreeper   *Dendrexetastes rufigula*
dendron (Gr) a tree   *exetastēs* (Gr) one who searches out, examines;
alluding to the bird's search for insects under the bark   *rufus* (L) red,
reddish   *gula* (L) the throat. Inhabiting South America, from
Colombia to Brazil.

Red-billed Woodcreeper   *Hylexetastes perrotii*
hulē (Gr) a wood, forest   *exetastēs*, see above: named after Jean Perrot
(1790-1858), a French taxidermist; he worked for the Paris Museum
in 1820. Inhabiting South America.

White-throated Woodcreeper   *Xiphocolaptes albicollis*
xiphos (Gr) a sword   *kolaptēr* (Gr) a chisel   *kolaptēs* (Gr) a chiseller
albus (L) white   *collum* (L) the neck; see notes above under *Glyph-*
orynchus spirurus. Inhabiting Venezuela, Brazil and Peru.

Barred Woodcreeper   *Dendrocolaptes certhia*
dendron (Gr) a tree   *colaptes*, see above   *kerthios* (Gr) a little bird, a
treecreeper. Inhabiting Central and South America.

Buff-throated Woodcreeper   *Xiphorhynchus guttatus*
xiphos (Gr) a sword   *rhunkhos* (Gr) the beak, bill; it has a powerful
beak, slightly curved, 25 to 38 mm (1 to 1½ in) long   *gutta* (L) a drop,
of a fluid   *guttatus* (L) spotted, speckled; it has very noticeable pale
spots and speckles on the buff plumage of the head and back of neck.
Inhabiting Central America from Guatemala to Panama, and South

America in Trinidad and Tobago, ranging south on the eastern side
of the Andes to Bolivia and the Amazon area.

Spix's Woodcreeper   *X. spixii*
Named after Dr J. B. von Spix (1781–1826), a German Professor of
Zoology at the Munich Museum; he was in Brazil from 1817 to 1820.
Inhabiting northern South America to the Amazon area.

Narrow-billed Woodcreeper   *Lepidocolaptes angustirostris*
lepidus (L) neat, graceful   *kolaptēr* (Gr) a chisel   *kolaptēs* (Gr) a
chiseller   *angustus* (L) narrow   *rostrum* (L) a beak, bill. Inhabiting
Brazil and Argentina.

Streak-headed Woodcreeper   *L. souleyetii*
Named after F. L. A. Souleyet (1811–1852), a French naval surgeon
and naturalist; he was in Peru in 1836. There are pale buff streaks on
the head and the back of the neck; it inhabits Central and South
America.

Black-billed Scythebill   *Campylorhamphus falcularius*
kampulos (Gr) bent, curved   *rhamphos* (Gr) a beak, bill, specially the
crooked beak of birds   *falx* (L) a sickle, genitive *falcis*   *-ulus* (L)
diminutive suffix   *-arius* (L) suffix meaning belonging to; by comparison with the beak of other woodcreepers the five scythebills in the
genus *Campylorhamphus* are quite remarkable; they are long, slender
and curved, a 7·5-cm (3-in) beak on a 25·5-cm (10-in) bird. Inhabiting Brazil, Paraguay, and Argentina.

Family   FURNARIIDAE   217 species
furnus (L) an oven   *furnarius* (L) a baker. The birds in the genus
Furnarius build a nest of mud that resembles an oven, and this has
given them the name Ovenbirds and the family as a whole the Latin
name Furnariidae, but it is a very big group with a wide variation of
nest-building and a number of different local and English names such
as Miner, Earthcreeper, Hornero, and Spinetail. They are small
brown forest birds living in Central and South America and ranging
from Mexico to southern Argentina.

Common Miner   *Geositta cunicularia*
ge (Gr) the ground, earth   *sittē* (Gr) a kind of woodpecker   *cuniculum*
(L) an underground passage, a cavity; spending all its time on the
ground, it burrows deep into a sandy bank for the nest, hence the
English name 'miner'. Inhabiting southern South America.

Rufous Hornero or Ovenbird *Furnarius rufus*
furnus (L) an oven *furnarius* (L) a baker *rufus* (L) red, ruddy. The
name Hornero is from the Spanish *hornero* (Sp) a baker. It inhabits
the southern part of South America.

Curve-billed Reedhaunter *Limnornis curvirostris*
limnē (Gr) a marsh, pond *ornis* (Gr) a bird *curvus* (L) curved
rostrum (L) the beak, bill; it searches among reeds for small crustaceans
and other small water animals. Inhabiting South America.

Grey-flanked Shaketail *Cinclodes oustaleti*
kinklos (Gr) a bird, perhaps the water ouzel *-odes* = *-oides* (New L)
from *eidos* (Gr) shape, form, resemblance; it is rather like the dipper,
or water ouzel, it lives near water, and has the habit of flicking its tail
up and down like the wagtail: named after Dr J. F. E. Oustalet (1844–
1905), a French Professor of Zoology; he was Curator of Mammals at
the Natural History Museum in Paris from 1900 to 1905. Inhabiting
southern South America.

Rush-loving Spinetail *Phleocryptes melanops*
phleōs (Gr) a water plant, a rush *kruptos* (Gr) hidden, secret; 'hidden
among rushes' *melas* (Gr), genitive *melanos*, black *ops* from *opsis*
(Gr) aspect, appearance. Known as a 'spinetail' because the tail
feathers end in bare spines. Inhabiting South America.

Pale-breasted Spinetail *Synallaxis albescens*
sunallaxis (Gr) exchange; a reference to the loud calls a pair make to
each other while foraging in the undergrowth, so establishing the pair
bond *albus* (L) white *-escens* (L) suffix meaning beginning to,
tendency; 'whitish'; it is chestnut coloured above with a pale buff
breast; 'spinetail', see above. Inhabiting Central and South America
from southern Mexico to southern Argentina.

Straight-billed Reedhaunter *Limnoctites rectirostris*
limnē (Gr) a marsh, pond *ktitēs* (Gr) an inhabitant *rectus* (L)
straight *rostrum* (L) the beak, bill. Inhabiting South America.

Pearled Treerunner *Margarornis squamiger*
margarōdēs (Gr) pearl-like *ornis* (Gr) a bird *squama* (L) a scale
gero (L) I bear, carry; the breast is covered with pearl-like marking
with black edges, giving the appearance of scales. Inhabiting
Venezuela and Colombia and ranging south to Peru and Bolivia.

Riker's Palmcreeper   *Berlepschia rikeri*
Named after H. Graf. von Berlepsch (1850-1915), a German zoologist who collected many thousands of specimens in South America,
and C. B. Riker (1863-1947), a zoologist who was in Brazil from 1884
to 1887. This bird, living in the Amazon area, spends almost its entire
life in the palm trees.

Red-fronted Thornbird   *Phacellodomus rufifrons*
phakelos (Gr) a bundle; sometimes written in Greek as *phakellos*, but
this is incorrect   *domos* (Gr) a house, structure; the nest is built with
bundles of sticks round the drooping limb of a tree, forming a column
sometimes 1·5 or 2 m (5 or 6 ft) high; several birds help to build the
structure which contains six or seven compartments, but probably
only one pair use it for breeding   *rufus* (L) red   *frons* (L) brow, forehead. Widespread in the northern half of South America on the
western side.

Spotted Barbtail   *Premnoplex brunnescens*
premnon (Gr) the trunk of a tree   *plexus* (L) a knitting, interweaving;
a reference to the form of nest-building, though the nests are usually
woven round an overhanging bough, or higher up in a fork; this bird
sometimes builds a globular structure   *brunneus* (New L) brown
-*escens* (L) suffix meaning beginning to, tendency, 'brownish'. Inhabiting southern Mexico to Bolivia.

Streak-breasted Treehunter   *Thripadectes rufobrunneus*
thrips (Gr) a woodworm   *dēktēs* (Gr) a biter   *rufus* (L) red   *brunneus*
(New L) brown. Inhabiting Central America.

Streak-capped Treehunter   *T. virgaticeps*
virga (L) a twig   *virgatus* (L) made of twigs; can mean striped   *ceps*
(New L) from *caput* (L) the head. Inhabiting western Venezuela,
Colombia, and Ecuador.

Plain Xenops   *Xenops minutus*
xenos (Gr) a stranger, foreigner   *ops* (Gr) the eye, face; a newly discovered bird and named in 1811; there are only four in this genus;
minutus (L) small, minute; at 13 cm (5 in) long it is one of the smallest
in this family which range from 13 to 28 cm (5 to 11 in). Inhabiting
Mexico to South America east of the Andes, including Bolivia,
Paraguay, northern Argentina, and on the west of the Andes to
Ecuador.

Tawny-throated Leafscraper   *Sclerurus mexicanus*
sklēros (Gr) hard, stiff   *oura* (Gr) the tail; the tail feathers have stiff
spines to help support the bird when climbing vertical tree trunks
-anus (L) suffix meaning belonging to; inhabiting Mexico and ranging
south through Central America to South America.

Streamcreeper   *Lochmias nematura*
lokhmē (Gr) a thicket, bush, specially one serving as a lair for an animal
lokhmaios (Gr) belonging to a bush; the name is probably misleading
as apparently the nest is usually a hole in a bank   *nēma* (Gr), genitive
nēmatos, thread   *oura* (Gr) the tail; a reference to the tail feathers
which end in bare spines. Sometimes known as the Sharptailed
Streamcreeper, it lives near mountain streams in eastern Panama,
ranging south through Central America and South America to
northern Argentina.

Family   FORMICARIIDAE   230 species
formica (L) an ant   *-arius* (L) suffix meaning belonging to; they probably do not eat ants (see below). The antbirds, which have been given
various names such as antshrike, antwren, and antthrush, form a large
group of about 230 species. The English names, and the Latin name
for the family, are misleading; it is not possible to say that they never
eat ants, but ants do not form part of their normal diet. However they
are sometimes seen following columns of army ants in order to feed on
the insects and small animals that are disturbed and are escaping from
the ants; this seems to be the reason for the name. Some species also
eat fruit and berries and some of the larger species are known to eat
very small snakes and lizards.

The birds range in size from about 7·5 to 18 cm (3 to 14 in), and
although related to the ovenbirds they are more colourful, especially
the males. Living in the dense forests of tropical Central and South
America they are not easily seen, and taken as a whole very little
accurate information about their habits is available.

Fasciated Antshrike   *Cymbilaimus lineatus*
kumbē (Gr) a cup, bowl; can mean a kind of bird, perhaps a tumbler
pigeon   *laimos* (Gr) the throat   *linea* (L) a line   *lineatus* (L) lined;
most of the plumage is marked with conspicuous pale or dark lines or
bands. Living on the eastern side of the Andes ranging south to Peru,
Bolivia, and the Amazon area. They have been given the English

name antshrike because the bill is hooked at the tip like that of the shrike, and used for killing and eating small animals.

Black-headed Antshrike   *Thamnophilus nigriceps*
thamnos (Gr) a bush, shrub   *philos* (Gr) loved, pleasing; some of the species in this genus come out of the dense forest and live in the bushes and bamboo thickets near human habitations   *niger* (L) black   *ceps* (New L) from *caput* (L) the head; for the name 'shrike' see above. Inhabiting Central and South America.

White-flanked Antwren   *Myrmotherula axillaris*
murmēx (Gr) the ant   *thēraō* (Gr) I hunt wild beasts; can mean I catch wild beasts   *-ulus* (L) diminutive suffix; 'catch small wild beasts'; as explained above in the notes under Family, the antwren does not eat ants   *axilla* (L) the armpit   *-aris* (L) suffix meaning belonging to; the male has a white patch of feathers under the wings which are exposed in courtship display or anger. It is a small wren-like bird only about 10 cm (4 in) long living in the forest undergrowth of Central and South America.

Klages's Antwren   *M. klagesi*
Named after S. M. Klages (fl 1900), a devoted and untiring naturalist and collector who sent many thousands of skins to the Carnegie Museum in Pittsburg; he was in Venezuela from 1898 to 1913 and Brazil in 1922. This antwren inhabits the forests in the northern part of South America.

Black-bellied Antwren   *Formicivora melanogaster*
formica (L) an ant   *voro* (L) I eat; the antwrens do not eat ants, as explained in the notes under Family opposite   *melas* (Gr), genitive *melanos*, black   *gastēr* (Gr) the belly. A small antbird resembling a wren, living in Brazil and Bolivia.

Striated Antbird   *Drymophila devillei*
drumos (Gr) a forest   *philos* (Gr) loved, pleasing: named after E. Deville (1824–1853), a French naturalist and taxidermist at the Natural History Museum in Paris; he was in South America from 1843 to 1847. Inhabiting South America.

Jet Antbird   *Cercomacra nigricans*
kerkos (Gr) the tail of an animal   *makros* (Gr) long   *nigricans* (L) swarthy, black; 'black as jet'; a black antbird with a long tail living in the forests of South America.

White-backed Fire-eye   *Pyriglena leuconota*
pyr (Gr) fire   *glēnē* (Gr) the eyeball   *leukos* (Gr) white   *nōtos* (Gr) the
back. Inhabiting Colombia and Argentina.

Goeldi's Antbird   *Myrmeciza goeldii*
murmēx (Gr) the ant   *murmēkizō* (Gr) I feel as if ants were creeping
about me; can mean I creep like ants: named after Dr E. A. Goeldi
(1859-1917), a Spanish-American zoologist who founded the Goeldi
Museum in Paraguay; he lived in Brazil from 1894 to 1905. This antbird inhabits the northern half of South America.

Black-headed Antthrush   *Formicarius nigricapillus*
formica (L) an ant   *-arius* (L) suffix meaning belonging to; they do not
eat ants, as explained in the notes under Family on p. 348   *niger* (L)
black   *capillus* (L) the hair of men, usually of the head; can mean the
hair of animals. Inhabiting Central America and ranging south
through Venezuela and Colombia to Bolivia and Brazil. The ant-
thrushes have rather big thrush-like heads.

Wing-banded Antbird   *Myrmornis torquata*
murmex = *murmos* (Gr) the ant   *ornis* (Gr) a bird; antbirds do not eat
ants (see above)   *torquatus* (L) wearing a twisted collar or necklace;
it has a collar of zig-zag white lines round the neck. Inhabiting Central
America and the northern part of South America.

Black-crowned Antpitta   *Pittasoma michleri*
pitta (Telugu or Tamil; these are Dravidian languages of India) a
small bird   *soma* (Gr) the body: named after Brig Gen N. Michler
(1827-1881) who did a survey in Panama during the years 1857 to
1861. The English name antpitta has been given because these birds
are similar to the pittas in the family Pittidae; this one lives in the
Panama area of Central America and ranges south to the northern
part of South America.

Rusty-breasted Antpitta   *Grallaricula ferrugineipectus*
grallae (L) stilts   *-arius* (L) suffix meaning belonging to   *-culus* (L)
diminutive suffix; 'of small stilts'; the antpittas and their close relatives, for example *Grallaria*, are mostly terrestrial and have longer legs
than the other antbirds, however this particular species is not often
seen on the ground   *ferrugo* (L) rust, of iron   *ferrugineus* (L) rust-
coloured   *pectus* (L) the breast. Inhabiting northern Venezuela,
Colombia, and northern Peru.

Great Antpitta   *Grallaria excelsa*
grallae (L) stilts   *-arius* (L) suffix meaning belonging to; it has long
legs (see above)   *excelsus* (L) high, elevated; a reference to its habitat
which is high up in the Andes in north-western South America. One
of the larger antbirds, it is about 25·5 cm (10 in) long.

Brown-banded Antpitta   *G. milleri*
Named after Leo E. Miller (fl 1910), he was in South America with
Dr F. M. Chapman, the well known ornithologist who wrote several
books about South Africa. They were there from 1910 to 1917; in 1918
Miller published his book *In the Wilds of South America*. This antpitta
lives in the northern part of South America.

Chestnut Crowned Antpitta   *G. ruficapilla*
rufus (L) red, ruddy   *capillus* (L) the hair of men, usually the hair of
the head; can mean the hair of animals. This is one of the large
ground-loving antbirds inhabiting the forests of the northern part of
South America.

Family   CONOPOPHAGIDAE   8 species
kōnōps (Gr), genitive *kōnōpos*, a gnat, mosquito   *phagein* (Gr) to eat.
The Gnat-eaters, as both their Latin and English names imply, are
insectivorous, their diet probably consisting entirely of small insects.
Related to, and in some respects resembling, the two previous families,
they are small ground-loving birds about 13 cm (5 in) long.

D'Orbigny's Gnat-eater   *Conopophaga ardesiaca*
Conopophaga, see above   *ardesco* (L) I gleam, glitter; possibly *-acus* (L)
relating to; this is a coined name and not to be taken too literally; the
plumage is brown above and grey below. The bird is named after
Prof A. D. d'Orbigny (1803-1857), a French zoologist who was Professor at the Museum of Natural History in Paris in 1854. He was in
South America from 1826 to 1833. It inhabits Bolivia and Peru.

Black-bellied Gnat-eater   *C. melanogaster*
melas (Gr), genitive *melanos*, black   *gaster* (Gr) the belly; the male has
a black cap on the head and black breast and underparts. Inhabiting
Bolivia and the Amazon area of Brazil.

Robert's Gnat-eater   *C. roberti*
Named after Alphonse Robert (fl 1900), a French naturalist and
collector; in 1901 he was in Brazil collecting for the Tring Zoological

Museum; this Gnat-eater was named after him in 1905. Inhabiting the northern part of South America.

Family RHINOCRYPTIDAE 29 species

rhis (Gr), genitive *rhinos*, the nose *kruptos* (Gr) hidden, covered; a reference to a peculiar anatomical feature of this family; they have a moveable flap to cover each nostril, thought to be a protection against dust or sand. They are ground-loving birds and similar to the Conopophagidae (above), ranging from 10 to 25·5 cm (4½ to 10 in) long. They have been given a variety of English names but the one most used is *Tapaculo*, a native name of Spanish origin, and often used by the native people with an amusing meaning which is 'Cover your bottom!'; this is a reference to their habit of strutting about with the tail cocked up and their rear end exposed. They spend almost their entire life on the ground, scratching about like domestic hens looking for insects and seeds; this has given them another English name, Gallito, derived from the Spanish and meaning a cockerel.

Black-throated Huet-huet *Pteroptochos tarnii*

pteron (Gr) a wing *ptōkhos* (Gr) one who crouches or cringes; 'a winged croucher'; these birds have been described as 'skulking about in the undergrowth'. They have secretive habits and are seldom seen though often heard: named after J. Tarn (1794–1877), a naturalist, and surgeon on H.M.S. *Adventure*, South America, from 1825 to 1830. It inhabits the forest undergrowth in south-western South America, most common in Chile and Patagonia; the name 'huet-huet' is an imitation of the call.

White-throated Tapaculo *Scelorchilus albicollis*

skelos (Gr) the leg *orkhilos* (Gr) a bird, probably a wren; perhaps 'a leggy wren'; with their cocked up tails they resemble wrens but have much longer and stouter legs *albus* (L) white *collum* (L) the neck. Inhabiting the western part of South America. For tapaculo see notes under Family, above.

Grey Gallito *Rhinocrypta lanceolata*

rhis (Gr), genitive *rhinos*, the nose *kruptos* (Gr) hidden, covered; see notes under Family, above *lancea* (L) a light spear or lance *lanceola*, a small lance *lanceolatus* (L) armed with a small lance; a reference to the pointed rather slender bill. Inhabiting Chile, Argentina, Paraguay, and Uruguay. For Gallito see notes under Family, above.

Sandy Gallito   *Teledromas fuscus*
tēle (Gr) far   *dromos* (Gr) running, escape; 'far-running'; when disturbed, like all members of the family it prefers running to flying; the
legs are long for a bird of this size, with a body of about 18 cm (7 in),
and it takes long strides; the flight muscles are weak and the leg
muscles strong   *fuscus* (L) dark, dark coloured; it is a sandy brown
colour. Most common in Argentina.

Andean Tapaculo   *Scytalopus magellanicus*
skutalē (Gr) a thick stick, a cudgel   *pous* (Gr) the foot; 'thick-footed';
a reference to the comparatively heavy legs and feet common in this
family, see above: named from the Strait of Magellan, which separates
the island of Tierra del Fuego from the mainland of Patagonia; however it is by no means confined to this area and ranges from northwest Venezuela through Colombia and along the line of the Andes to
Peru, Bolivia, Chile, Argentina to Cape Horn and the Falkland
Islands; in the Andes it is known to live as high as 4,000 m (13,000 ft).

Pale-throated Tapaculo   *S. panamensis*
One of the few that are found as far north as Central America, it
ranges from the mountains of Panama south to Colombia and Ecuador
-ensis (L) suffix meaning belonging to, usually applying to a locality.

Family   PITTIDAE   26 species
pitta is a name derived from a language spoken in Telingana, India,
which lies roughly north from Madras to the border of Orissa; it
means 'a small bird'. The fact that this little family are all in the same
genus, *Pitta*, tells us that their anatomy, habits, and diet, are all very
similar; minor differences are indicated by the specific name. They
are colourful birds, ranging in size from about 15 to 28 cm (6 to 11 in)
long, including a very short tail; the diet is varied, and includes
insects, worms, fruit and berries, and very small reptiles.

African Pitta   *Pitta angolensis*
Pitta, see above   *-ensis* (L) suffix meaning belonging to, usually applying to a locality; named from Angola, it is also found in Congo, Zaire,
Tanzania, and ranging south to Transvaal.

Blue-winged Pitta   *P. brachyura*
brakhus (Gr) short   *oura* (Gr) the tail; the tail is so short that it is
almost non-existent; the family as a whole have short tails. Widespread in south-eastern Asia and India, the Burma-Thailand-

Vietnam area, Malaysia, and ranging further east to Korea and Japan. A gaily-coloured bird with bright blue wings.

Red-breasted Pitta   *P. erythrogaster*
eruthros (Gr) red   *gaster* (Gr) the belly; actually the breast is partly green but the underparts or belly are bright red. Inhabiting the Philippines, and ranging south through Celebes, New Guinea and northern Australia.

Koch's Pitta   *P. kochi*
Named after Dr G. von Koch (1849–1914), a German zoologist and taxidermist; he was a Professor at the Darmstadt Museum in 1875. A rare pitta, and confined to certain areas in the mountains of Luzon.

Rainbow Pitta   *P. iris*
iris (Gr), genitive *iridos*, the rainbow; from *iridos* we get the English word iridescent; this is a brilliantly coloured bird with an iridescent plumage. Inhabiting Australia.

Phayre's Pitta   *P. phayrei*
Named after Lt Gen Sir A. P. Phayre (1812–1885), a naturalist who was in the Indian Army in 1828 and Commander of the British in Burma from 1862 to 1867. This pitta inhabits the Burma to Vietnam area.

Family   PHILEPITTIDAE   4 species
phileō (Gr) I love   *pitta*, a native name for a small bird (see notes under Family on p. 353). They have a superficial resemblance to pittas, of the family Pittidae; the Latin name for this family is taken to mean 'like a pitta'.

Velvet Asity   *Philepitta castanea*
castaneus (New L) chestnut-coloured, derived from *kastanos* (Gr) the chestnut tree; the male is black but in fresh plumage the feathers are tipped with yellow giving the appearance of chestnut-colour; as the tips wear off it becomes a rich black with a slight purple sheen that has been described as a 'velvety black'; the females are more green with some yellow. The asities and false-sunbirds are found only in Madagascar, and this one lives on the eastern side; they total only four species and range from 10 to 15 cm (4 to 6 in) in length; they eat insects, fruit and berries and some like nectar. Asity is the local native name.

Wattled False Sunbird   *Neodrepanis coruscans*
neos (Gr) new, recent   *drepanē* (Gr) a sickle   *drepanis* (Gr) a kind of
bird; a reference to its sickle-shaped bill; originally classified as a sunbird in the family Nectariniidae, it has recently been established for
various anatomical reasons that this was wrong, hence 'recent' and
now known as 'false sunbird'   *coruscus* (L) glittering, gleaming; a
reference to the iridescent blue plumage: it has a large wattle around
the eye though not present in the female. Confined to the forests of
Madagascar.

Small-billed False Sunbird   *N. hypoxantha*
hupo (Gr) under, beneath, less than usual   *xanthos* (Gr) yellow; the
breast and the underparts are yellow: it has a small sickle-shaped bill.
Inhabiting forests in eastern Madagascar, it is very rare and may now
be extinct due to the almost complete destruction of its forest habitat.

Family   ACANTHISITTIDAE   3 species
akanthis (Gr) a small bird; this is given in different translations as a
goldfinch, a linnet, and a siskin, but the New Zealand Wrens do not
appear to have any special relationship with these birds which are all
in the family Fringillidae   *sittē* (Gr) (Lat *sitta*) a woodpecker, and
sometimes translated as a nuthatch; however regardless of these translations *sitte* or *sitta* is frequently used for small creeping birds like nuthatches, e.g. Sittidae. The small family Acanthisittidae has been
variously placed taxonomically by different authors and some believe
they have a relationship with the Thornbill *Acanthiza*, of Australia.
They are small insect-eating birds only about 7·5 to 10 cm (3 to 4 in)
long.

Rock Wren   *Xenicus gilviventris*
xenos (Gr) a foreigner, a stranger   *-icus* (L) suffix meaning belonging
to; 'of foreign places'; at the time when this name was given New
Zealand would have seemed a very 'far-away' place   *gilvus* (L) pale
yellow   *venter* (L), genitive *ventris*, the belly, stomach; the underparts
are a pale grey-brown with some yellow on the flanks. This bird will
sometimes eat the fruit of alpine plants in addition to insects; it inhabits the slopes of mountains, and is confined to South Island, New
Zealand.

Bush Wren   *X. longipes*
longus (L) long   *pes* (L) the foot; the feet are large for a bird of this
size, particularly the hind toe and claw, which is useful for a bird that

spends a lot of its time climbing up and down trees searching for insects. Inhabiting both North and South Islands, New Zealand.

Rifleman *Acanthisitta chloris*
Acanthisitta, see notes under Family, above *khloros* (Gr) green or greenish yellow; the upperparts are greenish yellow and the flanks yellow. It has become known as a 'rifleman' because of the peculiar call that sounds like 'zing', and is similar to the sound of someone firing a rifle. It inhabits both Islands of New Zealand and some of the offshore islands, but not the northern part of North Island.

Family TYRANNIDAE 362 species
tyrannus (L) a monarch, a ruler, particularly a severe and cruel ruler; the tyrant flycatchers have earned this name because of the aggressive behaviour of many of the species, small birds attacking and harassing bigger birds such as birds of prey by diving at their backs. This behaviour is particularly noticeable in the kingbirds of the genus *Tyrannus*, but others in this group are equally aggressive when defending their territories.

This very large family, although similar in general appearance and feeding behaviour to the Old World flycatchers of the family Muscicapidae, anatomically they are different and not considered to be closely related. As their name implies they are insect eaters but the diet is by no means confined to insects; many species will also take fruit, small mammals, reptiles and amphibians; they range in size from about 7·5 to 36 cm (3 to 14 in) and are only found in the Americas, particularly South America.

White-tailed Shrike-Tyrant *Agriornis albicauda*
ager (L), genitive *agri*; a field, open country or *agrios* (Gr) living in the fields, of animals can mean wild, savage *ornis* (Gr) a bird, an aggressive bird as the English name shrike-tyrant suggests; the shrike is also known as the butcher bird *albus* (L) white *cauda* (L) the tail of an animal. Inhabiting South America.

Spot-billed Ground-Tyrant *Muscisaxicola maculirostris*
musca (L) a fly *saxum* (L) a large stone or rock *colo* (L) I inhabit; 'a rock-dwelling flycatcher' *macula* (L) a mark, a spot *rostrum* (L) the beak. Inhabiting western South America.

Say's Phoebe *Sayornis saya*
Named by Charles L. Bonaparte in honour of Thomas Say (1787-

1834), the famous American entomologist *ornis* (Gr) a bird: it is un-usual to have a personal name twice in the Latin name and is caused by a change of the generic name. It is said that phoebe, sometimes phebe, is an alteration of pewee influenced by the name Phoebe; pewee is a name used for birds in the genus *Contopus* (see p. 358), and is an imitation of their cry. Inhabiting North America and Mexico.

Vermilion Flycatcher   *Pyrocephalus rubinus*
pyr (Gr), genitive *pyros*, fire   *kephalē* (Gr) the head   *ruber* (L) red -*inus* (L) suffix meaning like; the male has a bright red head and under-parts. Inhabiting the southern part of the USA and ranging south through Central America to Argentina.

Cattle Tyrant   *Machetornis rixosus*
makhē (Gr) battle, combat   *makhētēs* (Gr) a fighter   *ornis* (Gr) a bird *rixa* (L) a fight, quarrel   -*osus* (L) suffix meaning full of, prone to; quarrelsome; although this is one of the aggressive tyrants it is best known for its habit of riding on the backs of cattle, sometimes for many miles, searching for insects or hopping to the ground for insects disturbed by the animal. It ranges from Colombia and Venezuela south to Argentina.

Fork-tailed Flycatcher   *Muscivora tyrannus*
musca (L) a fly   *voro* (L) I devour   *tyrannus* (L) a monarch, a ruler, particularly a severe and cruel ruler. An aggressive bird, widespread and ranging from southern Mexico south to Venezuela, Colombia and Brazil; a southern race (i.e. a sub-species) inhabits Paraguay, Argentina, and Uruguay, and in winter moves north as far as the islands of Trinidad and Curacao.

Thick-billed Kingbird   *Tyrannus crassirostris*
Tyrannus, see above   *crassus* (L) thick   *rostrum* (L) the beak; an aggressive bird inhabiting North and Central America. The name kingbird derives from the Latin *tyrannus*, a monarch.

Eastern Kingbird   *T. tyrannus*
One of the most aggressive of the tyrant flycatchers, it must not be thought that this has any connection with the tautonymous name, which could be translated as 'a tyrannical tyrant'; it is true, however, that the two authors, who gave each name separately, were doubtless influenced by the bird's behaviour; the fact that they are now in juxtaposition is because there has been a change in classification (see

p. 13 regarding tautonyms). Widespread, including southern Canada and south to Florida, along the Gulf Coast to Texas, New Mexico, Costa Rica, east to the Guyanas and south through western Brazil to Peru, Bolivia, and northern Argentina.

Grey-capped Flycatcher　*Myiozetetes granadensis*
myia (Gr) a fly　*zēteō* (Gr) I seek, search　*zētētēs* (Gr) a seeker, a searcher　*-ensis* (L) suffix meaning belonging to; named from Granada, in Nicaragua, Central America, it is not confined to that area and ranges south into northern South America.

Boat-billed Flycatcher　*Megarynchus pitangua*
megas (Gr) big　*rhunkhos* (Gr) the beak; it has a long heavy bill hooked at the tip: *pitangus* is a native South American name (Tupi language) for this large-billed flycatcher, inhabiting Mexico and ranging south through most of South America.

Cocos Island Flycatcher　*Nesotriccus ridgwayi*
nēsos (Gr) an island　*trikkos* (Gr) a small bird: named after R. Ridgway (1850–1929), an American ornithologist; he was Curator of Birds at the Smithsonian Institute in Washington from 1880 to 1929. Cocos is an island belonging to Costa Rica and lying just to the south in the Pacific Ocean.

Tropical Pewee　*Contopus cinereus*
kontos (Gr) short　*pous* (Gr) the foot　*cinereus* (L) ash-coloured, grey. Wide distribution in Central and South America including the West Indies and Trinidad; it is named from its call which is a distinct 'pee-wee'.

Eastern Wood Pewee　*C. virens*
virens (L) becoming green, greenish; in Canada it inhabits Manitoba, Ontario, Nova Scotia, and ranges south through eastern USA to Florida, west along the Gulf Coast to Texas, Costa Rica in Central America and south to Colombia and Peru in South America.

Yellow-bellied Flycatcher　*Empidonax flaviventris*
empis (Gr), genitive *empidos*, a gnat, mosquito　*onax = anax* (Gr) a king, lord or master　*flavus* (L) yellow　*venter* (L), genitive *ventris*, the belly; the diet is mostly insects, particularly mosquitos, and they are very skilled at taking them on the wing. Widespread in eastern Canada and ranging south through eastern USA, migrating in winter to Mexico and Panama.

Traill's Flycatcher *E. traillii*
Named by Audubon in honour of Dr Thomas S. Traill (1781–1862),
a Scottish doctor and Professor of Medicine at Edinburgh University
in 1832; he edited the 8th edition of *The Encyclopaedia Britannica*.
This flycatcher is widespread in Alaska and Canada and the USA
though not in the south-east; also in Mexico and ranging south to
Panama.

Golden-crowned Spadebill *Platyrinchus coronatus*
platus (Gr) flat, wide *rhunkhos* (Gr) the beak, snout; a mis-spelling;
it is usually spelt *rhynchus* in a Latin name; the bill is flattened and
spade-like *corona* (L) a crown *coronatus* (L) crowned. Inhabiting
Central and South America.

Black-headed Tody-Flycatcher *Todirostrum nigriceps*
todus (L) a small bird *rostrum* (L) the beak, snout; 'tody-beaked'; a
comparison with the real tody of the family Todidae, a small brightly-
coloured insectivorous bird rather like the smaller flycatchers; they
are not closely related *niger* (L) black *ceps* (New L) from *caput* (L)
the head. Inhabiting Costa Rica to northern South America.

Boat-billed Tody-Tyrant *Microcochlearius josephinae*
mikros (Gr) small *coclearum* (L) a spoon; a reference to the small
spoonshaped bill: named after Josephine V. McConnell (born 1906)
in 1914, when she was only eight years of age; the daughter of F. V.
McConnell (1868–1914), the zoologist, she married Paul L. de Laszlo.

Short-tailed Pygmy-Tyrant *Myiornis ecaudatus*
muia (Gr) a fly *ornis* (Gr) a bird; a reference to its insect-eating habits
ecaudis (L) without a tail *-atus* (L) provided with; 'not provided with
a tail'; a very small almost tailless flycatcher inhabiting Central and
South America.

Sharp-tailed Tyrant *Culicivora caudacuta*
culex (L), genitive *culicis*, a gnat, midge *voro* (L) I devour *cauda* (L)
the tail *acutus* (L) sharp, pointed. An insect-eating tyrant flycatcher
inhabiting South America.

Torrent Tyrannulet *Serpophaga cinerea*
serphos (Gr) a small insect, a gnat *phagein* (Gr) to eat *serpo-* is a misspelling of the Greek word *serphos* *cinereus* (L) ash-coloured, grey; a
small tyrant flycatcher living near mountain streams in Central and
South America.

Mountain Elaenia   *Elaenia frantzii*
elaenia (New L) from *elainos* (Gr) of the olive tree, olive-coloured; a
reference to the colour of the plumage: named after Dr A. von
Frantzius (1821–1877), a German naturalist who was in Brazil from
1849 to 1853, and Costa Rica from 1853 to 1868. Inhabiting mountainous areas in Central and South America.

Bolivian Tyrannulet   *Tyranniscus bolivianus*
-iscus (New L) from *-iskos* (Gr) diminutive suffix; 'a small tyrant'
-anus (L) suffix meaning belonging to; inhabiting Bolivia.

Family   OXYRUNCIDAE   1 species
oxus (Gr) sharp, pointed   *rhunkhos* (Gr) beak, bill; usually spelt
rhynchus in a Latin name; it has a strong pointed bill hence the English
name is also sharpbill. The classification of this bird has never been
satisfactorily settled as it appears to have no close relatives, but the
Tyrannidae seem to be the nearest and it has usually been placed in
that family; however some ornithologists now prefer to give it a family
on its own as in several ways it differs from the tyrant flycatchers, for
instance it does not eat insects and the diet seems to consist of fruit and
berries.

Sharpbill   *Oxyruncus cristatus*
Oxyruncus, see above   *cristatus* (L) crested; it has a yellow to scarlet
crest, sometimes described as orange. It inhabits Costa Rica and
Panama and ranges south through Amazonian Brazil and Paraguay,
but the distribution is not necessarily continuous.

Family   PIPRIDAE   53 species
pipra (Gr) a bird; some say possibly a woodpecker but these birds,
known as manakins, are not closely related to woodpeckers and are
not noticeably like them in habits and behaviour. The males are
brightly coloured and take part in spectacular aerobatic courtship
displays; the diet consists of insects and berries and sometimes small
invertebrates.

Golden-headed Manakin   *Pipra erythrocephala*
eruthros (Gr) red   *kephalē* (Gr) the head; the male has a golden-orange
crown; it inhabits Central America and the northern part of South
America. The English name manakin is from the Dutch *manneken*,
meaning a small man, a dwarf; they are very small birds, about the
size of the titmouse.

Red-headed Manakin   *P. rubrocapilla*
ruber (L) red   *capillus* (L) hair of men, usually the hair of the head,
can mean the hair of animals. Inhabiting Peru, Bolivia and Brazil.

Wire-tailed Manakin   *Teleonema filicauda*
teleos (Gr) entire, perfect   *nēma* (Gr) a thread   *filum* (L) a thread
cauda (L) the tail; a reference to the unusual tail which appears to consist of two long wires or threads as long as the whole body. Inhabiting
South America.

White-throated Manakin   *Corapipo gutturalis*
korē (Gr) a maiden, a girl; can mean a puppet, a doll   *pipō* (Gr) a
bird, possibly the woodpecker   *pipō* = *pipra* (see notes under Family
opposite) a 'puppet-bird' a 'manikin-bird'   *guttur* (L) the throat
-alis (L) suffix meaning relating to; a reference to the white throat.
Inhabiting northern South America.

White-bearded Manakin   *Manacus manacus*
manacus (New L) from *manneken* (Du) a small man, a dwarf; the white
throat gives it the name 'bearded', and this is prominent during the
courtship display dance when the white feathers are puffed out.
Widely distributed in the north-western part of South America as far
south as northern Argentina; also known in Trinidad.

Grey-headed Manakin   *Piprites griseiceps*
pipra (Gr) a bird (see notes under Family opposite)   *-ites* (Gr) suffix
denoting a group, belonging to a group   *griseus* (New L) from *greis*
(Ger) grey   *ceps* (New L) from *caput* (L) the head. Inhabiting Central
America.

Thrush-like Manakin   *Schiffornis turdinus*
ornis (Gr) a bird; named after Dr M. Schiff (1823–1896), a German
zoologist; he was Curator of Birds at the Senckenberg Museum from
1844 to 1853   *turdus* (L) a thrush   *-inus* (L) suffix meaning like; a
reddish-brown bird about the size of a thrush, it is widespread in
Central and South America from Mexico south to Brazil.

Family COTINGIDAE   79 species
cotinga is a name of Tupi origin, a language spoken by South American
Indians of the Amazon area; it is derived from two words   *coting*
(Tupi) wash, and *tinga* (Tupi) white; however the birds in this family
are by no means all white, and the name probably originates from the
White Bellbird,   *Procnias alba.* In fact it is a very varied group with all

kinds of different colours, unusual decorations, and sizes, ranging from about 10 to 46 cm (4 to 18 in); they eat insects, fruit and berries, and nearly all of them live in Central and South America.

Andean Cock-of-the-Rock *Rupicola peruviana*
rupes (L), genitive *rupis*, a rock, cliff *colo* (L) I inhabit, dwell in -*anus* (L) suffix meaning belonging to, 'of Peru'. A spectacular bird with a brilliant orange to red 'cape' surrounding the head and shoulders and a bushy crest. Ground-loving amongst rocks it lives in north-western South America, ranging through Venezuela, Columbia, Ecuador, Peru, and Bolivia.

Guianan Cock-of-the-Rock *R. rupicola*
Another brightly coloured bird with a crest similar to above; it is not confined to the Guianas and is also known in adjacent areas of Brazil and southern Venezuela.

Black-headed Berry-eater *Carpornis melanocephalus*
karpos (Gr) fruit *ornis* (Gr) a bird; a reference to its fondness of fruit and berries *melas* (Gr), genitive *melanos*, black *kephalē* (Gr) the head. Inhabiting south-east Brazil.

Purple-throated Cotinga *Porphyrolaema porphyrolaema*
porphuris (Gr) purple *laimos* (Gr) the throat. Inhabiting the northern part of South America.

Lovely Cotinga *Cotinga amabilis*
Cotinga, see notes under Family, above *amabilis* (L) lovely; a brightly coloured bird inhabiting Central America.

Pompadour Cotinga *Xipholena punicea*
xiphos (Gr) a sword *ōlenē* (Gr) an arm; can mean an armful, a bundle; probably a reference to the peculiarly twisted wing coverts *puniceus* (L) reddish, purple-coloured. Living in Venezuela and ranging south to the Amazon area. The male has brilliant ornamental plumage, and for this reason has been given the name Pompadour Cotinga after Madame de Pompadour, the famous French courtesan.

Yellow-billed Cotinga *Carpodectes antoniae*
karpos (Gr) fruit *dēktēs* (Gr) a biter: named in honour of Antonia Ridgway in 1884, the daughter of R. Ridgway (1850-1929), the Curator of Birds at the Smithsonian Institute in Washington. It inhabits Central America.

Green and Black Fruiteater   *Pipreola riefferii*
pipra (Gr) a bird   *-olus* (L) diminutive suffix: named after Rieffer in
1840, his initials and dates do not appear to be recorded; he formed a
collection from Bogata, Texas, in 1840. This fruiteater inhabits the
northern part of South America.

Red-crested Fruiteater   *Ampelion rubrocristatus*
ampelos (Gr) a vine, hence *ampeliōn* (Gr) a singing bird that frequents
vines   *ruber* (L) red   *cristatus* (L) crested. This bird, partial to grapes,
inhabits the northern part of South America.

Scaled Fruiteater   *Ampelioides tschudii*
ampelos (Gr) see above   *-oides* (New L) from *eidos* (Gr) apparent
shape, resemblance: named after J. J. Baron von Tschudi (1818–
1889), a Swiss ornithologist who was collecting in Peru from 1838 to
1842. Inhabiting the northern part of South America.

Black-capped Becard   *Pachyramphus marginatus*
pakhus (Gr) thick, can mean great, large   *rhamphos* (Gr) a beak,
specially a hooked beak as in birds of prey; a reference to the large
beak, hooked at the tip, which is useful for an insectivorous bird eating
large insects; this bird eats grasshoppers and beetles and similar
insects, rather than fruit   *margo* (L), genitive *marginis*, border, margin
marginatus (L) having a margin; a reference to the white-tipped tail
and the feathers tipped with iridescent blue; the female has tawny
margins on the wings. Becard is a name derived from *bécarde* (Fr) from
bec (Fr) a beak; it is usually applied to all the 15 species in the genus
Pachyramphus which are big-billed birds. They live in the northern
part of South America and in some cases also Central America.

Black-tailed Tityra   *Tityra cayana*
Tityrus (L) was the name of a shepherd in the 1st Eclogue of Virgil
cayana refers to Cayenne, in French Guiana, a name given by Linnaeus
in 1766. It ranges from Guyana, Venezuela, and Colombia in the
north, southwards through Peru, part of Brazil, and Bolivia and
Paraguay.

Long-wattled Umbrellabird   *Cephalopterus penduliger*
kephalē (Gr) the head   *pteron* (Gr) feathers; 'head-feathers'; it has a
spectacular canopy of feathers on the head which resembles a kind of
umbrella   *pendulus* (L) hanging down   *gero* (L) I carry; a reference
to the pendulus feathered wattle hanging from under the neck which

can be up to 46 cm (18 in) long; the length of the whole bird ranges from 41 to 48 cm (16 to 19 in). Inhabiting the western slopes of the Andes in Columbia and Ecuador.

Capuchin Monkbird   *Perissocephalus tricolor*
perissos (Gr) odd; in addition to meaning odd numbers it can also mean strange, remarkable   *kephalē* (Gr) the head; a reference to the naked head above a collar of feathers resembling a monk, hence Capuchin Monkbird   *tricolor* (L) three colours; the plumage is brown with a naked blue and grey head. Inhabiting the north-western part of South America.

White Bellbird   *Procnias alba*
According to the Greek legend Proknē, daughter of Pandion, was transformed by the gods into a swallow, but this does not mean that this bird is closely related to the swallows or resembles them; in the eighteenth and nineteenth centuries a number of names taken from classical mythology were used for scientific names without any particular significance for the animal concerned   *albus* (L) white. Inhabiting the north-eastern part of South America, from the Guyanas to Amazonia.

Three-wattled Bellbird   *P. tricarunculata*
tri- (L) three   *caruncula* (L) a small piece of flesh, a wattle   *-atus* (L) suffix meaning provided with; it has three remarkable long thin wattles growing round the mouth. Inhabiting most of Central America.

Family PHYTOTOMIDAE   3 species
phuton (Gr) a plant, tree   *tomos* (Gr) a cut, a piece cut off. This small family, all in the same genus, are closely related to the Cotingidae; they have a wasteful habit of cutting leaves, buds, fruit and twigs from trees or plants most of which remains uneaten, and to aid this habit they have a powerful finch-like beak with serrated edges; naturally they are unpopular with farmers and gardeners. Colourful birds with a harsh call, their length is about 18 cm (7 in).

Peruvian Plantcutter   *Phytotoma raimondii*
Named after Prof A. Raimondi (1825–1890), an Italian zoologist and explorer who lived in Peru from 1850 to 1870; he was appointed a Director of the Lima Museum in 1877. This plantcutter inhabits Peru.

Reddish Plantcutter   *P. rutila*
rutilus (L) red, ruddy; a fairly common and colourful bird inhabiting Argentina and Patagonia.

Family   MENURIDAE   2 species
mēnē (Gr) the moon   *oura* (Gr) the tail; the elaborate lyre-shaped tail is formed by the outer pair of feathers, about 50 cm (20 in) long, forming the frame of the lyre; these feathers, a light brown, are decorated with golden bars shaped like crescent moons. Superficially rather like pheasants, the lyrebirds appear to have no really close relatives, but can be connected in some respects with the scrub-birds (below); their diet is mostly insects and worms, but may include crustaceans and molluscs. Famous in Australia, where they live, they appear on one of the Australian postage stamps.

Albert's Lyrebird   *Menura alberti*
Named in honour of Prince Albert (1819-1861), Prince Consort of Queen Victoria. It inhabits the forests of south-eastern Australia.

Superb Lyrebird   *M. novaehollandiae*
novaehollandiae (New L) New Holland; this is a name that has occasionally been used for Java since the Dutch arrived there in 1596. In 1769 Australia, undiscovered by the Western World, was known as Terra Australis Incognitis based on reports from Marco Polo, the Venetian traveller (1254-1324). Later certain areas and possibly the whole continent, including Tasmania, became known as New Holland.

Family   ATRICHORNITHIDAE   2 species
atrikhos (Gr) hairless; a reference to the absence of bristles round the gape   *ornis* (Gr), genitive *ornithos*, a bird. Known as Scrub-Birds because they live mostly on the ground among the stunted bushes and brushwood where they hunt for insects, worms, crustaceans and seeds.

Noisy Scrub-Bird   *Atrichornis clamosus*
clamo (L) I cry out, I shout aloud   *clamosus* (L) full of clamour, noisy; as the name suggests, the call is loud and penetrating. Very rare, and possibly extinct, it inhabits south-western Australia.

Rufous Scrub-Bird   *A. rufescens*
rufus (L) red, ruddy   *-escens* (L) suffix denoting somewhat; reddish; the plumage is mostly rufous and pale brown. Inhabiting south-eastern Queensland and north-eastern New South Wales, Australia.

Family ALAUDIDAE 76 species

alauda (L) the lark. This is a family of rather inconspicuous almost drab brownish birds ranging from 12·5 to 23 cm (5 to 9 in) long; they have become famous because of their sweet song, particularly the sky-lark, which sings while hovering some 9 to 12 m (30 to 40 ft) above the ground. The distribution is almost world-wide, although only two species are known in America and two in Australia; their diet consists of insects, worms, small crustaceans and occasionally berries.

Rufous-naped Lark   *Mirafra africana*

mirus (L) wonderful; African 'wonder-bird'. Inhabiting Africa south of the Sahara, particularly the east and southern parts; not often seen to the west of Cameroon.

Singing Bush-lark   *M. javanica*

Named from Java, it is widespread including Africa and ranging through India, Burma, Malaysia, New Guinea, and Australia.

Desert Lark   *Ammomanes deserti*

ammos (Gr) sand, a sandy place   *manēs* (Gr) a type of cup; a reference to the form of the nest which is a cup scraped in the ground   *desertus* (L) a solitary place, the desert. Ranging from the Sahara in Africa to Afghanistan, southern Eurasia, and India.

Thick-billed Lark   *Rhamphocoris clot-bey*

rhamphos (Gr) a beak, bill   *corium* (L) leather; 'leather-billed': named after Dr A. Clot Bey (1795-1868) in 1850; he was an author and zoologist who spent some time in Egypt. A recent ruling states that hyphens may not be used in scientific names but this one will probably remain unchanged. This lark inhabits southern Eurasia including part of North Africa.

Calandra Lark   *Melanocorypha calandra*

melas (Gr), genitive *melanos*, black   *koruphos* (Gr) a small bird   *kalandra* (Gr) a kind of lark. Inhabiting Europe and Asia, and northern Africa.

Short-toed Lark   *Calandrella cinerea*

-ellus (L) diminutive suffix; 'a small lark'; one of the smaller larks, being only about 12·5 cm (5 in) in length   *cinereus* (L) ash-coloured; the plumage is usually pale brown with dark marking but there can be a variation with some grey, making it more ash-coloured. The birds in this genus have a shorter hind toe than other larks. This species is

widespread, ranging from Spain and North Africa east to Mongolia; South Africa on eastern side including Uganda, Ethiopia, part of Arabia; winter range can be the Sahara, Sudan, and east to India.

Stark's Short-toed Lark   *C. starki*
Named in 1902 after Dr A. C. Stark (1846-1899), a naturalist and author who was in South Africa from 1886 to 1899. This species inhabits Africa.

Dupont's Lark   *Chersophilus duponti*
khersos (Gr) dry land, barren land   *philos* (Gr) loved, pleasing; 'desert-loving': named after L. Dupont (c. 1820-1846), a French naturalist, collector, and dealer in animals. It inhabits arid areas of northern Africa.

Crested Lark   *Galerida cristata*
galerum (L) a helmet-like cap for the head   *galerita avis* (L) the crested lark   *cristatus* (L) crested; it has a prominent crest of brownish feathers. Widespread, ranging from Europe (though not Great Britain) east to Korea and south to the Himalayas and Arabia, northern Africa, Sudan, and east to Guinea.

Thekla's Lark   *G. theklae*
Named in memory of Thekla Brehm in 1858, who died that year at the age of 26; she was the daughter of Dr Christian Ludwig Brehm (1787-1864), the German zoologist. Inhabiting Spain, the Balearic Islands, North African coast, up the Nile to the mountains of Ethiopia, Somaliland and Kenya.

Wood Lark   *Lullula arborea*
Lullula is a coined word and said to be derived from the bird's call, which sounds like 'lu-lu-lu', a familiar phrase in the song   *arbor* (L) a tree   *arboreus* (L) pertaining to trees; this does not seem to be a suitable name for this bird, which nests on the ground, feeds on the ground, and prefers to live on hillsides where there are few trees, and other bare areas; however it frequents the edges of woods at times and has been seen to perch in the trees. It inhabits Europe, including Britain and Scandinavia, and ranges east to the Ural Mountains and south to Spain, Italy, and North Africa.

Skylark   *Alauda arvensis*
alauda (L) the lark   *arvum* (L) a field, cultivated land   *-ensis* (L) suffix meaning belonging to. Inhabiting Europe except the extreme

north and ranging from Britain across Asia to Japan and the Bering
Straits, south to North Africa and ranges east, south of the Caspian
Sea to Sinkiang. It has been introduced to other parts of the world.

Horned Lark   *Eremophila alpestris*
eremos (Gr) a desolate place, a wilderness   *philos* (Gr) loved, pleasing
alpestris (New L) pertaining to the Alps, high mountains; it frequents
desert areas and rocky alpine slopes up to 5,180 m (17,000 ft). It has
two tufts of black feathers that look like horns. The only lark in
America apart from the Skylark which has been introduced, it also
inhabits Eurasia from Scandinavia ranging east to Siberia, North
Africa east through to Middle East to mountain ranges of southern
Asia including the Himalayas, India, and Burma.

Family   HIRUNDINIDAE   74 species
hirundo (L) a swallow; genitive *hirundinis*. Superficially the swallows
resemble the swifts, both in appearance and manner of flying, but
they are not closely related; the swallows and martins, however, are
in the same family, and there is very little difference between them;
there is one martin in Britain that is called a swallow in America. The
House Martin can be distinguished from the swallows by its white
rump and the tail which is less forked. A very widespread family
known almost throughout the world, though not in New Zealand.
They eat insects, which they catch on the wing, and they only come
down to the ground to collect mud for their nests.

African River Martin   *Pseudochelidon eurystomina*
pseudēs (Gr) false, deceptive   *khelidōn* (Gr) a swallow   *eurus* (Gr) wide
stoma (Gr) the mouth   *-inus* (L) suffix meaning like, converting noun
to adjective; 'wide-mouthed'; it flies with a wide gape to assist in
catching flying insects. This swallow-like bird has been a problem for
ornithologists with regard to its place in classification, and at one time
it was placed with the wood swallows in the family Artamidae; some
authors give it a family on its own, Pseudochelidonidae, but recent
opinion seems to favour putting it with the true swallows, as it is given
here; because of this problem its Latin name was given as *Pseudochelidon*, 'deceptive swallow'. As recently as 1968 another species has
been found in Thailand. The African species frequents the banks of
the Lower Congo and Ubangi Rivers in West Africa.

White-eyed River Martin   *Ps. sirintarae*
This rare river martin was discovered in 1968 by the late Kitti Thon-

glongya (died 1974), who also discovered the Hog-nosed Bat *Crase-onycteris thonglongyai* and named in his honour by his friend J. Edwards Hill of the British Museum (Natural History). The river-martin was found in a marsh in the same area, Nakhon Sawan, in central Thailand, and named by Dr Boonsong Lekagul in honour of HRH Princess Sirindhorn Thepratanasuda, a daughter of King Bhumibol Adulyadej—'for her very keen interest in natural history'.

Tree Swallow   *Tachycineta bicolor*
takhus (Gr) fast, swift   *kineō* (Gr) I move   *kinētēs* (Gr) a mover; strictly, one that sets going a movement   *bicolor* (L) of two colours; the upper parts are blue and bronze-green. It frequents open wooded country and usually nests in tree hollows. It inhabits North America and ranges north as far as Alaska, and south to Panama and Cuba.

Caribbean Martin   *Progne dominicensis*
Prognē = *Proknē* (Gr); according to a legend, Proknē, daughter of Pandion, was transformed by the gods into a swallow   *-ensis* (L) suffix meaning belonging to, usually a locality; named from the Dominican Republic, it inhabits the Caribbean area and Central America.

Blue and White Swallow   *Notiochelidon cyanoleuca*
notios (Gr) southern   *khelidōn* (Gr) the swallow   *kuaneos* (Gr) blue *leukos* (Gr) white. Inhabiting Central and South America.

African Sand Martin   *Riparia paludicola*
ripa (L) the bank of a stream or river   *riparius* (L) one that frequents the banks of streams   *palus* (L), genitive *paludis*, a marsh, swamp *colo* (L) I inhabit; it makes its nest by tunnelling into the banks of streams or other sandbanks. Inhabiting Africa south of Sudan and Ethiopia, but not Zaire; Madagascar, southern India, the Burma-Vietnam area, Malaysia, and the Philippines.

Sand Martin   *R. riparia*
In the USA this bird is known as the Bank Swallow, which confuses the problem about the difference between a martin and a swallow (see notes under Family opposite). Inhabiting North, Central, and South America, Eurasia including the British Isles, Africa and Madagascar, India, the Burma-Vietnam area, Malaysia and Japan.

Crag Martin   *Hirundo rupestris*
hirundo (L) a swallow   *rupes* (L) a rock, cliff   *rupestris* (New L) living

among rocks. Inhabiting southern Europe, Mediterranean islands, the northern part of Africa, Asia Minor and ranging east to western China and southern India.

Common or Barn Swallow   *H. rustica*
rusticus (L) belonging to the country, rural. Probably the best known swallow with an almost world-wide distribution: North, Central and South America, Eurasia including Great Britain, Africa, India, Burma to Vietnam, Malaysia, New Guinea, and northern Australia.

Cliff Swallow   *Petrochelidon pyrrhonota*
petra (Gr) a rock   *khelidōn* (Gr) a swallow   *purrhos = purros* (Gr) red, flame-coloured   *noton* (Gr) the back; a reference to the upper tail coverts which are reddish. Although by nature it originally built the nest on rock and cliff faces, nowadays many cliff swallows prefer to build under the eaves of houses and barns. Inhabiting North, Central, and South America.

House Martin   *Delichon urbica*
Delichon is simply an anagram of *chelidon* (= *khelidōn* Gr) a swallow *urbicus* (L) pertaining to a city; it builds the nest under eaves of houses, on bridges, on sea cliffs, or inland cliffs. Inhabiting Eurasia including Great Britain, and Spain; ranging south to northern Africa, east to Asia Minor and the Himalayas, and the Thailand-Vietnam area.

Blue Rough-winged Swallow   *Psalidoprocne pristoptera*
psalis (Gr), genitive *psalidos*, a single bladed knife or razor, but can mean a pair of shears; a reference to the forked tail   *Prognē = Proknē* (Gr); according to a legend, Proknē, daughter of Pandion, was transformed by the gods into a swallow   *pristēs* (Gr) a saw or file   *pteron* (Gr) feathers, can mean wings; a reference to the serrated outer primary feathers on the wings; hence 'rough-winged' swallow; this is thought to assist the bird when clinging to vertical walls or cliffs. Inhabiting Eritrea and Ethiopia.

Family   MOTACILLIDAE   54 species
The family name, which is derived from the generic name *Motacilla*, is misleading; *Motacilla* does not mean wagtail, from *motator* (L) a mover, and *cilla*, a tail, because *cilla* does not mean a tail, and in fact there is no such word. There is a very rare word *cillo* (L) I move, but this would not make sense; Varro would not use a name that meant 'moving mover'. *Motacilla* is a coined word, using *-illa* (L) a diminu-

tive suffix, and means 'little mover'; when Varro explained this name
he said '*Quod semper movet caudam*', when he might have been expected
to say '*Quod semper movet cillam*', but he did not. This wrong interpretation probably started as long ago as the fourteenth century, and
still persists today among some ornithologists; for example we are
quite happy to interpret *Haliaeetus albicilla* as the White-tailed Eagle,
and in fact the specific name was supposed to mean 'white-tailed', but
it does not. Marcus Terentius Varro (116–27 B.C.) was a Roman
scholar and author of a great number of books; a genus of American
tropical shrubs, *Varronia*, is named after him.

The wagtails and pipits comprise a widespread family ranging from
the Americas through Eurasia and Africa to Australia. Sparrow-sized
birds, some with tails nearly as long as the body, their diet is varied
and consists mostly of insects but may include worms, small crustaceans and some seeds.

Forest Wagtail   *Dendronanthus indicus*   (or *Motacilla*)
dendron (Gr) a tree   *anthos* (Gr) a small bird like the yellow wagtail
-icus (L) suffix meaning belonging to, Indian; although named from
India, it also inhabits Siberia, Manchuria, Korea, northern China,
and Assam. The only wagtail that wags its tail sideways instead of up
and down.

White Wagtail   *Motacilla alba alba*
Motacilla, see notes under Family, above   *albus* (L) white: the forehead, sides of the head, and underparts are white, the upperparts are
grey with a black crown, wings and tail. Widespread in Eurasia including Iceland and Great Britain, and ranging east to Japan; the
northern part of Africa, India, Burma to Malaysia.

Pied Wagtail   *M. a. yarrellii*
This sub-species, and *M. a. alba*, above, are very similar; the Pied
Wagtail can be distinguished from the White Wagtail by its black
back: it is named after William Yarrell (1784-1856), the English
ornithologist and author; he wrote *The History of British Birds* which
was published in 1843. Distribution similar to White Wagtail.

Grey Wagtail   *M. cinerea*
cinereus (L) ash-coloured; although known as the Grey Wagtail it has
distinctly yellow underparts rather like the Yellow Wagtail, so it can
be mistaken for this bird, but the back is darker and the male has a
black throat. Inhabiting Eurasia including the British Isles but not

Iceland or northern Scandinavia; ranging east to Kamchatka and Japan and migrating south to northern Africa, the Burma-Vietnam area, Malaysia, Indonesia, and New Guinea.

Mountain Wagtail   M. clara
clarus (L) bright, shining. Inhabiting the mountains of Africa.

Yellow Wagtail   M. flava
flavus (L) yellow. Widespread in Eurasia including England and Wales but not Ireland, Scotland, Iceland, and northern Siberia; ranging east to the oriental region, Malaysia, Indonesia, and migrating south to northern Africa.

Rosy-breasted Longclaw   Macronyx ameliae
makros (Gr) long   onux (Gr) a claw; all birds in the family Motacillidae have a rather long hind claw, but those in the genus Macronyx have this claw unusually long: named in honour of Amélie, the Marquise de Tarragon (fl 1840); she married the Marquis Leone de Tarragon (1813-1896), a French zoologist. It inhabits East Africa ranging from Kenya through Tanzania to southern Natal.

Yellow-throated Longclaw   M. croceus
croceus (L) pertaining to saffron; can mean saffron-coloured, yellow, golden; the chin and throat are bright yellow, and the hind toe is exceptionally long, more than 4 cm ($1\frac{1}{2}$ in). Widespread in Africa, from Senegal and Angola in the west, and south to Natal.

Tawny Pipit   Anthus campestris
anthos (Gr) a small bird like the yellow wagtail   campus (L) open level land, a plain   campestris (L) pertaining to a plain, a field. The pipits in this genus also have long hind claws, like those in the genus Macronyx, above, but not so long as the latter. Living in open, dry and sandy country it inhabits Eurasia, but not Great Britain, Scandinavia and other northern parts; ranging south to Africa, and east to India and the Burma-Vietnam area.

Richard's or New Zealand Pipit   A. novaeseelandiae
novaeseelandiae (New L) New Zealand. The name probably refers to A. Richard (1794-1852) a botanist, and son of the famous botanist Prof L. C. M. Richard. Inhabiting eastern Eurasia, from Siberia south through India, Malaysia, New Guinea, Australia, Tasmania and New Zealand; also widespread in Africa.

Sprague's Pipit   *A. spragueii*
Named after I. Sprague (1811–1895) who also has the botanical genus
Spraguea named after him; he was an American botanist and worked
at the New York Botanical Gardens. Inhabiting North and Central
America.

Family   CAMPEPHAGIDAE   72 species
kampē (Gr) a caterpillar   *phagein* (Gr) to eat; the Cuckoo-Shrikes are
largely insect eaters, but they will also take berries and fruit, and have
been known to eat small lizards. Although in some respects they resemble cuckoos and also shrikes they are not closely related to either;
varying in size from 12·5 to 30 cm (5 to 12 in), they inhabit tropical
areas but not the Americas.

Ground Cuckoo-Shrike   *Pteropodocys maxima*
pteron (Gr) a wing   *pous* (Gr), genitive *podos*, a foot   *oxus* (Gr) sharp,
pointed; can mean quick, swift; 'swift on wing and foot'; largely
ground-living, where they hunt amongst the undergrowth for insects,
they also fly well and roost high in the trees   *maximus* (L) the largest;
it is the largest bird in the family, and may measure more than 30 cm
(12 in). Inhabiting Australia, but not in the north-western parts.

White-rumped Cuckoo-Shrike   *Coracina leucopygia*
korax (Gr), genitive *korakos*, a raven or crow; anything hooked like a
raven's beak   *-inus* (L) suffix meaning like; 'raven-like', a reference
to the hooked bill, similar to the hooked bill of the shrike, in the family
Laniidae   *leukos* (Gr) white   *pugē* (Gr) the rump, buttocks. Inhabiting Celebes.

Black-faced Cuckoo-Shrike   *C. novaehollandiae*
novaehollandiae (New L) New Holland; it was named from Tasmania
(see p. 365). This cuckoo-shrike has a black forehead, face, and throat.
Widespread in the oriental area, ranging from India to New Guinea,
some Pacific islands, Australia, Tasmania, and sometimes seen in New
Zealand.

Long-billed Greybird   *C. tenuirostris*
tenuis (L) thin, slender   *rostrum* (L) the beak, bill. The name Greybird is used for some of the cuckoo-shrikes because of their rather dull
grey, black, and white plumage. Inhabiting New Guinea, the
Solomon Islands, and Australia.

Black-and-White Triller   *Lalage melanoleuca*
Lalage (L) the name of a girl, possibly from *lallo* (L) I sing a lullaby;
this name and the English name Triller is a reference to the melodious
song   *melas* (Gr) black   *leukos* (Gr) white. Inhabiting part of the
Philippines.

White-winged Triller   *L. sueurii*
Named after C. A. Le Sueur (1778-1846), a French zoologist; he was
a Director of the Natural History Museum in Le Havre. This triller
inhabits the Lesser Sundas, New Guinea, and Australia.

Black Cuckoo-Shrike   *Campephaga flava*
kampē (Gr) a caterpillar   *phagein* (Gr) to eat   *flavus* (L) yellow; the
male is black and sometimes has some yellow plumage on the
shoulders; the female is olive-brown relieved with yellow margins on
the wings and tail and black and yellow bars on the underparts.
Inhabiting Sudan, Ethiopia, and Somalia and ranging south to
Angola.

Red-shouldered Cuckoo-Shrike   *C. phoenicea*
phoinix (Gr), genitive *phoinikos*, purple-red, crimson; the male is blue-
black with bright red shoulders. Widespread in Africa particularly
Congo.

Wattled Cuckoo-Shrike   *C. lobata*
lobos (Gr) a lobe   *lobatus* (New L) lobed; it has wattles consisting of
yellow lobes under the eyes. A rare and unusual cuckoo-shrike of
tropical Africa.

Short-billed Minivet   *Pericrocotus brevirostris*
peri (Gr) around, roundabout: hence *peri-* (New L) all around   *crocota*
(L) a saffron-coloured court dress   *brevis* (L) short   *rostrum* (L) the
beak, bill; the birds in this genus are not all yellow. The origin of the
name Minivet is obscure, but certainly it suggests something small,
and these birds are the smaller ones in the family. This one breeds in
the Himalayas and ranges south to India and the Burma-Vietnam
area.

Scarlet Minivet   *P. flammeus*
flammeus (L) flaming; can mean flame coloured, fiery red; the male
and female make a very colourful pair; the male with a black head
and upperparts and a bright red chest and belly, and the female with
a grey crown and upperparts and a bright yellow breast and belly.

Inhabiting southern India and Sri Lanka, Bangladesh, and the Burma-Vietnam area, and Malaysia.

Common Wood Shrike *Tephrodornis pondicerianus*
tephrōdēs (Gr) ash-coloured *ornis* (Gr) a bird *-anus* (L) suffix meaning belonging to; named from Pondicherry in south-east India, it also inhabits the Burma-Vietnam area and southern China.

Family PYCNONOTIDAE 118 species
puknos (Gr) compact; of plumage thick, close *nōton* (Gr) the back; the birds in this family, known as Bulbuls, have thick sometimes tangled-looking plumage on the back, but typically rather sparse on the nape of the neck where there are hair-like feathers. They are smallish birds, varying from the size of a sparrow to that of a blackbird, with a rather sombre plumage occasionally relieved by patches of white, yellow or red; the diet consists of insects, fruit and berries. The name bulbul is considered to be of Persian and Arabic origin and may be imitative, though the song is pleasant and in some cases melodious.

Finch-billed Bulbul *Spizixos canifrons*
spiza (Gr) a finch *ixos* (Gr) mistletoe; can mean the mistletoe berry *canus* (L) white, hoary *frons* (L) forehead, brow. Inhabiting Assam, the Burma-Vietnam area and southern China.

Slender-billed Greenbul *Andropadus gracilirostris*
anēr (Gr), genitive *andros*, a man, as opposed to a woman; however can be used to mean everyman, everyone *pados* (Gr) a tree; 'man of the trees' *gracilis* (L) slender, thin *rostrum* (L) the beak, bill. Inhabiting Africa.

Red-eyed Bulbul *Pycnonotus nigricans*
Pycnonotus, see above, under Family *nigricans* (L) blackish, swarthy; the upperparts are sooty brown with a blackish head and throat; the red eye stands out clearly against this background. Inhabiting the western part of southern Africa but not along the southern coastal belt.

Yellow-throated Bulbul *P. xantholaemus*
xanthos (Gr) yellow *laimos* (Gr) the throat. Inhabiting southern India.

Joyful Greenbul *Chlorocichla laetissima*
khloros (Gr) green *kikhlē* (Gr) a bird like the thrush *laeto* (L) I make

joyful, I gladden; *laetissima*, very joyful; a reference to the melodious song; most bulbuls are good singers, and for that reason are popular garden birds in Africa and the East, although they can be destructive. Inhabiting Africa.

Yellow-striped Greenbul   *Phyllastrephus flavostriatus*

phullon (Gr) a leaf   *strephō* (Gr) I twist; an obscure name, and not explained in the ornithological records in the British Museum (Natural History)   *flavus* (L) yellow   *stria* (L) a furrow, channel *striatus* (New L) striped. Inhabiting the central parts of Africa.

Terrestrial Bulbul or Brownbul   *P. terrestris*

terra (L) earth, ground   *terrestris* (L) belonging to the land; unlike the behaviour of bulbuls as a whole, it frequently descends to the ground in search of insects and grubs. Inhabiting south-eastern Africa.

Dwarf Bearded Bulbul   *Criniger finschi*

crinis (L) hair   *gero* (Gr) I bear, I carry; it has bristles under the chin and on the nape of the neck there are sparse hair-like feathers: named after Dr F. H. O. Finsch (1839-1917), a Dutch zoologist who worked in the Natural History Museum in Leyden from 1897 to 1904; he was in New Guinea from 1879 to 1882. This is a very small bulbul inhabiting Sumatra and Borneo.

Oriental Brown-eared Bulbul   *Hypsipetes flavala*

hupsipetēs (Gr) high-flying, soaring   *flavus* (L) yellow   *ala* (L) a wing; the upper parts including the crest are brown and there is a yellow patch on the wing. Inhabiting the Himalayas and ranging east to Yunnan and south through the Burma-Vietnam area to Malaysia.

McClelland's Bulbul   *H. mcclellandii*

Named after Dr J. McClelland (1805-1875), a Scottish zoologist who collected and studied in India. This bulbul inhabits the Burma-Vietnam area.

Family   IRENIDAE   14 species

named after Eirēnē, the goddess of peace, derived from *eirēnē* (Gr) peace, a time of peace. These birds are variously known as Ioras, Leafbirds, and Fairy Bluebirds; they are colourful birds of the oriental forests, the predominant colour being blue. The size ranges from 12·5 to 27 cm (5 to 10½ in) and the diet is mostly fruit and berries but also some insects and nectar.

Common Iora   *Aegithina tiphia*
aigithos (Gr) a hedge sparrow   *-inus* (L) suffix meaning like; the four
birds in this genus are small, about the size of a hedge sparrow, say
15 cm (6 in), but they are certainly more colourful   *tiphē* (Gr) an
insect, perhaps a kind of beetle. The origin of the name Iora is
obscure, but doubtless a local oriental native name. Inhabiting India,
the Burma-Vietnam area, and Malaysia.

Golden-fronted Leafbird   *Chloropsis aurifrons*
khloros (Gr) green   *opsis* (Gr) appearance, aspect; the predominant
colour is green   *aurum* (L) gold   *aureus* (L) golden, colour of gold
frons (L) forehead, brow; it has a bright orange-yellow patch on the
forehead. Inhabiting the Himalayan foothills, the southern part of
India including Sri Lanka, the Burma-Vietnam area and Malaysia.

Blue-winged Leafbird   *C. cochinchinensis*
-ensis (L) suffix meaning belonging to, usually a locality; what was
known as Cochinchina in the seventeenth century is now the central
and southern part of Vietnam; although named from that area it
ranges from India through the Burma-Vietnam area to Sumatra.

Fairy Bluebird   *Irena puella*
Named after Eirēnē, the goddess of peace, derived from *eirēnē* (Gr)
peace, a time of peace   *puella* (L) a girl, a maiden; in this case used to
suggest something pretty. A beautiful bird, as the English name suggests, and one of the largest in the family, at about 25·5 to 28 cm (10 to
11 in); it is widespread in the Oriental region ranging from India to
Malaysia.

Family   PRIONOPIDAE   9 species
priōn (Gr) a saw   *ops*, from *opsis* (Gr) appearance; a reference to those
in the genus *Prionops* which have a serrated edge to the bill, like a saw.
Some authors include this group, the Helmet Shrikes, with the
Shrikes or Butcher Birds in the family Laniidae, and there remains
some controversy among ornithologists as to how to place the shrikes
taxonomically. A third group, the Vanga Shrikes in the family
Vangidae, are restricted to Madagascar.

They are all rather like small birds of prey, ranging from 15 to 36 cm
(6 to 14 in) long, with the typical hooked beak, and they are all carnivorous, the diet including insects, small reptiles, small mammals
and birds. Some are known as Butcher Birds on account of their habit

of impaling their prey on a suitable spike or thorn which they will
return to later, and which also acts as a vice to hold the animal while
they dismember it for eating.

Rüppell's White-crowned Shrike   *Eurocephalus rueppelli*
euros (Gr) width, breadth   *kephalē* (Gr) the head; a reference to the
rather thick neck and head, the latter having a white crown: named
after Dr W. P. E. S. Rüppell (1794–1884), a zoologist who was in
North Africa from 1822 to 1827, and in Abyssinia (now Ethiopia)
from 1830 to 1834; he collected for the Senckenberg Museum. Inhabiting the eastern part of Africa.

Yellow-crested Helmet Shrike   *Prionops alberti*
priōn (Gr) a saw   *ops*, from *opsis* (Gr) appearance; a reference to the
serrated edge of the bill, like a saw: named after King Albert (1875–
1934), King of the Belgians from 1909 to 1934. Inhabiting Africa. The
name 'helmet' refers to the colouring of the head and neck.

Long-crested Helmet Shrike   *P. plumata*
pluma (L) a small soft feather   *plumatus* (L) feathered; a reference to
the white forward-pointing crest on the head. Widespread in Africa
as far south as Angola, Zimbabwe, and Mozambique.

Family   LANIIDAE   70 species
lanio (L) I tear, I rend in pieces   *lanius* (L) a butcher; a reference to
their unpleasant habit of impaling their prey on a spike or thorn, even
while still alive, to hold it while they dismember it for eating; this has
led to the name Butcher Bird which is sometimes used.

Pringle's Puff-back   *Dryoscopus pringlii*
drus (Gr), genitive *druos*, a tree   *skopeō* (Gr) I look; a reference to its
habit of hunting for larvae among the trees. Named after Col Sir J. W.
Pringle (1863–1938), who was Chief Inspector of Railways in Uganda
in 1891 and 1892. Inhabiting eastern Africa. It has a soft downy rump
and back and these feathers can be 'puffed out' by the male when
excited.

Boubou   *Laniarius aethiopicus*
laniarius (L) pertaining to a butcher; (see notes under Family above)
-icus (L) suffix meaning belonging to; although named from Ethiopia
it is widespread almost throughout Africa. The origin of the name
Boubou is obscure as it does not seem to be imitative and it is not
known to be a native name.

Fulleborn's Black Boubou   *L. fulleborni*
Named after Dr F. Fulleborn (1866-1933), a German zoologist. He
was in German East Africa from 1896 to 1900 and a Professor at Hamburg University in 1919. Inhabiting Africa.

Fiery-breasted Bush Shrike   *Malaconotus cruentus*
malakos (Gr) soft   *nōton* (Gr) the back; a reference to the soft downy
rump and back   *cruentus* (L) spotted or stained with blood; it has a
bright red breast and throat. Inhabiting Africa.

Lagden's Bush Shrike   *M. lagdeni*
Named after Sir G. Y. Lagden (1851-1934), an English naturalist
who was in the Gold Coast in 1883 and Commissioner in Basutoland
from 1893 to 1901; at that time a British Colony, it is now known as
Lesotho and is an independent nation of South Africa. This Bush
Shrike inhabits eastern Zaire.

Red-backed Shrike   *Lanius collurio*
lanius (L) a butcher (see notes under Family opposite)   *kolluriōn* (Gr)
a bird, possibly a thrush; the shrikes are not closely related to the
thrushes, in the subfamily Turdinae, but could be similar in appearance in some cases. This is one of the shrikes that keeps a 'larder' by
impaling its prey on spikes or thorns, which it can return to later; it is
widespread in Eurasia, ranging from Scandinavia and Russia south
to northern Spain, the Mediterranean area, and east through Asia
Minor, Palestine, Iran, to Siberia, Mongolia, Manchuria to western
China; some species migrate to Africa, Arabia, and India.

Northern Shrike   *L. excubitor*
excubitor (L) one who keeps watch, a sentinel; a reference to its habit
of sitting on a post or branch of a tree making no effort to be concealed, and watching for possible prey or for an intruder on its territory, which it would immediately attack and drive off; other shrikes
show similar behaviour. Widespread in both Old and New Worlds;
Alaska, Hudson Bay, Quebec, migrating south to California, and
Texas; Eurasia from Spain east through Siberia to Kamchatka,
south to Iran, Arabia and India, Eurasian birds winter in Mediterranean area and northern Africa.

Tibetan or Grey-backed Shrike   *L. tephronotus*
tephros (Gr) ash-coloured   *nōton* (Gr) the back. Named from Tibet,
it also inhabits Kashmir ranging east to Yunnan in China and south
to south-east Asia.

Bornean Bristle-head   *Pityriasis gymnocephala*
pityriasis (New L) from *pituron* (Gr) a skin disease of the head; a reference to some warts on the bare head   *gumnos* (Gr) naked   *kephalē*
(Gr) the head; the head is partly bare and has a patch of bristles, and
warts. Confined to Borneo, little is known about this curious bird, and
its proper place in classification has not been finally decided by
ornithologists.

Family   VANGIDAE   13 species
vanga (L) a mattock; this is a short pickaxe-like implement with a
curved blade or spike; a reference to the beak of the vanga shrikes
which is strong, slightly curved, and in some cases hooked at the tip.

Hookbilled Vanga   *Vanga curvirostris*
vanga (L) a mattock; see above   *curvus* (L) curved   *rostrum* (L) the
beak. The Vanga Shrikes are confined to Madagascar; they are
related to the other shrikes and their diet is similar, namely insects and
small reptiles, amphibians, and possibly small mammals.

Pollen's Vanga   *Xenopirostris polleni*
xenos (Gr) a stranger   *opsis* (Gr) appearance   *rostrum* (L) the beak;
'strange looking beak'; a reference to the heavy compressed and
hooked beak with an uptilted lower manbible: named after F. P. L.
Pollen (1842-1886), a naturalist and author who was in Madagascar
from 1863 to 1866.

Blue Vanga   *Leptopterus madagascarinus*
leptos (Gr) slender   *pteron* (Gr) feathers, can mean wings and usually
in plural   *-inus* (L) suffix meaning belonging to; like all vangas it is
found only in Madagascar.

Coral-billed Nuthatch   *Hypositta corallirostris*
hupo (Gr) under, less than usual; here used to mean 'not quite like'
sittē (Gr) (=*sitta*, Lat) a kind of woodpecker; has been translated as a
nuthatch   *corallinus* (L) coral-red (from *korallion* (Gr) coral, especially
red coral)   *rostrum* (L) the beak. Inhabiting forests of eastern Madagascar, up to 1,500 m (5,000 ft) in mountains.

Family   PTILOGONATIDAE   8 species
ptilon (Gr) a feather or wing; properly the soft feathers or down   *gonu*
(Gr), genitive *gonatos*, the knee; a reference to the soft feathers of the
underparts covering the knee; strictly speaking this is not the knee, as

it is the joint between the tibia and the tarsus, so anatomically speaking this is the ankle. This small family consists of the Silky Flycatchers, the Waxwings, and the strange Hypocolius; some authors put them in a family under the name Bombycillidae. They are smallish birds ranging from about 15 to 23 cm (6 to 9 in) long, having a mixed diet consisting of insects and fruit, and the Hypocolius probably only fruit.

Cedar Waxwing   *Bombycilla cedrorum*
bombyx (L), genitive *bombycis*, the silkworm   *bombycinus* (L) of silk, silky   *cilla* (New L) the tail; there is some doubt about this translation, see p. 370 under Family   *kedros* (Gr) the cedar tree   *-orum* (L) genitive plural ending; 'of cedar trees'; it inhabits open areas where there are scattered trees, but is not known to favour the cedar more than other kinds. The secondary feathers carry waxy red spangles, often plainly visible, hence the name waxwing. Inhabiting Quebec, Ontario and British Columbia, and ranging south to Georgia and California, migrating south in winter to West Indies and Panama.

Grey Silky Flycatcher   *Ptilogonys cinereus*
Ptilogonys, see notes under Family, above   *cinis* (L), genitive *cineris*, ashes   *cinereus* (L) ash-coloured; the plumage is soft and silky, similar to that of the waxwings. Inhabiting Mexico and Guatemala.

Phainopepla   *Phainopepla nitens*
phainō (Gr) I show   *phanos* (Gr) light, bright   *pepla* (Gr) a robe, a coat   *niteo* (L) I shine   *nitens* (L) shining; the male has glossy black plumage from the tall crest on the head to the long tail. Inhabiting California and Mexico and migrating south in winter to southern Mexico.

Black-and-Yellow Silky Flycatcher   *Phainoptila melanoxantha*
phainō, see above   *ptilon* (Gr) a feather or wing; properly the soft feathers or down   *melas* (Gr), genitive *melanos*, black   *xanthos* (Gr) yellow; the upperparts are a glossy black with a blue-green sheen, and the sides, flank, and rump are yellow. Inhabiting Costa Rica and western Panama.

Hypocolius   *Hypocolius ampelinus*
hupo (Gr) under, beneath   *colius* (New L) generic name for the coly, or mousebird; indicating that it somewhat resembles the African Coly   *ampelos* (Gr) a vine. Ampelis was the name of a singing bird in the satire *The Birds* by Aristophanes, and before about 1910 *Ampelis* was

in use for the waxwings, so *ampelinus* means waxwing-like; the behaviour is similar to the waxwings. The classification of this strange bird has been a problem for ornithologists; it inhabits an area around the Persian Gulf, including Afghanistan, Iraq, Iran, and Arabia.

Family DULIDAE 1 species
doulos (Gr) a slave; apparently the name was given to this rather unusual bird because Hispaniola Island had a reputation in the past as a place for slave trading. The Palm Chat, a starling-sized bird that probably only eats berries and some flowers, has been given a family on its own as there does not seem to be any certain relationship with other groups, although in some respects it resembles the waxwings, the silky flycatchers, and the hypocolius. The English name refers to its liking for the palm trees found in its natural habitat, and its habit of gathering in flocks that keep up a constant noisy chatter.

Palmchat *Dulus dominicus*
dulus (Gr) a slave; see above; *dominicus* refers to the Dominican Republic occupying the eastern part of the island of Hispaniola in the West Indies, the western part being Haiti; the palm chat is found only on this island and the nearby Gonave Island.

Family CINCLIDAE 4 species
kinklos (Gr) a kind of wagtail or water ouzel. These rather unusual birds, the only ones in the order Passeriformes that can swim and dive, are able to do this without the help of webbed feet; if they had webbed feet they could not be classified as 'perching birds', i.e. passerines. Known as Dippers, and sometimes Water Ouzels, the names are apt, as they live close to fast-flowing mountain streams and obtain most of their food from under the water; caddis flies, insect larvae, water beetles, snails and similar small water animals and they are known to catch very small fish like minnows.

Dipper *Cinclus cinclus*
Cinclus from *kinklos*, see above. Inhabiting mountain streams of Europe, including the British Isles, in the Atlas Mountains of northern Africa, in Asia Minor and Central Asia; they have been seen as high as 5,180 m (17,000 ft) in the mountains of Tibet.

American Dipper *C. mexicanus*
Named from Mexico, it has a large range from the Aleutian Islands and Alaska south to Guatemala and Panama and east to Dakota.

Family TROGLODYTIDAE 59 species

trōglē (Gr) a hole, a hollow *dutēs* (Gr) a burrower *trōglodutēs* (Gr) one who creeps into holes; the name refers to the form of the nest which is dome-shaped with an entrance hole, but not all wrens in this family build that type of nest; some build cup-shaped nests in holes and crevices or in bushes. They range in size from the little wren we know in Britain, about 9 cm ($3\frac{1}{2}$ in) long, to as much as 20 cm (8 in); there is only one wren living in Britain and Eurasia, *Troglodytes troglodytes*, all the others living in America, and mostly South America. They eat all kinds of insects and this is probably their only food.

Cactus Wren *Campylorhynchus brunneicapillus*

kampulos (Gr) bent, curved *rhunkhos* (Gr) the snout, beak; it has a fairly long slightly curved beak *brunneus* (New L) brown *capillus* (L) the hair of men, usually of the head; can mean the hair of animals; the predominant colour is brown. Known as the cactus wren because it usually builds the nest in a cactus or thorny bush and inhabits the cactus desert. Ranging from California, Utah and Texas south to Mexico.

Tooth-billed Wren *Odontorchilus cinereus*

odous (Gr), genitive *odontos*, a tooth *kheilos* (Gr) a lip, rim; of humans the lip, of birds the beak, bill; a reference to the serrated edge of the beak *cinis* (L), genitive *cineris*, ashes *cinereus* (L) ash-coloured. Inhabiting South America.

Canyon Wren *Salpinctes mexicanus*

salpinx (Gr) a war trumpet *salpinktēs* (Gr) a trumpeter; a reference to its loud call; although wrens, both male and female, have pleasant songs, a defiant male has a very loud call when proclaiming its territory. Named from Mexico, it ranges north, and south to Central America.

Sepia-Brown Wren *Cinnycerthia peruana*

kinnuris (Gr) a small bird *kerthios* (Gr) a little bird, perhaps a tree-creeper *-anus* (L) suffix meaning belonging to; inhabiting Peru.

Zapata Wren *Ferminia cerverai*

Named after Fermin Z. Cervera (dates not recorded), a naturalist who was in Cuba in 1926 when this wren was first discovered and named; it is very unusual for a bird to bear both the first name and the surname of a naturalist. It is confined to a marsh in southern Cuba on the Zapata Peninsula.

Black-throated Wren *Thryothorus atrogularis*
thruon (Gr) a rush, a reed *thouros* (Gr) rushing, impetuous; the male has a habit of building quite a number of nests that are never used, and seems to do this with a sense of urgency; one ornithologist described it as 'The male's frantic nest building'. Eventually the female chooses one of the nests and the male may use another one for roosting; they are often built in thick clumps of reeds *ater* (L) black *gula* (L) the throat *-aris* (L) suffix meaning pertaining to. Inhabiting Central America.

Spot-breasted Wren *T. maculipectus*
macula (L) a mark, a spot *pectus* (L) the breast; the white or pale underparts are heavily spotted and barred with black. Inhabiting Mexico and Central America.

House Wren *Troglodytes aedon*
Troglodytes, see notes under Family on p. 383 *aēdōn* (Gr) properly a songstress, but has been used to mean the nightingale; many wrens, both male and female, have a melodious song, and have been compared to the nightingale. Inhabiting North, Central, and South America.

Common Wren *T. troglodytes troglodytes*
Often known in Britain as the 'Jenny Wren', the Latin name unfortunately has become something of a joke; it is the Nominate Subspecies, so it has been saddled with the name *Troglodytes troglodytes troglodytes*. For an explanation of how this extraordinary name has come about see Chapter 1, Page 13, under Tautonyms; there are other cases in which an animal has what could be called a 'tautonymous trinomial name'. It is widespread in the western part of Europe, northern Africa, and ranging east through Asia Minor, to eastern China, Japan, and the Kuril and Aleutian Islands; it is also widespread in North America. It is the only true wren known that is not confined to the New World.

Shetland Islands Wren *T. t. zetlandicus*
More than 30 subspecies of the Common Wren are known; they seem able to colonise quite remote islands and become resident, and thus form a quite distinct subspecies; *zetland* (New L) Shetland *-icus* (L) suffix meaning belonging to; as the name tells us, it lives on the Shetland Islands.

Chestnut-breasted Wren   *Cyphorhinus thoracicus*
kuphos (Gr) curved   *rhis* (Gr), genitive *rhinos*, the nose; a reference to
the slightly curved beak   *thorax* (Gr), genitive *thorakos*, the breast
-icus (L) suffix meaning pertaining to; the throat and breast are rufous
chestnut in colour. Inhabiting Colombia, mostly west of the Andes,
and south to Ecuador and Peru.

Family   MIMIDAE   30 species
mimus (L) a mimic actor; the name has been given because of their
ability to mimic the songs of other birds, which of course accounts for
the English name Mocking Birds. They range in size from about 20
to 30 cm (8 to 12 in), and are similar in several ways to the thrushes,
to which they are related, and they are also related to the wrens.
Inhabiting the New World from Canada to South America, the diet
is probably mostly insects but they also take some fruit and berries,
and seeds.

Grey Catbird   *Dumetella carolinensis*
dumetum (L) a thorn brake, a thicket   *-ellus* (L) diminutive suffix;
they inhabit dense scrub and thorn bushes, usually near water: named
from Carolina, their range is widespread from southern Canada to
Florida, and west to Texas and New Mexico, migrating south in
winter to Cuba and Panama. The name catbird refers to one of the
calls which is a catlike mewing sound.

Brown-backed Mockingbird   *Mimus dorsalis*
mimus (L) a mimic actor   *dorsum* (L) the back   *-alis* (L) suffix meaning relating to; a reference to the brown back. Inhabiting Bolivia and
Argentina.

Mockingbird   *M. polyglottos*
polus (Gr) many   *glōssa* = *glōtta* (Attic) the tongue; 'many voiced';
mocking birds have a variety of songs, many quite melodious; some
ornithologists consider they are their own songs which happen to be
similar to other birds' songs, and this has given rise to the idea that
they are miming. Widespread in the USA from Wyoming east to Ohio
and Maryland, and from California south to the West Indies and
Mexico, and in some cases to Central and northern South America.

Galapagos Mockingbird   *Nesomimus trifasciatus*
nesos (Gr) an island   *mimus* (Gr) a mimic actor   *tria* (L) three
fascia (L) a band, girdle   *-atus* (L) suffix meaning provided with; a

reference to the white margins of the wing coverts which can be seen as three white bands. There are several different races of the mocking-bird confined to the Galapagos group of islands.

Curve-billed Thrasher   *Toxostoma curvirostre*
toxon (Gr) a bow for use with arrows; can mean anything bowed or arched   *stoma* (Gr) the mouth   *curvus* (L) curved   *rostrum* (L) the beak: in both names a reference to the slightly curved bill. Inhabiting Texas, New Mexico and Arizona and ranging south to Mexico. The English name thrasher is said to derive from thrush, but this has not been properly confirmed and it remains somewhat obscure; they are thrush-like birds, with a melodious song, and are certainly related to the thrushes.

Le Conte's Thrasher   *T. lecontei*
Named in honour of Dr John Lawrence Le Conte (1825-1883), an eminent zoologist; although he took a great interest in birds he made his name as an entomologist. This thrasher inhabits a similar area to the curve-billed thrasher, above, keeping mostly to arid parts.

Brown Thrasher   *T. rufum*
rufus (L) red, ruddy; the upperparts are reddish brown from the crown to the long tail. This thrasher often repeats a short phrase, like the Song Thrush *Turdus philomelos*. It inhabits a wide area from southern Canada south to Florida, the coast of the Gulf of Mexico, and Texas.

Brown Trembler   *Cinclocerthia ruficauda*
kinklos (Gr) a kind of wagtail   *kerthios* (Gr) a little bird, a treecreeper; an apparently meaningless name as it is neither a wagtail nor a tree-creeper, and it is not closely related to either   *rufus* (L) red, ruddy *cauda* (L) the tail of an animal: the English name has been given because of its peculiar habit of violent trembling. It is found only in the West Indies, Central America.

Pearly-eyed Thrasher   *Margarops fuscatus*
margaritēs (Gr) a pearl   *ops*, from *opsis* (Gr) aspect, appearance, also *ops* (Gr) the eye, face   *fuscus* (L) dark   *fuscatus* (L) dark-coloured. Inhabiting the West Indies.

Family   PRUNELLIDAE   13 species
prunus (L) a plum tree   *brunus* (New L) brown   *prunelle* (Fr) a sloe, from its dark colour; it is the fruit of the blackthorn; possibly with

-ellus (L) diminutive suffix. There is disagreement among authors about the origin of the name *Prunella* for the accentors, but it seems to be generally agreed that it refers to their brown colour; the one that is familiar to us in Britain, the Dunnock, is almost certainly named from its dun colour. *Accentor* is a Latin word, and means one who sings with another; it is a good songster but tends to be a soloist and is not reputed, as are some birds, to sing 'a duet' with its mate. They are sparrow-sized birds, or slightly larger, but the beak is thin and pointed, and the diet consists of insects, with more seeds and berries in the winter.

Alpine Accentor   *Prunella collaris*
Prunella, see above   *collum* (L) the neck   *collaris* (L) pertaining to the neck; it has a grey collar which shows against the background of brown. It inhabits Spain and northern Africa and ranges through the alpine areas of southern Europe and Asia to Japan, and is known to breed in alpine meadows and scrub from 1,500 to about 5,200 m (5,000 to about 17,000 ft).

Dunnock   *P. modularis*
modulator (L) one who measures, a director of music; can mean a musician; a reference to its high-pitched warble. Sometimes known as the hedge sparrow, it is not a sparrow, but is superficially somewhat like a sparrow and often builds its nest in hedges, particularly haw-thorn hedges; common in Britain, it is widespread in Europe though not in southern Russia and the Balkans; it is seen in Asia Minor and Iran, and winters in Israel, Greece, southern Italy, Egypt and northern Africa.

Family MUSCICAPIDAE   1,300 species
musca (L) a fly   *capio* (L) I take, seize. Under this main family head-ing, the thrushes and their kin number about 1,300 species, so it is usual to divide them into 11 subfamily groups; these are now almost universally recognised though authors may vary to some extent re-garding the subfamily names and the grouping. With such a large number of species it is not really possible to make any general remarks with regard to size, diet, and habits; anything of special interest will be mentioned under each subfamily heading. The reader should be reminded, as first mentioned in Chapter 4, that the suffix *-idae* denotes a family, and *-inae* a subfamily.

Subfamily TURDINAE   304 species
turdus (L) a thrush. This group includes a tremendous variety of

thrushes and their close relatives such as the wheatears, the robins, the redstarts, the nightingales, and the blackbirds; they are mostly insectivorous but also eat fruit and berries and some will take worms and snails when available. The group includes some of the finest songsters.

Forest Robin   *Stiphrornis erythrothorax*
stiphros (Gr) firm, sturdy   *ornis* (Gr) a bird   *eruthros* (Gr) red *thorax* (Gr) the breast; not to be confused with the robin we know in the British Isles, *Erithacus rubecula*, it is a bigger bird, though they are related; it inhabits Africa, from Cameroon to Uganda.

Rusty-bellied Shortwing   *Brachypteryx hyperythra*
brakhus (Gr) short   *pterux* (Gr) a wing   *hupo* (Gr) under, beneath *eruthros* (Gr) red. This shortwing inhabits India; there are five others, mostly oriental.

Wren-Thrush   *Zeledonia coronata*
Named after J. C. Zeledon (1846–1923), a naturalist from Costa Rica *corona* (L) a crown   *coronata* (L) wearing a crown; a reference to the tawny crown edged with black. Inhabiting Costa Rica and Panama.

Bearded Scrub Robin   *Cercotrichas barbata*
kerkos (Gr) the tail of an animal   *trikhas* (Gr) a thrush or field-fare; 'thrush-tailed'   *barba* (L) a beard   *barbata* (L) bearded. The English names are misleading because names such as robin, thrush, wheatear, chat and others are rather indiscriminately used in this large group. This scrub robin inhabits Africa.

Sooty Rock Chat   *Pinarornis plumosus*
pinaros (Gr) dirty   *ornis* (Gr) a bird; this name, and the English name 'sooty', indicate the rather drab grey plumage   *pluma* (L) a small soft feather   *-osus* (L) suffix meaning full of; known as a 'rock chat' from the habitat and the continuous chattering call. Inhabiting Africa.

Southern Scrub Robin   *Drymodes brunneopygia*
drumōdēs (Gr) of the woods   *brunneus* (New L) brown   *pugē* (Gr) the rump; a reference to the buff underparts. Inhabiting the interior part of south and south-west Australia.

Eurasian Robin   *Erithacus rubecula*
erithakos (Gr) a solitary bird; has been translated as the red-breast *ruber* (L) red   *-culus* (L) diminutive suffix; 'the little red one'. This is the robin that we know in the British Isles; it is confined to Europe and neighbouring countries such as Siberia, Asia Minor, part of

north-west Africa, the Canaries and Azores.

Thrush Nightingale   *Luscinia luscinia*
luscinius (L) a nightingale: this small bird is said to resemble *Erithacus rubecula*, above; it has a fine song considered to be more powerful than the nightingale *megarhynchos*, with pure bell-like notes. Inhabiting Eurasia from Denmark and Sweden to the River Ob area in Russia and ranging south to the Black Sea; it winters in the eastern part of Africa mostly south of the equator.

Nightingale   *L. megarhynchos*
megas (Gr) big, wide   *rhunkhos* (Gr) the beak; it would be difficult to discern that the beak is any bigger than other birds in the family. This is the nightingale that we know in Britain; it ranges across Eurasia and south to Africa.

Cape Robin   *Cossypha caffra*
kossuphos (Gr) a singing bird, like the blackbird   *caffra* is a coined name, from Kaffraria; this was a name applied to a large area in eastern South Africa, but is now obsolete. Inhabiting East Africa and ranging south through Tanzania to South Africa.

Redstart   *Phoenicurus phoenicurus*
phoinix (Gr), genitive *phoinikos*, red, dark, red   *oura* (Gr) the tail; a reference to the orange-chestnut coloured tail. This is the redstart well known in Britain, though it is not found in Ireland; it is widespread in Eurasia including Asia Minor and Iran, but local in north-west Africa and wintering south of the Sahara to the equator.

Eastern Bluebird   *Sialia sialis*
sialis (Gr) a kind of bird; a spectacular bright blue bird living in America; it ranges from eastern Canada south to Florida, the coast of the Gulf of Mexico, Texas, Mexico, and south to Honduras.

Townsend's Solitaire   *Myadestes townsendi*
muia (Gr) a fly   *edestēs* (Gr) an eater: named after John Kirk Townsend (1809-1851), an American ornithologist and author. The name solitaire has been given because it is a shy bird and tends to keep to itself; it inhabits the mountains of western North America and Alaska and winters south to Mexico.

Black-tailed Chat   *Cercomela melanura*
kerkos (Gr) the tail of an animal   *melas* (Gr) black   *oura* (Gr) the tail; inhabiting Arabia and Africa.

Stonechat   *Saxicola torquata*
saxum (L) a stone   *colo* (L) I inhabit; it lives in open country, heaths, moors, alpine meadows and mountains up to 3,600 m (12,000 ft); it has been said the name is taken from its alarm call which sounds like two stones being knocked together, and does not refer to a stony or rocky habitat   *torquatus* (L) wearing a collar made of twisted material; it has a partial white collar merging into a chestnut breast. It inhabits the south-western part of Europe including Denmark and the British Isles, East Africa and Madagascar; it ranges east through Asia Minor to China, Korea, and Japan.

Sooty Chat   *Myrmecocichla nigra*
murmēx (Gr), genitive *murmēkos*, an ant   *kikhlē* (Gr) a bird like a thrush; it has a thrush-like song but it is not known to eat ants; it often inhabits places where there are termite hills   *niger* (L) black; the male is a glossy black and the female a sooty brown. It is widely distributed in the central and western parts of Africa including Sudan, Cameroon, Zaire, Angola, and parts of East Africa.

Mountain Wheatear   *Oenanthe monticola*
oinē (Gr) poetic name for the vine   *oinanthē* (Gr) the first shoot of the vine; can mean a bird, one that appears in Greece at the same time as the first shoot of the vine, some say the wheatear   *mons* (L), genitive *montis*, a mountain   *colo* (L) I inhabit; it usually lives in rocky and stony areas but not always in the mountains. Inhabiting south-west Africa.

Wheatear   *O. oenanthe*
This is the most far-ranging of the wheatears, inhabiting most of Eurasia except the extreme north in Siberia; it is found in Alaska and Iceland and winters in tropical Africa and southern Arabia. It is the only wheatear that regularly inhabits the British Isles.

Short-toed Rock Thrush   *Monticola brevipes*
mons (L), genitive *montis*, a mountain   *colo* (L) I inhabit   *brevis* (L) short   *pes* (L) the foot. Inhabiting mountainous areas of Transvaal in South Africa.

Rock Thrush   *M. saxatilis*
saxum (L) a stone   *saxatilis* (L) one that dwells among rocks; the English and the Latin names must leave little doubt that it inhabits mountains; this is so and it also favours boulder-strewn rocky hillsides. It lives in southern Europe, Spain, part of north-west Africa, Asia

Minor and the mountains of south-west Asia; it winters in tropical Africa.

Ashy Ground Thrush   *Zoothera cinerea*

zōos (Gr) alive   *thēraō* (Gr) I hunt, can mean I catch; a reference to the diet of snails, worms, and grubs for which it hunts on the ground *cinereus* (L) ash-coloured. Inhabiting Philippines.

Wood Thrush   *Hylocichla mustelina*

hulē (Gr) a wood, forest   *kikhlē* (Gr) a bird like a thrush   *mustela* (L) a weasel   *mustelinus* (L) weasel-coloured; this is a reference to the chestnut-brown plumage with white underparts. It inhabits the central and eastern parts of the USA ranging south to Florida, Texas, and in winter Mexico and Panama.

White-collared Blackbird   *Turdus albocinctus*

turdus (L) a thrush   *albus* (L) white   *cinctus* (L) a girding; can mean a girdle. Inhabiting India and the Burma-Vietnam area, Tibet and China.

Blackbird   *T. merula*

merula (L) the European blackbird. Inhabiting nearly the whole of Europe including the British Isles, and part of northern Africa, and ranging east through Asia Minor, the Caucasus, the mountains of southern Asia to southern China.

American Robin   *T. migratorius*

migrator (L), genitive *migratoris*, a wanderer; a reference to its rare visits to western Europe in the autumn and winter; quite unlike the Eurasian Robin *Erithacus rubecula* which we know in Britain, it is much bigger, about 25 cm (10 in) long compared to 14 cm ($5\frac{1}{2}$ in); it is more like and closely related to the blackbird *T. merula*, above, and it is only the chestnut red breast that has given rise to the name 'robin'. It inhabits Alaska, most of Canada, and ranges south to Mexico and Guatemala.

Olive Thrush   *T. olivaceus*

oliva (L) an olive   *olivaceus*, like an olive; a reference to the dark olive-coloured plumage, both upper and underparts. Inhabiting open hill-sides and mountains and has been seen as high as 3,000 m (10,000 ft) on Kilimanjaro; widespread in most of Africa.

Song Thrush   *T. philomelos*

philos (Gr) beloved, dear   *melos* (Gr) a song; 'a song-lover'; there could be a connection with the Greek legend in which Philomela was changed by the gods into a nightingale, another bird with a fine song. Widespread in Europe including the British Isles and ranging east through Asia Minor to the Lake Baikal area; it winters in southern Europe, northern Africa and Arabia.

Fieldfare   *T. pilaris*
pilus (L) a hair   *-aris* (L) suffix meaning pertaining to; apparently this was a translation from the Greek *thrix*, genitive *trikhos*, hair, but this was a mistake because the word used was *trikhas* (Gr) a kind of thrush, the fieldfare. Inhabiting Europe including the British Isles and Iceland and ranging east into Asia to the Lake Baikal area, and south to Arabia in winter. Fieldfare is from an Old English word meaning 'field-dweller', or 'fell-dweller'.

Mountain Blackbird   *T. poliocephalus*
polios (Gr) hoary, grey   *kephalē* (Gr) the head; this grey-headed thrush lives in the mountainous areas of Malaysia, Indonesia, New Guinea, and the neighbouring Pacific Islands.

Ring Ouzel   *T. torquatus*
torquatus (L) wearing a necklace made of twisted material; a reference to the white breast-band which is prominent in the male. Inhabiting southern Europe and isolated parts in northern Scandinavia and northern Britain, and wintering in Africa and Asia Minor. Ouzel is from an Old English word *osle*, the blackbird.

Mistle Thrush   *T. viscivorus*
viscum (L) the mistletoe   *voro* (L) I devour; Aristotle recorded that it was fond of mistletoe berries; in addition to fruit and berries it will also take earthworms and insects. Widespread in Europe including the British Isles but not the northern part of Scandinavia, ranging east through Asia Minor and the mountains of south-west Asia and up to 3,350 m (11,000 ft) in the Himalayas; some migration in winter to Mediterranean areas.

Subfamily ORTHONYCHINAE   20 species
orthos (Gr) straight   *onux* (Gr) a claw, genitive *onukhos*; Prof Temminck (1778–1858), the Dutch zoologist, says so named from the straightness of its claws. This group includes the Logrunners, Whipbirds, Quail-Thrushes, and their kin; they eat a variety of insects and some will also take worms, fruit, and berries. They all live in the Oriental and Australasian regions.

Northern Logrunner   *Orthonyx spaldingi*
Named after E. Spalding (fl 1867–1894), a naturalist and taxidermist
of Australia; he worked at the Queensland Museum from 1880 to
1894. It spends a lot of its time on the ground, running about over
fallen tree trunks and forest debris hunting for insects; it also eats
berries. Inhabiting forests along the coast of northern Queensland in
Australia.

Eastern Whip-bird   *Psophodes olivaceus*
psophōdēs (Gr) noisy   *oliva* (L) an olive   *olivaceus* (L) similar to, olive-
coloured; a reference to the dark olive-green upperparts; a noisy bird,
it has a long, loud whistling call which increases in strength to finish
with a sort of 'whip-crack'. Inhabiting eastern Australia from northern
Queensland south through New South Wales to Victoria.

Cinnamon Quail-Thrush   *Cinclosoma cinnamomeum*
kinklos (Gr) a kind of ouzel, or thrush   *soma* (Gr) the body; 'thrush-
bodied'; it is about the size and colour of the song thrush *Turdus philomelos*; *cinnamomeum* (L) cinnamon; a reference to the chestnut brown
upperparts. Inhabits arid stony deserts of south-western Australia.

Malay Rail Babbler   *Eupetes macrocercus*
eu- (Gr) a prefix meaning well, nicely   *petes* (New L) a flier, derived
from *petomai* (Gr) I fly; 'flying well'   *makros* (Gr) long   *kerkos* (Gr)
the tail of an animal; a reference to the unusually long tail. Inhabiting
Malaysia and Borneo.

Subfamily   TIMALIINAE   252 species
timalia (New L) a bird name, probably from an Indian native name.
The correct classification of some of the birds in this large subfamily
has been a problem for ornithologists; they range in size from about
10 to 46 cm (4 to 18 in), they range in colour from rather drab to very
colourful; as a general rule they have rather short rounded wings,
longish tails, and a bill like that of a thrush, and the diet may be
insectivorous or vegetarian. They are mostly known as Babblers on
account of their continuous loud chattering and calling.

Ashy-headed Babbler   *Trichastoma cinereiceps*
trikhas (Gr) a kind of thrush, or fieldfare   *stoma* (Gr) the mouth;
'thrush-billed'   *cinereus* (L) ash-coloured   *ceps* (New L) from *caput*
(L) the head. Inhabiting Malaysia.

Tickell's Babbler   *T. tickelli*
Named after Col S. R. Tickell (1811–1875), a naturalist who was in

India and Burma from 1834 to 1865. This babbler inhabits India and ranges east through Burma to Philippines.

White-throated Babbler   *Malacopteron albogulare*
malakos (Gr) soft   *pteron* (Gr) feathers, wings   *albus* (L) white   *gula* (L) the throat. Inhabiting Malaysia.

Rusty-cheeked Scimitar Babbler   *Pomatorhinus erythrogenys*
pōma (Gr), genitive *pōmatos*, a lid, a cover   *rhis* (Gr), genitive *rhinos*, the nose; a reference to the corneous covering of the nares which is one of the distinguishing features of this genus   *eruthros* (Gr) red   *genus* (Gr) the jaw, the cheek. Inhabiting the Himalayas and ranging east from Bangladesh to southern China and Taiwan. It has a long curved bill, hence 'scimitar' babbler.

Black-throated Wren-Babbler   *Napothera atrigularis*
napē (Gr) a woody dell   *thēraō* (Gr) I hunt, I catch; 'a hunter in woods'   *ater* (L) black   *gula* (L) the throat   *-aris* (L) suffix meaning pertaining to. Inhabiting Borneo. The name 'wren' refers to the size, one of the smallest babblers.

Barred-Wing Wren-Babbler   *Spelaeornis troglodytoides*
spēlaios (Gr) a grotto, a cave   *ornis* (Gr) a bird; a reference to the form of the nest which is dome-shaped with an entrance hole at the side   *trōglē* (Gr) a hole, a hollow   *trōglodutēs* (Gr) one who creeps into holes   *-oides* (New L) from *eidos* (Gr) apparent shape, resemblance; a reference to it being somewhat like the Common Wren *Troglodytes troglodytes*, hence the English name Wren-Babbler; the wren itself takes its Latin name from the form of its nest. Inhabiting the Himalayas and the northern part of India and ranging south and east to the Burma-Vietnam area.

Wedge-billed Wren-Babbler   *Sphenocichla humei*
sphen (Gr) a wedge   *kikhlē* (Gr) a bird like a thrush: named after A. O. Hume (1829-1912), a naturalist and author who was in India in 1873. This wren-babbler lives in Assam and Burma.

Deignan's Babbler   *Stachyris rodolphei*
Stachyris (New L) spike-like, from *stakhuēros* (Gr) with ears of corn; a reference to the short spiky bill. Named after Rodolph M. de Schauensee (born 1901), an American zoologist and author of Philadelphia who specialised in taxonomy; he was in Siam (now Thailand) from 1932 to 1938; and H. G. Deignan (born 1906), a zoologist and author

who was also in Siam from 1928 to 1937. This babbler inhabits Thailand and neighbouring countries.

Red-capped Babbler   *Timalia pileata*
timalia (New L) a bird name, probably from an Indian native name *pileus = pilleus* (L) a felt cap   *pilleatus* (L) wearing the *pilleus*; a reference to the red plumage on the crown. Inhabiting India, the Burma-Vietnam area and Java.

Wren-Tit   *Chamaea fasciata*
khamai (Gr) on the earth, on the ground; it spends most of its time in thick cover hopping about looking for insects, and rarely flying *fascia* (L) a band or girdle   *fasciata* (L) banded, can mean striped; a reference to the buff brown underparts marked with long dark streaks. Inhabiting Oregon, California and south to Lower California.

Arrow-marked Babbler   *Turdoides jardinei*
turdus (L) a thrush   *-oides* (New L) from *eidos* (Gr) apparent shape, resemblance; at 24 cm (9$\frac{1}{2}$ in) it is about the same size as a thrush but the plumage is darker; some of the birds in this genus, comprising 26 species, are thrush-like in colour, others less so; this one is flecked with small white sharp-pointed marks, hence 'arrow-marked': named after Sir W. Jardine (1800-1874), an English naturalist and author. It is widespread in Africa, particularly central areas centred on Zaire and Kenya.

Spiny Babbler   *T. nipalensis*
-ensis (L) suffix meaning belonging to, usually of places; it inhabits Nepal, north-east India, and surrounding areas. It has spiny shaft tips to its feathers, hence 'spiny babbler'.

White-crested Laughing Thrush   *Garrulax leucolophus*
garrulus (L) talkative   *-ax* (L) suffix denoting an aggressive tendency; the 44 species in this large genus are known for their loud cackling sort of 'laughter', and they seem to attempt to outdo each other, each one calling louder than the one before   *leukos* (Gr) white   *lophos* (Gr) a crest. Widespread in the Oriental Region from India to Malaysia.

Collared Laughing Thrush   *G. yersini*
Named after Dr A. E. J. Yersin (1863-1943), a Swiss-French bacteriologist and Director of the Pasteur Institute of Indo-China in 1919; this country has now been broken up and in 1951 became Laos, Cambodia, and Vietnam. This laughing thrush inhabits Annam.

Peking Robin   *Leiothrix lutea*
leios (Gr) smooth, soft   *thrix* (Gr) hair, in this case taken to mean
feathers   *luteus* (L) golden-yellow; the throat is yellow merging into
an orange coloured breast; this also gives it the English name 'robin'.
From the southern Himalayas it ranges east across northern Burma
to China, remaining south of the Yangtze River.

Chestnut-headed Tit-Babbler   *Alcippe castaneceps*
Alkippē was the daughter of Ares, a Greek god of war; there is probably no special connection with Alkippē; a number of generic and
specific names have been given that were taken from classical mythology without any particular reason   *castaneus* (New L) chestnut-
coloured, from *kastanos* (Gr) the chestnut tree   *ceps* (New L) from
caput (L) the head. Inhabiting the Himalayas and ranging south
through India and Burma to Malaysia, Indonesia, and the Philippines.

Chapin's Flycatcher-Babbler   *Lioptilus chapini*
leios (Gr) smooth, soft   *ptilon* (Gr) a feather or a wing, but usually
means the soft feathers or down of a bird: named after Dr J. P. Chapin
(1889–1964), an American ornithologist and author who spent many
years in the Belgian Congo, now Zaire, and collected for the American
Museum of Natural History. Inhabiting Zaire.

Bare-headed Rockfowl   *Picathartes gymnocephalus*
pica (L) a magpie   *kathartēs* (Gr) a cleanser, a purifier   *Cathartes* the
Turkey Vulture; a combination of *pica* and *cathartēs*; it is partly corvine
(resembling a crow) and bald like a vulture   *gumnos* (Gr) naked
kephalē (Gr) the head; the front of the head and the cheeks are naked
and the skin is bright yellow. During the past 100 years, since the bird
was first discovered, there has been continuous discussion among
ornithologists about the taxonomic position; the consensus of opinion
now seems to be that its nearest relative is the babbler. It is confined
to some small areas in Sierra Leone and Ghana, western Africa.

Subfamily PARADOXORNITHINAE   19 species
paradoxos (Gr) contrary to opinion, strange   *ornis* (Gr), genitive
ornithos, a bird; like the Bare-headed Rockfowl, above, this small
group of birds has been a puzzle for the ornithologists. The beak is
heavy and rounded, like that of a parrot, which has given rise to the
name Parrotbill, but they are not closely related to the parrots; it has
been suggested that they are related to the babblers, and also to the

titmice, and this problem gave them the name *Paradoxornis*, 'strange bird'. They vary in length from about 12·5 to 30 cm (5 to 12 in), and their food is mainly insects, but they will take seeds and other vegetarian food when necessary.

Bearded Tit or Reedling   *Panurus biarmicus*
panu (Gr) all, altogether   *oura* (Gr) the tail; in this case meant to stress the length of the tail; this bird is about 16·5 cm (6½ in) long of which 7·5 cm (3 in) is tail; *biarmicus* (obscure); a name given by Linnaeus who often failed to give explanations of his names: 'Perhaps a corruption of *beardmanica* ("bearded manikin"), the barbarous name invented by Albin and referred to by Linnaeus with reference to the male's black moustache.' (Macleod). E. Albin (fl 1713–1759) was an English artist and author of *The Natural History of Birds*. The male has a black stripe on either side of the bill that gives the appearance of drooping moustaches. The main distribution is in the Black Sea area and it inhabits the reeds around lakes and swamps, hence 'reedling'; it ranges east to the River Ob, and west to England and Spain.

Ashy-throated Parrotbill   *Paradoxornis alphonsianus*
Paradoxornis, see notes above under Subfamily; named after Prof Alphonse Milne-Edwards (1835–1900), a French zoologist who was a Professor at the Paris Museum of Natural History in 1876 and a Director from 1891 to 1900. This parrotbill inhabits the southern part of Eurasia including part of the Oriental Region.

Greater Red-headed Parrotbill   *P. ruficeps*
rufus (L) red   *ceps* (New L) from *caput* (L) the head. Inhabiting India and the Burma-Vietnam area.

Subfamily   SYLVIINAE   339 species
sylva = *silva* (L) a wood, a forest. The members of this very big subfamily are known as Old World Warblers, nearly all living in Eurasia and Africa and ranging south-east to Australasia; a small group of about 12 species known as Gnatcatchers live in the Americas, ranging from the USA to Brazil. They are mostly smallish birds ranging from about 9 to 20 cm (3½ to 8 in), though some are a bit larger; the diet is mostly insects but they will also take worms and other invertebrates and some fruit and berries when necessary. Many are very good and melodious songsters which has given them the popular name Warblers.

Blue-grey Gnatcatcher   *Polioptila caerulea*
polios (Gr) hoary, grey   *ptilon* (Gr) a feather or wing   *caeruleus* (L)
blue. Inhabiting the southern part of the USA and wintering in
Mexico, Guatemala, Cuba and Bahamas.

Cetti's Warbler   *Cettia cetti*
There are 12 species in the genus *Cettia*, named after François Cetti
(1726–1780); he was an Italian Jesuit, and a zoologist and author,
and published a book *The Natural History of Sardinia* in 1774. This
warbler inhabits south-western Europe including Corsica and Sardinia and other Mediterranean islands, and the northern coastal strip
of Africa; recently it has been known to visit England; it ranges east
through Asia Minor, Iraq, to Pakistan.

Knysna Scrub Warbler   *Bradypterus sylvaticus*
bradus (Gr) slow   *pteron* (Gr) wings; a reference to the mode of flight
sylvaticus = *silvaticus* (L) belonging to a wood, or trees. Inhabiting
dense scrub and bramble thickets at the forest edge, it is named from
the town of Knysna on the south coast of South Africa, and is found
only in the extreme south ranging from Table Mountain east to Natal.

Chinese Bush Warbler   *B. tacsanowskius*
Named after Dr W. Taczanowski (1819–1890), a Polish zoologist and
author who worked at the Zoological Museum in Warsaw from 1855
to 1890. It inhabits China, India, and the Burma-Vietnam area.

Grasshopper Warbler   *Locustella naevia*
locusta (L) a locust, a grasshopper   *-ellus* (L) diminutive suffix; grasshoppers probably form part of the diet, but it has been suggested that
the high-pitched rasping call, very like the sound of a stridulating
grasshopper, has given rise to the English name   *naevus* (L) a mole or
wart on the body; can mean a spot, a blemish; the pale buff throat
and underparts are usually marked with spots and streaks. Inhabiting
western Europe including the British Isles but not northern Scandinavia; it winters in southern Europe, northern Africa, and ranges east
to India.

Great Reed Warbler   *Acrocephalus arundinaceus*
akron (Gr) a point, top   *kephalē* (Gr) the head; the crown of the head
is more pointed than in other Sylviinae   *arundo* = *harundo* (L) the
reed, cane   *harundinaceus* (L) like a reed, pertaining to reeds. One of
the largest birds in this subfamily, about 19 cm ($7\frac{1}{2}$ in) long, it builds
its nest in the reeds. Widespread in Europe but not Britain or Sweden

and Norway, it ranges east through southern Eurasia to Burma and western China, and south to Africa.

Sedge Warbler   *A. schoenobaenus*
skhoinos (Gr) the reed, rush   *bainō* (Gr) I walk, step; can also mean to stand, to be in a place. Inhabiting Europe including the British Isles but not Spain; east to India and western China, and wintering in Africa as far south as Natal.

Melodious Warbler   *Hippolais polyglotta*
hippolais (New L) from *hupolais* (Gr) a small bird; the Liddell and Scott Lexicon gives the Chiffchaff or 'the singing hedge-sparrow' *polus* (Gr) many   *glōtta* = *glōssa* (Gr) the tongue; 'many voiced'; it has a variety of musical songs and is sometimes imitative. Inhabiting France, Spain, Italy and the northern part of Africa.

Blackcap   *Sylvia atricapilla*
sylva = *silva* (L) a wood, a forest   *ater* (L) black   *capillus* (L) the hair of men, usually the hair of the head; can mean the hair of animals; the male has a sharply defined black crown and the female a reddish brown crown. Widespread in Eurasia including the British Isles, south to northern Africa, east to Asia Minor and part of Arabia, and western Siberia.

Whitethroat   *S. communis*
communis (L) common, general: widespread in Europe including the British Isles but not northern Scandinavia; ranging south to northern Africa, and east to Arabia, south-west Asia and India. The white throat is very noticeable against the brown of the head and upperparts.

Sardinian Warbler   *S. melanocephala*
melas (Gr), genitive *melanos*, black   *kephalē* (Gr) the head; unlike the blackcap (above) the head is entirely black to below the eye. Named from Sardinia it is widespread in the Mediterranean region in a wide belt all round the coast, including northern Africa; it ranges east through Asia Minor, part of Arabia, the Persian Gulf, and probably somewhat further east.

Chiffchaff   *Phylloscopus collybita*
phullon (Gr) a leaf   *skopos* (Gr) a watchman, a lookout man   *skopeō* (Gr) I see; a reference to its habit of searching amongst foliage for insects: *collybita* is probably a corruption of *kollubistēs* (Gr) a money-

changer, as the monotonous song is said to resemble the jingle of coins being counted; in France the bird is sometimes known as *Compteur d'Argent* (MacLeod). Inhabiting Eurasia including the British Isles, parts of northern Africa, Asia Minor, Arabia, and ranging east to India and Burma.

Wood Warbler   *P. sibilatrix*
sibilator (L) a whistler   *-atrix* (L) suffix meaning feminine; a reference to the song which is a series of notes, often given in flight, and repeated at a gradually increasing rate. Inhabiting Europe including England and Scotland but not Ireland or Spain, and ranging east to the Black Sea area and the Ural Mountains.

Goldcrest   *Regulus regulus*
rex (L), genitive *regis*, a king   *regulus* (L) a little king; in ancient times it was called 'king of the birds', probably by Aristotle, because of the golden stripe on the crown, and hence 'goldcrest'. Widespread in Eurasia, including the British Isles, Spain, and also northern Africa; ranging east to Black Sea area, Iran, Turkestan, Himalayas and Tibet.

Golden-crowned Kinglet   *R. satrapa*
satrapa (L) a wealthy ruler, a governor of a Persian province. Inhabiting North America from Alaska and in the east from Nova Scotia ranging south through Mexico to Guatemala.

Brown Emu-tail   *Dromaeocercus brunneus*
dromaios (Gr) swift, running at full speed; giving rise to the Latin name of the emu *Dromaius*, on account of this bird being a very fast runner *kerkos* (Gr) the tail of an animal; 'emu-tailed'   *brunneus* (New L) brown. Inhabiting Africa.

Black-throated Prinia   *Prinia atrogularis*
prinya (Javanese) a bird, probably this particular warbler as they are found in Java   *ater* (L) black   *gula* (L) the throat   *gularis*, pertaining to the throat. Inhabiting southern Asia and the Oriental Region to Malaysia and Indonesia.

Red-winged Warbler   *P. erythroptera*
eruthros (Gr) red   *pteron* (Gr) feathers, wings. Inhabiting Africa.

Carruthers' Cisticola   *Cisticola carruthersi*
cista (L) a wooden box or basket, often woven of twigs   *colo* (L) I inhabit, I dwell in; a reference to the nest, which is a ball of twisted grass with a side entrance, sometimes with leaves stitched in; has been

called a 'purse nest': named after A. Douglas M. Carruthers (1882–1962), a naturalist who was exploring in Uganda in 1906. This warbler lives in Africa.

Golden-headed Cisticola   *C. exilis*
exilis (L) thin, small; one of the smallest warblers, it is only about 9 cm ($3\frac{1}{2}$ in) long: the male has a golden crown during the breeding season. Inhabiting the Indian subcontinent, China, the Burma-Vietnam area, ranging south through Indonesia to the northern and eastern parts of Australia, and north-eastern Tasmania.

Common Tailorbird   *Orthotomus sutorius*
orthos (Gr) straight   *tomē* (Gr) a cutting off; a reference to the cutting of the fibres that are used to sew the nest together   *sutilis* (L) sewed together   *sutor* (L) a shoemaker   *sutorius* (L) actually means 'belonging to a shoemaker'; it is intended as a reference to the nest which is a remarkable structure made of a leaf folded over or leaves sewn together, the nest itself being made inside this 'purse'. Widespread in the Oriental region including Sri Lanka and southern China.

Bar-throated Apalis   *Apalis thoracica*
Apalis, an obscure name, probably from an African native name; it is not given in reference books, and not explained in the ornithological records in the British Museum (Natural History)   *thōrax* (Gr), genitive *thōrakos*, the chest   *-icus* (L) suffix meaning belonging to; it has a black bar across the throat at the top part of the chest.. Inhabiting eastern and southern Africa south of the Equator but not found in coastal areas.

Yellow-backed Eremomela   *Eremomela icteropygialis*
erēmos (Gr) lonely, solitary; of places, a wilderness, desert   *melos* (Gr) a song; a 'desert songster'; a reference to its habitat which is dry open bush and desert areas   *ikteros* (Gr) the jaundice; also a bird of a yellowish green colour, by looking at which a jaundiced person was cured—the bird died!   *pugē* (Gr) the rump, buttocks   *-alis* (L) suffix meaning relating to. Inhabiting Africa south of the Sahara and ranging south through eastern Africa to south Africa but not in the eastern and most southern coastal areas.

Long-billed Crombec   *Sylvietta rufescens*
sylva = *silva* (L) a wood, a forest   *-etta*, a diminutive suffix   *rufus* (L) red, ruddy   *-escens* (L) suffix meaning beginning to, approaching, i.e. 'reddish'; the predominant colour of the plumage is a reddish

brown: the name Crombec comes from the Old English word *crump*, or *crumb*, meaning crooked, and *bec*, a beak; a reference to the beak which is long and curved. Inhabiting south-eastern Zaire, Zambia, and Malawi and ranging south to South Africa.

Spinifex Bird   *Eremiornis carteri*

erēmos (Gr) lonely, solitary; of places, a wilderness, desert   *ornis* (Gr) a bird; an allusion to its habitat which is arid and often desert country with low scrub: named after T. C. Carter (1863-1931), an English naturalist who explored and collected specimens in Australia. The English name derives from a type of Australian grass known as spinifex. This warbler inhabits Australia from central Queensland ranging west through Northern Territory to part of Western Australia, and also Barrow Island.

Fernbird   *Bowdleria punctata*

Named after Dr R. Bowdler Sharp (1847-1909), an English zoologist and author who worked in the Bird Room at the British Museum (Natural History) from 1872 to 1909   *punctum* (L) a hole, a puncture *punctatus*, spotted as with punctures; a reference to the spotted dark brown throat and breast. The English name refers to the habitat which is wasteland and ferny scrub; it inhabits New Zealand and nearby islands.

Subfamily   MALURINAE   29 species

malos (Gr) has various meanings, and some consider it is the same as *malakos* (Gr) soft; here it is taken to mean slender   *oura* (Gr) the tail; a reference to the long slender tail.

This group of small birds about 10 to 16·5 cm (4 to 6½ in) long, mainly insectivorous, have been given various English names such as Wren, Wren-Warbler, and Grass-Wren; they live in Australia and some neighbouring islands.

Superb Wren-Warbler   *Malurus cyaneus*

kuaneos (Gr) dark blue, glossy blue; of the swallow; it is not entirely blue, but the head and shoulders are blue, with the hind neck, the throat, and the lower back black, making a striking contrast. Inhabiting eastern and southern Australia, Tasmania, and some islands in the Bass Strait.

Red-backed Wren-Warbler   *M. melanocephalus*

melas (Gr), genitive *melanos*, black   *kephalē* (Gr) the head; the head and neck are glossy black and the back is red. Inhabiting western part of Australia and ranging east to northern Queensland and south to New South Wales.

Blue Wren-Warbler   *Todopsis cyanocephala*
todus (L) a small bird, a tody   *opsis* (Gr) aspect, appearance; indicating that it is rather like a tody, of the family Todidae, a small brightly coloured insectivorous bird   *kuaneos* (Gr) dark blue, glossy blue *kephalē* (Gr) the head. Inhabiting New Guinea.

Black Grass-Wren   *Amytornis housei*
Named after Amytis, daughter of Astyages, the last king of the Median Empire   *ornis* (Gr) a bird; and Dr F. M. House (1865-1936), a zoologist and sheep breeder from Western Australia. During a long period in which this bird was not seen by any observers it was found again in the Kimberley district of Western Australia in 1968. The plumage is mainly black; habitat woodland and probably spinifex grass.

Striated Grass-Wren   *A. striatus*
stria (L) a furrow, a channel   *striatus* (New L) striped. Inhabiting Australia.

Subfamily   ACANTHIZINAE   59 species
akantha (Gr) a thorn   *-iza* is a mutilated form of *spiza* (Gr) a finch or any small bird, derived from *spizō* (Gr) I pipe, I chirp, from the shrill note of small birds; the name is an allusion to the short fine bill, hence the English name 'thornbill'.
   This group of Australasian Warblers have been given various names such as Thornbill, Bristle-Bird, and Gerygone; they are small, from about 9 to 18 cm ($3\frac{1}{2}$ to 7 in) long, and mainly insectivorous.

Bristle-Bird   *Dasyornis broadbenti*
dasus (Gr) hairy, shaggy   *ornis* (Gr) a bird; a reference to the rictal bristles: named after Kendall Broadbent (1837-1911) a naturalist and specially a taxidermist who worked at the Queensland Museum. Inhabiting Australia.

Brown Weebill   *Smicrornis brevirostris*
smikros (Gr) an Ionic form of *mikros* (Gr) small   *ornis* (Gr) a bird; an allusion to the size of this tiny bird, about 9 cm ($3\frac{1}{2}$ in) long   *brevis* (L) short   *rostrum* (L) the beak; a reference to the small beak, hence also 'weebill'. Inhabiting Queensland and Victoria and ranging across southern Australia to Western Australia.

Yellow-tailed Thornbill   *Acanthiza chrysorrhoa*
Acanthiza, see notes under Subfamily, above   *khrusos* (Gr) gold
orhos (Gr) the rump, bottom; a reference to the rump and adjacent
part of the tail which is yellow. Inhabiting Australia.

Pilot-Bird   *Pycnoptilus floccosus*
puknos (Gr) thick, compact; of a bird's plumage thick, close   *ptilon*
(Gr) a feather or wing; properly the soft feathers or down of a bird
floccus (L) a flock of wool   *floccosus* (L) full of flocks of wool; very
woolly; a reference to the thick plumage. The name 'pilot bird' has
been given because it has a reputation for guiding the Superb Lyrebird *Menura novaehollandiae* to food, but it is also said that it is the Lyrebird that is the 'pilot'. Inhabiting the mountains of the south-eastern
part of Australia.

Chestnut-tailed Ground-Wren   *Hylacola pyrrhopygia*
hula (Doric) = *hulē* (Gr) a wood, a forest   *colo* (L) I inhabit   *purros*
(Gr) red; flame-coloured   *pugē* (Gr) the rump, buttocks. Inhabiting
south-eastern Australia.

White-throated Gerygone Warbler   *Gerygone olivacea*
gērugonē (Gr) born of sound; an echo; the name is from *The Idylls of
Theocritus*, poems of idealised rustic life; a reference to the sweet song
of these warblers, 'a distinctive and beautiful warble'   *olivaceus* (L)
like an olive, olive coloured; the upperparts are ashy-brown to olive,
and the chin and throat are white. Inhabiting the northern part of
Western Australia, ranging east to Queensland and south to Victoria;
also found in New Guinea.

Yellow-breasted Gerygone Warbler   *G. sulphurea*
sulfur (L) brimstone, sulphur   *sulfureus* (L) like sulphur; a reference
to the yellow breast. Inhabiting the Burma-Vietnam area and
Malaysia.

Spotted Scrub-Wren   *Sericornis maculatus*
sērikos (Gr) silken   *ornis* (Gr) a bird   *macula* (L) a mark, a spot
maculatus, spotted; a reference to the black marks on the white throat
and breast. Inhabiting south-western Australia and some neighbouring islands.

Black-and-Green Sericornis   *S. nigroviridis*
niger (L) black   *viridis* (L) green. Inhabiting New Guinea.

Rufous Songlark   *Cinclorhamphus mathewsi*
kinklos (Gr) a small bird, probably a water ouzel or dipper   *rhamphos* (Gr) the crooked beak of birds; a reference to the slightly curved beak: named after G. M. Mathews (1876-1949), an Australian naturalist and author; he founded the Mathews Ornithological Library and published his book *The Birds of Australia* in 1910. Widespread on the Australian mainland apart from Cape York Peninsula in Queensland.

Subfamily   MUSCICAPINAE   134 species
musca (L) a fly   *capio* (L) I take, seize. This subfamily, known as Old World Flycatchers, is widespread in Europe, Africa, southern Asia, and ranging east to Australasia; as the name tells us, they are not known in the New World, i.e. the Americas, and it also suggests, correctly, that their main food is insects.

Pale Flycatcher   *Bradornis pallidus*
bradus (Gr) slow   *ornis* (Gr) a bird   *pallidus* (L) pale; the throat, belly and underparts are pale brown to white, the upperparts darker. Widespread through most of Africa.

South African Black Flycatcher   *Melaenornis pannelaina*
melas (Gr), genitive *melanos*, black   *ornis* (Gr) a bird   *panu* (Gr) altogether; can mean very, exceedingly   *elaia* (Gr) an olive   *elainos* (Gr) of the olive tree; the plumage is entirely black with a bluish-green gloss. Inhabiting Kenya and ranging south to South Africa and across to West Africa south of Congo.

White-browed Forest Flycatcher   *Fraseria cinerascens*
Named after Louis Fraser (fl 1819-1866) of the Zoological Society of London; he went on an expedition to Niger during 1841 and 1842 *cinereus* (L) ash-coloured   *-ascens = -escens* (L) suffix meaning beginning to, approaching, greyish. Inhabiting Africa.

Pied Flycatcher   *Ficedula hypoleuca*
ficedula (= *ficecula* and *ficetula*) (L) a garden warbler or similar small bird, a *biccafico* (It) from *biccare*, to peck, and *fica*, a fig, 'a fig pecker'; they are not known to eat figs but might have been seen pecking insects from figs; insects, larvae, and spiders are their main food   *hupo* (Gr) under, less than usual   *leukos* (Gr) white; the underparts are a dullish white, 'off white'. Inhabiting large areas of Europe including western Britain, parts of Spain, and northern Africa, ranging east to

Siberia and south to Asia Minor; wintering in Africa, south to Tanzania.

Palawan Flycatcher   *F. platenae*
Named after Dr Margarete Platen, wife of Dr C. C. Platen (1843–1899); they were in the Celebes and Philippines from 1878 to 1894. The English name refers to Palawan Island in the Philippines where the bird was found.

Brown Flycatcher   *Muscicapa latirostris*
musca (L) a fly   *capio* (L) I take, seize; the diet consists of various types of insects including beetles and grasshoppers   *latus* (L) wide   *rostrum* (L) the beak; a reference to the unusually broad bill: the upperparts are a blackish brown. Inhabiting eastern Siberia and ranging south to Manchuria and Korea, parts of India south-east through Burma to Indonesia and the Philippines.

Spotted Flycatcher   *M. striata*
stria (L) a furrow   *striatus* (New L) striped; it has dark stripes on the forehead and breast, but on the forehead these appear like broken lines, hence spotted. Widespread in Europe including the British Isles, and part of northern Africa, ranging east into Asia to the borders of Manchuria, and wintering in north-west India and Africa.

Grey Tit-Flycatcher   *Myioparus plumbeus*
muia (Gr) a fly   *parus* (L) a titmouse   *plumbum* (L) lead   *plumbeus* (L) leaden, lead-coloured. Inhabiting Africa.

Southern Yellow Robin   *Eopsaltria australis*
ēōs (Gr) dawn, the red colour at dawn; can mean the east   *psaltria* (Gr) a harpist; 'a harp-player from the east'   *auster* (L) the south   *australis* (L) southern; it does not necessarily mean Australia. Inhabiting the eastern part of Australia.

Yellow-bellied Robin   *E. flaviventris*
flavus (L) yellow   *venter* (L), genitive *ventris*, the belly. 'Robin' is a local name; although related to the robins it is not a true robin, as in the subfamily Turdinae. Inhabiting the Pacific Islands to the east of New Guinea.

White-faced Robin   *Tregallasia leucops*
Named after T. H. Tregallas (1864–1938), an Australian ornithologist from Victoria   *leukos* (Gr) white   *ops* (Gr) the eye, the face. Inhabiting Australia and New Guinea.

Rock Robin Flycatcher   *Petroica archboldi*
petra (Gr) a rock   *-icus* (L) suffix meaning belonging to: named after
R. Archbold (born 1907); an American zoologist who worked in the
Mammals Department at the American Museum of Natural History;
he was in New Guinea in 1933 and 1934. This flycatcher lives in
Papua New Guinea.

Red-capped Robin Flycatcher   *P. goodenovii*
Named after Bishop Samuel Goodenough (1743-1829), an English
naturalist who was at one time Vice-President of the Linnean Society
of London. Widespread in Australia although not found in desert
areas or humid forests.

Grey-headed Flycatcher   *Culicicapa ceylonensis*
culex (L), genitive *culicis*, a gnat, a fly   *capio* (L) I take, seize   *-ensis*
(L) suffix meaning belonging to, usually of places; 'of Ceylon', now
known as Sri Lanka. From India it ranges east through the Oriental
area to Malaysia and Indonesia.

Subfamily   RHIPIDURINAE   38 species
rhipis (Gr), genitive *rhipidos*, a fan   *oura* (Gr) the tail; the birds in this
subfamily have fan-shaped tails that can be spread into quite a large
fan which they can also wave up and down, which accounts for the
English name Fantail Flycatcher.

Grey Fantail   *Rhipidura fuliginosa*
fuligo (L), genitive *fuliginis*, soot   *-osus* (L) suffix meaning full of,
prone to; 'very sooty'; a reference to the plumage which is mainly
grey. Widespread in Australia, also found in Tasmania, New Zealand,
New Caledonia, New Guinea; and also occurs on some Pacific
islands.

Red-backed Fantail Flycatcher   *R. opistherythra*
opisthe (Gr) behind   *eruthros* (Gr) red. Inhabiting New Guinea.

Red-tailed Fantail   *R. phoenicura*
phoinix (Gr), genitive *phoinikos*, red   *oura* (Gr) the tail. Inhabiting the
Indonesian area in the Oriental region.

Rufous-fronted Fantail   *R. rufifrons*
rufus (L) red   *frons* (L) forehead, brow. Inhabiting Celebes and ranging east to New Guinea, Solomon and other Pacific islands, and
Australia.

Subfamily MONARCHINAE   133 species
monarkhos (Gr) a monarch, a king; the describers N. A. Vigors and Dr
T. Horsfield wrongly supposed that these birds were allied to the
Kingbirds of the family Tyrannidae, genus *Tyrannus*, named from
their aggressive behaviour (see notes under Family on p. 356). N. A.
Vigors, FRS (1785–1840) was Editor of the *Zoological Journal* from
1827 to 1834; Dr T. Horsfield (1773–1859) was a scientist who was in
Sumatra from 1796 to 1818; he named many mammals and birds in
the East.

Cape Flycatcher   *Batis capensis*
batis (Gr) a ray or skate; can also mean a bird that frequents bushes;
living in woodland areas it builds its nest in trees and bushes   *-ensis*
(L) suffix meaning belonging to, usually of a place; from the Cape of
Good Hope, South Africa; it ranges through the Cape Province as far
north as Natal.

Little Yellow Flycatcher   *Erythrocercus holochlorus*
eruthros (Gr) red   *kerkos* (Gr) the tail of an animal   *holos* (Gr) whole,
entire; can mean on the whole, in general   *khloros* (Gr) green; can
also mean honey-coloured, yellow. Inhabiting Africa.

Black Paradise Flycatcher   *Terpsiphone atrocaudata*
terpsis (Gr) enjoyment, delight   *phōnē* (Gr) a sound, the voice, of men
or animals; a reference to the joyful song, although some of the birds
in this genus are rather indifferent songsters   *ater* (L) black   *cauda*
(L) a tail   *caudata* (L) having a tail; the black tail may be as much as
25 cm (10 in) long. Inhabiting southern Asia and ranging south
through Burma to Malaysia and Indonesia.

Seychelles Paradise Flycatcher   *T. corvina*
corvus (L) a rook   *-inus* (L) suffix meaning like; it is not as big as a
rook, the body, without tail, being about 20 cm (8 in), but the colour
is a glossy blue-black; the central tail feathers may be 25 to 30 cm (10
to 12 in) long. It is confined to a small island in the east of the Seychelles group known as La Digue.

African Paradise Flycatcher   *T. viridis*
viridis (L) green; the head neck and breast is a metallic blue-green.
Widespread in Africa extending to south-western Arabia, but not
known in Madagascar.

Black Myiagra Flycatcher   *Myiagra atra*
muia (Gr) a fly   *agra* (Gr) a catching, a hunting; can mean that which

is taken in hunting, the prey *ater* (L) black. Inhabiting New Guinea.

Restless Flycatcher *M. inquieta* (formerly *Seisura*)
inquietus (L) restless, unquiet; a reference to the bird's habit of making constant small movements, as though nervous. Inhabiting northern Australia and ranging along the coastal area to the north-east and the east; it is also known in the south-west.

Yap Island Monarch *Monarcha godeffroyi*
Monarcha, see notes under subfamily opposite: J. C. Godeffroy (1813–1885) was a German zoologist; he founded the Godeffroy Museum in Hamburg and organised various collectors who went to the South Pacific. Yap Island, a small Pacific island, lies about 1,288 km (800 miles) north of New Guinea in the group known as the Caroline Islands.

Black-faced Monarch Flycatcher *M. melanopsis*
melas (Gr), genitive *melanos*, black *opsis* (Gr) appearance, face; the colour is mainly grey, with a sharply defined black throat and forehead which surround the beak. Inhabiting Timor, New Guinea, and the coastal belt in eastern Australia from Cape York south to Victoria.

Versicolour Flycatcher *Mayrornis versicolor*
Mayr + *ornis* (Gr) a bird; the name refers to Dr Ernst Mayr (born 1904), a Professor of Zoology at Harvard University, and a Director of the Museum of Comparative Zoology; he has written several authoritative books on birds *versicolor* (L) of various colours, partly coloured. Inhabiting the islands of the South Pacific.

Subfamily PACHYCEPHALINAE 48 species
pakhus (Gr) thick, stout *kephalē* (Gr) the head; sometimes known by the name Thickheads, on account of the large head and rather shrike-like bills; they are also known as Whistlers, a reference to the pleasant call notes, and also sometimes called Shrike-Thrushes. They range in size from 15 to about 30 cm (6 to about 12 in), and the diet is mainly insects though some berries may also be taken.

Crested Bellbird *Oreoica gutturalis*
oros (Gr), genitive *oreos*, a mountain *-icus* (L) suffix meaning belonging to *guttur* (L) the throat *-alis* (L) suffix meaning relating to; a reference to the black gorget round the white face and throat; it has a black crest on the forehead and the name 'bellbird' refers to the bell-like note heard at the end of the song. Widespread in Australia but mostly away from coastal areas.

Golden-faced Pachycare   *Pachycare flavogrisea*
pakhus (Gr) thick, stout   *karē* (Gr) the head; see above under Subfamily   *flavus* (L) yellow, golden   *griseus* (New L) grey. Inhabiting
New Guinea.

Lorentz's Whistler   *Pachycephala lorentzi*
Named after Dr H. A. Lorentz (1871–1944), a zoologist who was
exploring in New Guinea during the years 1903 to 1910, and later
was appointed as Dutch Consul in South Africa. The birds in this
genus are usually known as Whistlers on account of the pleasant
whistling call notes. Inhabiting New Guinea.

Golden Whistler   *P. pectoralis*
pectus (L), genitive *pectoris*, the breast   *-alis* (L) suffix meaning relating to; a reference to the prominent band of black across the golden
breast and the white throat which makes a contrast. It is widespread
in the Australasian area, from Indonesia to the Bismarck Archipelago,
and Fiji, and along the southern part of Australia including Tasmania.

Rufous Shrike-Thrush   *Colluricincla megarhyncha*
kolluriōn (Gr) a bird like a thrush   *kinklos* (Gr) a bird, perhaps the
water ouzel; this bird in the genus *Cinclus* is considered by some to be
related to the thrushes; it is more often known as a Dipper   *megas* (Gr)
big   *rhunkhos* (Gr) the beak; a reference to the shrike-like beak.
Inhabiting Indonesia and Australia.

Family   AEGITHALIDAE   7 species
aigithalos (Gr) the tit, titmouse; sometimes classified as one of three
subfamilies under Paridae, in which case the name becomes Aegitha-
linae; the other two are Parinae and Remizinae. This little group are
known as Long-tailed Tits.

Long-tailed Tit   *Aegithalos caudatus*
cauda (L) the tail of an animal   *-atus* (L) suffix meaning provided
with; a reference to the tail which may be even longer than the body,
the whole bird being about 12·5 to 15 cm (5 to 6 in). Widespread in
Europe including the British Isles, ranging east through Asia Minor,
Iran, and the Caucasus to Kamchatka, Japan, and China.

White-cheeked Long-tailed Tit   *A. leucogenys*
leukos (Gr) white   *genus* (Gr) the cheek, the chin. Inhabiting southern
Asia.

Common Bush Tit *Psaltriparus minimus*
psaltria (Gr) a female harpist   *parus* (L) a titmouse, or tit; a reference
to the high-pitched buzzing notes of the call   *minimus* (L) smallest;
it is only 11·5 cm (4½ in) long. Inhabiting the western part of North
America, and Mexico.

Family   REMIZIDAE   9 species
remiz is probably a Polish name for these birds; its use as a generic
name was given by Dr F. P. Jarocki in an obscure Warsaw periodical
in 1819; it replaced the former name *Anthoscopus*. Sometimes classified
as a subfamily, see above under Family Aegithalidae; they are
generally known as Penduline Tits.

Mouse-coloured Penduline Tit   *Remiz musculus*
mus (L) a mouse, diminutive *musculus*, a small mouse; in this case
taken to mean like a mouse; the name penduline is from the Latin
pendulus, hanging, pendent, and refers to the type of nest which is built
suspended from twigs of willow or other trees and usually hanging
over water. Inhabiting Africa.

Penduline Tit   *R. pendulinus*
Inhabiting the central part of Europe and ranging east across Eurasia,
and south to southern France, Italy, Greece, Asia Minor to China.

Verdin   *Auriparus flaviceps*
aurum (L) gold   *parus* (L) a tit   *flavus* (L) yellow, golden   *ceps* (New
L) from *caput* (L) the head; the head, face, and throat are golden
yellow   *verdin* is a French word for a bird, usually used to mean the
yellowhammer. Inhabiting the south-western part of the USA and
northern Mexico.

Fire-capped Tit   *Cephalopyrus flammiceps*
kephalē (Gr) the head   *pur* (Gr), genitive *puros*, fire   *flamma* (L) a
blaze, a flame   *ceps* (New L) from *caput* (L) the head. The range is
from India to China.

Family   PARIDAE   46 species
parus (L) a titmouse, a tomtit. Sometimes classified as a subfamily;
see under Family Aegithalidae opposite. These small birds, variously known as tits, tomtits, titmice, and chickadees, are popular
wherever they come in contact with humans and are regular visitors
to the garden bird table. Mainly insectivorous, they can manage on

other foods when necessary, and will take bread and other scraps put out in the garden, and of course nuts; this can be a real life-saver for them during a hard winter.

Coal Tit   *Parus ater*
parus (L) a titmouse   *ater* (L) black; not entirely black, the colour is mostly grey, and the black head has a white patch on the back of the neck and the cheeks. Inhabiting Eurasia including the British Isles, Africa north of the Sahara, and ranging east to Kamchakta to Japan.

Black-capped Chickadee   *P. atricapillus*
ater (L) black   *capillus* (L) the hair of men, usually the hair of the head; can mean the hair of animals: 'chickadee' is imitative of the call. Inhabiting Canada, the USA, and ranging south to Texas and northern Mexico.

Blue Tit   *P. caeruleus*
caeruleus (L) blue; this particularly refers to the blue cap. Widespread in the western part of Eurasia including the British Isles, Africa north of the Sahara, ranging east through Asia Minor, to Caucasus, and Iran.

Crested Tit   *P. cristatus*
crista (L) usually means a tuft on the head of animals   *cristatus* (L) crested. Inhabiting Europe but not the British Isles except northern Scotland, and not Italy, but ranging east to western Siberia.

Elegant Titmouse   *P. elegans*
elegans (L) neat, elegant. Inhabiting the Philippines.

Great Tit   *P. major*
major (L) greatest; one of the largest tits at about 14 cm ($5\frac{1}{2}$ in). Inhabiting Eurasia including the British Isles, Africa north of the Sahara, and ranging east to India, the Burma-Vietnam area, Malaysia, Indonesia and China.

Willow Tit   *P. montanus*
mons (L), genitive *montis*, a mountain   *-anus* (L) suffix meaning belonging to; it is known to inhabit mountain shrubs at 2,133 m (7,000 ft) altitude in Asia: although it often breeds in marshy areas it does not have a special liking for willow trees. It is widespread in Eurasia including the southern half of Britain, but not Spain or Africa, and ranges east to Siberia, Kamchatka, and Japan.

Marsh Tit *P. palustris*
palus (L) a swamp, a marsh *palustria* (L) swampy or marshy places;
a misleading name as it does not usually frequent marshy places.
Inhabiting Europe including England and Wales but not Ireland and
Scotland, and not northern Scandinavia or Spain. A separate population inhabits eastern Asia.

Chestnut-backed Chickadee *P. rufescens*
rufus (L) red, ruddy *-escens* (L) suffix meaning approaching, beginning to; i.e. 'reddish': the name 'chickadee' is imitative of the call.
Inhabiting North America.

Sultan Tit *Melanochlora sultanea*
melas (Gr), genitive *melanos*, black *khloros* (Gr) green, can mean
honey-coloured, yellow *sultaneus* (New L) like a sultan, from the
colourful exotic plumage; mainly black offset by a bright golden-
yellow crest, chest, and underparts. Probably the largest tit with a
length up to 20 cm (8 in), inhabiting the eastern Himalayas from
Nepal south to Burma and Malaysia.

Family SITTIDAE 21 species
sittē (Gr) a kind of woodpecker; also used to mean the nuthatch. The
Nuthatches are small insectivorous birds closely related to the tits and
often seen with them in parties hunting for insects and spiders; they
will also take seeds and nuts, wedging a nut into a crevice and hammering at it with their strong beaks like a woodpecker; they walk
about tree trunks, up, down, and across just as though they were on
the ground.

Red-breasted Nuthatch *Sitta canadensis*
sittē (Gr) = *sitta* (New L) the nuthatch *-ensis* (L) suffix meaning
belonging to, usually of a place; it is not confined to Canada, and
ranges from Alaska to Newfoundland and south to California, Florida,
and northern Mexico.

Nuthatch *S. europaea*
Widespread in Europe including the southern half of Britain and the
southern half of Scandinavia, ranging east to include Asia Minor, the
Caucasus, Iran, Siberia, Japan, India, China and the Burma area.

Corsican Nuthatch *S. whiteheadi*
Named after John Whitehead (1860-1899), an English explorer and

naturalist who made an expedition to Corsica, and later to Borneo in 1885, and the Philippines in 1893.

Pink-faced Nuthatch   *Daphoenositta miranda*
daphoinos (Gr) has been used to mean, of wild beasts, blood-reeking, bloody; probably more correctly of their colour, red, tawny   *sittē* (Gr) = *sitta* (New L) the nuthatch; a reference to the colour of the face   *miror* (L) I wonder, I admire   *mirandus* (L) wonderful, marvellous. Inhabiting Papua New Guinea.

Wall Creeper   *Tichodroma murina*
teikhos (Gr) a wall   *dromos* (Gr) running about   *murus* (L) a wall -*inus* (L) suffix meaning belonging to; a reference to its habit of climbing about on rocks looking for insects, but it also frequents trees. Inhabiting mountains in Spain, northern Africa, in the Alps and other European ranges, Asia Minor, the Caucasus, Iran, and east through the Himalayas to north-west China; in inhabited areas it climbs about on walls.

Family   CERTHIIDAE   6 species
kerthios (Gr) a little bird, a treecreeper. These small birds, known as Treecreepers, are related to the Nuthatches; as the name suggests, they spend a lot of their time creeping about on trees looking for small invertebrate animals and often in parties with tits or nuthatches; unlike the latter which have soft tails, they have stiff tails that they use for support when climbing. Although essentially they are birds of the Old World, it will be seen that one species has managed to establish itself in America.

Short-toed Treecreeper   *Certhia brachydactyla*
brakhus (Gr) short   *daktulos* (Gr) a finger, can mean a toe; a name which tells us that the toes are short compared to other creepers which all have longish toes and claws. Widespread in Europe but not known in the British Isles or Scandinavia; it ranges south to the northern part of Africa and east to parts of Asia.

Common Treecreeper   *C. familiaris*
familiaris (L) a servant; can mean familiar, habitual. Very widespread in both America and Eurasia; it ranges from Alaska south to Nicaragua in Central America, and northern Siberia south to parts of China, the Himalayas and Japan; in Europe it is well known in the British Isles, the southern part of Scandinavia, but not Spain or the northern part of Africa.

Spotted Creeper   *Salpornis spilonota*
salpinx (Gr) a war trumpet   *ornis* (Gr) a bird; an allusion to the shrill
call note   *spilos* (Gr) a blemish, a spot   *nōton* (Gr) the back; the
plumage is mainly brown, the upperparts and back being liberally
spotted with white. Inhabiting most of Africa ranging south to
Zambia and Zimbabwe, and northern South Africa; also the Himalayan foothills and India.

Family   RHABDORNITHIDAE   only 2 species
rhabdos (Gr) a rod or stick   *rhabdōtos* (Gr) striped; of animals streaked
ornis (Gr), genitive *ornithos*, a bird; a reference to the striped head.
Closely related to the Certhiidae, above, these two birds known as
Philippine Creepers are confined to some of the Philippine Islands.

Plain-headed Creeper   *Rhabdornis inornatus*
inornatus (L) undecorated, plain; a reference to the absence of the
stripes on the head. Found only in the Philippine Islands.

Stripe-headed Creeper   *R. mystacalis*
mustax (Gr), genitive *mustakos*, a moustache   *-alis* (L) suffix meaning
relating to, like; a reference to the white stripes on the head and above
the beak which give the appearance of a moustache. Like the Plain-
headed Creeper, above, it is found only in the Philippine Islands.

Family   CLIMACTERIDAE   6 species
klimaktēr (Gr) the rung of a ladder; a reference to their manner of
climbing trees as though on a ladder.

White-throated Treecreeper   *Climacteris leucophaea*
leukos (Gr) white   *phaios* (Gr) dark, grey; a reference to the white
throat which is in contrast to the mainly grey plumage. Eating mainly
insects and spiders, the treecreepers work their way up tree trunks in
a spiral climb searching for their prey. Inhabiting Australia and New
Guinea.

Black-tailed Treecreeper   *C. melanura*
melas (Gr), genitive *melanos*, black   *oura* (Gr) the tail. Inhabiting
Australia.

Family   DICAEIDAE   58 species
dicaeum (New L) a coined name meaning the flowerpecker. They do
not actually eat flowers but they take nectar from flowers with a

specially shaped tongue; they also eat berries and some species are particularly fond of mistletoe berries; probably all species also eat insects and spiders.

Olive-backed Flowerpecker   *Prionochilus olivaceus*

prion (Gr) a saw   *kheilos* (Gr) a lip, rim; of humans the lip, of birds the beak; a reference to the serrated or saw-like edge of the beak *oliva* (L) an olive   *olivaceus* (L) similar to, olive-coloured. Inhabiting the Philippines.

Anna's Flowerpecker   *Dicaeum annae*

dicaeum (New L) a coined name meaning the flowerpecker: named after Dr Anna A. Weber-Van Bosse (1852-1942), a Dutch botanist who was in New Guinea from 1888 to 1890; she was the wife of Prof M. W. C. Weber, the Dutch zoologist. This flowerpecker lives in Sumbalra and Flores.

Scarlet-backed Flowerpecker   *D. cruentatum*

cruento (L) I make bloody   *cruentatum*, stained with blood, red. Inhabiting India, the Burma-Vietnam area, Sumatra and Borneo.

Nehrkorn's Flowerpecker   *D. nehrkorni*

Named after Adolf Nehrkorn (1841-1916), a German ornithologist who collected for the Berlin and Brunswick museums. Inhabiting the mountains of Celebes.

Crested Berrypecker   *Paramythia montium*

paramuthia (Gr) encouragement, exhortation; a reference to the behaviour of the male during nest-building; he takes no part in the work but accompanies the female while she is collecting the material and building, as though to give encouragement   *mons* (L), genitive *montis*, a mountain; it inhabits the mountains and forests of New Guinea and is one of the flowerpeckers that is fond of the mistletoe berries.

Spotted Pardalote   *Pardalotus punctatus*

pardalōtus (Gr) spotted like a leopard   *punctus* (L) a puncture   *punctatus* (L) spotted as with punctures; these flowerpeckers have come to be known as 'pardalotes' on account of their spotted plumage. Inhabiting the eastern and southern part of Australia and Tasmania.

Forty-spotted Pardalote   *P. quadrigintus*

quadriginta (L) forty. Inhabiting Tasmania and Banks Island.

Family NECTARINIIDAE 118 species

nektar (Gr) nectar; in Greek mythology the drink of the gods; also *nectar* (L) nectar *-inus* (L) a suffix meaning belonging to; a reference to their liking for the honey from the glands of plants. This family, known as Sunbirds, have a similar diet to the Flowerpeckers, and in addition to nectar they eat insects and spiders, and in some cases berries.

Violet-backed Sunbird *Anthreptes longuemarei*
anthos (Gr) a flower *threptēr* (Gr) a feeder, a rearer; a reference to their habit of taking nectar from flowers; named after G. de Longuemare (fl 1831–1841), a French amateur collector. Inhabiting Africa.

Blue-headed Sunbird *Nectarinia alinae*
Named in honour of Lady Aline Jackson, the wife of Sir F. J. Jackson (1860–1929), naturalist and author; he was Governor of Uganda from 1911 to 1917. Inhabiting East Africa.

Sumba Sunbird *N. buettikoferi*
Named after Dr J. Büttikofer (1850–1927), a Dutch zoologist who was Director of the Zoological Gardens in Rotterdam from 1897 to 1924. Sumba is one of the chain of islands in Indonesia that lie to the east of Java.

Malachite Sunbird *N. famosa*
famosus (L) famous, much talked of: malachite is a mineral from which a green pigment is made; this refers to the upperparts, including the head and neck, which are a metallic green. Inhabiting Sudan, Ethiopia, and ranging south through East Africa to South Africa.

Loten's Sunbird *N. lotenia*
Named after J. G. Loten, FRS (1710–1789), a British naturalist who was Governor of Ceylon (now known as Sri Lanka) from 1752 to 1757. This sunbird inhabits India and Sri Lanka.

Beautiful Sunbird *N. pulchella*
pulcher (L) beautiful *-ellus* (L) diminutive suffix; 'a little beauty'; most sunbirds are small, less than $12\frac{1}{2}$ cm (5 in) including the tail, and have spectacular colouring with a metallic sheen; they can be compared to the humming birds of South America but they are not related. Inhabiting Senegal and Sierra Leone in the western part of Africa, Sudan to Somalia in the east, and south to Tanzania.

Mrs Gould's Sunbird   *Aethopyga gouldiae*
aithos (Gr) burnt; can mean a red-brown colour   *pugē* (Gr) the rump;
a reference to the dark colour of the plumage under the tail: named
after Mrs E. Gould (1804-1841) a bird artist, and the wife of J. Gould,
FRS (1804-1881) an English zoologist. Inhabiting southern Asia,
India, the Burma-Vietnam area and China.

Streaked Spider Hunter   *Arachnothera magna*
arakhnēs (Gr) a spider   *thēraō* (Gr) I hunt, chase, wild beasts; can
mean I catch wild beasts; it has been seen to hover and pick a spider
out of its web, but like other birds in this family it also takes insects,
nectar, and mistletoe berries   *magnus* (L) great; it is one of the largest
species, measuring 21·5 cm (8½ in) from the beak to the tip of the tail;
the beak itself may be 5 cm (2 in) long. The English name refers to the
colour of the upperparts which are olive to yellow and streaked with
black. Inhabiting the Himalayan area, Sikkim, Assam, and the
Burma-Vietnam area.

Family   ZOSTEROPIDAE   79 species
zōstēr (Gr) a girdle, a band   *ops* (Gr) the eye; a reference to the white
ring round the eye, and they are usually known as White-Eyes. Almost
without exception the birds in this family have a white ring round the
eye; they are small birds, from 10 to 12·5 cm (4 to 5 in) long, and
related to the Sunbirds; the diet is similar consisting of nectar, insects
and berries.

Mangrove White-Eye   *Zosterops chloris*
khloros (Gr) green; a yellowish green plumage is predominant in this
family, and nearly all have the white ring round the eye. Inhabiting
Indonesia, Papua New Guinea, and Australia.

Pale White-Eye   *Z. pallida*
pallidus (L) pale, pallid; as the English and Latin names suggest, the
plumage is pale, with more yellow than other species. Inhabiting
Ethiopia, Kenya, Tanzania, and South Africa.

Oriental White-Eye   *Z. palpebrosa*
palpebra (L) the eyelid; can mean the eyelashes   *-osus* (L) suffix meaning prone to, full of; 'prominent eyelashes'; a reference to the ring of
small white feathers round the eye. Inhabiting south-eastern Asia,
including Afghanistan, India, the Burma-Vietnam area, south-
western China, and Indonesia.

African Yellow White-Eye   *Z. senegalensis*
-ensis (L) suffix meaning belonging to, usually a place; in addition to
Senegal it ranges south through East and Central Africa to Mozambique and South Africa. The plumage is mainly yellow; the name
does not refer to the ring round the eye, but see *Z. wallacei*, below.

Yellow-spectacled or Wallace's White-Eye   *Z. wallacei*
The only species known that has a yellow eye-ring instead of white:
named after Dr Alfred R. Wallace, FRS (1823–1913), the British
zoologist, who was in Malaysia from 1848 to 1852; he is well known
for establishing the Wallace Line, an imaginary line separating Bali,
Borneo, and Mindanao, from Lombok, Celebes, and Talaut Island,
and this is zoologically important. This white-eye inhabits the Lesser
Sundas.

Woodford's White-Eye   *Woodfordia superciliosa*
Named after C. M. Woodford (1852–1927), the Resident Commissioner in the Solomons Protectorate from 1897 to 1915   *supercilium* (L) the eyebrow   *-osus* (L) suffix meaning prone to, full of;
'prominent eyebrows'; a reference to the ring of feathers round the
eye. Inhabiting the Solomon Islands.

Javan Crested White-Eye   *Lophozosterops javanica*
lophos (Gr) a crest, the crest of birds   *-icus* (L) belonging to, 'of Java';
inhabiting Java and neighbouring islands.

Black-capped Speirops   *Speirops lugubris*
speira (Gr) anything wound or wrapped round a thing   *ops* (Gr) the
eye; a reference to the ring of small white feathers that surround the
eye   *lugubris* (L) of or belonging to mourning, mourning apparel; a
reference to the black cap. Inhabiting Cameroon.

Family   EPTHIANURIDAE   5 species
phthinas (Gr) decreasing, wasting   *oura* (Gr) the tail; *Epthianura* is a
coined name and supposed to mean 'decreased tail', a reference to the
unusually small tail. This small family are known as Australian Chats,
the diet is insects hunted on the ground though they will take nectar
if available.

White-fronted Chat   *Epthianura albifrons*
albus (L) white   *frons* (L) forehead. Inhabiting Queensland, New
South Wales, Victoria, and ranging west along the southern part of
Australia.

Crimson Chat   *E. tricolor*
tricolor (New L) of three colours; it has a bright plumage consisting of crimson, brown, and white. Inhabiting western and southern parts of Australia and partly central.

Gibber-bird   *Ashbyia lovensis*
Named after Edwin Ashby (1861–1941), an Australian ornithologist; he was President of the Royal Australian Ornithologists Union in 1926   *-ensis* (L) suffix meaning belonging to; named from Love's Creek, near Alice Springs in central Australia. Gibber is a native name for the desert stones and pebbles polished by sandblast; this chat lives in desert areas of central Australia.

Family   MELIPHAGIDAE   169 species
meli (Gr) honey   *phagein* (Gr) to eat. Known in general as Honey-eaters, they have a tongue that is semi-tubular and bristle-tipped; adaptations for feeding on nectar from flowers, but they also take insects, fruit, and berries. Ranging in size from 10 to 36 cm (4 to 14 in), they show a great diversity of structure and habits, but the one thing they all have in common is the specially adapted tongue.

Long-billed Honeyeater   *Melilestes megarhynchus*
meli (Gr) honey   *lēstēs* (Gr) a robber; 'a honey robber'   *megas* (Gr) big   *rhunkhos* (Gr) the beak, bill. Inhabiting New Guinea.

Slaty-chinned Longbill   *Toxorhamphus poliopterus*
toxon (Gr) a bow for use with arrows; can mean anything bowed or arched   *rhamphos* (Gr) a beak, bill, specially the curved beak of birds; a reference to the long downward curved bill   *polios* (Gr) hoary, grey, usually of hair   *pteron* (Gr) feathers; a reference to the slate-coloured plumage. Inhabiting New Guinea.

Brown Honeyeater   *Lichmera indistincta*
lichmērēs (Gr) playing with the tongue, licking, as of snakes; the tongue is specially shaped and bristled-tipped for taking nectar and insects from flowers   *indistinctus* (L) unpretentious, indistinct; a reference to the sombre brown plumage. Inhabiting Australia and ranging west to Bali and north to New Guinea.

Red-headed Honeyeater   *Myzomela erythrocephala*
muzō (Gr) I suck in   *meli* (Gr) honey   *eruthros* (Gr) red   *kephalē* (Gr) the head. One of the smallest birds in this family, only $12\frac{1}{2}$ cm (5 in) long, it lives in the mangrove swamps of Australia.

Dainty Honeyeater   *M. pulchella*
pulcher (L) beautiful   *-ellus* (L) diminutive suffix, 'a little beauty'.
Inhabiting New Ireland.

Yellow-faced Honeyeater   *Meliphaga chrysops*
meli (Gr) honey   *phagein* (Gr) to eat   *khrusos* (Gr) gold   *khruseos* (Gr)
gold-coloured, golden-yellow   *ops* (Gr) the eye, also the face. Inhabiting Australia from northern Queensland and ranging along the
coastal belt to the southern parts.

Lewin's Honeyeater   *M. lewini*
Named after J. W. Lewin (1770-1819), a naturalist, author, and
artist, of New South Wales; he wrote and published the book *Birds of
New Holland*. Inhabiting eastern Australia.

White-eared Mountain Meliphaga   *M. montana*
mons (L), genitive *montis*, a mountain   *montanus* (L) of a mountain. In
some cases honeyeaters are known as meliphagas; this one lives in the
mountains of New Guinea.

Orange-cheeked Honeyeater   *Oreornis chrysogenys*
oros (Gr), genitive *oreos*, a mountain   *ornis* (Gr) a bird   *khrusos* (Gr)
gold   *khruseos* (Gr) gold-coloured, golden-yellow   *genus* (Gr) the
jaw, cheek. Another honeyeater inhabiting the mountains of New
Guinea.

White-throated Honeyeater   *Melithreptus albogularis*
meli (Gr) honey   *threptos* (Gr) fed, brought up   *albus* (L) white   *gula*
(L) the throat   *-aris* (L) suffix meaning pertaining to. Inhabiting
Australia and New Guinea.

Strong-billed Honeyeater   *M. validirostris*
validus (L) strong   *rostrum* (L) the beak. Inhabiting Tasmania.

White-fronted Melidectes   *Melidectes leucostephes*
meli (Gr) honey   *dektēs* (Gr) a beggar, a receiver   *leukos* (Gr) white
stephos (Gr) a crown, a garland. Inhabiting New Guinea.

Bar-breasted Honeyeater   *Ramsayornis fasciatus*
Named after Dr E. P. Ramsay (1842-1916), an Australian zoologist;
he was Curator of the Australian Museum in Sydney from 1874 to
1894   *ornis* (Gr) a bird   *fascia* (L) a band or girdle   *fasciatus* (L)
banded, can mean striped; a reference to the stripes across the breast.
Inhabiting Australia.

Rufous-throated Honeyeater   *Conopophila rufogularis*
kōnōps (Gr), genitive *kōnōpos*, a gnat, a mosquito   *philos* (Gr) loved,
pleasing; in addition to nectar the honeyeaters like the insects that are
attracted by the nectar, and the specially adapted and bristle-tipped
tongue assists in their collection; it is also thought that the birds help
with the pollination of the flowers   *rufus* (L) red   *gularis* (L) pertaining to the throat. Inhabiting northern Australia.

Eastern Spinebill   *Acanthorhynchus tenuirostris*
akantha (Gr) a thorn, a prickle   *rhunkhos* (Gr) the beak   *tenuis* (L)
thin, slender   *rostrum* (L) the beak; these are references to the unusually long spinelike bill. Ranging from northern Queensland to
southern Australia, including Kangaroo Island and Tasmania.

Bellbird   *Anthornis melanura*
anthos (Gr) a flower, a blossom   *ornis* (Gr) a bird; a reference to its
fondness of flowers where it can obtain nectar   *melas* (Gr) black
oura (Gr) the tail; the tail is really dark brown but it varies and is
occasionally almost black. These birds sometimes sing in chorus and
it sounds like the pealing of bells; they are confined to New Zealand
and neighbouring islands.

Tui or Parson Bird   *Prosthemadera novaeseelandiae*
prosthema (Gr) an addition, an appendage   *deirē* (Gr) the neck, throat;
a reference to the two white tufts of feathers on each side of the throat
which resemble the two white 'lapels' sometimes worn by clergymen,
and this has given rise to the name 'parson bird'; it also has a white
lacy collar   *novaeseelandiae* (New L) New Zealand. *Tui* is a Maori
name for the bird which is confined to New Zealand and neighbouring
islands.

Cape Sugarbird   *Promerops cafer*
pro- (L) for   *merops* (L) a bird, the bee-eater; *promerops* is a coined
name, and the reason for the Latin prefix is obscure; the bee-eater, in
the genus *Merops*, eats bees but does not eat honey, whereas the sugarbird of South Africa does eat honey; why there are two species of
Promerops living in Africa, a long way from the other honeyeaters of
the east, is something of a mystery, and may be an example of con-
vergent evolution   *kafir* (Ar) the name of a South African people.
Inhabiting the southern part of South Africa.

Family EMBERIZIDAE
emberiza (New L) from Swiss-German *emmeritz*, a bunting or yellow

hammer. This very big group of about 550 species can be divided into 5 subfamilies: the Emberizinae (the Buntings and American Sparrows), the Catamblyrhynchinae (the Plush-capped Finch), the Cardinalinae (the Cardinal-Grosbeaks), the Thraupinae (the Tanagers and Honeycreepers) and the Tersinae (the Swallow-Tanager).

Ornithologists have various opinions about the divisions, and the naming of the subfamilies, as correct anatomical distinctions are by no means obvious; this book follows the system used by E. S. Gruson in his book *A Checklist of the Birds of the World*.

They range in size from 8 to about 30 cm (3 to about 12 in), and their general appearance and beak structure is very varied; the food is mainly insects, berries, seeds and fruit, and the honeycreepers also take nectar.

Subfamily EMBERIZINAE   281 species
Crested Bunting   *Melophus lathami*
melas (Gr) black   *lophos* (Gr) a crest; the upperparts are a glossy black including the pointed crest: named after Dr J. Latham, FRS (1740–1837) an English zoologist and author. Inhabiting the Himalayas, Yunnan, and Burma.

Yellow Bunting or Yellowhammer   *Emberiza citrinella*
Emberiza, see above under Subfamily   *citrus* (L) the citron-tree *citrinus* (New L) citron or lemon-coloured   *-ellus* (L) diminutive suffix. Inhabiting Europe including the British Isles and ranging east through Eurasia to the Sea of Okhotsk.

Golden-breasted Bunting   *E. flaviventris*
flavus (L) yellow, golden   *venter* (L), genitive *ventris*, the belly, stomach; the yellow plumage extends up the breast to the throat. Inhabiting Africa from Nigeria to Ethiopia and ranging south to South Africa.

Black-headed Bunting   *E. melanocephala*
melas (Gr), genitive *melanos*, black   *kephalē* (Gr) the head. Inhabiting south-eastern Europe, Asia Minor, Georgia, the Caucasus, Iran, and the mountains of Turkestan.

Reed Bunting   *E. schoeniclus*
skhoinos (Gr) a rush, or reed   *skhoiniklos* (Gr) a water-bird, a name used by Aristotle, probably to mean the white water-wagtail; it is not related to the White Wagtail *Motacilla alba* and in this case is only used to indicate its habitat among the reeds. Wide distribution throughout

Europe and Central Asia, but not Iceland; it includes northern Africa and Asia Minor and extends east along a narrower band to Kamchatka and northern Japan.

Socotra Mountain Bunting   *E. socotrana*
-anus (L) suffix meaning belonging to; Socotra is an island in the Indian Ocean lying about 322 km (200 miles) to the east of the Horn of Africa; the Horn of Africa is not marked in maps; it is the northeastern part of Somalia.

Lapland Bunting or Longspur   *Calcarius lapponicus*
calcar (L), genitive *calcaris*, a spur   *-icus* (L) belonging to; 'of Lapland'; the generic name refers to the unusually long hind toe and claw. Inhabiting the northern coasts of Eurasia from Norway to America and Greenland but not Iceland, and migrating south in winter.

Snow Bunting   *Plectrophenax nivalis*
plēktron (Gr) a cock's spur   *phenax* (Gr) a cheat, an impostor; for technical reasons the name was changed by Dr Stejneger from *Plectrophanes*, 'spur-showing', to *Plectrophenax*, but the name does not seem to have any meaning except that it took the place of *Plectrophanes*, i.e. 'an impostor'; it does have a long hind claw   *nix* (L), genitive *nivis*, snow *nivalis* (L) of snow, snowy. Inhabiting northern American and Eurasian coasts and arctic islands, ranging south to the British Isles and parts of Eurasia as far east as Kamchatka.

American Tree Sparrow   *Spizella arborea*
spiza (Gr) a finch   *-ellus* (L) diminutive suffix   *arbor* (L) a tree *arboreus* (L) pertaining to a tree: the Greek word *spiza*, a finch, 'a small piping bird', is fairly widely used for small birds; the American sparrows are not related to the sparrows of the Old World (Family Ploceidae). This tree sparrow inhabits North America ranging from Alaska, northern Canada, and Newfoundland, to southern USA.

Patagonian Sierra-Finch   *Phrygilus patagonicus*
phrugilos (Gr) a bird, perhaps a finch   *-icus* (L) suffix meaning belonging to, 'of Patagonia'; this is the southern part of Argentina, South America   *sierra* is Spanish for a ridge of mountains.

White-throated Seedeater   *Sporophila albogularis*
spora (Gr) seed   *philos* (Gr) loved, pleasing   *albus* (L) white   *gula* (L) the throat   *-aris* (L) suffix meaning pertaining to. Inhabiting Brazil.

Slate-coloured Seedeater *S. schistacea*
skhistos (Gr) divided, cleft, as a piece of slate; hence *schist* (New L) slate, and *schistaceus* (New L) slaty, slate-coloured. Inhabiting Central and South America.

Blue Seedeater *Amaurospiza concolor*
amauros (Gr) dark *spiza* (Gr) a finch, 'a small piping bird'; a reference to the blue plumage *concolor* (L) similar in colour, of one colour. Inhabiting Central and South America.

Cuban Bullfinch *Melopyrrha nigra*
melas (Gr) black *purrhos* (Gr) red, flame-coloured; a reference to the black plumage with the flame-coloured breast; it is not related to the Old World bullfinches in the family Fringillidae. Inhabiting Cuba and nearby islands.

St Lucia Black Finch *Melanospiza richardsoni*
melas (Gr) black *spiza* (Gr) 'a small piping bird', a finch: named after W. B. Richardson (fl 1919), an American naturalist and author; he was in Central America on several expeditions during the period 1887 to 1917. St Lucia Island is one of the group known as the Windward Islands in the Caribbean Sea.

Large Ground Finch *Geospiza magnirostris*
gē (Gr) the earth, ground *spiza* (Gr) 'a small piping bird', a finch *magnus* (L) great *rostrum* (L) the beak; one of the larger ground finches with a big strong beak. This is one of a number of finches that are confined to the Galapagos Islands, and are known as Darwin's Finches; C. R. Darwin, FRS (1809-1882) was the zoologist who became famous for his work on the theory of evolution by natural selection, which was based on the Galapagos finches.

Cocos Island Finch *Pinaroloxias inornata*
pinaros (Gr) dirty *loxos* (Gr) crosswise; *Loxia* was used for the crossbills but later was used for other finch-like birds; *pinaros* indicates the dull plumage; cf. Sooty Rock Chat *Pinarornis plumosus* *inornatus* (L) plain, unadorned; another reference to the rather drab greyish-brown plumage. It is confined to Cocos Island, some 966 km (600 miles) to the north-east of Galapagos, where all the other Darwin finches live (see above).

Rufous-sided Towhee *Pipilo erythrophthalmus*
pipio (L) a young piping or chirping bird *pipilo* (L) I chirp *erythros*

(Gr) red *ophthalmos* (Gr) the eye; they have a bright red eye and rufous plumage on the flanks. Inhabiting Canada and ranging through most of the USA to northern Mexico. The name towhee is imitative of the two-note call which sounds like to-whee.

Pectoral Sparrow   *Arremon taciturnus*
arrhēmōn (Gr) silent, without speech   *taciturnus* (L) not talkative, quiet; a reference to the absence of a song; it only makes a repeated high call: 'pectoral' from *pectus* (L) the breast; refers to a black patch on each side of the breast. Inhabiting Venezuela and Colombia.

Black-headed Brush-Finch   *Atlapetes atricapillus*
Atlapetes, obscure, possibly a corruption of *altus* (L) high + *petes* (New L) a flier, from *petomai* (Gr) to fly, of birds, bees etc.; the name is not explained in the ornithological records in the British Museum (Natural History)   *ater* (L) black   *capillus* (L) the hair, usually the hair of the head, can mean the hair of animals. Inhabiting Central and South America.

Rufous-capped Brush-Finch   *A. pileatus*
pileus = *pilleus* (L) a felt cap   *pilleatus* (L) wearing a felt cap. Inhabiting southern Mexico.

Subfamily   CATAMBLYRHYNCHINAE   1 species
katamblunō (Gr) I make blunt   *rhunkhos* (Gr) the beak; an allusion to the stubby beak. Little is known about this unusual finch-like bird and so for the present it is placed in a family on its own.

Plush-capped Finch   *Catamblyrhynchus diadema*
diadema (Gr) a band or fillet, usually worn round the head; a reference to the golden brown forecrown which is erect and plush-like to the touch. Inhabiting Venezuela and Colombia and ranging south through Ecuador and Peru to Bolivia.

Subfamily   CARDINALINAE   37 species
cardinalis (L) a chief, a principal; here taken to mean red on account of the red robes worn by cardinals; although all the birds in this subfamily are not red, the name is an allusion to the red plumage of the birds in the genus *Cardinalis* from which the subfamily takes its name.

Cardinal   *Cardinalis cardinalis*   (formerly *Pyrrhuloxia*)
Inhabiting southern Canada and ranging south through the USA to Florida, the Gulf Coast, Mexico, and part of Central America.

Yellow-shouldered Grosbeak   *Caryothraustes humeralis*
karuon (Gr) a nut, can be almost any kind of nut *thrauō* (Gr) I break
in pieces   *thraustos* (Gr) broken; a reference to its ability to break open
the toughest nuts   *humerus = umerus* (L) the upper arm   *umerale* (L)
a cape for the shoulders; a reference to the yellow plumage on the
'shoulders': the name grosbeak has been given because of the large,
powerful, conical bill. Inhabiting South America.

Blue Grosbeak   *Passerina caerulea*
passer (L) a sparrow, or other small bird   *-inus* (L) suffix meaning like;
it is similar to a sparrow in size and shape, but not in colour   *caeruleus*
(L) dark blue. Inhabiting North and Central America.

Painted Bunting   *P. ciris*
keiris (Gr) a bird mentioned in Greek mythology: 'painted' is a suitable name for this bird with a plumage of gaudy colours including
blue, green, yellow, and various shades of red in a distinct pattern.
Inhabiting southern USA, Mexico, and Panama.

Subfamily   THRAUPINAE   233 species
thraupis (Gr) a little bird. This large group consists of smallish birds
measuring from 10 to 20 cm (4 to 8 in); the diet is varied and may
include seeds, fruit, insects, and in some cases honey. They are mostly
known as Tanagers and Honeycreepers though some unusual names
are used such as Euphonias and Flower-Piercers.

Yellow-green Bush Tanager   *Chlorospingus flavovirens*
khloros (Gr) green, pale green, to yellow   *spingos = spinos* (Gr) a small
bird   *flavus* (L) yellow   *virens* (L) becoming green, or green. Inhabiting Ecuador.

Grey-capped Hemispingus   *Hemispingus reyi*
hēmi- (Gr) prefix meaning half   *spingos = spinos* (Gr) a small bird; in
this case *hēmi-* is taken to mean like, rather than half, and as spinos has
been used to mean a finch, the name would mean 'like a finch': named
after Dr J. G. C. E. Rey (1838-1909), a German zoologist who made
a special study of oology, i.e. birds' eggs. This tanager lives in South
America.

Summer Tanager   *Piranga rubra*
piranga (New L) from a native name for this bird   *rubra* (L) red; it is
almost entirely rosy red. Fairly widespread in certain parts of the
USA from Wisconsin south to Florida and west to New Mexico and

California, in summer, hence the name Summer Tanager; in winter migrating south to Mexico, through Central America, to Peru.

Azure-shouldered Tanager   *Thraupis cyanoptera*
thraupis (Gr) a little bird, sometimes taken to mean a goldfinch *kuaneos* (Gr) glossy blue   *pteron* (Gr) feathers, can mean wings. Inhabiting South America.

Golden-crowned Tanager   *Iridosornis rufivertex*
iris (Gr), genitive *iridos*, the rainbow; can mean any bright-coloured circle, as a halo; a reference to the circular golden patch on the top of the black head   *ornis* (Gr) a bird   *rufus* (L) red   *vertex* (L) a top, a summit; a misleading name as the crown is usually golden, but it can vary in different localities. Inhabiting Venezuela, Colombia, and Ecuador.

Thick-billed Euphonia   *Euphonia laniirostris*
euphonia (Gr) goodness of voice; indicating a rather better song than other tanagers which taken as a whole are poor songsters   *lanius* (L) a butcher   *rostrum* (L) the beak; 'butcher-beaked'; a reference to the powerful beak. Inhabiting Central America and north-western South America, ranging from Costa Rica south to Bolivia and neighbouring parts of Brazil.

Glistening-green Tanager   *Chlorochrysa phoenicotis*
khlōros (Gr) green   *khruseos* (Gr) golden-yellow   *phoinix* (Gr), genitive *phoinikos*, purple-red, crimson   *ous* (Gr), genitive *ōtos*, the ear; an allusion to the small blue and red patches just above the ear coverts. Inhabiting the western part of Pacific Colombia and north-western Ecuador.

Black-headed Tanager   *Tangara cyanoptera*
Tangara is a Tupi native name for a brightly-coloured bird   *kuaneos* (Gr) glossy blue   *pteron* (Gr) feathers, can mean wings; it has a black head and the wings are black with blue flight feathers. Inhabiting Guyana, Venezuela, north-eastern Colombia, and northern Brazil.

Black and Green Tanager   *T. nigroviridis*
niger (L) black   *viridis* (L) green; the plumage appears to be alternate streaks of black and green, with the green changing almost to blue in certain parts and with changing light. Inhabiting the mountains of Venezuela, Colombia, and Ecuador, and ranging south to the northern part of Bolivia.

Purple Honeycreeper *Cyanerpes caeruleus*
kuaneos (Gr) dark blue, glossy blue *herpēstēs* (Gr) a creeping thing
caeruleus (L) dark blue; the plumage of the male is generally purple
and black, while the female is quite different being olive green. In
addition to insects and fruit the honeycreepers take nectar. Inhabiting
Trinidad and the north-western part of South America including the
Amazon Basin.

Bluish Flowerpiercer *Diglossa caerulescens*
di- (Gr) a prefix meaning two, double *glōssa* (Gr) the tongue; the tip
of the tongue is divided and has feather-like ends to facilitate taking
the nectar, and the beak is specially adapted to actually pierce the
corolla of the flower *caeruleus* (L) dark blue *-escens* (L) suffix meaning approaching, beginning to be, i.e. bluish; the plumage is rather a
nondescript greyish blue. Inhabiting the Andes Mountains in
Colombia, Ecuador, Peru, and northern Bolivia.

Subfamily TERSININAE 1 species

Swallow-Tanager *Tersina viridis*
Tersina, a name coined by Vieillot, evidently from the French vernacular name *La Tersine*; further information is not given in Vieillot's
description in the ornithological records in the British Museum
(Natural History) *viridis* (L) green; the plumage of the male is really
a torquoise blue but this changes to an emerald green, according to
changes of light; the female is bright green. Inhabiting Trinidad and
most of the northern half of South America, ranging from Panama in
Central America to northern Argentina including Brazil and Peru.
The Swallow-Tanager is difficult to classify and so has been given a
subfamily on its own; some authors give it a family. Anatomically it
is very different from a tanager and is not related to the swallows, but
it has a swallow-like bill and catches flying insects in a similar manner
to swallows. L. J. P. Vieillot (1748–1831) was a well-known French
zoologist and author.

Family PARULIDAE 120 species
parula (New L) from *parus* (L) a titmouse *-ulus* (L) diminutive suffix;
a misleading name as this family consists of birds that are mostly larger
than the tits, ranging from 10 to 18 cm (4 to 7 in), whereas the tits are
mostly about 10 to $12\frac{1}{2}$ cm (4 to 5 in). Known as Wood Warblers or
American Warblers they are not related to the Old World Warblers

in the subfamily Sylviinae; the diet is mostly insects and fruit though some will take nectar.

Blue-winged Warbler   *Vermivora pinus*
vermis (L) a worm   *voro* (L) I devour; it may include worms in the diet but insects are the main source of food   *pinus* (L) a pine; it is known to frequent pine forests but this is not the only habitat. The plumage is mostly yellow with blue wings; it inhabits the central and eastern USA ranging south to Georgia and wintering further south to Nicaragua.

Virginia's Warbler   *V. virginiae*
Named in honour of Mrs M. Virginia C. Anderson, the wife of Dr W. W. Anderson (1824–1901), a surgeon in the United States Army from 1849 to 1861. The range is western USA and Mexico.

Northern Parula   *Parula americana*
Parula, see under Family, above: a small greyish blue warbler that inhabits the eastern part of North America.

Yellow-throated Warbler   *Dendroica dominica*
dendron (Gr) a tree   *-icus* (L) suffix meaning belonging to; it favours mature woodland areas: named after the Dominican Republic, one of the islands in the West Indies, it is fairly widespread in the USA and ranges south to Central America in winter.

Kirtland's Warbler   *D. kirtlandii*
Named after Dr J. P. Kirtland (1793–1877), an American zoologist from Ohio. A very rare warbler, it is restricted apparently by choice to an area of pine forest in north central Michigan, but migrates south in winter.

Ovenbird   *Seiurus aurocapillus*
seiō (Gr) I shake, I move to and fro   *oura* (Gr) the tail; it does not actually wag the tail but raises it when walking over a pile of leaves and then lowers it again   *aurum* (L) gold   *aureus* (L) golden, colour of gold   *capillus* (L) the hair of men, usually the hair of the head, can mean the hair of animals; it has a reddish-gold stripe on the crown. The name ovenbird comes from the form of the nest, a domed structure with an entrance at the side. Inhabiting Canada and the USA and ranging south in winter to the Gulf Coast, the Lesser Antilles, and Mexico to Colombia, South America.

Worm-eating Warbler *Helmitheros vermivorus*
helmins (Gr) a worm *thēraō* (Gr) I hunt, I take; it eats various small invertebrates *vermis* (L) a worm *voro* (L) I devour. Inhabiting the USA and moving south in winter to the Bahamas, the West Indies, and Central America as far as Panama.

Prothonotary Warbler *Protonotaria citrea*
protos (Gr) first *notarius* (L) a secretary; this refers to the Chief Secretary of the Chancery at Rome, who wears yellow robes *kitrea* (Gr) the citron tree; indicating the yellow plumage of the bird. Inhabiting the south-eastern part of the USA, Central America, and the northern part of South America.

Grey-crowned Yellowthroat *Geothlypis poliocephala*
gē (Gr) the earth, ground *thlupis* (Gr) a small bird; an obscure Greek word probably used to mean a finch *polios* (Gr) hoary, grey *kephalē* (Gr) the head; the birds in this genus usually have yellow plumage on the throat and upper breast. Inhabiting North and Central America.

Canada Warbler *Wilsonia canadensis*
Named after Alexander Wilson (1766-1813), a Scottish ornithologist who settled in the USA and became known as the 'Father of American Ornithology'; a number of other birds bear his name *-ensis* (L) suffix meaning belonging to; it is not confined to Canada, and ranges south in eastern USA, and winters in Guatemala, Central America, south to Ecuador and Peru in South America.

Golden-fronted Redstart *Myioborus ornatus*
muia (Gr) a fly *boros* (Gr) devouring *ornatus* (L) decorated, adorned; a reference to the brilliant golden face and breast in contrast to the dark olive-grey upperparts. Inhabiting the eastern Andes in Venezuela and Colombia, South America.

Golden-crowned Warbler *Basileuterus culicivorus*
basileutōr = *basileus* (Gr) a king *culex* (L), genitive *culicis*, a gnat, a midge *voro* (L) I devour. Inhabiting Central and South America.

Rufous-capped Warbler *B. rufifrons*
rufus (L) red *frons* (L) forehead, brow; the crown and ear-coverts are chestnut-coloured. Inhabiting Mexico and Central America and ranging south to Colombia and Venezuela.

White-eared Conebill　*Conirostrum leucogenys*
conus (L) a cone　*rostrum* (L) the beak; a reference to the cone-shaped
bill　*leukos* (Gr) white　*genus* (Gr) the cheek. Inhabiting Central and
South America.

Bananaquit　*Coereba flaveola*
coereba (Braz) a small bird　*flavus* (L) yellow　*-olus* (L) diminutive
suffix; 'small and yellow': it is not known specially to favour bananas
but is fond of various fruits and nectar. Inhabiting Central America
and the West Indies, Peru, Bolivia, Paraguay, Brazil and Argentina.

Family　DREPANIDIDAE　15 species
drepanon (Gr) a sickle; a reference to the sickle-shaped bill, though
some species have a short parrot-like bill. Usually known as Hawaiian
Honeycreepers they are confined to the Hawaiian Islands.

Crested Honeycreeper　*Palmeria dolei*
Named after Henry C. Palmer (fl 1890); he collected for Rothschild's
Tring Museum on Laysan, one of the islands in the Hawaiian group,
from 1890 to 1893: also S. B. Dole (1844–1926), a judge in Hawaii
and President of the Hawaiian Republic from 1894 to 1900.

Iiwi　*Vestiaria coccinea*
vestis (L) clothing, a covering　*-arius* (L) suffix meaning pertaining
to　*coccinus* (L) scarlet garments　*coccineus* (L) scarlet coloured; a
reference to the brilliant red plumage which covers most of the body,
and which was at one time used to make feather cloaks. *Iiwi* is a
Hawaiian native name.

Amakihi　*Viridonia virens*
viridis (L) green　*virens* (L) becoming green; referring to the bright
green plumage. *Amakihi* is a Hawaiian native name.

Akepa　*Loxops coccinea*
loxos (Gr) cross-wise　*ops* (Gr) the eye, face; can also mean aspect,
appearance; it refers to the beak which has the tips of the lower and
upper mandibles slightly crossed　*coccineus* (L) scarlet coloured; the
plumage is more a bright orange colour than scarlet. *Akepa* is a
Hawaiian native name.

Parrotbill　*Psittirostra psittacea*
psittakos (Gr) a parrot　or *psittacus* (L) a parrot　*rostrum* (L) the beak
-aceus (L) suffix meaning similar to, like; this is one of the Drepani-

didae that has a bill more like that of a parrot, not long and curved like a scimitar. Inhabiting the Hawaiian Islands.

Family VIREONIDAE   39 species
vireo (L), genitive *vireonis*, a kind of bird, according to some the green-finch; they are similar in size and colour to the greenfinch, but are only distantly related. Usually known as Vireos and Greenlets, the diet consists of insects and fruit.

Bell's Vireo   *Vireo belli*
Named after J. G. Bell (1812–1899); an American naturalist and taxidermist of New York. Inhabiting North America and Central America.

Red-eyed Vireo   *V. olivaceus*
oliva (L) an olive   *olivaceus* (L) similar to; 'olive-coloured'; it is olive-green above and whitish below. One of the best known Vireos it ranges from Canada, south through the USA but mostly in the west central part, to the Gulf of Mexico and south to Brazil.

Golden-fronted Greenlet   *Hylophilus aurantifrons*
hulē (Gr) a wood, forest   *philos* (Gr) beloved, dear; a reference to the habitat which is mostly forest and woodland areas   *aurum* (L) gold *auratus* (L) golden   *aurantium* (New L) an orange   *frons* (L) the fore-head; the crown is bright orange. Inhabiting Central and South America.

Tawny-crowned Greenlet   *H. ochraceiceps*
ōkhra (Gr) yellow-coloured earth, yellow-ochre   *ochraceus* (New L) like ochre, pale yellow   *ceps* (New L) from *caput* (L) the head; it has a yellowish to orange crown. Inhabiting Central America, the Guyanas, Venezuela, Brazil, Colombia, Ecuador and Bolivia.

Family ICTERIDAE   92 species
ikteros (Gr) the jaundice; also a bird of a yellowish-green colour, by looking at which a jaundiced person was cured; the bird then died! In this group there is great diversity of feeding habits, nest building, mating behaviour, and size, which ranges from about 18 to 53 cm (7 to 21 in) in length.

Dusky-green Oropendola   *Psarocolius atrovirens*
psar (Gr), genitive *psaros*, a starling   *kolios* (Gr) a kind of woodpecker. Wagler does not fully explain the name; the Icteridae are analagous

to the Old World starlings and one species is known as the Military Starling, but they are not starlings and they are not woodpeckers   *ater* (L) black   *virens* (L) becoming green. The name *Oropendola* is derived from *oro* (Sp) gold, and *pendola* (Sp) a feather, a plume; the wings have bright yellow outer feathers. Inhabiting the forests of the northern part of South America. Dr Joannes G. Wagler (1800–1832) was a German naturalist and Professor of Zoology at Munich University from 1827 to 1832.

Green Oropendola   *Ps. viridis*
viridis (L) green. Inhabiting the northern tropical areas of South America.

Chestnut-headed Oropendola   *Ps. wagleri*
Named after Dr J. G. Wagler (see above). Inhabiting the forests of Central America and the northern part of South America.

Scarlet-rumped Cacique   *Cacicus uropygialis*
cacicus (New L) from cacique, a native Indian chief, particularly in areas where there is a Spanish culture, as in the West Indies and Central and South America   *oura* (Gr) the tail   *pugē* (Gr) the rump, buttocks   *-alis* (L) suffix meaning relating to; a reference to the red plumage of the tail and rump. Inhabiting the West Indies, Central America, and northern South America.

Northern Oriole   *Icterus galbula*
Icterus, see under Family, above   *galbina* (L) diminutive *galbula*, a small bird, probably an oriole; also *galbina* (L) pale green garments; the male is a glossy black with some orange parts, but the female is olive yellow above with yellow underparts and tail. Inhabiting Canada including Nova Scotia, and ranging south through eastern USA to Georgia and Texas.

White-edged Oriole   *I. graceannae*
Named after Grace Anna Lewis (1821–1912), an American naturalist, specially a botanist, and teacher of biology. This oriole inhabits the western tropical area of South America; these orioles must not be confused with the Old World Orioles of the family Oriolidae, and they are not closely related.

Jamaican Blackbird   *Nesopsar nigerrimus*
nesos (Gr) an island   *psar* (Gr) a starling   *niger* (L) black   *nigerrimus*, very black. The American Blackbirds must not be confused

with our blackbirds in the family Muscicapidae, and they are not closely related; this one lives on the island of Jamaica.

Oriole Blackbird   *Gymnomystax mexicanus*
gumnos (Gr) naked   *mustax* (Gr) the upper lip; there is a patch of bluish white bare skin at the base of the bill and on the cheeks   *-anus* (L) suffix meaning belonging to; although named from Mexico by Linnaeus, this was an error; it inhabits north-western South America. It is one of the largest birds in this family, the male being up to 48 cm (19 in) long.

Yellow-headed Blackbird   *Xanthocephalus xanthocephalus*
xanthos (Gr) yellow   *kephalē* (Gr) the head; the head, neck, and upper breast are bright yellow, making a remarkable contrast with the rather drab colour of the body. Inhabiting the southern part of western Canada, the western USA, and ranging south in winter to southern Mexico.

Red-winged Blackbird   *Agelaius phoenicus*
agelaios (Gr) gregarious; a gregarious bird, breeding socially in groups, though sometimes only in scattered pairs   *phoinix* (Gr), genitive *phoinikos*, purple red, crimson; there is a red patch on the upper part of the wing. Inhabiting southern Canada, widespread in the USA, and ranging south in winter to southern Mexico.

Eastern Meadowlark   *Sturnella magna*
sturnus (L) a starling   *-ellus* (L) diminutive suffix; it is bigger than a starling but is one of the smaller birds in this family   *magnus* (L) large; this only indicates that it is bigger than the other birds in the genus *Sturnella*. Inhabiting Canada and ranging south through the eastern part of the USA to Florida, west to Mexico, Central America, Colombia and Venezuela.

Western Meadowlark   *S. neglecta*
neglectus (L) a neglecting, neglect; this species is so like *S. magna*, above, that for nearly 100 years it was thought to be the same bird; when ornithologists discovered and agreed that it was a separate species it was given the specific name *neglecta*, and the new name has been internationally recognised since 1844. Although the two species often live in overlapping territories they do not hybridise; how do they know? The two quite different songs are thought to be the answer. Inhabiting south-western Canada and ranging south through the western half of the USA to Mexico.

Greater Antillean Grackle   *Quiscalus niger*
quiscalus (New L) a quail; they do not look like quails and are not
related to them, but the flight may be similar; their close relatives the
Meadowlarks, above, are noted for their quail-like flight   *niger* (L)
black; the plumage is black with a glossy sheen of steel blue. Inhabiting the Greater Antilles; this is the group of islands that includes Cuba,
Jamaica, and the Haiti and Dominican Republics.

Brown-headed Cowbird   *Molothrus ater*
molobros (Gr) a greedy beggar, a lazy fellow; in this case intended to
mean a parasite; *molothros* is not a Greek word, it is a mistake and
should be *molobros*. Cowbirds are parasitic, but not in the same way as
the cuckoo; they lay several eggs, each one in a different nest, and a
great variety of different species; the host's eggs or nestlings are not
thrown out and may grow up quite amicably with the intruder. In
ancient times when bison were abundant cowbirds followed them to
feed on the insects on their backs or those flushed from the grass by the
trampling feet, now they follow the herds of cattle which has given
rise to the name 'cowbirds'   *ater* (L) black; the plumage is dark
glossy blue-black with brown head and neck. Inhabiting southern
Canada, widespread in the USA and ranging south to southern
Mexico.

Family   FRINGILLIDAE
Subfamily   FRINGILLINAE   3 species
fringilla (L) a small bird, according to some the chaffinch.

Chaffinch   *Fringilla coelebs*
coelebs = *caelebs* (L) unmarried, single, whether of a bachelor or a
widower; sometimes chaffinches gather in flocks of one sex only, and
at one time it was thought that the females migrated south leaving the
males behind, and they became known as 'bachelor birds'. Widespread in Europe including Scandinavia and the British Isles, ranging
east through the central part of Asia to the western half of Mongolia;
also northern Africa, Asia Minor, the Middle East, the Canary Islands
and the Azores.

Brambling   *F. montifringilla*
mons (L), genitive *montis*, a mountain; it breeds mainly in the mountains of northern Europe and northern Asia. The distribution is
similar to *F. coelebs*, above, but extending into more northern areas
and to the east as far as Kamchatka and south to Japan.

Subfamily CARDUELINAE   122 species
carduus (L) the wild thistle   *carduelis* (L) the thistle-finch, goldfinch;
a reference to their liking for seeding thistle heads but they will take
other seeds and occasionally insects on the ground.

Canary   *Serinus canaria*
serinus (New L) pertaining to the bird known as the serin; also *serin*
(Fr) the canary: the name canary is said to originate from *canis* (L) a
dog, because of the large dogs that were kept on one of the islands in
Roman times, and was called Canaria; eventually the name was used
for the whole group of islands. Wild canaries live in the Canary
Islands, the Azores, and Madeira; they are greenish yellow streaked
with brown and black, but the pure yellow ones seen in captivity are
the result of selective breeding.

Yellow Canary   *S. flaviventris*
flavus (L) yellow   *venter* (L), genitive *ventris*, the belly; the upperparts
are olive green with dark streaks and the forehead, breast and belly
a bright yellow. Inhabiting the western part of southern Africa.

Serin   *S. serinus*
See above under Canary. Inhabiting western Europe but only recently
seen in southern Sweden and Britain; north-western Africa and up
to 2,450 m (8,000 ft) in the Atlas Mountains, Asia Minor and
Palestine.

Eurasian Goldfinch   *Carduelis carduelis*
See above under Subfamily. Inhabiting Eurasia including southern
Scandinavia, the British Isles, northern Africa, Asia Minor, and
ranging east to western Mongolia and western Tibet; it is found also
in the Canary Islands and the Azores, and has been introduced to
New Zealand and the USA.

Greenfinch   *C. chloris*
khlōros (Gr) green; the plumage is generally a fairly dark olive green
with bright yellow patches on the wings and tail. Widespread in
Europe including southern Scandinavia, the British Isles, northern
Africa, the Azores, Asia Minor, and northern Iran.

Eurasian Siskin   *C. spinus*
spingos = *spinos* (Gr) a small bird, probably a kind of finch. Inhabiting
Europe including the British Isles and the southern half of Scandinavia, and ranging east to China, Korea and Japan.

Linnet   *Acanthis cannabina*
akanthis (Gr) a bird fond of thistles, the siskin, or the linnet   *cannabis*
(L) hemp   *cannabinus* (L) pertaining to hemp; a reference to its liking
for the seeds of the hemp plant, but it also eats other seeds, specially
thistles, and some insects. Widespread in Europe including southern
Scandinavia, the British Isles, the Canary Islands and Africa north of
the Sahara, Asia Minor, ranging east to Kazakhstan and south to
parts of the Oriental Region.

Redpoll   *A. flammea*
flammeus (L) flaming, fiery; can mean fiery-red, a reference to the red
forehead. Inhabiting the northern part of Europe including the
British Isles, and north to Baffin Land, Greenland, and Siberia; introduced to New Zealand.

House Finch   *Carpodacus mexicanus*
karpos (Gr) fruit   *daknō* (Gr) I bite   *dakos* (Gr) a biting animal; a
'fruit-biter'; the Greek word *dakos* really means a poisonous biting
animal such as a snake   *-anus* (L) a suffix meaning belonging to;
named from Mexico, it ranges from British Columbia in Canada
south through the western part of the USA to southern Mexico.

Red-mantled Rosefinch   *C. rhodochlamys*
rhodon (Gr) a rose   *rhodoeis* (Gr) rose-coloured   *khlamus* (Gr) a short
cloak. Inhabiting Asia including India and the Burma-Vietnam
area.

Pine Grosbeak   *Pinicola enucleator*
pinus (L) a pine-tree or fir-tree   *colo* (L) I inhabit, dwell in; a reference to one of their favourite trees   *enucleo* (L) I take out the kernel;
with its powerful beak it can dig into the hard cones of pine trees. The
name grosbeak indicates 'big-beak'. Inhabiting the northern part of
North America, Scandinavia, and Asia, and ranging south to the
Burma-Vietnam area.

Red Crossbill   *Loxia curvirostra*
loxos (Gr) cross-wise   *curvus* (L) curved   *rostrum* (L) the beak; the
plainly curved and rather parrot-like bill is crossed at the tip though
this is not always easily seen; the male is red to orange-red but the
female is a yellowish green. Widespread along the coastal part of
North America from Alaska south to Panama in Central America,
ranging east in Canada to Newfoundland; fairly widespread in certain
areas of Eurasia including Britain, Spain, northern Africa, Asia

Minor, and east to Manchuria and Japan and south to India, Burma, the Philippines and Indonesia.

Bullfinch   *Pyrrhula pyrrhula*
purrhos = purros (Gr) red, flame-coloured   *purroulas* (Gr) a red-coloured bird, perhaps the bullfinch; the cheeks and the underparts are red, with upperparts grey, black, and white markings. Inhabiting Eurasia including the Azores, northern Spain, the British Isles, northern Asia Minor, and ranging east to Kamchatka and Japan.

Hawfinch   *Coccothraustes coccothraustes*
kokkos (Gr) a kernel   *thrauō* (Gr) I break in pieces   *-istes* (Gr) suffix meaning the one who acts; 'kernel breaker'; a reference to the powerful beak which can actually crack open a cherry stone. Inhabiting Europe including the southernmost part of Scandinavia, and England and Wales and northernmost part of Africa, ranging east across Asia to the Pacific, Japan, and south to India.

Family   ESTRILDIDAE   124 species
estrilda (New L) from *astrild*, an Afrikaans name for this bird; it is also known in South Africa by the Afrikaans name *Rooibekkie*; the usual English name is Waxbill.

Red-headed Blue-bill   *Spermophaga ruficapilla*
sperma (Gr) seed   *phagein* (Gr) to eat; it is mostly a seed-eater but also takes insects   *rufus* (L) red   *capillus* (L) hair of men, usually the hair of the head, can mean the hair of animals. Inhabiting Africa.

African Firefinch   *Lagonosticta rubricata*
lagōn (Gr), genitive *lagonos*, the flank, of men or animals   *stiktos* (Gr) dotted, dappled; a reference to white spots on the side of the breast   *rubrica* (L) red ochre   *-atus* (L) suffix meaning provided with; the plumage is red becoming brown to grey towards the tail. Northern part of western Africa ranging east to Ethiopia, and south on the eastern side to South Africa.

Blue-breasted Waxbill or Angolan Cordon-Bleu   *Uraeginthus angolensis*
oura (Gr) the tail   *aeginthus* (New L) derived from *aigithos* (Gr) a hedge-sparrow; 'sparrow-tailed'; the bird is similar in size and shape to the Cape Sparrow of Africa   *-ensis* (L) suffix meaning belonging to; named from Angola it is also found in the eastern part of South Africa specially the Kruger Park.

Waxbill   *Estrilda astrild*
estrilada   see Family, p. 439. Most of the waxbills have a pink or red
bill with a waxy appearance suggesting sealing wax which has given
rise to the English name. This waxbill is fairly common and widespread in Africa south of the Sahara.

Black-cheeked Waxbill   *E. erythronotos*
eruthros (Gr) red   *noton* (Gr) the back; the plumage is black on the
sides of the face, and red on the back and underparts. Inhabiting
Africa south of the Sahara but not further south than the Orange
River in South Africa.

Avadavat (or Amadavat)   *Amandava amandava*
Avadavat and *Amandava* are corruptions of the name of the Indian city
Ahmadabad, in the province of Bombay; it was from here that the
first of these birds were sent to Europe. A small red to orange bird no
bigger than a wren, with a typical waxy-looking red bill, it is much
favoured as a cage bird. Inhabiting India, the Burma-Vietnam area,
Malaysia, and Indonesia.

Zebra Waxbill   *A. subflava*
sub (L) under, below   *flavus* (L) yellow; a reference to the yellow
throat which merges into scarlet, and the sides of the body have yellow
bars which gives rise to the English name Zebra Waxbill. Inhabiting
Africa south of the Sahara.

Black Chinned Quail-Finch   *Ortygospiza atricollis*
ortux (Gr), genitive *ortugos*, the quail   *spiza* (Gr) a finch; an allusion
to the quail-like flight, rising steeply with whirring wings and then
dropping to cover   *ater* (L) black   *collum* (L) the neck; the cheeks
and throat are black. Inhabiting the Senegal area and ranging east to
Ethiopia and south through the eastern part of Africa to South Africa.

Zebra Finch   *Poephila guttata*
poa (Gr) grass, can mean a grassy place, a meadow; becomes *poē* when
used in compounds   *philos* (Gr) loved, pleasing; the habitat is usually
riverside grasslands   *guttatus* (L) spotted, speckled; it also has bands
of black and white which gives it the English name. Common in
Australia it also inhabits Lesser Sundas and Moluccas.

Green-faced Parrot-Finch   *Erythrura viridifacies*
eruthros (Gr) red   *oura* (Gr) the tail; a reference to the crimson tail
viridis (L) green   *facies* (L) form, face; not necessarily meaning only

the face as the plumage is mostly grass-green. Known only from Luzon.

Gouldian Finch *Chloebia gouldiae*
khloë (Gr) the shoots of young green plants, grass *bios* (Gr) living, manner or means of living; this finch and its relatives are sometimes known as grassfinches, and although they have a finch-like bill they are not true finches; they seldom come to the ground but climb about on the stems of tall grasses and eat the seeds and insects they find there. This finch is named in honour of Mrs E. Gould (1804-1841), a bird artist and the wife of J. Gould, FRS, a zoologist and the author of books on birds. Inhabiting northern areas of Australia and sometimes migrating south to breed.

Chestnut-breasted Finch *Lonchura castaneothorax*
lonkhē (Gr) a spear-head *oura* (Gr) the tail; a reference to the pointed tail *castanea* (L) the chestnut tree *castaneus* (New L) chestnut-coloured *thōrax* (Gr) the breast. Inhabiting northern Australia to New South Wales, also New Guinea.

Java Sparrow *Padda oryzivora*
The generic name *Padda* is a reference to the paddy fields where they find rice, their favourite food *oruza* (Gr) rice *voro* (L) I eat; a hybrid word formed from Greek and Latin; they will also take other seeds but can become quite a pest in the rice fields. Inhabiting Burma ranging south to Malaysia and Indonesia.

Family PLOCEIDAE 150 species
plokē (Gr) a twining, a weaving *plokeus* (Gr) a plaiter, a weaver; a reference to the form of the nest, which in some cases is a huge complicated woven nest, a communal place housing many birds.

Subfamily VIDUINAE 9 species
The generic name *Vidua* is derived from Whydah (sometimes Ouidah) a small town on the coast of Dahomey in western Africa where these birds were first discovered. They are parasitic, and choose certain species of birds where the eggs are laid, and only one in each nest; the stranger does not throw out the host's eggs or fledglings as does the cuckoo. They are seed-eaters and also take insects.

Pintailed Whydah *Vidua macroura*
Vidua, see above *makros* (Gr) long *oura* (Gr) the tail; the very long tail can be up to about 20 cm (8 in) which looks strange on a bird with

a body of 11·5 to 12·75 cm (4½ to 5 in). Widespread in Africa south of the Sahara including South Africa.

Paradise Whydah   *V. paradisea*
paradeisos (Gr) a park or pleasure ground, often used to mean a garden; an Oriental word meaning the garden of Eden; the name is probably used for this whydah on account of the quite remarkable and spectacular tail which is lyre-shaped in flight. Only the males have the long tail and in some species even the male does not have it and may look similar to the female. Widespread in Africa south of the Sahara.

Subfamily   PLOCEINAE   141 species
plokeus (Gr) see under Family, p. 441.

Donaldson Smith's Sparrow Weaver   *Plocepasser donaldsoni*
plokeus (Gr) a weaver; a reference to the form of the nest   *passer* (L) a sparrow; the sparrows in this family are true sparrows and include our House Sparrow: named after Dr A. Donaldson Smith (1864–1939), an American zoologist who was in East Africa in 1894 and 1895. Inhabiting Africa south of the Sahara.

Social Weaver   *Philetairus socius*
philos (Gr) loved, pleasing   *etairos* (Gr) a companion   *socius* (L) sharing, joining in; this species is famous for the enormous communal nests which they all help to build, woven from grasses to make a structure that can accommodate two or three hundred pairs, each in its own compartment. Inhabiting the western and central part of southern Africa.

House Sparrow   *Passer domesticus domesticus*
passer (L) a sparrow   *domus* (L) a house   *domesticus* (L) belonging to the house, familiar, domestic. Sometimes known as the English Sparrow, it is widespread in Europe, parts of northern Africa, southwestern Asia, India, Burma, and east to Manchuria, but it has been introduced to so many other countries that its original range is hard to define; for example introduced to Canada, the USA, West Indies, parts of South America, Australia, Tasmania, and New Zealand.

Italian Sparrow   *P.d. italiae*
This subspecies is similar to the House Sparrow, having a brown cap and a small black bib. It is not known except in Italy and Corsica. (For notes about subspecies, see Chapter 2, page 18).

Cape Sparrow   *P. melanurus*
melas (Gr), genitive *melanos*, black   *oura* (Gr) the tail; it has black tail
feathers with buff margins. Common in Cape Town and the western
part of southern Africa.

Tree Sparrow   *P. montanus*
montanus (L) belonging to a mountain; it is not restricted to mountainous areas and inhabits Europe, including the British Isles, Asia,
including India, and south to the Burma-Vietnam area, Indonesia,
and east to Japan. The appearance is similar to the House Sparrow
though it is slightly smaller; it has been introduced to North America,
Australia, and New Zealand.

Swahili Sparrow   *P. suahelicus*
-icus (L) suffix meaning belonging to; there is no such place as Swahili;
it is a language originally from Zanzibar, and now spoken and used as
a common tongue by many tribes in large areas of East Africa and
Zaire.

Rock Sparrow   *Petronia petronia*
petra (L) a rock, a crag   *petronius* (L) of a rock, or crag; a reference to
the habitat of dry rocky slopes with grass and low scrub. Inhabiting
the Canary Islands and the Mediterranean area and ranging east
through Asia Minor to the mountains and desert areas of south-west
Asia to Tibet and China.

Snow Finch   *Montifringilla nivalis*
mons (L), genitive *montis*, a mountain   *fringilla* (L) a small bird,
according to some authorities the chaffinch   *nix* (L), genitive *nivis*,
snow   *nivalis* (L) of snow, snowy; a reference to the habitat of rocky
slopes and screes, high mountains and borders of glaciers at 6,000 m
(20,000 ft) or more in the Alps. Inhabiting the mountains, though not
all, ranging from the Pyrenees to the Altai, and some of the mountains
of Tibet.

Golden Palm-Weaver   *Ploceus bojeri*
plokeus (Gr) a weaver; a reference to the form of the nest: named after
Wenzel Bojer (1800–1856), a Czechoslovakian naturalist from
Prague; he collected in Madagascar and Zanzibar and was at one
time Curator of the Mauritius Museum. The males in the genus
Ploceus are golden yellow and the woven nests are frequently hung in
palm trees. Inhabiting Somalia and Kenya.

Black-headed Weaver   *P. cucullatus*
cucullus (L) a hood   *cucullatus* (L) hooded; a reference to the black head and neck which gives the impression that it is wearing a hood. Widespread in Africa except the south-western part and The Cape.

Baya Weaver   *P. philippinus*
-inus (L) suffix meaning belonging to; named from the Philippines, it inhabits India, the Burma-Vietnam area, Malaysia, Indonesia, and the Philippines. *Bayā* is a Hindi word of East India for the weaver.

Red-billed Quelea   *Quelea quelea*
quelea (New L) derived from an African native name; it has a striking red bill. Inhabiting a part of the west coast of Africa south of the Sahara and ranging east along a belt of country to Sudan and Ethiopia, and south through eastern Africa to northern South Africa.

Yellow-shouldered Widow Bird   *Euplectes macrourus*
eu- (Gr) prefix meaning well, nicely   *plektos* (Gr) plaited, twisted; 'well plaited'; a reference to the nest which is made of woven grasses. The range is very similar to the Red-billed Quelea, above, but only extending south as far as Mozambique, Malawi, and Angola.

Family   STURNIDAE   106 species
sturnus (L) a starling; some of the birds in this group are known as Mynahs, and have become familiar to us because they are often kept as cage birds on account of their clever imitation of the human voice; they are, however, true starlings, and some other starlings are clever imitators.

Glossy Starling   *Aplonis metallicus*
Aplonis, obscure, but may have some connection with Apollo, the Greek sun-god, i.e. 'shining'. H. Wolstenhome, BA, of Wahroonga, Sydney, in the Official Checklist of the Birds of Australia, referred to the MS of Dr Herbert Langton (deceased), expert on the derivation of bird names; he says: 'Like myself he could not find the origin of certain obscure names, such as *Aplonis*, *Biziura*, and *Epthianura*' *metallicus* (L) belonging to metal, metallic; a reference to the shining plumage, and hence the English name. Inhabiting the southern Pacific on the many small islands, and Papua New Guinea, and Australia.

Burchell's Glossy Starling   *Lamprotornis australis*
lamprotēs (Gr) brilliance   *ornis* (Gr) a bird; a reference to the metallic

green plumage with a purple sheen *australis* (L) southern; this does not mean Australia, it means the southern part of Africa: named after Dr W. J. Burchell (1782-1863), a naturalist and author who explored South Africa in 1811; he is better known through having Burchell's Zebra named after him. This starling inhabits much of southern Africa particularly Angola and Botswana.

Wattled Starling *Creatophora cinerea*
kreas (Gr), genitive *kreatos*, a piece of flesh *phoreō* (Gr) I bear, I have: a reference to the black wattles which grow on the side of the face and under the beak in the breeding season *cinis* (L), genitive *cineris*, ashes *cinereus*, ash-coloured; the plumage is mostly grey with some black on the flight and tail feathers that have a greenish gloss. Inhabiting south-western Arabia, Somalia, and ranging west to Zaire, Angola, and south to The Cape.

Rose-coloured Starling *Sturnus roseus*
sturnus (L) a starling *roseus* (L) red, rose-coloured; the colour of the body is rose pink. Inhabiting south-eastern Europe, Asia Minor, Iran, Turkestan, the Caspian region and Altai, and migrating south to India in winter.

Common Starling *S. vulgaris*
vulgaris (L) ordinary, commonplace. Known now in many parts of the world, it inhabits Iceland and most of Europe except Spain and southern France; ranging east across western and central Asia. It has been introduced to North America, South Africa, Australia and New Zealand.

Rothschild's Mynah *Leucopsar rothschildi*
leukos (Gr) white *psar* (Gr) a starling; a remarkable starling almost entirely white except for some bare skin on the face which is blue: named after Lionel Walter, Baron Rothschild, FRS (1868-1937), a zoologist and author; he founded the Tring Museum in 1889. It is very rare, and confined to the island of Bali, Indonesia.

Chinese Mynah *Acridotheres cristatellus*
akris (Gr), genitive *akridos*, a locust *thēraō* (Gr) I hunt, I catch; it will take insects other than locusts, and also eats fruit and grain *crista* (L) a crest, usually means a tuft on the head of animals *cristatus* (L) crested *-ellus* (L) diminutive suffix; it has a small tuft of rather ragged feathers on the head just behind the bill. Inhabiting southern China south of the Yangtze River, and Burma.

Common Mynah *A. tristis*
tristis (L) sad, gloomy; here used to mean dull colouring compared to other mynahs; although glossy the plumage is mostly black with some brown. Inhabiting Kazakhstan in the USSR and ranging south to Iran, Afghanistan, India, the Burma-Vietnam area, and Indonesia; it has been introduced to South Africa, the Seychelles, Mauritius, Australia, Tasmania, and New Zealand.

Golden-crested Mynah *Mino coronatus*
mino (New L) from *myna* = *mynah*, from the Hindi *mainā corona* (L) a crown *coronatus*, having a crown; it has a yellow head and a long yellow crest lying flat on the head. Inhabiting India and the Burma-Vietnam area.

Yellow-billed Oxpecker *Buphagus africanus*
bous (Gr) an ox, a bull *phagein* (Gr) to eat; a reference to its habit of clinging to the backs of large animals like buffalo, rhino, and antelopes, where they climb about searching for ticks, flies, and other insects on which they feed *africanus* (L) of Africa. Inhabiting Eritrea and ranging south to South Africa, though not eastern Kenya and Tanzania; it tends to keep to the areas that are the habitat of the host animals.

Red-billed Oxpecker *B. erythrorhynchus*
eruthros (Gr) red *rhunkhos* (Gr) the snout, beak. The range is very similar to *B. africanus*, above, and this bird has been seen at about 2,437 m (8,000 ft) in Kenya.

Family ORIOLIDAE 28 species
oriolus (New L) from *aureolus*, the diminutive of *aureus* (L) of gold, golden; the name originates from the Golden Oriole which has a brilliant golden head and upper part of the body with black wings and tail. The diet is insects and fruit, and they range in size from about 18 to 30·5 cm (7 to 12 in); known as Old World Orioles, they must not be confused with the American orioles of the genus *Icterus*, family Icteridae, to which they are not closely related.

Crimson-breasted Oriole *Oriolus cruentus*
Oriolus, see above *cruentus* (L) spotted or stained with blood; a reference to the red breast. Inhabiting Malaysia and Indonesia.

Golden Oriole *O. oriolus*
Oriolus, see above; this oriole is said to have a rather musical call that

sounds like 'ori-ole'. Inhabiting north-western Africa, Iberia, wide-spread across most of Eurasia but not the northern part, east as far as Tibet and India.

Yellow Figbird   *Sphecotheres flaviventris*
sphēx (Gr), genitive *sphekos*, a wasp   *thēraō* (Gr) I hunt, can mean I catch, wild beasts; it is not recorded that they eat wasps, but certainly various large insects such as cockchafers and grasshoppers, and also fruit especially wild figs   *flavus* (L) yellow   *venter* (L), genitive *ventris*, the belly; a reference to the bright yellow underparts. Inhabiting the northern coastal area of Australia.

Figbird   *S. vieilloti*
Named after L. J. P. Vieillot (1748-1831), the well-known French zoologist and author. Inhabiting Papua New Guinea, and Australia as above.

Family DICRURIDAE   20 species
dikroos = dikros (Gr) forked   *oura* (Gr) the tail; nearly all species in this family have a forked tail. Usually known as Drongos, they range in length from about 18 to 38 cm (7 to 15 in), though including the very long tail of the Racket-tailed Drongos this could be 63 cm (25 in); the diet consists of insects and small frogs and lizards, and sometimes nectar.

Mountain Drongo   *Chaetorhynchus papuensis*
khaitē (Gr) long flowing hair, giving rise to *chaeta* (New L) a bristle *rhunkhos* (Gr) the beak; all drongos have rictal bristles around the beak but in this case the bristles are longer than the beak   *-ensis* (L) suffix meaning belonging to; named from the then Territory of Papua, it inhabits the mountain forests of Papua New Guinea.

New Ireland Drongo   *Dicrurus megarhynchus*
Dicrurus, see above   *megas* (Gr) big, wide   *rhunkhos* (Gr) the beak; a reference to the big strong beak. It is named from New Ireland, one of the islands of the Bismarck Archipelago, lying to the north-east of Papua New Guinea.

Square-tailed Drongo   *D. ludwigii*
Named after C. F. H. Baron von Ludwig (1784-1847), a South African botanist from Cape Town. One of the drongos, which usually have forked tails, that has a square tail. Inhabiting the southern part of South Africa. *Drongo* is a native name in Madagascar.

Greater Racket-tailed Drongo   *D. paradiseus*

paradeisos (Gr) a park; the specific name in this case refers to the similarity of the bird to the Birds of Paradise, on account of the long
ornamental tail; in this bird the tail consists of two long shafts devoid
of feathers and ending with two small 'flags' or 'rackets'. Inhabiting
India, the Burma-Vietnam area, and Malaysia.

Family CALLAEIDAE   3 species

kallaion (Gr) a cock's comb; can mean wattles; a reference to the
brightly coloured fleshy wattle on each side of the lower mandible.
Usually known as Wattlebirds, the smallest is about 25 cm (10 in) and
the largest 50 cm (20 in) long; the diet is mainly vegetable food but
sometimes they also take insects and nectar.

Kokako   *Callaeas cinerea*

cinerea (L) ash-coloured; the plumage is mostly a bluish-grey. Kokako
is a Maori name for this bird; confined to the South Island of New
Zealand, it has bright blue and yellow wattles; although widespread
it is very rare.

Saddleback Wattlebird   *Creadion carunculatus*

kreas (Gr) flesh, a piece of meat   *kreadion* (Gr diminutive) a small
piece of meat   *caruncula* (L) a small piece of flesh   *-atus* (L) suffix
meaning provided with; a reference to the small orange-coloured
wattles on each side of the lower mandible. The name 'saddleback'
refers to the wide brown band across the glossy black back. It is now
found only on Hen Island, off the North Island of New Zealand, and
three small islands to the south-west of Stewart Island at the southern
end of South Island, New Zealand.

Family GRALLINIDAE   4 species

grallae (L) stilts   *-inus* (L) suffix meaning like, belonging to; a reference to the legs which are long for a bird of this size, about 25 cm (10
in), and influenced by it often being seen walking about in shallow
water, thus keeping the body dry. This small family have been given
various names such as chough and magpie-lark, but they are not
closely related to any of these birds. The diet consists of insects and
small invertebrates found in or near water such as worms and snails.

Magpie-Lark   *Grallina cyanoleuca*

Grallina, see above   *kyaneos* (Gr) dark blue   *leukos* (Gr) white; the
plumage is a piebald mixture of black and white, rather than dark

blue. Widespread in Australia, but not Tasmania; usually seen near water.

White-winged Chough *Corcorax melanorhamphos*
corvus (L) a crow *korax* (Gr) a crow: it is not a chough or a crow though the appearance is similar, and it is about the same size as a rook or a carrion crow, namely 46 cm (18 in) *melas* (Gr), genitive *melanos*, black *rhamphos* (Gr) a beak, bill, especially a crooked beak as in birds of prey; it has a black curved beak and the plumage is black except for a white patch on the wings. Inhabiting Australia in the eastern part from southern Queensland to New South Wales, and much of the southern region.

Family ARTAMIDAE 10 species
artamos (Gr) a butcher; known as Wood-Swallows, these birds were originally confused with the shrikes, sometimes known as Butcher-birds, in the genus *Lanius* (L) a butcher; Linnaeus described the first known species as a shrike in 1771. Many of the shrikes were well known to have the habit of impaling their prey on any convenient spikes, where they could remain as though in a larder until required for eating. The wood-swallows feed mostly on insects of various kinds, some being hawked in the air.

White-breasted Wood-Swallow *Artamus leucorhynchus*
Artamus, see above *leukos* (Gr) white *rhunkhos* (Gr) the snout, beak; the beak is white, tinged with very pale blue, and the breast and underparts are white. The birds in this small family are swallow-like in some respects but are not closely related. Inhabiting Indonesia, the Philippines, New Guinea, and ranging east to the Fiji Islands; in Australia the northern coastal region and in the east ranging south to New South Wales.

White-browed Wood-Swallow *A. superciliosus*
super (L) over, above *cilium* (L) the eyelid, giving rise to *cilia* (New L) a hair *ciliosus* (New L) full of hairs; it has a prominent white eye-brow. Widespread in Australia in eastern areas.

Family CRACTICIDAE 11 species
kraktikos (Gr) noisy; an allusion to the strident alarm call but they do have a pleasant song. Known variously as Butcherbirds, Bell Magpies, and Currawongs, they have the same trick as the shrikes, in the genus *Lanius,* of impaling their prey on a suitable spike (see above, under Family Artamidae).

Grey Butcherbird   *Cracticus torquatus*
Cracticus, see above   *torquatus* (L) wearing a collar; a reference to the
white band round the neck; the plumage of the upper parts is grey.
Fairly widespread in Australia except the northern coastal region.

Pied Currawong   *Strepera graculina*
strepo (L) I make a noise   *streperus* (New L) noisy; a reference to the
strident alarm call which sounds like a loud 'currawong', and no doubt
this has given rise to the Australian native name   *graculus* (L) a jackdaw   *-inus* (L) suffix meaning like; the Latin word originates from
the jackdaw's call, which is supposed to sound like 'gra gra', though
some consider it is 'jack jack'; the currawongs are not closely related
to the jackdaw. The Pied Currawong is mostly black with some white
on the wings, rump, and tail; it inhabits the eastern coastal area of
Australia, and Lord Howe Island which lies about 805 km (500 miles)
to the east of the mainland.

Family   PTILONORHYNCHIDAE   17 species
ptilon (Gr) a feather   *rhunkhos* (Gr) the snout, beak; a reference to the
feathers that partly cover the beak. Known as Bowerbirds from their
habit of building an elaborate bower of grasses, sometimes roofed over
like a sort of wigwam, and sometimes just an avenue; in front there is
a circular court carefully decorated with coloured stones and other
coloured objects, and some species even paint the interior, making
their own paint with a mixture of saliva and the juice of berries, and
apply this with a 'brush' consisting of a piece of fibrous bark; however
some species do not build a bower. The diet consists of fruit, insects,
worms, and other small invertebrates.

Tooth-billed Bowerbird   *Scenopoeetes dentirostris*
skēnē (Gr) a covered or sheltered place   *poiētēs* (Gr) one who makes,
a maker; a perfect description of the male bird and his bower   *dens*
(L), genitive *dentis*, a tooth   *rostrum* (L) the beak; an allusion to the
serrated edges of the beak used by the male for cutting the leaves to
decorate his bower. Inhabiting the north-eastern part of Australia.

Archbold's Bowerbird   *Archboldia papuensis*
Named after R. Archbold (born 1907), a zoologist who was in the
Mammals Department of the American Museum of Natural History,
and was in New Guinea in 1933 and 1934   *-ensis* (L) suffix meaning
belonging to; named from Papua New Guinea, where it lives.

New Guinea Regent Bowerbird *Sericulus bakeri*
sērikos (Gr) silken *-culus* (L) diminutive suffix, here taken to mean
somewhat; a reference to the glossy plumage (cf Satin Bowerbird,
below): named after G. F. Baker, junior (1878-1937), a naturalist
and a New York banker; he was a Trustee of the American Museum
of Natural History. Inhabiting Papua New Guinea.

Satin Bowerbird *Ptilonorhynchus violaceus*
ptilon (Gr) a feather *rhunkhos* (Gr) the beak; a reference to the
feathers that partly cover the beak *violaceus* (L) violet-coloured; the
plumage of the male is a silky black, the feathers having specially
formed edges that give a glossy violet blue effect in sunlight. Inhabiting eastern Australia from Queensland south to Victoria.

Spotted Bowerbird *Chlamydera maculata*
khlamus (Gr) a short cloak, or mantle *derē = deirē* (Gr) the neck,
throat; a reference to the mantle or neck frill that can be raised or
lowered *macula* (L) a mark, a spot *maculatus*, spotted; the brownish
plumage is mottled with golden buff spots. Inhabiting eastern
Australia from Queensland south to northern Victoria.

Family PARADISAEIDAE 40 species
paradeisos (Gr) a park, a garden; in a legendary sense it has come to
mean a garden as in Heaven. The Birds of Paradise first became known
to the western world in the sixteenth century when a ship from the
Spanish fleet, the *Vittoria*, returned to Spain from the Molucca Islands
bringing a gift of some of the skins of these birds from a sultan in the
Moluccas to the King of Spain; the feathers and plumes were so
beautiful and spectacular that the Spaniards said they must be from
paradise, and so gave them the name Birds of Paradise. Closely related
to the Bowerbirds, they also have a spectacular mating display; the
diet is similar.

Multi-crested Bird of Paradise *Cnemophilus macgregorii*
knemos (Gr) the shoulder of a mountain, can mean the lower part of a
mountain *philos* (Gr) loved, pleasing: named after Sir W. Mac-
gregor (1846-1919), Governor of Queensland, Australia, from 1909
to 1914; he was in New Guinea for several years about 1890. This bird
lives on the lower slopes of the mountains of Papua New Guinea.

Paradise Crow *Lycocorax pyrrhopterus*
lukos (Gr) a daw or a chough; a daw is a kind of crow, cf. jackdaw

korax (Gr) a raven or crow; the birds of paradise are similar to crows in some respects and are closely related to them   *purros* (Gr) red *pteron* (Gr) wings; a reference to the wings, although they are usually a light brown rather than red. Inhabiting the Molucca Islands.

Curl-breasted Manucode   *Manucodia comrii*
Manucodia from *manucodiata* (New L) derived from a Malay name for the bird, *manuqdewata*, literally 'a bird of the gods': named after Dr P. Comrie (1832–1882), a Naval Surgeon on *HMS Basilisk* from 1871 to 1874: the English name is an allusion to the crinkled feathers of the body which finish as a 'topknot' on the head; it is sometimes known as the Curl-crested Manucode. It lives on the Trobriand and D'Entre-casteaux Islands which lie just to the east of Papua New Guinea.

Victoria Riflebird   *Ptiloris victoriae*
ptilon (Gr) a wing or feather   *os* (L), genitive *oris*, the mouth, pertaining to the mouth; a reference to the feathers that extend over the base of the upper mandible: named in honour of Queen Victoria (1819–1901), Queen of Great Britain and Ireland from 1837 to 1901. The Riflebirds derive their English name from the high-pitched call which is said to resemble the whine or whistle of a rifle bullet. Inhabiting the Cape York Peninsula of Australia.

Twelve-wired Bird of Paradise   *Seleucides melanoleuca*
seleukos (Gr), genitive *seleukidos*, a bird that eats locusts; this is a reference to its liking for insects, which would probably include locusts *melas* (Gr), genitive *melanos*, black   *leukos* (Gr) white, can mean grey; the plumage of the male is black and the underparts are pale but not white, and usually a pale lemon-yellow. The English name refers to six plumes that grow from each flank and terminate with thin shafts up to $30\frac{1}{2}$ cm (12 in) long that resemble wires. It lives in central and western Papua New Guinea.

Black-billed Sicklebill Bird of Paradise   *Drepanornis albertisii*
drepanon (Gr) a curved sword, a sickle   *ornis* (Gr) a bird: named after Luigi M. d'Albertis (1841–1901), an Italian zoological-ethnologist; he was in New Guinea from 1871 to 1877. It inhabits the high mountains of Papua New Guinea.

Princess Stephanie's Bird of Paradise   *Astrapia stephaniae*
astrapē (Gr) lightning, can mean the brightness of flowers; a reference

to the brightly coloured iridescent plumage: named in honour of Princess Stephanie (1864–1945) of Belgium. Inhabiting Papua New Guinea.

Queen Carola's Parotia   *Parotia carolae*
parōtis (Gr) a gland beside the ear; can also mean a lock of hair beside the ear; a reference to the crest on the head which extends down beside the ear; in this species the crest is kept constantly on the move: named in honour of Queen Carola of Saxony (born 1833), the wife of King Albert of Saxony (1828–1902); he was the King of Saxony from 1873 until his death in 1902. Inhabiting Papua New Guinea.

King of Saxony Bird of Paradise   *Pteridophora alberti*
pteris (Gr), genitive *pteridon*, a kind of fern, so called from its feathery leaves; a reference to the plumes of feathers on the crown that may reach a length of 46 cm (18 in)   *phoreō* (Gr) I have, I bear: named in honour of King Albert of Saxony (see above). Inhabiting the mountains of Papua New Guinea.

King Bird of Paradise   *Cicinnurus regius*
kikinnos (Gr) a curled lock of hair, a ringlet   *oura* (Gr) the tail; an allusion to the two wirelike central tail feathers which have tips consisting of green circular feathers   *rex* (L), genitive *regis*, a king; although named a 'king bird' it is the smallest member of the family, being only about 15 cm (6 in) long, but the extraordinary tail feathers of the male may increase this by another 18 cm (7 in). Inhabiting Papua New Guinea and the Kepulaun Aru Islands.

Greater Bird of Paradise   *Paradisaea apoda*
Paradisaea (see under Family on p. 451)   *a-* (Gr) prefix meaning not, without   *pous* (Gr), genitive *podos*, a foot; 'without feet'; the name was given by the first European naturalists to see the skins of these birds, who found that they had no legs, as they had been cut off during the process of preparing the skins for shipping. Without more careful examination they assumed that the birds had no legs, and never came down to land, and they thought up all kinds of bizarre explanations about their mode of living, including a hollow in the back of the male where the egg was deposited in flight, and the female remaining sitting on it during this 'airborne incubation'! One of the largest birds of paradise, it inhabits southern Papua New Guinea and the Kepulauan Aru Islands, which lie to the south-west.

Family CORVIDAE 103 species

corvus (L) a raven; this family consists of the Jays, Magpies, Ravens, and Crows, and of course the Rook, which is a type of crow, in the same way that the Jackdaw is a type of crow. There are some big birds in this family, for instance the raven *Corvus corax* can be as much as 61 cm (24 in) long; they eat a great variety of different foods including carrion, vegetable matter, insects, the eggs of birds and reptiles, and even small reptiles.

Steller's Jay *Cyanocitta stelleri*

kuaneos (Gr) dark blue, glossy blue   *kitta* (Gr) a jay; the front part of the body is black, the wings and tail a purplish blue, and altogether less colourful than some of the other jays: named after G. W. Steller (1709-1769), a German zoologist and explorer, particularly of northern areas such as Alaska and Siberia; several other birds have been named in his honour. This jay ranges over a large area of North America from southern Alaska, south on the western side, to Nicaragua in Central America.

Scrub Jay *Aphelocoma coerulescens*

aphelēs (Gr) smooth   *komē* (Gr) the hair, usually of the head; an allusion to the sleek plumage   *caeruleus* (L) dark blue   *-escens* (L) suffix meaning approaching, beginning to, i.e. 'almost dark blue'; the colour of the feathers is a blend of light blue and dark blue, the hind parts and tail being light blue. The range is across the USA from Washington to the coast of the Pacific and south to southern Mexico; there is also an isolated race in Florida. It is known as the Scrub Jay from its habitat, the dense scrub of oaks, sand pines, and similar forest trees.

Azure-hooded Jay *Cyanolyca cucullata*

kuaneos (Gr) dark blue, glossy blue   *lukos* (Gr) a kind of chough   *cucullus* (L) a hood   *cucullatus* (L) hooded. Inhabiting Central America.

Purplish Jay *Cyanocorax cyanomelas*

kuaneos (Gr) dark blue, glossy blue   *korax* (Gr) a raven or crow   *melas* (Gr) black; a mixture of blue and black to suggest purple. Inhabiting central South America.

Eurasian Jay *Garrulus glandarius*

garrulus (L) talkative; although shy and usually remaining hidden it has a great variety of cries and calls and is a good mimic   *glans* (L), genitive *glandis*, an acorn   *glandarius* (L) of or belonging to acorns; the diet is very varied but in autumn they eat a lot of acorns and also bury them in the ground for future use. Inhabiting Europe, including the British Isles, but not northern Scandinavia or northern Russia; north-west Africa, Asia Minor, and ranging east to Manchuria, Japan, China, and northern Laos and Vietnam.

Green Magpie   *Cissa chinensis*
kissa (also Attic *kitta*) (Gr) a chattering greedy bird, like a jay   *-ensis* (L) suffix meaning belonging to, 'of China'; in addition to China it inhabits large areas of south-eastern Asia including the Burma-Vietnam area, the Malay Peninsula, Sumatra and Borneo. The colour is mostly a bright blue-green.

Azure-winged Magpie   *Cyanopica cyanus*
kuaneos (Gr) dark blue, glossy blue   *pica* (L) a magpie; a hybrid word of Greek and Latin; the wings and the long tail feathers are blue. One group inhabits a large area of south-western Spain and Portugal, but the main population is in China, Korea, and Japan.

Andaman Treepie   *Dendrocitta bayleyi*
dendron (Gr) a tree   *kitta* (Attic) = *kissa* (Gr) a chattering greedy bird like a jay: named after Sir E. C. Bayley (1821-1894), an English statesman and an archeologist, he was in India from 1842 to 1878. Inhabiting the Andaman Islands which lie to the south of Burma in the Bay of Bengal; there are nine magpies that are known as treepies.

Common Magpie   *Pica pica*
pica (L) a magpie. Very widespread in the Northern Hemisphere, from Alaska, Canada, the western USA, to northern Africa, Eurasia including the British Isles, Asia Minor, southern China and Japan.

Abyssinian or Stresemann's Bush-Crow   *Zavattariornis stresemanni*
Named in 1938 by Dr Moltoni for Prof E. Zavattari (born 1883), an Italian Professor of Zoology in Pavia, northern Italy   *ornis* (Gr) a bird: 'Zavattari's bird'; also Prof Dr E. Stresemann (born 1889), a German zoologist and Director of Ornithology at the Zoological Museum of Berlin in 1921. It is unusual for both the generic and specific names to refer to zoologists. Dr E. Moltoni (born 1896), was an Italian ornithologist of Milan, and author of the book *Birds of Italian Somaliland*. This rare bush-crow was discovered in southern Abyssinia, now known as Ethiopia. It is not recorded elsewhere as yet.

Clark's Nutcracker   *Nucifraga columbiana*
nux (L), genitive *nucis*, a nut   *frango* (L) I break, I dash to pieces; part
of the diet consists of pine seeds and acorns which it extracts from the
shell with great skill: named from British Columbia, it inhabits the
mountains from southern Alaska through British Columbia, Alberta,
South Dakota, New Mexico, and California; it has been seen as high
as 3,600 m (12,000 ft) in the mountains. Named after Brigadier-
General W. Clark (1770–1828), American Army; he led the Lewis-
Clark River Missouri and River Columbia Expedition in 1804 to
1806; he became Governor of Missouri in 1813. Captain M. Lewis
became Governor of Louisiana in 1907.

Red-billed Chough   *Pyrrhocorax pyrrhocorax*
purrhos (Gr) flame-coloured, red; a reference to the slender curved red
bill and red legs and feet   *korax* (Gr) a raven or crow: it is usually
found in high mountainous areas, and inhabits the western part of
Scotland, Ireland, and England, mountain areas of Spain, northern
Africa, Switzerland, Italy and Greece; it ranges east through Asia
Minor, Iran and Tibet to north-western China and Manchuria. The
Alpine Chough *Pyrrhocorax graculus* has been seen on Mount Everest
at about 8,200 m (27,000 ft) which is probably as high as any bird
can fly.

Raven   *Corvus corax*
corvus (L) a raven   *corax* (L) = *korax* (Gr) a raven. Widespread round
the world in the Northern Hemisphere.

Carrion Crow   *C. corone corone*
korōnē (Gr) a kind of sea-bird, a sea-crow; can mean a kind of ordinary
crow, a jackdaw; the second specific name, properly known as the
sub-specific name, indicates that it is the nominate sub-species (see
Chapter 2, p. 19); the Hooded Crow, a sub-species, is given below.
The Carrion Crow is widespread in Eurasia including England and
Scotland but not the northern coastal area of the USSR; it ranges east
to the Pacific but probably never south of the Tropic of Cancer.

Hooded Crow   *C. c. cornix*
cornix (L) a crow; this is a sub-species of the Carrion Crow, and has a
grey upperpart of the body which makes the black head and neck
stand out as a 'hood'. It inhabits Scotland and Ireland and a large
area of central Eurasia roughly between the River Elbe and the River
Yenisei; where the area overlaps that of the Carrion Crow these two
sub-species often interbreed producing a mixed race, a hybrid.

Thick-billed Raven   *C. crassirostris*
crassus (L) thick, heavy   *rostrum* (L) the beak; the ravens tend to have big heavy beaks. Inhabiting north-eastern Africa.

Rook   *C. frugilegus*
frux (L), genitive *frugis*, fruits of the earth, vegetables grown underground; can be used to mean fruits in general; it has a very mixed diet which includes earthworms, insects, fruit, and grain   *lego* (L) I gather, I collect   *frugilegus* (L) fruit-gathering. Widespread in Europe including the British Isles but only the southern part of Scandinavia; the Azores and Madeira and parts of northern Africa, and most of Asia.

Jackdaw   *C. monedula*
monedula (L) a jackdaw, a daw; daw is an English word meaning a bird of the crow kind. Widespread in Europe including the British Isles but not northern Spain, south-west France, northern Scandinavia and northern Russia; it is found in the northern part of Africa and ranges east through Asia Minor, India, the Burma-Vietnam area, part of the Himalayas, and in Mongolia.

Fish Crow   *C. ossifragus*
os (L), genitive *ossis*, a bone   *frango* (L) I break, I dash to pieces; prominent in the varied diet of this crow, as the English name suggests, are marine animals, and it is particularly fond of shellfish; the Latin name refers to its ability to break open the hard shells. It inhabits the coastal regions, rivers and lakes of the Atlantic coast from southern Massachusetts south to Florida and west along the Gulf coast to eastern Texas.

PART FOUR
The Mammal Orders

41 Mammals in General—the Class Mammalia

This group of animals—the class Mammalia—includes tiny creatures like the harvest mouse, weighing about 7 g (only $\frac{1}{4}$ oz), and the whales which may weigh over 100 tonnes. Between these two extremes are all the animals most familiar to us—such as cats, dogs, horses, lions, elephants and ourselves. They are the only animals that have true hairs and, by means of the mammary glands produce milk to feed their young; hence the name 'mammals', the word being derived from the Latin *mamma*, a breast.

For purposes of classification the class Mammalia is divided into three subclasses: the Prototheria or 'first animals', the Metatheria or 'later animals', and the Eutheria, 'typical' or 'well-made animals'. The echidnas and the duck-billed platypus are the only ones in the subclass Prototheria, the kangaroos and other pouched animals form the subclass Metatheria, and all the remainder form the subclass Eutheria.

The first group, the Monotremes, are primitive animals that lay eggs. The second group, the Marsupials, are a more advanced form of life in which the young are born after a very short gestation period, but are not fully developed and have to start life in the pouch. All the animals in the third group give birth to young in an advanced state of development, though in most cases the young are quite helpless and dependent on the care of the mother. For instance, a lioness may continue to teach her cubs to hunt for about a year.

The placenta, which unites the foetus to the mother's womb and supplies it with nourishment and oxygen, is not found in the egg-laying mammals, and only in a rudimentary form in the marsupials. All the remaining mammals have the placenta, and it is not found in any other animals (except for a rather similar organ in one or two fishes and lizards). It has almost certainly evolved from the reptiles.

42 Duck-Billed Platypus and Echidnas
MONOTREMATA

Subclass PROTOTHERIA
prōtos (Gr) first; *thēr* (Gr) a wild animal.

Order MONOTREMATA
monas (Gr) single; *trēma* (Gr) a hole.

Family ORNITHORHYNCHIDAE one species
ornis (Gr) genitive *ornithos*, a bird; *rhunkhos* (Gr) a beak, a bill.

Duck-billed Platypus *Ornithorhynchus anatinus*
anas (L), genitive *anatis*, a duck *-inus* (L) suffix meaning like, belonging to. Platypus is from *platus* (Gr) flat, broad and *pous* (Gr) a foot. This is the only known species; it is a most peculiar animal, and has some similarity to birds as well as mammals, and even to reptiles. It is one of the very few mammals that is truly venomous; the male has sharp spurs on the hind legs that are connected by a duct to poison glands. They have only one aperture for the elimination of liquid and solid waste, for copulation, and for the birth of the young; hence the name 'monotreme'. They lay soft-shelled eggs in a nest like the reptiles. Found only in Australia and Tasmania.

Family TACHYGLOSSIDAE five species
takhus (Gr) fast, swift *glossa* (Gr) the tongue.

Echidna or Australian Spiny Anteater *Tachyglossus aculeatus*
aculeus (L) a sting, a point; *aculeatus* thus means provided with prickles

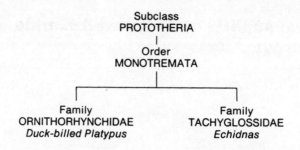

Subclass
PROTOTHERIA
|
Order
MONOTREMATA

Family
ORNITHORHYNCHIDAE
Duck-billed Platypus

Family
TACHYGLOSSIDAE
Echidnas

or stings. This probably refers to the body, not to the tongue, or to the spurs on the hind legs. These spurs are equipped with a poison duct, hence *echidna* (Gr) a viper, an adder (see Platypus above). With the echidna, unlike the platypus, the eggs remain in a pouch which develops after mating; later the mammary glands discharge milk into this pouch.

Tasmanian Spiny Anteater   *T. setosus*
saeta (=seta) (L) a bristle   *-osus* (L) suffix meaning full of.

New Guinea or Three-toed Spiny Anteater   *Zaglossus bruijni*
za (Gr) prefix with intensive sense, meaning very, much   *glossa* (Gr) the tongue; 'a long tongue'. A. A. Bruijn (fl. 1875-1885) was an officer in the Dutch navy; he collected specimens in the Malay Archipelago, especially Celebes and New Guinea.

43 **Marsupials or Pouched Animals**
MARSUPIALIA

In this group of mammals the young are born very small, and as little more than partly developed embryos; but they quickly find their way to the mother's pouch. Although collectively known as 'pouched mammals', in some cases the pouch is only a fold of skin and in others is completely absent. The young then attach themselves to a teat and remain attached for several weeks.

Subclass METATHERIA

meta (Gr) among, between; of place, after; of time, after, later *thēr* (Gr) a wild animal (see page 460).

Order MARSUPIALIA

marsupium (L) a purse, a pouch.

Family DIDELPHIDAE   about 65 species
di- (Gr) prefix meaning two, double *delphus* (Gr) the womb; an allusion to the pouch as a 'secondary womb' in which the young develop after birth.

Woolly Opossum   *Caluromys lanatus*
kalos (Gr) beautiful, fair *oura* (Gr) the tail *mus* (Gr) a mouse *lana* (L) wool *-atus* (L) suffix meaning provided with, i.e. 'woolly'. Inhabiting tropical South America.

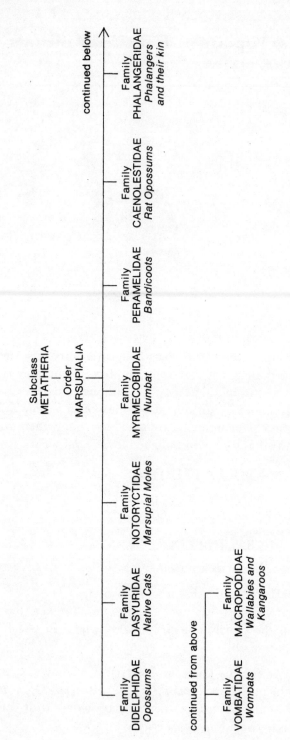

Subclass
METATHERIA

Order
MARSUPIALIA

Family
DIDELPHIDAE
Opossums

Family
DASYURIDAE
Native Cats

Family
NOTORYCTIDAE
Marsupial Moles

Family
MYRMECOBIIDAE
Numbat

Family
PERAMELIDAE
Bandicoots

Family
CAENOLESTIDAE
Rat Opossums

Family
PHALANGERIDAE
*Phalangers
and their kin*

continued below

continued from above

Family
VOMBATIDAE
Wombats

Family
MACROPODIDAE
*Wallabies and
Kangaroos*

Common or Virginia Opossum *Didelphis marsupialis*
-*alis* (L) suffix meaning pertaining to, like. Inhabiting North and South America.

Murine or Mouse Opossum *Marmosa murina*
marmose (Gr) a name of obscure origin, given by the Comte de Buffon in the 18th century *mus* (L), genitive *muris*, a mouse *ina* (L) suffix meaning like, belonging to. Inhabiting Brazil. This is one of a very large genus of about 40 species.

Water Opossum or Yapok *Chironectes minimus*
kheir (Gr) the hand *nēo* (Gr) I swim *nēktēs*, a swimmer; 'a hand for swimming'. They have webbed feet and are good swimmers *minimus* (L) smallest; it was first described as an otter, and in comparison it was very small. The name Yapok is derived from the South American river Oyapok, between Brazil and Guyana. Inhabiting Guatemala and southern Brazil.

Family DASYURIDAE about 45 species.
dasus (Gr) hairy *oura* (Gr) the tail; they have a bushy tail.

Broad-footed Marsupial Mouse *Antechinus stuarti*
anti- (Gr) prefix meaning against, opposed to, but can also mean equal to, like, resembling the word that follows *echinos* (Gr) a hedgehog; the head is similar to a hedgehog but it does not have prickly spines. J. McDouall Stuart (1815-1866) was a Scottish-Australian explorer who made several expeditions to the interior; Mount Stuart is named after him. The small marsupials are indeed mouse-sized, but they are not mice. Widespread in Australia.

Flat-skulled Marsupial Mouse *Planigale ingrami*
planus (L) level, flat *galea* (L) a helmet; the skull is so flattened that it can creep through 6 mm ($\frac{1}{4}$ in) crevices. It is the smallest marsupial and is named after Sir William Ingram, Bart (1847-1924), an English newspaper proprietor who shared in the financing of a collecting expedition in Australia. Named by Oldfield Thomas in 1906, it inhabits Western Australia. Oldfield Thomas (1858-1929) was at one time curator of mammals at the Natural History Museum, London. His classic paper of 1911 identified the Linnaean mammals of 1758 and is a landmark in mammalian nomenclature. He probably wrote more papers on and named more genera and species of mammals than any other zoologist, during the years 1880 to 1929. He

came to a tragic end, as he took his own life through grief at his wife's death.

Crest-tailed Marsupial Mouse   *Sminthopsis crassicaudata*
sminthos (Gr) an old Cretan word for a field mouse   *opsis* (Gr) aspect, appearance   *crassus* (L) thick   *cauda* (L) the tail of an animal   *-atus* (L) suffix meaning provided with. Inhabiting central and southern parts of Australia.

Pouched Jerboa   *Antechinomys spenceri*
Antechinus, see above   *mus* (Gr) a mouse. Named in honour of Professor Sir W. B. Spencer (1860–1929), a Director of the Natural History Museum in Melbourne and an explorer. He discovered *Notoryctes* (below). Inhabiting the desert areas of Australia.

Eastern Dasyure or Quoll   *Dasyurus quoll*
dasus (Gr) hairy   *oura* (Gr) the tail; it has a bushy tail. Quoll is an Aborigine word for this animal; it lives in Australia and Tasmania.

Western Native Cat   *Dasyurinus geoffroii*
-inus (L) suffix meaning like, belonging to. Named after Etienne Geoffroy Saint-Hilaire (1772–1844) the well known French zoologist. His son Isidore was also a zoologist and wrote a book about the life and work of his father. This dasyure inhabits the western part of Australia.

Little Northern Dasyure or Native Cat   *Satanellus hallucatus*
Satan (Gr) satan, from a Hebrew word meaning 'the enemy'   *-ellus* (L) diminutive suffix; 'little devil'; probably referring to its destructive nature   *hallex* (L), genitive *hallicis*, the thumb or big toe   *-atus* (L) suffix meaning provided with. Unlike other dasyures this one has the big toe. Inhabiting northern Australia.

Tiger Cat or Spotted Native Cat   *Dasyurops maculatus*
Dasyurus, see above; *ops* from *opsis* (Gr) aspect, appearance; 'like a dasyure'   *macula* (L) a spot, a mark   *-atus* (L) suffix meaning provided with, i.e. spotted, speckled; it has white spots on the body. Widespread in Australia.

Tasmanian Devil   *Sarcophilus harrisii*
sarx (Gr), genitive *sarkos*, flesh   *philos* (Gr) loved, pleasing; 'a meat lover'. Named after Deputy Surveyor Robert Harris; his notes on this animal were published in London in 1808. It is probably now confined to Tasmania.

Thylacine or Pouched Wolf   *Thylacinus cynocephalus*
thulakos (Gr) a bag, pouch   *cinus* (= *cynus*) from *kuōn* (Gr), genitive
kunos, a dog   *kephalē* (Gr) the head; it has a head like a dog or wolf.
Inhabiting the mountains of Tasmania, and now very rare, possibly
extinct.

Family NOTORYCTIDAE   2 species

Marsupial Mole   *Notoryctes typhlops*
notos (Gr) the south   *oruktēs* (Gr) one who digs; 'a southern digger'
tuphlos (Gr) blind   *ops* from *opsis* (Gr) appearance, 'apparently
blind'; the eyes are almost useless and hidden under the fur. It lives
in Australia, hence 'the south'.

Family MYRMECOBIIDAE   1 species

Numbat or Marsupial Anteater   *Myrmecobius fasciatus*
murmēx (Gr), genitive *murmēkos*, an ant   *bios* (Gr) living, also manner
of living; 'one that lives on ants'   *fascia* (L) a band, a girdle   *-atus*
(L) suffix meaning provided with; it has dark bands round the body.
Numbat is a name used by the Aborigines of Australia, where this
animal lives.

Family PERAMELIDAE   about 22 species

Long-nosed Bandicoot   *Perameles nasuta*
pera (L) a bag, a pouch   *meles* (L) a badger   *nasutus* (L) having a
large nose; it inhabits the eastern part of Australia. Pandikokku,
which became 'bandicoot', is the Telugu word for pig-rat, and used
in southern India and Ceylon for a large rat. The bandicoot-rat of
India almost certainly was familiar to some of the early settlers who
arrived in Australia. Thus there was the transfer of the name to
similar animals that they encountered, even though they were
marsupials and not placentals.

Rabbit Bandicoot   *Thylacomys lagotis*
thulakos (Gr) a bag, a pouch   *mus* (Gr) a mouse   *lagos* (Gr) a hare
ous (Gr), genitive *ōtos*, the ear; 'a pouched mouse with hare's ears'.
Inhabiting Australia.

Brindled Short-nosed Bandicoot   *Isoodon marcourus*
isos (Gr) equal, similar   *odōn* (Gr) a tooth; the teeth are similar in

size and shape   *makros* (Gr) long   *oura* (Gr) the tail; it has a short nose and a long tail. Inhabiting the eastern part of Australia.

Pig-footed Bandicoot   *Chaeropus ecaudatus*
khoiros (Gr) a young pig   *pous* (Gr) a foot; alluding to the striking resemblance of the forefeet to those of a pig   *ex* (L) from out of, without   *cauda* (L) the tail of an animal   *-atus* (L) suffix meaning provided with; 'not provided with a tail'. The name is not correct as the bandicoot *has* a tail, but the first one discovered and named had no tail, a loss to which this animal is apparently prone. In 1857 J. L. G. Krefft, the Australian zoologist, found that he had great difficulty in obtaining specimens. He showed a drawing of the original one with no tail to the natives, who failed to recognise it as their tailed 'landwang' and brought to him examples of the common bandicoot minus the tail which they had removed, hoping to satisfy his requirements. The Pig-footed Bandicoot inhabits southern Australia and Tasmania, but is now extremely rare, and possibly extinct.

Family   CAENOLESTIDAE   7 species

Rat Opossum   *Caenolestes fuliginosus*
There tends to be confusion about the two names 'opossum' and 'possum'; there is no significance in the different names, though 'possum' was probably first used in America for the opossums of the family Didelphidae. Captain James Cook was said to have first used this abbreviated form and it is sometimes used in Australia for those in the family Phalangeridae. The expression 'playing possum' alludes to the animal's habit of lying on its back and pretending to be dead when in danger from predators.
kainos (Gr) new, recent   *lēstēs* (Gr) a robber   T. S. Palmer, in his *Index Generum Mammalium*, quotes—'The affix *lestes* is connected in mammalogy with small and ancient fossil marsupials, . . . so that the above may be considered to represent an existing animal with ancient fossil relatives. (Thomas)'   *fuligo* (L) genitive *fuliginis*, soot   *-osus* (L) suffix meaning full of, prone to; the coat is dark grey. Inhabiting Ecuador, South America. Dr Theodore Sherman Palmer (1868–1958) was at one time Secretary of the American Ornithologists Union; his *Index Generum Mammalium*, published in 1904, took twenty years to complete. The work was begun in 1884 by Dr C. Hart Merriam (1855–1942), Chief of the U.S. Biological Survey, and so pressure of other

work forced him to hand over the task to Palmer. Assisted by a large number of researchers he finally completed the work in 1904, its year of publication.

Chilean or Fat-tailed Rat Opossum   *Rhyncholestes raphanurus*
rhunkhos (Gr) beak, snout   *lēstēs* (Gr) a robber; see above   *raphanos* (Gr) a radish   *oura* (Gr) the tail; the tail is thickened at the base giving it somewhat the appearance of a radish. Very little is known about this rare opossum. It inhabits Chiloe Island, Chile, and the neighbouring mainland.

Family   PHALANGERIDAE   about 46 species
phalanx (Gr), genitive *phalangos*, a line or order of battle; also, in the biological sense, the bone of a finger. Commonly known as phalangers, these marsupials have an unusual dexterity with their fingers, and the big toe is opposed to the other toes as our thumb is opposed to the fingers. However the name is not derived from this; it is an allusion to the peculiarity of the hind foot, in which the second and third digits are webbed together. It has been stated that the name is derived from the Greek *phalanx*, meaning a web. *Phalanx* does not mean a web, but there is a Greek word *phalangion*, a venomous spider, and after the classical period this may also have been used to mean a web.

New Guinea Spotted Cuscus   *Phalanger maculatus*
Phalanger, see above   *macula* (L) a spot   *-atus* (L) suffix, meaning provided with. Cuscus is a native name from the Moluccas islands in Indonesia.

Celebes Cuscus   *P. celebensis*
-ensis (L) suffix meaning belonging to, usually local names; in this case from Celebes.

Brush-tail Opossum   *Trichosurus vulpecula*
thrix (Gr), genitive *trikhos*, the hair, of man or beast   *oura* (Gr) the tail   *vulpes* (L) a fox   *-culus* (L) diminutive suffix; 'a little fox'. Inhabiting Australia, Tasmania, and New Zealand.

Pygmy Flying Possum or Pygmy Glider   *Acrobates pygmaeus*
akros (Gr) at the top   *bates* (Gr) from *baino*, I walk, step; 'an acrobat'   *pugmaios* (Gr) dwarfish. Flying possums have a web of skin stretching from the forepaws to the hind paws which enables them to glide quite

long distances: see notes about possums and opossums on page 468. Inhabiting Australia and New Guinea.

Pentailed Phalanger   *Distoechurus pennatus*
dis (Gr) twice, double   *stoikhos* (Gr) a row, a line   *oura* (Gr) the tail; 'a two-lined tail'   *penna* (L) a feather or wing   *pennatus* (L) feathered; the tail has two lines of long stiff hairs, on both sides, and only very short fur above and below, which gives it the appearance of a feather. Inhabiting New Guinea.

Pygmy Possum or Dormouse Phalanger   *Cercartetus nanus*
kerkos (Gr) the tail   *artaō* (Gr) I fasten to, I hang one thing upon another   *-etus* (L) a suffix usually used to designate a place; it has a prehensile tail, it can hang itself up on its tail   *nanus* (L) a dwarf. Inhabiting Australia, Tasmania, and New Guinea.

Dormouse Possum   *Burramys parva*
Burra is a town in south-eastern Australia where fossil remains of this animal were found in 1890; it was not until 1966 that one was caught alive. *mus* (Gr) a mouse   *parvus* (L) minor.

Leadbeater's Possum   *Gymnobelideus leadbeateri*
gumnos (Gr) naked   *belos* (Gr) anything thrown, a dart   *-ideus* (New L) suffix denoting similarity; 'naked dart' because it lacks the flying membranes of the Flying Possum (see page 469). J. Leadbeater (fl. 1852–1875) was an assistant at the National Museum in Melbourne; he discovered this possum in a forest near Melbourne, Australia.

Sugar Squirrel or Sugar Glider   *Petaurus australis*
petaurum (L) a springboard used by acrobats   *australis* (L) southern; it does not necessarily mean Australia. This Sugar Glider inhabits Tasmania and eastern parts of Australia.

Striped Possum   *Dactylopsila trivirgata*
daktulos (Gr) a finger; can mean a toe   *psilos* (Gr) naked; the toes of this possum are naked   *tres*, or *tria* (L) three   *virgatus* (L) striped. Inhabiting Australia.

Scaly-tailed Possum   *Wyulda squamicaudata*
Wyulda is the West Australian Aborigine name for the brush-tailed possum   *squama* (L) a scale of an animal   *cauda* (L) the tail of an animal. Inhabiting Australia.

Koala Bear   *Phascolarctos cinereus*
phaskōlos (Gr) a leather bag, a pouch   *arktos* (Gr) a bear; 'a bear

with a pouch' *cinis* (L), genitive *cineris*, ashes; cinereus, ash-coloured. These harmless little animals have been sadly persecuted for their fur, but a few colonies are still living in eastern Australia. Koala is an Aboriginal name and is said to mean 'no drink animal'.

Common Ringtailed Opossum *Pseudocheirus laniginosus*
pseudēs (Gr) false, deceptive *kheir* (Gr) the hand; an allusion to the hand-like forefeet, the two inner toes being opposable to the other three *lana* (L) wool and hence *laniger* (L) wool-bearing *-osus* (L) suffix meaning full of; 'woolly'. Inhabiting Australia.

Rock-haunting Ringtailed Opossum *Petropseudes dahlii*
petra (Gr) a rock *pseudēs* (Gr) false, deceptive; when naming this opossum in 1923 Thomas used *petra* because of its rocky habitat, and 'tacked' this word on to *pseudes* to show its relationship with *Pseudocheirus* (above). Named in honour of Professor K. Dahl (1871–1951), a Norwegian zoologist who was in Australia from 1894 to 1896. This opossum is widespread in Australia.

Great Glider *Schoinobates volans*
skhoinobatēs (Gr) a rope-dancer *volo* (L) I fly; *volans*, flying; a large glider that may measure more than 1 m (3 ft) in length. Inhabiting the forests of eastern Australia.

Family VOMBATIDAE 2 or possibly 3 species

Common Wombat or Ursine Wombat *Vombatus ursinus* (formerly *Phascolomis*)
Vombatus is derived from wombat, a native name for this animal in New South Wales *ursus* (L) a bear *-inus* (L) suffix meaning like. Inhabiting Australia and Tasmania.

Soft-furred or Hairy-nosed Wombat *Lasiorhinus latifrons*
lasios (Gr) hairy, shaggy *rhis* (Gr) genitive *rhinos*, the nose *latus* (L) broad, wide *frons* (L) brow, forehead. Inhabiting the southern part of Australia.

Family MACROPODIDAE about 47 species

makros (Gr) long, large *pous* (Gr), genitive *podos*, a foot; alluding to their long and powerful hind legs.

Hare Wallaby *Lagorchestes leporoides*
lagōs (Gr) a hare *orchēstēs* (Gr) a dancer *lepus* (L) a hare *-oides*

(New L) from *eidos* (Gr) shape, form. The hare wallabies are hare-like in colour and habits; they live in the southern and western parts of Australia and some islands off the coast.

Banded Hare Wallaby   *Lagostrophus fasciatus*
lagōs (Gr) a hare   *strophis* (Gr) a band, a girdle   *fascia* (L) a band, a girdle   *-atus* (L) suffix meaning provided with; it has dark cross-bands down the back. Inhabiting south-western Australia.

Brush-tailed Rock Wallaby   *Petrogale penicillata*
petra (Gr) a rock   *gale* (Gr) a marten-cat or weasel   *penicillus* (L) a painter's brush   *-atus* (L) provided with. It lives amongst rocks and boulders and is widespread in Australia.

Ringtailed Rock Wallaby   *P. xanthopus*
xanthos (Gr) yellow   *pous* (Gr) a foot; 'yellow-footed'; it has dark rings along the tail and yellow marking on the limbs. Widespread in Australia in rocky habitat.

Little Rock Wallaby   *Peradorcus concinna*
pēra (Gr) a pouch, a wallet   *dorkas* (Gr) a gazelle or antelope; a reference to its agility among rocks, like the chamois   *concinnus* (L) pleasing proportion, harmony of form. Widespread in Australia in rocky habitat.

Nail-tail Wallaby   *Onychogalea frenata*
onux (Gr), genitive *onukhos*, a nail, a claw; the tip of the tail ends in a horny spur   *gale* (Gr) a marten-cat or weasel: *freno* (L) I bridle, I curb; this refers to the face marking which resembles a bridle. Inhabiting grassy plains in Australia.

Red-legged Pademelon   *Thylogale stigmatica*
thulakos (Gr) a bag, a pouch   *gale* (Gr) a marten-cat or weasel: *stigma* (L), genitive *stigmatis*, a brand put upon slaves   *-ica* (New L) a suffix sometimes used to emphasise a certain characteristic; an allusion to a mark like a brand on the neck: pademelon is from paddymalla, an old native name in Australia. Inhabiting eastern parts of Australia.

Red-necked Pademelon   *T. thetis*
Halmaturus thetis (now *Thylogale thetis*). 'M. F. Cuvier has named *Kangourou thetis* a new species, brought from Port Jackson, in 1826, by the frigate *La Thetis*, commanded by M. de Bougainville, and

which M. Busseuil is to describe in the narrative which is to be published of this voyage.' (R. P. Lesson, 1827.) This refers to Admiral H. Y. P. Baron de Bougainville (c. 1781–1846), son of the famous vice-Admiral L. A. Baron de Bougainville. This pademelon inhabits a coastal strip of eastern Australia.

Red-bellied Pademelon   *T. billardieri*
Named by Professor Desmarest, of the Paris Museum, in 1822, in honour of the naturalist and navigator Jacques J. H. de La Billardière (1755-1834). He obtained the original Tasmanian specimen while on a voyage in search of Captain J. F. Comte de La Pérouse (1741–1788) who was lost off the New Hebrides in 1788. This pademelon inhabits the southern part of Australia, and Tasmania.

Bruijn's Pademelon   *T. bruijni*
Named after the Dutch painter C. de Bruijn who provided the first exact description of a kangaroo in 1714. It was the first kangaroo known to Europeans. In 1711 it had been seen at Batavia (now Djakarta) in Java, in the Dutch Governor's garden, where he kept one as a pet or perhaps more as a curiosity. It must have been brought there by travellers as this pademelon lives in New Guinea.

Brush Wallaby   *Wallabia rufogrisea*
Wallabia is derived from a native Australian word *wolobā   rufus* (L) red, ruddy   *griseus* (New L) derived from the German *greis*, grey. Sometimes known as the Red-necked Wallaby, it has a reddish-brown coat and grey chest and belly. It lives among brushwood and small trees in the eastern part of Australia, and Tasmania.

Pretty-faced Wallaby   *W. elegans*
elegans (L) neat, elegant. Living in Queensland and possibly New South Wales, it is now very rare.

The Euro or Wallaroo   *Macropus robustus*
makros (Gr) long, large   *pous* (Gr) the foot   *robustus* (L) strong. *Euro* and *walaru* are native names for this wallaby. It inhabits rocky areas of Australia.

Red Kangaroo   *M. rufus*
rufus (L) red, ruddy. Found in most parts of Australia.

Great Grey Kangaroo   *M. canguru*
Kanguru is a native name for the animal and is said to mean 'I don't know'. Inhabiting eastern Australia.

Quokka or Short-tailed Pademelon   *Setonix brachyurus*
seta (= *saeta*) (L) a bristle, stiff hair   *onux* (Gr) a claw   *brakhus* (Gr) short   *oura* (Gr) the tail. *Quokka* is an Australian native name for a pademelon. Inhabiting south-western Australia.

Lumholtz's Tree Kangaroo   *Dendrolagus lumholtzi*
dendron (Gr) a tree   *lagōs* (Gr) a hare; it is a tree-climber   Dr C. S. Lumholtz (1851–1922) was a Norwegian zoologist and author and was in Queensland during the years 1881 to 1893. These kangaroos live in Queensland.

Bennett's Tree Kangaroo   *D. bennettianus*
Named after Dr G. Bennett (1804–1893) a surgeon and zoologist who spent most of his life in Australia. This kangaroo inhabits Queensland and the island of New Guinea.

Goodfellow's Tree Kangaroo   *D. goodfellowi*
W. Goodfellow (1866–1953) was a zoologist who collected these kangaroos in New Guinea during the years 1909 to 1911; he was perhaps the greatest bird collector ever, chiefly for the British Museum (Natural History). He made many expeditions from 1898 to 1925 and is said to have brought back skins by the thousand. This kangaroo lives in New Guinea.

New Guinea Forest Wallaby   *Dorcopsis muelleri*
dorkas (Gr) a gazelle   *opsis* (Gr) appearance; suggesting a small kangaroo, in the same way that in Africa a gazelle is a small antelope, wallabies being small kangaroos. Named after Salomon Müller (1804–1863), a Dutch zoologist. He collected in Netherlands New Guinea and worked with Professor H. Schlegel (1804–1884) of Leiden Museum.

Short-nosed Rat Kangaroo or Bettong   *Bettongia cuniculus*
Bettong is a native name for the rat kangaroo   *cuniculus* (L) a rabbit; can also mean an underground passage. These small kangaroos burrow in the ground and often live in rabbit warrens with the rabbits that were brought to Australia.

Rufous Rat Kangaroo   *Aepyprymnus rufescens*
aipus (Gr) high   *prumnos* (Gr) the hindmost, endmost; alluding to the long hind legs   *rufus* (L) red, ruddy   *-escens* (L) suffix meaning beginning to, slightly; in this case 'reddish'. Inhabiting the Queensland area of Australia.

Desert Rat Kangaroo   *Caloprymnus campestris*
kalos (Gr) beautiful, fair   *prumnos* (Gr) hindmost, endmost; an allusion to the well developed hind legs   *campestris* (L) a plain, level country; it lives in the desert areas of central Australia.

Long-nosed Rat Kangaroo or Potoroo   *Potorous tridactylus*
Potoroo is a native name in New South Wales for this small kangaroo *treis, tria* (Gr) three   *daktulos* (Gr) a finger, a toe; this rat kangaroo has one toe absent and two others fused together, giving the appearance of three toes. Potoroos are now extinct in some parts of Australia but can still be found in Tasmania.

Musky Rat Kangaroo   *Hypsiprymnodon moschatus*
hupsos (Gr) height, also the top, summit   *prumnos* (Gr) the hindmost, endmost; alluding to the disproportionate development of the hind legs   *odous* (Gr) genitive *odontos*, a tooth; the reason for 'tooth' is obscure, and not given in T. S. Palmer's standard work *Index Generum Mammalium* (1904)   *moskhos* (Gr) musk and *moschatus* (New L) musky; it has musk glands that produce a strong odour. Inhabiting limited areas of Queensland.

This is a strange assortment of rather primitive little animals, inhabiting almost the entire world, except Australia, the polar regions, and most of South America. They are to be found everywhere, not only in the country but in city parks and suburban areas, and yet very few people have actually seen a live shrew or mole, except possibly in captivity; hedgehogs are most often seen dead on the roadside after being hit by traffic.

They are generally of flesh eating habits, and in addition to insects they eat grubs, worms and snails; to obtain their prey some burrow in the earth, some hunt on the surface, and some, such as the water shrews and desmans, are good swimmers and catch their prey in the water.

Subclass EUTHERIA

eu- (Gr) prefix meaning well, nicely; sometimes used to mean the typical animals in a group *thēr* (Gr) a wild animal (see page 32 for notes about the placenta).

Order INSECTIVORA

voro (L) I devour.

Family SOLENODONTIDAE 2 species.

Cuban Solenodon *Solenodon cubanus*
sōlēn (Gr), genitive *sōlēnos*, a channel, a pipe *odous* (Gr), genitive

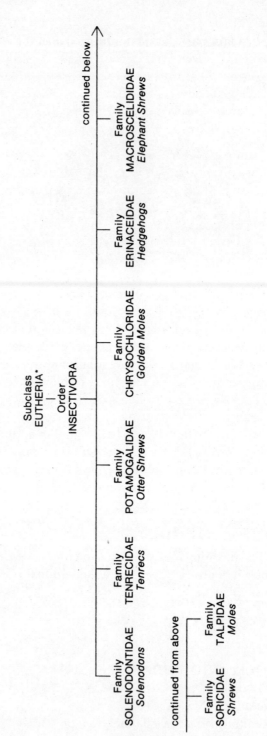

Subclass
EUTHERIA*
—
Order
INSECTIVORA

Family
SOLENODONTIDAE
Solenodons

Family
TENRECIDAE
Tenrecs

Family
POTAMOGALIDAE
Otter Shrews

Family
CHRYSOCHLORIDAE
Golden Moles

Family
ERINACEIDAE
Hedgehogs

Family
MACROSCELIDIDAE
Elephant Shrews

continued below

continued from above

Family
SORICIDAE
Shrews

Family
TALPIDAE
Moles

*Hereafter, all the described Orders are of the Subclass Eutheria and so only the Orders and Families are included in the tables accompanying Chapters 44 to 60.

odontos, a tooth, a fang; an allusion to the groove in the second incisor on each side of the lower jaw; it was originally supposed to be a channel for poisonous saliva, but tests have shown the poison to be extremely weak. Some now consider the groove may simply be to give strength to the tooth. *-anus* (L) suffix meaning belonging to; it lives in the Bayama Mountains in Cuba.

Haitian Solenodon *S. paradoxus*
para (Gr) contrary to, against *doxa* (Gr) opinion *paradoxos* (Gr) unexpected, strange. This species is found only in Haiti.

Family TENRECIDAE about 30 species
These animals are also known as tanrecs, or tendracs, from the Malagasy word *tandraka*; they are found only in Madagascar.

Common Tenrec *Tenrec ecaudatus*
ex (=*e*) (L) a prefix meaning from, out of; can mean without, like the Greek prefix *a-* *cauda* (L) the tail of an animal: *-atus* (L) suffix meaning provided with; 'not provided with a tail'; the common tenrec does not have a tail. It has sharp spines on the body rather like a hedgehog.

Hedgehog Tenrec *Setifer setosus*
saeta (=*seta*) (L) a bristle *fero* (L) I bear, I carry *-osus* (L) suffix meaning full of; 'a very bristly bristle carrier'.

Pygmy Hedgehog Tenrec *Echinops telfairi*
ekhinos (Gr) a hedgehog *ops*, from *opsis* (Gr) aspect, appearance: named after C. Telfair (1777–1833), a zoologist who founded the Botanical Gardens in Mauritius.

Banded or Streaked Tenrec *Hemicentetes semispinosus*
hēmi- (Gr) half, + *Centetes* (*Tenrec* formerly *Centetes*); indicating that this genus *Hemicentetes* differs from *Tenrec* in the presence of a third upper incisor, smaller canines, and the form of the skull *semi-* (L) half *spina* (L) a thorn, the prickles of animals *-osus* (L) suffix meaning full of; 'half-prickly'; this tenrec has bands of prickles alternating with bands of coarse black hair.

Rice Tenrec *Oryzorictes hova*
oruza (Gr) rice *oruktēs* (Gr) one who digs; they burrow in the rice fields *Hova* a dominant race in Madagascar, specially of the middle class and a name often used for animals inhabiting Madagascar.

Long-tailed Tenrec   *Microgale longicaudata*
mikros (Gr) small   *galē* (Gr) a marten-cat or weasel   *longus* (L)
long   *cauda* (L) the tail of an animal   *-atus* (L) suffix meaning
provided with.

Web-footed or Water Tenrec   *Limnogale mergulus*
limnē (Gr) a marshy lake   *galē* (Gr) a marten-cat or weasel; it is a
good swimmer, and one specimen was collected in a marsh . *mergo*
(L) I dip, I plunge   *-ulus* (L) adjectival ending denoting tendency;
the hind feet are fully webbed.

Family   POTAMOGALIDAE   3 species
Giant Water Shrew or Giant Otter Shrew   *Potamogale velox*
potamos (Gr) a river   *galē* (Gr) a marten-cat or weasel; a good
swimmer, it is a shrew, and not an otter or a marten-cat   *velox* (L)
fast, swift. Inhabiting western and central Africa.

Dwarf Otter Shrew   *Micropotamogale lamottei*
mikros (Gr) small   *potamogale* (see above) named after the French
zoologist Dr M. Lamotte who collected in West Africa. This shrew
inhabits the coastal regions of West Africa.

Family   CHRYSOCHLORIDAE   about 20 species
khrusos (Gr) gold   *khlōros* (Gr) green, can mean honey-coloured or
yellow; a reference to the beautiful iridescent hair.

Cape Golden Mole *Chrysochloris asiatica*
asiatica, of Asia. This is a misleading name as it does not belong to
Asia. It was given this name by Linnaeus because it was mistakenly
thought to come from Siberia; in fact it lives in South Africa.

De Winton's Golden Mole   *Cryptochloris wintoni*
kruptos (Gr) secret, hidden   *khlōros*, see above; they live a 'hidden'
life, almost entirely underground. W. E. de Winton (1856–1922) was
at one time Superintendent of the Zoological Gardens, London. This
mole inhabits Africa south of the equator.

Hottentot Golden Mole   *Amblysomus hottentotus*
amblus (Gr) blunt, point taken off   *sōma* (Gr) the body; 'blunt-body';
it has brilliant iridescent golden fur and a bulb-shaped body. The
Hottentots are a native people of Africa, and the name Hottentot is
from the Dutch, meaning a stutterer, on account of their peculiar
language. Range similar to above.

Family   ERINACEIDAE   about 15 species
erinaceus, or *ericius* (L) a hedgehog.

Moon Rat   *Echinosorex gymnurus*
ekhinos (Gr) a hedgehog   *sorex* (L) a shrew-mouse   *gumnos* (Gr)
naked   *oura* (Gr) the tail; the tail is naked and scaly. Inhabiting
southeast Asia.

Lesser Gymnure   *Hylomys suillus*
hulē (Gr) a wood, a forest   *mus* (Gr) a mouse: *sus* (L) a pig, hence
suillus (L) of swine; it looks rather like a small hog, hence 'hedgehog'.
Inhabiting Burma, Thailand, Borneo, Sumatra and South China.

Hedgehog   *Erinaceus europaeus*
erinaceus (L) a hedgehog   *europaeus* (L) belonging to Europe. This is
the common hedgehog of the British Isles and Europe. It also ranges
across parts of Asia.

Family   MACROSCELIDIDAE   14 species, possibly more.

Short-eared Elephant Shrew or Trumpet Rat   *Macroscelides
proboscideus*
makros (Gr) long   *skelis* (Gr), genitive skelidos, the leg; it has long
hind legs rather like the kangaroo   *proboscis* (L), genitive *proboscidis*,
a proboscis; derived from *pro* + *boskō* (Gr) I feed   *-ideus* (New L)
suffix meaning similarity. Inhabiting northern Africa.

Chequered Elephant Shrew   *Rhynchocyon cirnei*
rhunkhos (Gr) the snout, beak   *kuōn* (Gr) a dog; 'dog-nosed'; the
small proboscis is more like that of an elephant. Peters named this
shrew after Herr Cirne who stayed for two months in the Bororo
district of Mozambique, where this shrew was found. Inhabiting
central areas of Africa.

Family   SORICIDAE   more than 200 species; a large but uncertain number.
sorex (L), genitive *soricis*, the shrew-mouse.

European Pygmy Shrew   *Sorex minutus*
minutus (L) very small, minute. However, it is not the smallest shrew,
this honour probably belongs to Savi's Pygmy Shrew *Suncus etruscus*.

Common European Shrew   *S. araneus*
aranea (L) a spider, hence *araneus*, relating to a spider; this is a reference to the old belief that both the shrew and the spider are poisonous.

Smoky Shrew   *S. fumeus*
fumeus (L) smoky; the fur is greyish to brown on the back and a dirty white beneath. Inhabiting North America.

Pacific Water Shrew   *S. bendirei*
Major C. E. Bendire (1836–1897) was an American zoologist and author. Inhabiting North America. A good swimmer.

Northern Water Shrew   *S. palustris*
paluster (L), genitive *palustris*, marshy, boggy. Inhabiting Alaska and the northern part of North America.

North American Pygmy Shrew   *Microsorex hoyi*
mikros (Gr) small   *sorex* (L) the shrew-mouse: named after Dr Philip R. Hoy (1816–1892), an American explorer and naturalist. It is the smallest mammal in the New World and inhabits the northern part of North America.

Water Shrew   *Neomys fodiens*
neō (Gr) I swim   *mus* (Gr) a mouse   *fodio* (L) I dig; they dig holes in the banks of rivers. Inhabiting Europe, including most of the British Isles, and also parts of Asia.

Mediterranean Water Shrew   *N. anomalus*
anomalos (Gr) different, irregular; a reference to the absence of the keel of stiff hairs under the tail as possessed by *fodiens*. It inhabits not only Mediterranean areas, but can also be found in quite high mountains in Asia Minor.

Short-tailed Shrew   *Blarina brevicauda*
Blarina is a coined name for certain shrews given by Gray in 1838 who was an inveterate 'coiner'; it has been suggested that it derives from Blair, Nebraska, and *-inus* (L) pertaining to; this shrew can be found in that part of North America, but it is now considered that this explanation is an invention, and is not valid   *brevis* (L) short *cauda* (L) the tail of an animal. Inhabiting central parts of North America.

Least Shrew   *Cryptotis parva*
kruptos (Gr) hidden   *ous* (Gr), genitive *ōtos*, the ear; the ears are very

small and hidden under the fur   *parvus* (L) small; it is not so small as *Microsorex hoyi* (see above), but smaller than a mouse. Inhabiting North America.

Savi's Pygmy Shrew   *Suncus etruscus*
suncus from 'far sunki' (Arabic), a shrew   *etruscus*, an Etruscan, i.e. an inhabitant of Etruria, a district in north-western Italy, now known as Tuscany. Paolo Savi (1798-1871) was an Italian geologist and ornithologist and a Professor of Zoology at the University of Pisa. It inhabits large areas of southern Europe as well as Italy, and is considered to be the world's smallest mammal.

House Shrew   *S. murinus*
mus (L), genitive *muris*, a mouse   *murinus* mouselike. It has a vast range in continental Europe, central and southern Asia, and ranging to Indonesia.

Lesser White-toothed Shrew   *Crocidura suaveolens suaveolens*
krokus (Gr) nap, pile of cloth, a flock of wool   *oura* (Gr) the tail; the tail has short bristles interspersed with longer projecting hairs *suavis* (L) sweet   *olens* (L) smelling: *suave-olens* (L) sweet-smelling. Sometimes known as musk shrews, a reference to the musky odour which is less potent than that of common shrews. The white-toothed shrews have less red pigmentation on the teeth than other shrews; this shrew inhabits southern Europe, North Africa, and parts of Asia.

Scilly White-toothed Shrew   *C. s. cassiteridum*
kassiteros (Gr) tin   *kassiterides* (Gr) the tin islands, or Scilly Isles off the Cornish coast of England. There is no evidence that tin was found on the Scilly Isles but in the Early Iron Age tin was worked in Cornwall. This shrew was found on one of the islands comparatively recently, about 1924.

Armoured or Girder-backed Shrew   *Scutisorex congicus*
scutum (L) an oblong shield   *sorex* (L) the shrew-mouse   *-icus* (L) suffix meaning belonging to; *congicus*, of Congo, now Zaire. This little animal has the most remarkable backbone, resembling in miniature a massive girder designed to carry enormous weights; the reason for the elaborate structure is not known. This shrew is found only in Zaire.

Someren's Girder-backed Shrew   *S. somereni*
Dr R. A. L. Someren (1880-1955) was a naturalist who lived in

Uganda for many years from about 1905 onwards. This shrew was first discovered in Uganda.

Family   TALPIDAE   about 20 species
talpa (L) a mole.

Russian Desman   *Desmana moschata*
'Desman' is a Swedish word meaning musk   *moskhos* (Gr) musk; the desmans have glands under the tail which produces an overpowering musky smell. Inhabiting the central part of Russia.

Pyrenean Desman   *Galemys pyrenaicus*
galē (Gr) a marten-cat or polecat   *mus* (Gr) a mouse   *-icus* (L) suffix meaning belonging to; 'of the Pyrenees'. The desmans have webbed feet and are good swimmers. This one inhabits the northern part of Spain and Portugal in addition to the Pyrenees.

European Mole   *Talpa europaea*
Inhabiting Europe and parts of Asia.

Mediterranean Mole   *T. caeca*
caecus (L) blind; the moles, living underground, are usually almost completely blind. Inhabiting south-western parts of Europe.

Eastern Mole   *T. micrura*
mikros (Gr) small   *oura* (Gr) the tail; the very short tail is hidden under the fur. Inhabiting eastern Asia including Japan.

American Shrew Mole   *Neurotrichus gibbsi*
neos (Gr) new   *oura* (Gr) the tail   *thrix* (Gr), genitive *trichos*, the hair of man or beast; 'new hairy-tail'. George Gibbs (1815-1873) was a scientist from Oregon, U.S.A. This shrew mole is found on the west coast of North America from British Columbia to central California.

Townsend's Shrew Mole   *Scapanus townsendi*
scapanē (Gr) a spade or hoe; a reference to the claws, specially adapted for digging. J. K. Townsend (1809-1851) was an author who travelled and collected in the U.S.A. This mole lives in the northern coastal regions of North America.

Broad-footed Mole   *S. latimanus*
latus (L) broad   *manus* (L) the hand. Inhabiting the southern regions of North America.

Hairy-tailed Mole   *Parascalops breweri*
para (Gr) beside, near; i.e. resemblance   *skalops* (Gr) the digger, the mole. Dr T. M. Brewer (1814–1880) was a zoologist and author of Boston, Massachusetts. This mole lives in the eastern region of North America and parts of northern Carolina.

Eastern American Mole   *Scalopus aquaticus*
skalops (Gr) the digger, the mole   *pous* (Gr) the foot; 'a foot for digging'   *aqua* (L) water, hence *aquaticus* (L) living in water; although this mole is a good swimmer it does *not* live in the water. Inhabiting the south-eastern part of North America.

Star-nosed Mole   *Condylura cristata*
kondulos (Gr) knob of a joint, a knuckle   *oura* (Gr) the tail; a misleading name based on a faulty drawing by De La Faille   *crista* (L) the crest of an animal or bird, hence *cristatus* (L) crested; it has a 'star' or crest of 22 pink fleshy fingers growing on the snout. Inhabiting eastern North America.

45 **Bats** CHIROPTERA

The bats are divided into two main groups; the suborder Mega-chiroptera, meaning 'large hand-wing', and the suborder Micro-chiroptera 'small hand-wing'. The scientific name of the order Chiroptera is apt, as their wings are attached to their 'hands' and their method of flight is peculiar in that they appear to 'claw their way through the air'. Slow motion pictures show this clearly, and it is quite different from the wing action of a bird. The group names 'large' and 'small' are not so apt, as some bats in the large group, Megachiroptera are smaller than some of the big ones in the small group Microchiroptera. However, the division is a good one having regard to the structure of the skeleton and their eating habits.

Insect-eating bats of the suborder Microchiroptera have the remark-able ability to fly in the dark using ultra high frequency sound waves to avoid obstacles and even track down tiny insects in flight and catch them. This is all done without using their eyes and it is known as echolocation. They emit these sound waves continuously during flight, which are reflected by any object in the flight path and picked up by the bat's ears. This is a quite different phenomenon from that of cats, which are supposed to be able to see in the dark. In fact, cats cannot see in complete darkness, although their eyes are sensitive to

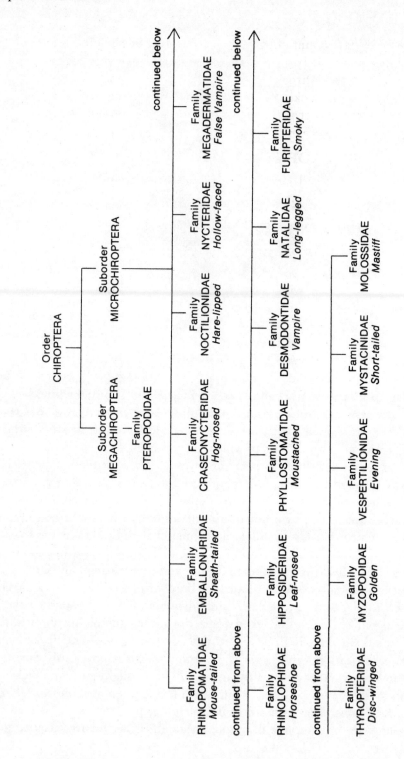

a very small amount of light, a characteristic of all animals that like to hunt at night. My own cat on one occasion, on a very dark winter's night, came in through the kitchen window and jumped down straight into a bucket of water! No self-respecting bat using echo-location would ever make such a ridiculous and humiliating mistake.

Subclass EUTHERIA (see pages 460 and 477)
Order CHIROPTERA

kheir (Gr) the hand *pteron* (Gr) wings (see introductory notes)

Suborder MEGACHIROPTERA fruit-eating bats

megas (Gr) big fruit-eating bats; 'big hand-wing'.

Family PTEROPODIDAE about 150 species

pteron (Gr) wings *pous* (Gr), genitive *podos*, the foot; 'wing-footed'; an allusion to the wing membrane which arises from the side of the back and the back of the second toe.

Indian Short-eared Fruit Bat *Cynopterus brachyotis*
kuōn (Gr), genitive *kunos*, a dog *pteron* (Gr) wings; 'a winged dog'; it has a dog-like head *brakhus* (Gr) short *ous* (Gr), genitive *ōtos*, the ear. Inhabiting Sri Lanka and south-east Asia, and ranging from southern China to Celebes.

Fruit Bat or Flying Fox *Pteropus vampyrus*
Pteropus, see above *vampir* (Slavonic). In eastern European folklore a vampire was a dead man who rose from the grave to prey upon the living; a 'blood-sucker'. The name is not appropriate as this bat is not a blood-sucker like the true vampire bats (see page 494). Inhabiting Burma and ranging east to Java, the Philippines, and Timor.

Grey-headed Flying Fox *P. poliocephalus*
polio (L) I polish; can mean I whiten, I adorn *kephalē* (Gr) the head. Inhabiting coastal and northern Australia.

Hammer-headed Bat *Hypsignathus monstrosus*
hupsos (Gr) height *gnathos* (Gr) the jaw; an allusion to the 'deeply arched mouth': monstrum (L) a monster *-osus* (L) a suffix meaning full of, or intensive; 'very monstrous'; it is exceptionally hideous. Inhabiting forest areas from Uganda to West Africa.

Straw-coloured Fruit Bat *Eidolon helvum*
eidolon (Gr) an image, a phantom; evidently an allusion to its move-

ments; *eidolon* is also the name for a small winged figure, human or combining human with animal elements, and found in Greek vase painting   *helvus* (L) light bay, yellow. Widespread in Africa south of the Sahara.

Arabian Straw-coloured Fruit Bat   *E. sabaeum*
Sabaean (L) a native of Saba, in southern Arabia.

Queensland Blossom Bat   *Syconycteris australis*
sukon (Gr) a fig   *nukteris* (Gr) a bat; it is not known that this bat eats figs though many fruit bats do. Its main diet is the nectar of flowering trees, hence the name 'blossom bat'   *auster* (L) the south   *-alis* (L) suffix meaning pertaining to. This is one of the smallest bats in the suborder Megachiroptera, or 'big bats', being only about 75 mm (3 in) long. Inhabiting the Queensland area of Australia.

Suborder   MICROCHIROPTERA   insect-eating bats
mikros (Gr) small   *kheir* (Gr) the hand   *pteron* (Gr) wings; 'small hand-wing'; (see introductory notes on page 487).

Family   RHINOPOMATIDAE   4 species
rhis (Gr), genitive *rhinos*, the nose   *pōma* (Gr), genitive *pomatos*, a lid, a cover; these bats have a flange of skin connecting the ears, a small pad on the nose, and valvular nostrils opening through a narrow slit.

Mouse-tailed Bat   *Rhinopoma microphyllum*
Rhinopoma, see above   *mikros* (Gr) small   *phullon* (Gr) a leaf; this could be interpreted as 'small leaf nose cover': they have a tail similar to that of a mouse. Ranging from West Africa to India, and possibly to Burma, Thailand and Sumatra.

Family   EMBALLONURIDAE   about 48 species
emballō (Gr) I put in   *oura* (Gr) the tail; the tail is partially sheathed in the tail membrane and emerges through the membrane to lie on the upper surface; it is used as a rudder when flying.

Sheath-tailed Bat   *Emballonura monticola*
Emballonura, see above   *mons* (L), genitive *montis*, a mountain   *colo* (L) I inhabit; it lives in the mountains of Malaya, Sumatra, Borneo, Java and Celebes.

Tomb Bat   *Taphozous perforatus*
taphos (Gr) a tomb   *zōos* (Gr) alive, living; 'one that lives in a tomb';

great numbers of these bats were found in the tombs by the French expedition which collected them during investigations in Egypt at the beginning of the nineteenth century *perforatus* (L) perforated; the interfemoral membrane is perforated to accommodate the tail. Inhabiting Egypt, Kenya, Arabia, and north-western India.

Pouch-bearing Bat *T. saccolaimus*
sakkos (Gr) a coarse cloth of hair, a sack or bag *laimos* (Gr) the throat; this tomb bat, as well as some others, has a small sac on the throat the purpose of which is not definitely established. Ranging from India and Sri Lanka to Borneo, Sumatra and Java.

Yellow-bellied Tomb Bat *T. flaviventris*
flavus (L) yellow *venter* (L), genitive *ventris*, the belly; the usual colour of bats is brownish but this one is black with pale yellow beneath. Inhabiting eastern Australia.

Family CRASEONYCTERIDAE 1 species
A new genus and species created as recently as 1974 by John Edwards Hill of the British Museum (Natural History). As in all cases, the name for the family is constructed from the generic name.

Hog-nosed Bat *Craseonycteris thonglongyai*
krasis (Gr), genitive *kraseōs*, a mixing, a blending *nukteris* (Gr), genitive *nukteridos*, a bat; an allusion to this bat having composite characters. Named by J. Edwards Hill in honour of his friend the late Kitti Thonglongya (died 1974) who discovered the animal in Thailand.

Family NOCTILIONIDAE 2 species
nox (L), genitive *noctis*, night; there is no such word in Latin as *noctilio* or *noctilionis*, although there is a Latin word *vespertilio* (genitive *vespertilionis*) from *vesper* (L) evening, and meaning 'an animal of the evening', a bat. *Noctilio* and *noctilionis* appear to be coined words from *nox* and supposed to mean 'an animal of the night'.

Hare-lipped or Fish-eating Bat *Noctilio leporinus*
Noctilio, see above *lepus* (L), genitive *leporis*, a hare *-inus* (L) suffix meaning like; it has full, swollen-looking lips, the upper lip being divided by a fold of skin; it patrols the sea and fresh waters and catches fish in its claws. Inhabiting Mexico, Central and South America, and the West Indies.

Family NYCTERIDAE   about 20 species
nukteris (Gr), genitive *nukteridos*, a bat.

Egyptian Hollow-faced Bat   *Nycteris thebaica*
Thebai (Gr) Thebes, an ancient city of Upper Egypt, surviving today
in the ruins of Luxor. Also known as Slit-faced Bats, this refers to a
groove on the face extending from the nose to between the eyes; there
is a deep hollow in the forehead. This bat inhabits Egypt and ranges
south to Angola.

Family MEGADERMATIDAE   5 species
megas (Gr) large   *derma* (Gr), genitive *dermatos*, the skin of man or
animals; from the large wing and interfemoral membrane.

Australian False Vampire Bat   *Macroderma gigas*
makros (Gr) large   *derma* (Gr) the skin   Gigas was a giant. Known
as 'false vampire bat' because although carnivorous it is not a 'blood
drinker'. Inhabiting Australia.

Family RHINOLOPHIDAE   75 species, possibly more
rhis (Gr), genitive *rhinos*, the nose   *lophos* (Gr) a crest; this refers to
the nose-leaf, part of which forms a sort of crest.

Greater Horseshoe-nosed Bat   *Rhinolophus ferrumequinum*
ferrum (L) iron, or can mean almost any instrument made of iron
equinum (L) relating to horses; a horseshoe; part of the nose-leaf is in
the form of a horseshoe. It inhabits Europe, North Africa, northern
India, China, and Japan.

Lesser Horseshoe-nosed Bat   *R. hipposideros*
hippos (Gr) a horse   *sidēros* (Gr) iron, or anything made of iron; 'a
horse-iron' or horseshoe. Range as *R. ferrumequinum*; both these bats
inhabit parts of the British Isles.

Family HIPPOSIDERIDAE   about 57 species
see *hipposideros* above.

Large Malayan Leaf-nosed Bat   *Hipposideros diadema*
diadema (L) a royal headband, a decoration; a reference to the nose-
leaves. These bats are related to the horseshoe-nosed bats, above,
and have a horseshoe-shaped nose-leaf. Inhabiting Burma, Australia
and the Solomon Islands.

Flower-faced Bat   *Anthops ornatus*
anthos (Gr) a flower   *ops* (Gr) the eye, face   *ornatus* (L) decorated; a
reference to the nose-leaves. From the Solomon Islands.

Family PHYLLOSTOMATIDAE   about 136 species
phullon (Gr) a leaf   *stoma* (Gr), genitive *stomatos*, the mouth; these
bats have a spear-shaped leaf on the nose just above the mouth, but
not on it.

Moustached or Leaf-lipped Bat   *Chilonycteris rubiginosa*
kheilos (Gr) the lip, brim; of animals the muzzle, beak   *nukteris* (Gr)
a bat   *rubigo* (L) rust   *-osus* (L) suffix meaning full of, prone to;
the lower lip has plate-like outgrowths and the body is rust-coloured.
Inhabiting Central America, northern South America, the West
Indies and common in Trinidad.

Naked-backed Bat   *Pteronotus davyi*
pteron (Gr) wings   *nōtos* (Gr) the back; the wing membrane is connected with the middle line of the back instead of the sides of the
body as in closely related species. Named after Dr John Davy (1790–
1868), brother of Sir Humphry Davy, 'well known for his physio-
logical papers'. The range is similar to *Chilonycteris*, above.

Leaf-chinned Bat   *Mormoops megalophylla*
mormō (Gr) a hideous monster, a hobgoblin   *ops*, from *opsis* (Gr)
aspect, appearance   *megas* (Gr), genitive *megalon*, big   *phullon* (Gr)
a leaf; 'a big-leafed monster'; the chin has leaf-like projections.
Usually found in caves in southern North America, Central America,
Venezuela, Trinidad and the West Indies.

Mexican Big-eared Bat   *Macrotus mexicanus*
makros (Gr) long   *ous* (Gr), genitive *ōtos*, the ear   *-anus* (L) suffix
meaning belonging to; of Mexico. It ranges south to Guatemala.

Spear-nosed Bat   *Phyllostomus hastatus*
phullon (Gr) a leaf   *stoma* (Gr) the mouth   *hastatus* (L) armed with
a spear; it has a spear-shaped leaf on the mouth. Inhabiting Central
and South America.

Jamaican Long-tongued Bat   *Monophyllus redmani*
monos (Gr) alone, single   *phullon* (Gr) a leaf; it has only one nose leaf.
Named by Dr Leach after Dom R. S. Redman, of Jamaica, from whom
a specimen was obtained. Inhabiting the West Indies. Dr W. E. Leach

(1790–1836) was an author and zoologist at the British Museum (Natural History) from 1813 to 1821.

White-lined Bat   *Vampyrops vittatus*
vampir (Slavonic); in eastern European folklore a monster, 'a blood-sucker'   *ops*, from *opsis* (Gr) aspect, appearance; it is not a true vampire bat and does not feed on blood   *vittatus* (L) bound with a ribbon or can mean striped; it is coffee-coloured with white stripes along the back. Inhabiting Costa Rica and ranging eastwards to Venezuela.

Family   DESMODONTIDAE   3 species
desmos (Gr) a bundle   *odous* (Gr), genitive *odontos*, a tooth; this refers to the two large curved, cone-shaped incisors in the upper jaw, apparently pressed together (i.e. bundled) and occupying the entire space between the canines.

Common Vampire Bat   *Desmodus rotundus*
rotundus (L) round; the body is spherical in shape. These are the true vampire bats, the 'blood-suckers'. They are unable to feed on anything except blood which they obtain from live animals. The specially adapted teeth, extremely sharp, enable them to make a 'scooping' incision to obtain the blood, reputedly without waking a sleeping animal. Inhabiting Central and South America.

Hairy-legged Vampire Bat   *Diphylla ecaudata*
di- from *dis* (Gr) two   *phullon* (Gr) a leaf; they do not have true nose-leaves, but have two small pads on the snout   *e-* (= *ex*) (L) out; can mean without   *cauda* (L) the tail   *-atus* (L) provided with; 'not provided with a tail'; they are tailless. Inhabiting tropical areas of Central and South America.

Family   NATALIDAE   about 10 species (see below)

Long-legged Bat   *Natalus stramineus*
This bat has a floating glandular disc in the front part of the head, known as the natalid organ. It is the only creature having such a process and its purpose has not been established. The term 'natalid organ' derives from its generic name *Natalus*, but the origin of the name remains obscure, and is not given by T. S. Palmer in his standard work *Index Generum Mammalium* (1904). *Stramineus* (L) made

of straw; sometimes known as Straw-coloured Bats, and also Funnel-eared Bats, referring to the unusual shape of the ears. Inhabiting Mexico, Central and South America, and the West Indies.

Family FURIPTERIDAE   2 species
furo (L) I rage, I rave   *pteron* (Gr) wings; 'winged furies'. The Furies, or Erinyes, in Greek mythology, were hideous avenging deities. Aeschylus represents them as 'daughters of the night', and Sophocles as 'daughters of darkness'. Several bats have been named after similar disagreeable mythical characters.

Smoky Bat   *Furipterus horrens*
horreo (L) I bristle; it has a very bristly muzzle. The body colour is grey. Inhabiting Panama and Brazil.

Family THYROPTERIDAE   2 species
thureos (Gr) a door-shaped shield   *pteron* (Gr) wings; this refers to a suction disc where the wing is attached to the wrist and another at the ankle. These discs enable the bat to attach itself to a smooth surface.

Disc-winged or Tricoloured Bat   *Thyroptera tricolor*
tres or *tria* (L) three   *color* (L) colour. Inhabiting Central and South America.

Family MYZOPODIDAE   only 1 species
muzō (Gr) I suck in   *pous* (Gr), genitive *podos*, the foot; 'sucker-footed one'; it has suction discs similar to *Thyroptera* (see above).

Golden Bat   *Myzopoda aurita*
aurum (L) gold, colour of gold   *-itus* (L) suffix meaning provided with—but more likely *aurita* (L) long-eared; it does have unusually long ears. A very rare bat and found only in Madagascar.

Family VESPERTILIONIDAE   about 280 species
vesper (L) evening   *vespertilio* (L), genitive *vespertilionis*, a bat; an animal of the evening. These bats are found in all temperate and tropical regions throughout the world including the British Isles.

Common or Brown Bat   *Myotis myotis*
mus (Gr) a mouse   *ous* (Gr), genitive *ōtos*, the ear; an allusion to the

large ears. Ranging from France to south-western Asia and to southern China.

Whiskered Bat   *M. mystacinus*
mustax (Gr), genitive *mustakos*, a moustache; this refers to the whiskers *-inus* (L) suffix meaning like. Ranging from Ireland to Japan, and India to Borneo.

Fishing Bat   *Pizonyx vivesi*
piezō (Gr) I squeeze   *onux* (Gr) a claw; an allusion to the toes and claws, which are compressed. The describer, Ménégaux, gives no clue as to the identity of Vives who apparently did not discover this bat. H. A. Ménégaux himself (1857–1937) was a French zoologist who was at the Paris Museum of Natural History from 1901 to 1926. It lives on the western side of North America.

Common Pipistrelle   *Pipistrellus pipistrellus*
pipistrello (Italian) a bat. This is the small bat seen in the evening in the British Isles. It is widespread and ranges from Ireland east to Japan, Kashmir and Taiwan.

American Western Pipistrelle   *P. hesperus*
hesperos (Gr) evening; can mean western, land of the setting sun. Inhabiting the western part of North America and Mexico.

American Eastern Pipistrelle   *P. subflavus*
sub- (L) under, beneath   *flavus* (L) yellow; it has a reddish-brown coat and pale underparts. Inhabiting eastern and central parts of North America and ranging south to Honduras.

Noctule   *Nyctalus noctula*
nux (Gr), genitive *nuktos*, night   *nuktalos* (= *nustalos*) (Gr) drowsy; said to be an allusion to its habit of coming out at dusk: *nox* (L), genitive *noctis*, night   *-ulus* (L) suffix denoting tendency; 'usually at night'. It has a wide range including Great Britain, Europe, Asia, and ranging eastwards to Burma and Malaya.

Leisler's Bat   *N. leisleri*
Dr L. P. A. Leisler (1771–1813) was a German scientist and author. This bat ranges from Ireland to south-western Asia as far as Punjab.

Serotine Bat   *Eptesicus serotinus*
Eptesicus is a coined word, derived from *eptēn* (Aorist tense of *petomai*) (Gr) I fly, and *oikos* (Gr) a house, and meaning 'a house-flyer'; they

are often seen flying near houses and roosting under the eves   *serotinus*
(L) late, backward; this name does not appear to have any meaning
with regard to the bat's development or behaviour; it does not come
out particularly late in the evening. It is quite a large bat, known in
the British Isles, and inhabits Eurasia and North Africa.

Club-footed Bat   *Tylonycteris pachypus*
tulē (Gr) a swelling or lump, a pad   *nukteris* (Gr) a bat   *pakhus* (Gr)
thick   *pous* (Gr) the foot; 'a pad-footed bat'; the under surface of the
base of the thumbs and the soles of the feet are expanded into fleshy
pads. Inhabiting southern Asia, Sumatra, Java, Borneo, the Philippines and Celebes.

Frosted Bat   *Vespertilio murinus*
vespertilio (L) a bat   *murinus* (L) mouse-like   the name 'frosted' refers
to the basic fur colour being obscured by intermingled white or white-
tipped hairs. Widespread in Europe and Asia, it has been reported in
both Great Britain and Japan.

Hinde's Bat   *Scotoecus hindei hindei*
skotos (Gr) darkness   *oikeō* (Gr) I dwell; 'dwelling in darkness'.
Named after Dr S. L. Hinde (1863–1931) who was in East Africa from
1896 to 1901. This bat is widespread in East Africa and has been
recorded in Somalia, Ethiopia, southern Sudan, Kenya, Uganda,
Zaire and Tanzania.

Falaba Bat   *S. h. falabae*
falabae, of Falaba; this does not refer to the town of Falaba in Sierra
Leóne, West Africa, but to a ship named 'Falaba'. Dr J. C. Fox, who
collected this bat, was nearly lost when the ship was 'barbarously'
sunk by natives. The bat has been recorded in Nigeria and Cameroun,
but the distribution will probably prove to be more widespread in
Africa.

Butterfly Bat   *Glauconycteris poensis*
glaukos (Gr) bluish-green, silvery-grey; the fur is light grey at the tips
nukteris (Gr) a bat   *-ensis* (L) suffix meaning belonging to; *poensis*
refers to the Island of Fernando Po, off the coast of Cameroun, West
Africa; it is found in other areas of Africa. It has a butterfly-like flight.

Red Bat   *Lasiurus borealis*
lasios (Gr) hairy   *oura* (Gr) the tail; sometimes known as the Hairy-
tailed Bat   *boreas* (L) the north wind, hence *boreus* northern   *-alis* (L)

suffix meaning pertaining to. Inhabiting North and South America
and the West Indies.

Hoary Bat   *L. cinereus*
cinerea (L) ash-coloured; hoary can mean white, or greyish-white.
The range is similar to *L. borealis*.

Barbastelle   *Barbastella barbastellus*
barbastelle (Fr) a little beard; they are sometimes known as Hairy-
lipped Bats. Inhabiting Europe, north-eastern Africa and Asia.

European Long-eared Bat   *Plecotus auritus*
plekō (Gr) I twine, I twist   *ous* (Gr), genitive *ōtos*, the ear; referring
to the unusual form of the junction at the base of the ears   *auris* (L)
the ear, hence *auritus* (L) long-eared. Found in Europe, Asia and
North Africa.

American Long-eared Bat   *P. macrotis*
makros (Gr) long   *ous* (Gr), genitive *ōtos*, the ear. Living in the south-
eastern part of North America.

Spotted Bat   *Euderma maculata*
eu- (Gr) prefix meaning well, nicely but sometimes used to mean
typical   *derma* (Gr) skin; in this case the interpretation would be
'nicely coloured skin'   *macula* (L) a spot   *-atus* (L) suffix meaning
provided with; spotted. It has a unique colour pattern with white
spots. A rare bat and found only in certain localities in western U.S.A.
and north-western Mexico. Also known as the Death's Head Bat.

Tube-nosed Bat   *Murina suilla*
mus (L), genitive *muris*, a mouse, hence *murinus* mouse-like   *sus* (L),
genitive *suis*, a pig and so *suillus* pig-like; the nostrils are at the end of
tubes. Inhabiting south-east Asia, Sumatra, Borneo, the Philippines
and other islands in that area.

Little Tube-nosed Bat   *M. aurata*
aurum (L) gold, hence *auratus* (L) golden, gilded; this refers to the
golden-brown coat. Inhabiting south-eastern Siberia, Japan, Burma
and northern India.

Painted Bat   *Kerivoula picta*
Kehelvoulha is a Singalese name for this bat   *pingo* (L) I paint and
pictus, painted; the brightest of all bats, it has long woolly hair and
the colour is orange or scarlet with black wings, and said to look like

a leaf in autumn. Inhabiting India, Sri Lanka, southern China, and Indonesia ranging east to Bali Island.

Pale or Desert Bat   *Antrozous pallidus*
antron (Gr) a cave   *zōon* (Gr) an animal; *zōos* alive, living; it usually roosts in caves   *pallidus* (L) pale. Found only in the west of North America.

Family   MYSTACINIDAE   1 species
mustax (Gr), genitive *mustakos*, a moustache; can mean the upper lip
inus (L) suffix meaning pertaining to.

Short-tailed Bat   *Mystacina tuberculata tuberculata*
tuber (L) a knob and so *tuberculum* (L) a small knob   *-atus* (L) suffix meaning provided with; it has small knobs, or pimples, on the upper lip. Widespread in the North Island of New Zealand, and the northern area of South Island, and on the Barrier Islands. It is becoming very rare and possibly is now extinct.

Short-tailed Bat   *M. t. robusta*
robustus (L) strong, robust. Inhabiting Stewart Island off the south coast of South Island, New Zealand.

Family   MOLOSSIDAE   88 species, possibly more
molossos (Gr) a kind of wolf-dog used by shepherds; these bats have dog-like faces.

Celebes Naked or Hairless Bat   *Cheiromeles parvidens*
kheir (Gr), genitive *kheiros*, the hand   *melos* (Gr) a limb; 'a hand limb'; T. S. Palmer, in his standard work *Index Generum Mammalium* (1904) says 'Possibly in allusion to the first toe, which is separated from the others like a thumb and probably opposable to them, thus giving the foot the appearance of a hand'   *parvus* (L) small   *dens* (L) a tooth. Probably the most unattractive of the bats in appearance, and furthermore possessing glands that can produce a quite revolting smell. Inhabiting Celebes and the Philippines.

Philippine Naked or Hairless Bat   *C. torquatus*
torquatus (L) having a collar; the skin round the neck is in folds, giving the appearance of a collar. This bat and *C. parvidens* differ only in small details. Inhabiting the Malay States, Indonesia and the Philippines.

Free-tailed Bat   *Tadarida taeniotis*

Tadarida is a dubious coinage by C. S. Rafinesque (1783-1840) a scholar of strange origin who died in poverty, though later he was re-interred in Lexington, Kentucky, U.S.A., where he had previously taught botany at the University. He outlined a pre-Darwinian theory of evolution. *taenia* (L) a head-band   *ous* (Gr), genitive *ōtos*, the ear; 'ear-bands'. The name 'free-tailed' refers to the tail being independent of the posterior portion of the flight membrane. Inhabiting southern Europe and ranging to eastern Siberia; also north-western Africa, Egypt, Iran, Japan and Taiwan.

Railer Bat   *T. thersites*

Named after Thersites 'The ugliest man in the Greek camp before Troy'; a relative of Diomedes, he was a railing demagogue, and this bat is ugly and chatters constantly. It is a far-fetched and interesting connection, but considered authentic. Inhabiting West Africa, Zaire, Rwanda and possibly Mozambique and Zanzibar.

46 **Flying Lemurs or Colugos** DERMOPTERA

These rather extraordinary animals are somewhat of a puzzle for zoologists. They have no close relatives, and so have been given an order all to themselves. At one time they were thought to be bats, and at another time primates, but their anatomy shows that in neither case was this correct. One remarkable feature is the structure of the lower front teeth, formed like small combs, each having from 6 to 11 teeth, probably to facilitate leaf-eating and fur-cleaning. They have a membrane of skin extending from the forelegs to the hind legs and out to the tail, used when gliding; they cannot fly, in a true sense, but are probably the most expert of all mammals that are able to glide.

Subclass **EUTHERIA** (see pages 460 and 477)
Order **DERMOPTERA**
derma (Gr) the skin, hide of animals *pteron* (Gr) a feather; can mean wings.

Family **CYNOCEPHALIDAE** 2 species
(formerly GALEOPITHECIDAE)
kuōn (Gr), genitive *kunos*, a dog *kephalē* (Gr) the head; they have a somewhat dog-shaped head; they are not lemurs in spite of the name.

Order
DERMOPTERA
|
Family
CYNOCEPHALIDAE
Flying Lemurs

Malayan Flying Lemur   *Cynocephalus variegatus*
(formerly *Galeopithecus*)
varius (L) variegated; the coat is a mottled grey, fawn and buff.
Inhabiting south-east Asia, including Sumatra, Java and Borneo.

Philippine Colugo or Flying Lemur   *C. volans*
volo (L) I fly and *volans* flying. Sometimes called the Cobego; inhabiting the Philippines.

47 Tree Shrews, Lemurs, Monkeys, Apes and Man PRIMATES

This large group consists of the Apes, the Monkeys and Humans, but it also includes a number of less monkey-like creatures such as the Lorises and the Lemurs. So the Order is divided into two Suborders; Lemuroidea the 'lemur-like', and Anthropoidea the 'man-like'.

Then there are the Tree-shrews, anatomically quite a problem, but zoologists do not all agree that these should be included with the primates. There is a tendency now to revert to their previous classification as insectivores. Scientific research continues, so until some agreement is reached they can remain with the primates.

The reader will probably experience some difficulty regarding the English names. For example, langurs, baboons, gibbons, orang-utans, gorillas and chimpanzees, to mention only a few; how do we know which of these are monkeys and which are apes? As a general rule it can be said that monkeys have tails and apes do not. Thus, the apes are gibbons, orang-utans, gorillas and chimpanzees.

With the exception of the human race, and a few monkeys that reach Europe and the north-eastern parts of Asia, the primates live in tropical countries.

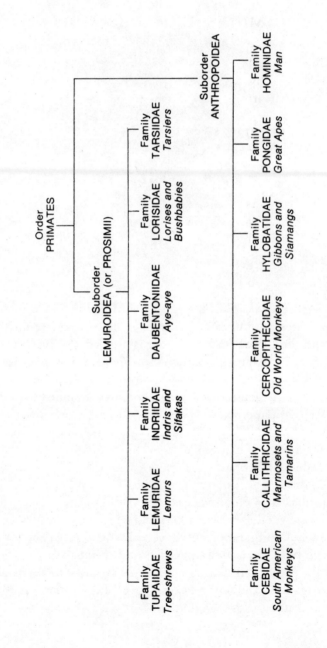

Subclass EUTHERIA (see pages 460 and 477)

Order PRIMATES
primus (L) first, foremost

Suborder LEMUROIDEA (or PROSIMII)
lemures (L) ghosts, spectres; in ancient Rome the spirits of the evil dead; supposed to be so named because of the animal's nocturnal habits and stealthy movements *-oides* (New L) from *eidos* (Gr) apparent shape, form, a kind, sort.

Family TUPAIIDAE about 20 species
tupai is a Malayan name that is used for various small squirrel-like creatures.

Common Tree Shrew *Tupaia glis*
Tupaia see above *glis* (L) a dormouse; it is not a dormouse though rather similar. Inhabiting India, Burma, China and the East Indies.

Madras Tree Shrew *Anathana ellioti*
anatheō (Gr) I run up Dr D. G. Elliot (1835-1915) was an American zoologist, at one time Curator of the Field Museum of Natural History, Chicago. Inhabiting southern Asia.

Smooth-tailed Tree Shrew *Dendrogale murina*
dendron (Gr) a tree *galē* (Gr) a marten-cat or weasel; the word *galē* has been used in nomenclature for various small mammals *mus* (L), genitive *muris*, a mouse, hence *murinus* (L) mouse-like, but they are bigger than a mouse. Inhabiting Vietnam and the Kmer Republic.

Mindanao Tree Shrew *Urogale everetti*
oura (Gr) the tail *galē* (Gr) see above; the tail is a distinctive feature as it is close-haired and not bushy as in the genus *Tupaia*. A. H. Everett (1848-1898) was a zoologist resident in Sarawak from 1869 to 1890. Mindanao is one of the Philippine Islands where these tree shrews live.

Pen-tailed Tree Shrew *Ptilocercus lowi*
ptilon (Gr) a wing *kerkos* (Gr) the tail of an animal; 'wing-tailed'; the tail has feather-like fringes of long white hairs. Sir Hugh Low (1824-1905) was Resident in Perak, Malaya, from 1877 to 1889. This tree shrew inhabits Borneo and the Malayan Peninsula.

Family LEMURIDAE   about 15 species

lemures (L) ghosts, spectres, (see page 505). Lemurs are found only in Madagascar, now known as the Malagasy Republic.

Grey Gentle Lemur   *Hapalemur griseus*
hapalos (Gr) soft, gentle   *lemures* (L) see above; an allusion to the long soft fur   *griseus* (New L) grey, derived from the German greis.

Broad-nosed Gentle Lemur   *H. simus*
simus (L) flat-nosed, snub-nosed. This species now very rare.

Ring-tailed Lemur   *Lemur catta*
catta (New L) a cat. Probably the best known of the lemurs, it has striking black and white rings round the tail.

Ruffed Lemur   *L. variegatus*
variegatus (L) variegated; it usually has a black and white pattern on the body and a ruff of long yellowish hair around the neck, but the colour variation is quite remarkable.

Black Lemur   *L. macaco*
macacus (New L) derived from *macaco* (Port) a native name for a monkey.

Weasel or Sportive Lemur   *Lepilemur mustelinus*
lepos (L) pleasantness, charm   *mustela* (L) a weasel and so *mustelinus* (L) of a weasel; 'weasel-like'.

Hairy-eared Dwarf Lemur   *Cheirogaleus trichotis*
kheir (Gr) the hand   *galē* (Gr) a marten-cat or weasel; an allusion to the long fingers and freely movable thumb, adapted for prehension *thrix* (Gr), genitive *trikhos*, the hair of both man and animals   *ous* (Gr), genitive *ōtos*, the ear. This lemur is now very rare.

Greater Dwarf Lemur   *C. major*
major (= *maior*) (L) greater.

Fat-tailed Dwarf Lemur   *C. medius*
medius (L) middle, intermediate. Now very rare.

Lesser Mouse Lemur   *Microcebus murinus*
mikros (Gr) small   *kēbos* (Gr) a monkey   *mus* (L), genitive *muris*, a mouse, hence *murinus*, mouselike; probably the smallest living primate, and not much bigger than a large mouse.

Fork-crowned Lemur   *Phaner furcifer*
phaneros (Gr) visible, evident   *furca* (L) a fork   *fero* (L) I bear, I carry;
it has black fork-shaped markings on the head. Now very rare.

Family   INDRIIDAE   4 species; subspecies are not included in the
count of the number of species in a family
The name 'indri' is said to originate from a Malagasy word meaning
'look' or 'there it is'; indris and sifakas are found only in Madagascar.

Woolly Indri   *Avahi laniger*
Avahi is a native name for the woolly lemur   *lana* (L) wool   *gero* (L)
I carry, hence *laniger* (L) wool-bearing. Like the lemurs, indris live
in Madagascar and are closely related.

Diadeemed Sifaka   *Propithecus diadema diadema*
pro (Gr) before, in front of; this probably refers to its relative age in
evolution, 'before apes' *pithēkos* (Gr) an ape   *diadema* (L) a royal
headband; it has a headband of white fur   *sifac* is a Malagasy word.
The nominate subspecies (see pages 18 and 19).

Perrier's Sifaka   *P. d. perrieri*
M. Perrier de la Bathie was a French botanist in Madagascar; the
name was given quite recently, in 1931. This is a subspecies (see pages
18 and 19).

Black Sifaka   *P. d. edwardsi*
Professor Alphonse Milne-Edwards (1835–1900) was a French zoologist; in collaboration with another French zoologist, A. Grandidier
(1836–1921) he wrote a classical work on what was then Madagascar,
and its ecology; a subspecies (see pages 18 and 19).

Verreaux's Sifaka   *P. verreauxi verreauxi*
Named after J. B. E. Verreaux (1810–1868) a French zoologist.
Sifakas are found only in Madagascar where they live in the forests
round the coast. This sifaka, the nominate subspecies (see pages 18 and
19) is now very rare.

Coquerel's Sifaka   *P. v. coquereli*
Named after Dr C. Coquerel (1822–1867), a French zoologist who
lived for a time in Madagascar; a subspecies (see pages 18 and 19)
and now very rare.

Crowned Sifaka   *P. v. coronatus*
coronatus (L) provided with a crown. A rare subspecies (see pages 18 and 19).

Van der Decken's Sifaka   *P. v. deckeni*
Baron Van der Decken (1833–1865) was a Dutch explorer and naturalist. A rare subspecies.

Indri   *Indri indri*
Indri, see above, under Family. Sometimes known as the Endrina, it lives in the tree-tops of the forests of eastern Madagascar; it is very rare.

Family   DAUBENTONIIDAE   1 species
Named after Louis J. M. Daubenton (1716-1799), the French zoologist, who discovered this animal in the year 1780.

Aye-aye   *Daubentonia madagascariensis* (*Daubentonia* formerly *Cheiromys*)
This peculiar animal presented a problem for the zoologists, as the anatomical features were most unusual, although they showed it to be a primate. It has an extraordinary long, thin third finger, probably used for digging the larvae of wood-boring insects out of the holes in the wood, and the teeth are similar to those of a rodent. It was first described as a squirrel, and was given eight different generic names between 1795 and 1846. It has now been given a family to occupy on its own; it is very rare. The name aye-aye is derived from a Malagasy name *aiay*; it inhabits eastern Madagascar.

Family   LORISIDAE   11 species
The name loris is derived from the Dutch *loeris*, meaning a clown, a booby; the lorises have a comical appearance.

Slender Loris   *Loris tardigradus*
tardus (L) slow   *gradus* (L) a step, a pace; all lorises take slow, deliberate steps. This loris lives in southern India and Sri Lanka.

Slow Loris   *Nycticebus coucang*
nux (Gr), genitive *nuktos*, night   *kēbos* (Gr) a monkey; 'night-monkey'; they are nocturnal and the very large eyes enable them to see in a poor light: *kukang* is the native name in Malaya. The range is from Assam to Tongking, south to Singapore, Sumatra, Java and Borneo.

Lesser Slow Loris   *N. pygmaeus*
pugmē (Gr) a fist and so *pugmaios* (Gr) about a foot long, or tall, i.e.
'dwarfish'. It lives in Vietnam.

Angwantibo   *Arctocebus calabarensis*
arktos (Gr) a bear   *kēbos* (Gr) a monkey   Calabar is in Nigeria
-ensis (L) suffix meaning belonging to   *angwantibo* is an Efik name for
this animal, the language of a people of south-eastern Nigeria; it
inhabits central West Africa.

Potto   *Perodicticus potto*
pēros (Gr) maimed, disabled   *deiktikos* (Gr) showing, proving; 'showing maimed'; this refers to the second fingers and toes, which are very
short, and look like stumps after an amputation; potto is a word of
Niger–Congo origin meaning a tailless monkey; this loris has a very
short tail. It lives in the forests of East, Central and West Africa.

Thick-tailed Bushbaby   *Galago crassicaudatus*
Galago is probably from *golokh* (Wolof) a monkey; the Wolof are a
people of the western Sudan   *crassus* (L) thick   *cauda* (L) the tail of
an animal   *-atus* (L) suffix meaning provided with. The range is
Gambia and surrounding areas in West Africa, and large areas in
East Africa down to Rhodesia.

Senegal Bushbaby   *G. senegalensis*
-ensis (L) suffix meaning belonging to; the type species came from
Senegal but it inhabits other large areas of Africa.

Allen's Bushbaby or Moholi Galago   *G. alleni*
Rear-Admiral W. Allen F.R.S. (1793–1864) was an English zoologist
who led an expedition to Niger in 1841   *moholi*, or *maholi*, is an African
native name. Inhabiting large areas of Africa, from Kenya in the east
to Gambia in the west.

Demidoff's Bushbaby   *G. demidovi*
Paul Demidov (1738–1821) was a Russian scientist and traveller;
his name first appeared in a Moscow publication in 1808. This bushbaby lives in a small area in West Africa, mostly confined to Cameroun
and Gabon.

Family   TARSIIDAE   3 species

Philippine Tarsier   *Tarsius syrichta*
tarsos (Gr) the flat of the foot, the part between the toes and the heel;

an allusion to the tarsal bones of the hind legs which are exceptionally long   *suriktas* (Gr) a player on Pan's pipe, to make a whistling sound; tarsiers have a very high-pitched call. The Latin name was given by Linnaeus in 1758, more than 200 years ago! Inhabiting the Philippines.

Horsfield's or Western Tarsier   *T. bancanus*
Banka is an island in the East Indies lying just off the coast of Sumatra *-anus* (L) suffix meaning belonging to. Dr T. Horsfield (1773–1859) was a scientist who was in Sumatra from 1796 to 1818; he named many mammals and birds of the East. This tarsier lives in Sumatra and Borneo, but was found first on Banka Island.

Celebes or Eastern Tarsier   *T. spectrum*
spectrum (L) an image, an apparition; tarsiers are nocturnal and have very large eyes, which would suggest 'an apparition'. Celebes is a large island in Indonesia lying to the east of Borneo.

Suborder   ANTHROPOIDEA (or SIMIAE, PITHECOIDEA)
anthrōpos (Gr) man   *-oides* (New L) from *eidos* (Gr) shape, resemblance; 'man-like'.

Family   CEBIDAE   about 35 species
kēbos (Gr) a monkey.

Douroucouli or Night Monkey   *Aotus trivirgatus*
(formerly *Nyctipithecus*)
a- (Gr) prefix meaning not, or there is not   *ous* (Gr), genitive *ōtos*, the ear; they have very small ears hidden under the fur   *tria* (L) three *virga* (L) a twig   *-atus* (L) suffix meaning provided with, so *virgatus* (L) can mean striped; they have black and brown lines on the face. Douroucouli is a South American native name; it has also been known as the Owl Monkey on account of its 'owl-like' face, and 'night monkey' because it is active at night. Inhabiting central South America.

Collared Titi or Widow Monkey   *Callicebus torquatus*
kallos (Gr) a beauty; *kalos* (Gr) beautiful   *kebos* (Gr) a monkey *torquatus* (L) wearing a collar or necklace; it has a collar of white fur. Possibly named 'widow' because the coat is dark-coloured or even black. Inhabiting a large area north of the Amazon in South America.

Orabassu Titi   *C. moloch*
Moloch was a Semitic god to whom children were sacrified; the name
probably has no significance (see notes about Linnaeus under Common Marmoset, page 513), but may be associated with its ugly face.
Orabassu is derived from the Tupi native name *oyapussa; titi* is a
Spanish South American name meaning 'little cat'. Inhabiting parts
of the Amazon basin in South America.

Bald Uakari   *Cacajao calvus*   (formerly *Brachyurus*)
cacajao (Tupi) a South American native name   *calvus* (L) bald; the
face and head of this monkey are almost hairless. It lives in the forests
of the upper Amazon Basin area, South America.

Black-headed Uakari   *C. melanocephalus*
melas (Gr), genitive *melanos*, black, dark   *kephalē* (Gr) the head.
Inhabiting the upper Amazon Basin. Uakari is a Tupi South American
native name.

Hairy or Monk Saki   *Pithecia monachus*
pithēkos (Gr) an ape   *monakhos* (Gr) solitary, a monk; it has a cape of
long hair on the head and shoulders resembling a monk's habit.
Inhabiting the Guyana area and the Amazon Basin.

Pale-headed or White-faced Saki   *P. pithecia*
Saki is a name derived from a Tupi South American word. This saki
has a remarkable white face; it inhabits the Guyana area and part of
the northern Amazon Basin.

Black Saki   *Chiropotes satanus*
kheir (Gr) the hand   *potēs* (Gr) a drink; it drinks from the back of its
hand by soaking the fur in water and licking it   *Satanas* (Gr) Satan,
a Hebrew word meaning the enemy; it is sometimes used to indicate
a black or an ugly animal. Inhabiting the Amazon area.

White-nosed Saki   *C. albinasus*
albus (L) white   *nasus* (L) the nose. Inhabiting a southern part of the
Amazon Basin.

Guatemalan Howler Monkey   *Alouatta villosa*
(*Alouatta* formerly *Mycetes*)
alouatta (New L) derived from the French alouate   *villus* (L) shaggy
hair   *-osus* (L) suffix meaning full of, prone to; 'a lot of shaggy hair'.
Inhabiting Central America.

Mantled or Panamanian Howler Monkey   *A. pailiata*
pallium (L) a type of Greek mantle, hence *palliata*, clad in a pallium; the hair is longer on the sides and of a different colour giving the impression of a cloak. Howler monkeys have a special formation of the throat, a kind of sound-box, which increases the power of the voice; they live in southern Mexico and the northern part of South America.

Red Howler Monkey   *A. seniculus*
senex (L) old, aged; *seniculus* (L) a little old man; the name is an allusion to its appearance. Inhabiting South America.

Brown Howler Monkey   *A. fuscus*
fuscus (L) dark, dusky. Inhabiting south-eastern Brazil.

Red-handed Howler Monkey   *A. belzebul*
belzebul, another spelling of Beelzebub, god of flies, from the Hebrew *baal*, a lord, and *zeebub*, a fly; suggesting black, evil, an allusion to the rather ugly face. It inhabits a large area south of the Amazon.

Black Howler Monkey   *A. caraya*
The Caraya are a people living in the central part of South America where this monkey is found.

White-throated Capuchin   *Cebus capucinus*
kēbos (Gr) a monkey   the Franciscan Monks were known as Capuchins and the dark patch of hair on the crown of this monkey's head resembles a monk's cap. Inhabiting Central America and the north-western tip of South America.

Weeper or Black-capped Capuchin   *C. nigrivittatus*
niger (L) black   *vitta* (L) a ribbon, a band   *-atus* (L) suffix meaning provided with; 'having a black band'; they communicate by making a continuous 'weeping cry'. Inhabiting the Guyana and Surinam area and the northern Amazon Basin; there are other species of capuchin monkeys all living in South America; they are often sold as pets, and have been used by organ-grinders.

Common Squirrel Monkey   *Saimiri sciureus*
(formerly *Chrysothrix*)
Saimiri is from a Brazilian-Portuguese word for a small monkey *sciurus* (L) a squirrel; see notes about the suborder Sciuromorpha on page 113. Living in the forests of the Orinoco, Guyana and the Amazon area.

Black Spider Monkey *Ateles paniscus*
a- (Gr) a prefix meaning not, or there is not *teleios* (Gr) complete, entire; 'not complete'; it usually has no thumb though sometimes it is vestigial *paniscus* (L) a sylvan deity, 'a little pan', from Pan and the diminutive suffix *-iscus* (L). Inhabiting two large areas in the Amazon Basin.

Woolly Spider Monkey *Brachyteles arachnoides*
brakhus (Gr) short *teleios* (Gr) complete, entire; it has very short thumbs or none at all *arakhnēs* (Gr) a spider *-oides* (New L) derived from the Greek *eidos*, apparent shape, form. Inhabiting the forest area to the south-east of Brazil.

Humboldt's Woolly Monkey *Lagothrix lagothricha*
lagos (Gr) a hare *thrix* (Gr), genitive *trikhos*, hair of man or beast, wool: an allusion to the woolly, hare-like fur: Baron v. Humboldt (1769–1859) was a German scientist and explorer who lived in South America from 1799 to 1804. The coat of these monkeys is thick and woolly; they inhabit the north-western part of South America.

Peruvian Mountain Woolly Monkey *L. flavicauda*
flavus (L) yellow *cauda* (L) the tail of an animal; it has a yellow streak on the underside of the tail. This is a very rare monkey found only in the mountains of northern Peru.

Family CALLITHRICIDAE (formerly HAPALIDAE)
about 35 species
kallos (Gr) a beauty *kalos* (Gr) beautiful *thrix* (Gr), genitive *trikhos*, hair; they have elaborate 'hair styles' and colourful coats.

Common Marmoset *Callithrix jacchus*
(*Callithrix* formerly *Hapale*)
Jacchus (= Iacchus) a Roman god associated with Bacchus, the god of wine. In 1758 Linnaeus used several names from classical mythology for the specific names of certain species, and it is probable that they were given without thought of any physical significance. Marmoset is from *marmouset* (Fr) a grotesque figure, a young monkey (a little boy); it may have derived from *marmor* (L) marble, and later *marmoretum*, a little marble figure. This marmoset inhabits the eastern part of Brazil.

White-eared Marmoset   *C. aurita*
auris (L) the ear and hence *auritus* (L) long-eared. Inhabiting the
eastern part of Brazil.

Silvery Marmoset   *C. argentata*
argentum (L) silvery and hence *argentatus* (L) ornamented with silver.
Inhabiting central parts of South America south of the Amazon.

Pygmy Marmoset   *Cebuella pygmaea*
kēbos (Gr) a monkey   *-ellus* (L) diminutive suffix; 'a small monkey';
it is the smallest of the Anthropoidea   *pugmaios* (Gr) dwarfish, from
pugmē (Gr) a fist. Inhabiting the northern part of Peru.

Golden Lion Marmoset   *Leontideus rosalia*
leo (L), genitive *leontos*, a lion   *-ideus* (New L) from *eidos* (Gr) the
shape, resemblance   *rosa* (L) a rose   *-alis* (L) suffix meaning like:
referring to the colour, a reddish gold. It inhabits the south-eastern
area of Brazil.

Cotton-headed Tamarin   *Saguinus oedipus*   (formerly *Tamarin*)
sagouin (Fr) a squirrel-monkey   *-inus* (L) suffix meaning like   *oidipous*
(Gr) swollen-footed; possibly no direct connection with this meaning.
There is a legend about King Polybus whose shepherd found a baby
abandoned on Mount Cithaeron, with injuries to his feet. The King
reared him as his own son and named him Oedipus, because of his
swollen feet. Inhabiting the north-western part of South America.

Pied Tamarin   *S. bicolor*
bicolor (L) of two colours; it is brown and white. Tamarin is a Caribbean word meaning a squirrel-monkey; this tamarin inhabits an area
north of the Amazon.

Red-handed Tamarin   *S. midas*
Possibly an allusion to Midas, the legendary King of Phrygia, who
was supposed to have had 'ass's ears'; this tamarin has unusually large
ears. Inhabiting Brazil in an area around the mouth of the Amazon.

Brown-headed Tamarin   *S. fuscicollis*
fuscus (L) dark-coloured   *collum* (L) the neck. Inhabiting an area
north of the Amazon.

Emperor Tamarin   *S. imperator*
imperator (L) a leader, a chief. A taxidermist, so the story goes, had
never seen a live tamarin and twisted the white 'moustache' upwards

to look like the Emperor of Germany, instead of letting it droop in the natural position. It thus acquired the name Emperor Tamarin as a joke, but the name stuck, and the Latin name became established as *Saguinus imperator*. Inhabiting the central part of South America.

Family CERCOPITHECIDAE   about 60 species
kerkos (Gr) the tail of an animal   *pithēkos* (Gr) an ape. In fact this is a family of monkeys, not apes (see page 503).

Toque Monkey   *Macaca sinica*
Macaca is derived from *macaque* (Fr) from Portuguese, a monkey *sinica* is a coined word meant to mean 'of China'; it was given by Linnaeus in 1771 because he thought this monkey came from China. In fact, it inhabits Sri Lanka. It has a whorl of hair on the head.

Bonnet Monkey   *M. radiata*
radiatus (L) provided with spokes or rays; the hair on the head radiates to form a circular cap, or bonnet. From southern Asia.

Lion-tailed Monkey   *M. silenus*
Silenus was a companion of the Roman god Bacchus and Sileni (plural) were gods of the woods. It has a tuft at the tip of the tail like that of a lion. Inhabiting the forests in the Western Ghats of south-western India.

Pig-tailed Monkey   *M. nemestrina*
Nemestrinus was the god of groves: 'pig-tail' is said to be a reference to the hair on the head, but more likely refers to the tail which is short and 'pig-like'. Ranging from the peninsular part of southern Thailand through Malaya to Borneo, Sumatra and some small adjacent islands.

Crab-eating Macaque   *M. fascicularis*   (formerly *irus*)
fascia (L) (kindred with *fascis*) a band, a filet, hence *fasciculus* (L), the diminutive, a small band   -*aris* (L) suffix meaning pertaining to; Sir Stamford Raffles does not explain the significance of the specific name, but it is probable that he had in mind the coloration of the animal. It inhabits Malaysia, the Philippines, and Indonesia as far east as Flores and Timor. Sir T. Stamford Raffles (1781-1826) was Lieutenant Governor of Java in 1811 and later Sumatra. He founded Singapore in 1819.

Rhesus Monkey   *M. mulatta*
A mulatto is the offspring of a black person mating with a white, from

mulus (L) a mule; this monkey is a yellowish sandy-brown colour
Rhēsos (Gr) a king of Thrace; it gives the name to the rhesus blood
group because experiments in blood transfusion were carried out with
this monkey. Inhabiting India, and ranging as far north as the
Himalayas, and eastwards to China.

Assamese Macaque   *M. assamensis*
-ensis (L) suffix meaning belonging to, usually applied to localities;
belonging to Assam, India.

Formosan or Round-headed Macaque   *M. cyclopis*
kuklos (Gr) round, a circle   *ops*, from *opsis* (Gr) aspect, face; 'round-
faced' or 'round-headed'. Inhabiting Formosa.

Stump-tailed Macaque   *M. arctoides*   (formerly *speciosa*)
arktos (Gr) a bear   *-oides* (New L), from *eidos* (Gr) apparent shape,
form; 'bear-like'; it is a heavily built very hairy monkey, and has a
stumpy tail about 5 cm (2 in) long. Inhabiting south-east Asia.

Japanese Macaque   *M. fuscata*
fuscus (L) dark coloured, or black; it has a dark grey coat. Inhabiting
Japan.

Barbary Ape   *M. sylvana*
silva (L) a wood   *-anus* (L) suffix meaning belonging to; Sylvanus
was a god of the woods. Although known as apes, they are actually
monkeys that have no tail (see page 503). They are the famous monkeys
which live on the Rock of Gibraltar and their numbers are
maintained at a certain level by the British Government. In their wild
state they live in Morocco and Algeria, and it is not known for certain
whether they were originally imported or are the last survivors of
European monkeys. Barbary is the name for the belt of land north of
the Sahara stretching from Egypt to the Atlantic.

Moor Macaque   *M. maurus*
mauros (Gr) dark; it is almost black, and often confused with the Black
Ape. Inhabiting the Island of Celebes, Indonesia.

Celebes Black Ape   *Cynopithecus niger*
kuōn (Gr), genitive *kunos*, a dog   *pithēkos* (Gr) an ape   *niger* (L) black:
the tail is only vestigial but it is not an ape (see page 503). It has a
dog-like head.

Grey-cheeked Mangabey   *Cercocebus albigena*
kerkos (Gr) the tail   *kēbos* (Gr) a monkey; the mangabeys have long

tails   *albus* (L) white; can mean pale or whitish   *genus* (Gr) the jaw, cheek. Mangabey is a Malagasy word, derived from the port town of Mangabe; the name is misleading and due to a mistake as these monkeys do not live in Madagascar; they live in Uganda, Congo, Cameroun and other parts of central West Africa.

Black Mangabey   *C. aterrimus*
ater (L) black, hence *aterrimus* (L) very black. Inhabiting forests of southern Congo.

Agile Mangabey   *C. galeritus galeritus*
galeritus (L) wearing a hood or skull cap; it has a distinct crest of hair on the head. Inhabiting forests of central Africa. The nominate subspecies.

Golden-bellied Mangabey   *C. g. chrysogaster*
khrusos (Gr) golden   *gaster* (Gr) the belly. A subspecies of *C. galeritus*, it lives in the forests of central and western Africa.

White-collared or Sooty Mangabey   *C. torquatus*
torquatus (L) having a collar; it has a collar of white fur like a ruff, and the coat is usually grey but shows some variation. Inhabiting the forests of central western Africa.

Sacred or Hamadryas Baboon   *Papio hamadryas*
papio (New L) a baboon, from the French *papion*: a hamadryad is a wood nymph from *hama* (Gr) together with, and *drus* (Gr) a tree. Inhabiting hillsides in Arabia, Ethiopia and Sudan.

Yellow Baboon   *P. cynocephalus*
kuōn (Gr), genitive *kunos*, a dog   *kephalē* (Gr) the head; all baboons have a rather dog-like face and muzzle. It inhabits the central part of East Africa.

Chacma Baboon   *P. ursinus*
ursus (L) a bear   *-inus* (L) suffix meaning like   *chacma* is a Hottentot native name. It is found in Rhodesia and South Africa.

Anubis Baboon   *P. anubis*
Supposed to resemble the pictorial representation of Anubis, a god in ancient Egyptian religion. Inhabiting Chad, Sudan, Ethiopia and the northern part of Kenya.

Guinea Baboon   *P. papio*
Inhabiting Guinea and neighbouring areas in western Africa.

Mandrill   *Mandrillus sphinx*
Mandrill is probably a combination of 'man' and 'drill', a West
African native name for a baboon; a man-like ape   the sphinx was
a monster of Egyptian origin figuring also in Greek mythology; there
is probably no special reason for giving it this name (see page 513). It
inhabits tropical rain forests in Gabon and neighbouring areas.

Drill   *M. leucophaeus*
Drill is a West African native name   *leukos* (Gr) white   *phaios* (Gr)
dusky, dark; there is sometimes white hair surrounding the face and
on the chest but most of the coat is dark brown. It lives in the forests
of the Gabon and Cameroun areas in West Africa.

Gelada Baboon   *Theropithecus gelada*
thēr (Gr) a wild animal   *pithēkos* (Gr) an ape   *gelada* is an Ethiopian
native name. It lives in the hills of Ethiopia, and has been seen at
heights of over 2,000 m (7,000 ft) above sea level.

Green or Grass Monkey   *Cercopithecus aethiops*
The monkeys of the genus *Cercopithecus* are known as Guenons; they
have very long tails and colourful fur. The Latin name is misleading as
they are monkeys and not apes (see page 503).
kerkos (Gr) the tail of an animal   *pithēkos* (Gr) an ape   *aithos* (Gr)
burnt, red-brown colour   *opsis* (Gr) aspect, appearance; hence
aethiops = Ethiopian, or negro, i.e. 'burnt-face'. It has a black face
and the coat sometimes has a green tinge. Widespread in Africa south
of the Sahara.

Moustached Monkey   *C. cephus*
cephus is a Linnaean name, a variant of the Greek *kēbos* and similar to
the Arabic *keb*, a monkey, and a name still used in the East African
coastal area. It has a prominent white moustache. Inhabiting central
western Africa.

Diana Monkey   *C. diana*
A white crescent on the forehead is supposed to resemble the bow of
the goddess Diana but this may not be the origin of the name. Inhabiting the central part of western Africa.

Owl-faced Monkey   *C. hamlyni*
R. I. Pocock says that the first specimen of this monkey was 'procured
alive from the Ituri Forest, Congo, for the Hon. Walter Rothschild
by J. D. Hamlyn'. It is possible that Hamlyn was not a collector but

a dealer in animals. The Hon. Walter Rothschild later became Lord Rothschild (see page 558). R. I. Pocock F.R.S. (1863–1947) was Superintendent of the Zoological Gardens in London from 1904 to 1923. The Owl-faced Monkey has large round eyes like those of an owl; it is a rare species living in eastern Congo.

Blue or Diadem Monkey   *C. mitis*
mitis (L) mild, gentle; it is a quiet and shy monkey   *diadema* (L) a royal headband; it has a light coloured band on the forehead. Living in central areas of Africa including Congo, Uganda and Tanzania.

De Brazza Monkey   *C. neglectus*
neglectus (L) neglected, not chosen; so named because the specific rank had been overlooked for some time. J. C. S. de Brazza (1859-1888) was a French naturalist and explorer who spent some time in Congo where this monkey lives. It is also found in parts of Uganda and Kenya.

Talapoin Monkey   *C. talapoin*
Talapoin is the name for a Buddhist monk, particularly of Pegu, in Lower Burma; from *tala poi* (Old Peguan) meaning 'my lord'; it has a dark crown on the head like a monk's cap. Inhabiting some large areas on the west coast of central Africa.

Red-tailed Monkey   *C. ascanius*
Ascanius was the legendary son of Aeneas and Creusa in Greek mythology. This is an olive-green monkey with a red tail living in a small area in the western part of Africa.

Allen's Swamp Monkey   *C. (Allenopithecus) nigroviridis*
The subgenus is named after Dr Joel Asaph Allen (1838-1921) the eminent American zoologist, who was curator of birds and mammals at the American Museum of Natural History, New York, during the years 1885 to 1908   *pithēkos* (Gr) an ape   *niger* (L) black   *viridis* (L) green; it is black and yellowish with a speckled effect. Inhabiting the central part of Africa. The name in brackets indicates that it is a subgenus of *Cercopithecus*.

Patas Monkey   *Erythrocebus patas*
eruthros (Gr) red   *kēbos* (Gr) a monkey; it is a reddish-brown colour *patas*, from *pata* (Wolof); the Wolof are a people of the western Sudan where this monkey lives. It·is also found in western parts of Uganda and Kenya.

Hanuman or Entellus Langur   *Presbytis entellus*

presbus (Gr) an old man and so *presbuteros* (Gr) an elder, a priest; can mean greater, more important. The leader, or dominant male of a pack, has a distinct behaviour pattern when issuing orders to his subordinates. Entellus was a massively built veteran hero in Virgil's Aeneid. Hanuman is a Hindu word for a monkey god; this langur is sacred in India. Inhabiting Sri Lanka and peninsular India, ranging north to Sikkim and Kashmir, and to southern Tibet.

Purple-faced Langur   *P. senex*

senex (L) an old man, person; from its appearance, the male having copious whiskers, often white, and the face is reddish to purple. Inhabiting southern India and Sri Lanka.

Spectacled or Dusky Langur   *P. obscurus*

obscurus (L) dark; it has a dark coat and white rings round the eyes, hence 'spectacled'. Inhabiting southern Burma, southern Thailand, Malaga and some small adjacent islands.

Maroon Langur   *P. rubicundus*

rubicundus (L) red, ruddy. Inhabiting Borneo.

White-headed Langur   *P. leucocephalus*

leukos (Gr) white   *kephalē* (Gr) the head; it is possible that this langur will be established as a subspecies; it lives in a small area in Kwangsi, to the north of Tongking.

Douc Langur   *Pygathrix nemaeus*

pugē (Gr) the rump, buttocks   *thrix* (Gr) hair; alluding to the long hair on the rump   *nemus* (L) a grove, a forest   *douc* is a French Cochinchina name. Cochin-china was a name used by the French in the 17th century for the central and southern areas of Vietnam and the capital was Saigon. This langur inhabits southern Vietnam and Laos, and is now very rare.

Golden Snub-nosed Monkey   *Rhinopithecus roxellanae*

rhis (Gr), genitive *rhinos*, the nose   *pithēkos* (Gr) an ape. This snub-nosed monkey is named after the consort of the great Turkish sultan Sulaiman I; she was a famous Russian lady of doubtful repute, named Roxellane; like the monkey she had golden-red hair and a turned-up nose. This is a very rare monkey discovered in 1870, and living in the mountains of eastern Tibet and north-eastern China.

Pig-tailed Langur   *Simias concolor*
simias (L) an ape, a monkey   *concolor* (L) the same colour. This langur's tail is short and naked; the coat is entirely brown but the face, hands and feet are black. Inhabiting the mountains of the Mentawei Islands, off the west coast of Sumatra. It is very rare.

Proboscis Monkey   *Nasalis larvatus*
nasus (L) the nose and so *nasalis* (New L) pertaining to the nose   *larva* (L) a ghost; can mean a mask   *-atus* (L) suffix meaning provided with, 'having a mask'; this is a reference to the nose, which is long and quite remarkable. It lives in the swamp forests of Borneo.

Black Colobus   *Colobus polykomos*
kolobos (Gr) docked, mutilated; they have very small thumbs or the thumb completely absent, giving the appearance that it has been cut off   *polus* (Gr) many   *komē* (Gr) the hair; the coat is handsomely marked in black and white and long and silky. Widespread in the tropical forests of Africa.

Red Colobus   *C. badius*
badius (L) chestnut coloured; it has a black body and chestnut coloured head, arms and legs. Range as above.

Olive Colobus   *C. verus*
verus (L) true, genuine; meaning a true or typical colobus. It is confined to an area in and around Ghana on the west coast of Africa.

Family   HYLOBATIDAE   7 species
hulē (Gr) a wood, a forest   *bainō* (Gr) I walk, I step   *batēs* (Gr) one that treads; can mean a climber.

White-handed or Lar Gibbon   *Hylobates lar*
Lar is an honorary title in Rome, equivalent to the English Lord. The Lar Gibbon lives in Thailand, the Malay Peninsula, Sumatra, Java and Borneo.

Dark-handed Gibbon   *H. agilis*
agilis (L) nimble, agile; the gibbons are probably the most expert acrobats in the animal kingdom. This gibbon has a dark upper surface on the hands and feet; it inhabits the Malay Peninsula and Sumatra.

Hoolock Gibbon   *H. hoolock*
Hulluk is a Burmese native name for the gibbon; it inhabits Burma and Thailand.

Black or Harlan's Gibbon   *H. concolor concolor*
concolor (L) the same colour   Dr Richard Harlan (1796-1843) was
an American physician, naturalist and author. This gibbon inhabits
Cambodia, Vietnam and Laos; the nominate subspecies.

White-cheeked Gibbon   *H. c. leucogenys*
leukos (Gr) white   *genus* (Gr) the jaw, cheek; the cheeks are white or
pale yellow. This is a subspecies of *H. concolor* living in Vietnam, Laos
and Thailand.

Grey Gibbon   *H. moloch*
Moloch was a Semitic god to whom children were sacrificed; the
specific name *moloch*, given by Audebert in 1797, probably has no
special significance. The custom of naming monkeys and apes after
mythical creatures and classical and heathen beings probably derives
from Linnaeus (see page 513). Jean Baptiste Audebert (1759-1800) was
a distinguished naturalist, bird artist and engraver. This gibbon inhabits
Java and Borneo.

Siamang   *Symphalangus syndactylus*
sym- (=*syn-*) (Gr) together   *phalanx* (Gr), genitive *phalangos*, soldiers
in line of battle; in a biological sense the small bones between the
joints of the fingers, and also the toes; in the siamang the second and
third toes are joined together by a web of skin   *daktulos* (Gr) a finger;
also a toe; so this name means 'the toes-joined one, with joined toes'.
Inhabiting Sumatra.

Dwarf Siamang   *S. klossi*
C. B. Kloss (1877-1949) was a zoologist living in Singapore during
the years 1903 to 1932   *siamang* is a Malay native name for the gibbon.
Some authorities classify this small gibbon in the genus *Hylobates*.
It was discovered in the early part of the twentieth century and is
found only in the Mentawai Islands off the west coast of Sumatra.

Family PONGIDAE   4 species
mpongi is a Congolese name, probably originally used to mean a
gorilla, but later used to mean the orang-utan; this is a Malay word
meaning 'forest-man'.

Orang-Utan   *Pongo pygmaeus*   (*Pongo* formerly *Simia*)
pugmaios (Gr) small, dwarfish; a misleading name as it is quite a large
ape, standing about four feet high; the comparison was probably
made with a man. Inhabiting Borneo and Sumatra.

Chimpanzee   *Pan troglodytes*   (*Pan* formerly *Anthropopithecus*)
pan (Gr) all, the whole; in Greek mythology Pan was the rural god
of Arcadia, of pastures and woods   *trōglē* (Gr) a hole   *dutēs* (Gr) a
burrower, a diver; a peculiar name for the chimpanzee as a troglodyte
is a cave-dweller, whereas chimpanzees spend some of their time in
trees and make nests there where they sleep, and some of their time
on the ground; they do not live in caves. The name was given nearly
200 years ago by J. F. Blumenbach, in 1779, and in those days it was
probably thought to be some kind of cave man. It inhabits widely
scattered areas of central parts of western Africa north of the Zaire
River (formerly Congo River), particularly where there are tall
deciduous forests.

Pygmy Chimpanzee   *P. paniscus*
-*iscus* (L) dim, suffix, 'a small Pan'; it is also known as the Dwarf
Chimpanzee, and is slender and much smaller than *P. troglodytes*. The
name chimpanzee is from a Zaire native name *kimpenzi*. It lives in the
swamp forests of Zaire, south of the Zaire River and ranging southwards to the Lukenie River.

Western Gorilla   *Gorilla gorilla gorilla*
gorillai (Gr) gorillas or hairy humans; the origin of the word is obscure.
Hanno the Carthaginian said it was a tribe of hairy women; the name
almost certainly is from western Africa. This gorilla inhabits the
forests of Cameroun and Gabon in central western Africa. (See
Tautonyms on page 13.) It is the nominate subspecies.

Mountain Gorilla   *G. g. beringei*
Discovered by Capt Oskar von Beringe, an officer in what was then
German East Africa, in 1902. It was named by Dr Paul Matschie
(1862-1926) the German zoologist, and originally misspelled as
beringeri. It lives high up in the mountains to the west of Lake Kivu
in eastern Zaire, and contiguous regions of Uganda, Rwanda,
Burundi and Tanzania. A subspecies.

Family   HOMINIDAE   1 species
homo (L), genitive *hominis*, a man.

Man   *Homo sapiens*
sapiens (L) wise, sensible(!). Widespread throughout the world.

THE PRIMATES
Their relationship to humans and other animals, showing the Phylum, Subphylum, Class,
Subclass, Order, Suborders, and Families, in sequence.

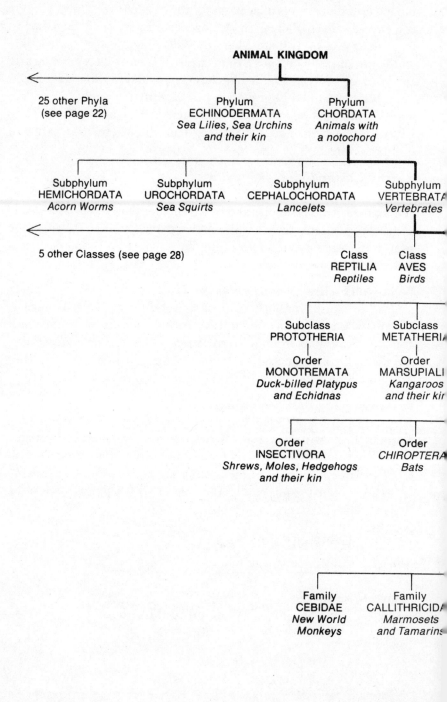

ANIMAL KINGDOM

25 other Phyla
(see page 22)

Phylum
ECHINODERMATA
*Sea Lilies, Sea Urchins
and their kin*

Phylum
CHORDATA
*Animals with
a notochord*

Subphylum
HEMICHORDATA
Acorn Worms

Subphylum
UROCHORDATA
Sea Squirts

Subphylum
CEPHALOCHORDATA
Lancelets

Subphylum
VERTEBRATA
Vertebrates

5 other Classes (see page 28)

Class
REPTILIA
Reptiles

Class
AVES
Birds

Subclass
PROTOTHERIA

Subclass
METATHERIA

Order
MONOTREMATA
*Duck-billed Platypus
and Echidnas*

Order
MARSUPIALI
*Kangaroos
and their kin*

Order
INSECTIVORA
*Shrews, Moles, Hedgehogs
and their kin*

Order
CHIROPTERA
Bats

Family
CEBIDAE
*New World
Monkeys*

Family
CALLITHRICIDA
*Marmosets
and Tamarins*

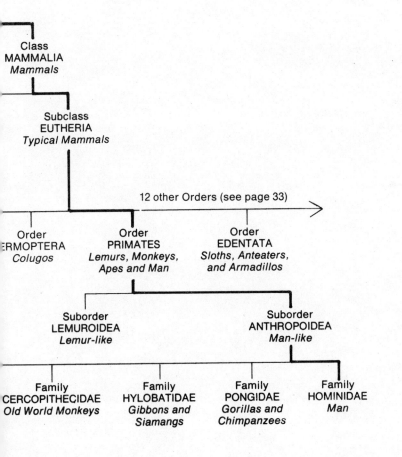

Class
MAMMALIA
Mammals

Subclass
EUTHERIA
Typical Mammals

12 other Orders (see page 33)

Order
ERMOPTERA
Colugos

Order
PRIMATES
*Lemurs, Monkeys,
Apes and Man*

Order
EDENTATA
*Sloths, Anteaters,
and Armadillos*

Suborder
LEMUROIDEA
Lemur-like

Suborder
ANTHROPOIDEA
Man-like

Family
CERCOPITHECIDAE
Old World Monkeys

Family
HYLOBATIDAE
*Gibbons and
Siamangs*

Family
PONGIDAE
*Gorillas and
Chimpanzees*

Family
HOMINIDAE
Man

48 Sloths, Armadillos and Anteaters
EDENTATA

This is a curious little group of mammals coming under the order Edentata, and even this name makes no sense as it means 'without teeth', whereas in fact most of them do have teeth, and the Giant Armadillo has up to ninety! The name came about because it was first applied to the toothless anteaters. These have no teeth and feed exclusively on termites and other insects. (They should not really be known as 'anteaters' because, strictly speaking, termites are not ants. They are distantly related to ants, but are not classed as the same order, termites being in the order Isoptera, meaning 'equal or similar wing', and ants in the order Hymenoptera, meaning 'membrane wing'.) With regard to the sloths, there are only two known types, although these are divided into several species; the nine neck vertebrae of the Three-toed Sloth is unique, being more neck vertebrae than even the giraffe.

Subclass EUTHERIA
Order EDENTATA

e (= *ex*) (L) prefix meaning out of; can mean without *dens* (L), genitive *dentis*, a tooth -*atus* (L) suffix meaning provided with; 'not provided with teeth'.

Family MYRMECOPHAGIDAE 4 species
murmex (Gr) genitive *murmēkos*, an ant *phagein* (Gr) to eat.

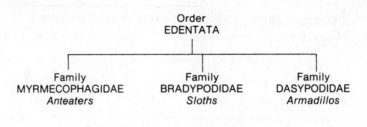

Order
EDENTATA

| Family
MYRMECOPHAGIDAE
Anteaters | Family
BRADYPODIDAE
Sloths | Family
DASYPODIDAE
Armadillos |

Giant Anteater   *Myrmecophaga tridactyla*
treis or *tria* (Gr) three   *daktulos* (Gr) a finger, a toe. Inhabiting
Central and South America.

Silky or Two-toed Anteater   *Cyclopes didactylus*
kyklops (Gr) round-eyed, or possibly *kuklos* (Gr) round, a circle   *pes*
(L) a foot; may refer to this anteater's unusual feet which have a
jointed sole that enables the claws to bend round and almost completely encircle the branch of a tree   *dis* or *di-* (Gr) twice, double
daktulos (Gr) a finger, a toe. Inhabiting Central America and the
northern part of South America.

Tamandua   *Tamandua tetradactyla*
tamandua is a Brazilian word for an 'ant-trap'   *tetras* (Gr) four
daktulos (Gr) a finger, a toe. Inhabiting Central and South America.

Family   BRADYPODIDAE   probably 7 species
bradus (Gr) slow   *pous* (Gr), genitive *podos*, a foot; hanging from the
branches of trees, they usually move very slowly, hence the name
sloth.

Three-toed Sloth or Ai   *Bradypus tridactylus*
treis, tria (Gr) three   *daktulos* (Gr) a finger, a toe. It has three toes on
the forefeet and hind feet. It inhabits Central America and most of
the northern half of South America.

Two-toed Sloth or Unau   *Choloepus didactylus*
khōlos (Gr) lame, maimed   *pous* (Gr) a foot; this name has been given
because it has only two digits on the forefeet; the name is misleading
because it has three toes on the hind feet   *dis* or *di-* (Gr) twice, double
daktulos (Gr) a finger, a toe. Unau and Ai (above) are Tupi native

names for the sloth. It inhabits Central America and the north-western part of South America.

Hoffmann's Sloth   *C. hoffmanni*
Dr Carl Hoffmann (1823-1859) was in Costa Rica, Central America, from 1854 until his death in 1859. This two-toed sloth inhabits Costa Rica and ranges south to Ecuador in South America.

Family   DASYPODIDAE   21 species
dasus (Gr) hairy, rough   *pous* (Gr), *genitive* podos, a foot; in this case meaning 'rough-footed'.

Long-haired Armadillo   *Chaetophractus vellerosus*
khaitē (Gr) long flowing hair   *phraktos* (Gr) protected   *vellus* (L), genitive *velleris*, wool, hair   *-osus* (L) suffix meaning full of. Inhabiting South America.

Giant Armadillo   *Priodontes giganteus*
priōn (Gr) a saw   *odous* (Gr), genitive *odontos*, a tooth; 'saw-toothed'; they have up to ninety small teeth: giganteus (L) very big. Inhabiting forested areas of the Amazon basin.

Central American Armadillo   *Cabassous centralis*
Cabassous is probably from *capacou* (Galibi) an armadillo; the native language of a people of French Guiana   *-alis* (L) suffix meaning relating to. From Central America.

La Plata Three-banded Armadillo   *Tolypeutes matacus*
tolupē (Gr) wool made into a ball, a ball wound up   *-tes* (Gr) suffix meaning in connection with; these small armadillos are the only ones that roll themselves up to form a ball when in danger; the Mataco are a people of Bolivia, Paraguay and Argentina   *-acus* (L) suffix meaning relating to. It lives in Paraguay and parts of Bolivia and Argentina.

Brazilian Three-banded Armadillo   *T. tricinctus*
tria (L) three   *cinctus* (L) a girdle; there are three belts of bony scales round the body. Inhabiting the eastern part of Brazil.

Nine-banded Armadillo   *Dasypus novemcinctus*
dasus (Gr) hairy, rough   *pous* (Gr) a foot; in this case it means 'rough-footed'   *novem* (L) nine   *cinctus* (L) a girdle; it has nine conspicuous belts of bony scales round the body. It inhabits the southern part of North America and is widespread in South America.

Fairy Armadillo or Pichiciego   *Chlamyphorus truncatus*
khlamus (Gr) a short cloak or mantle   *phora* (Gr) a carrying; can also
mean that which is carried, a burden; referring to the armour plating
of bony scales on the armadillo   *trunco* (L) I maim, I cut off, hence
truncatus (New L) cut off; the armour plating on the back stops
abruptly above the rump and the rear end of the body is protected
by a bony shield. It is the smallest of the twenty one species, being
only about 21 cm (6 in) long; it inhabits Argentina and Bolivia.
Pichiciego is the local Allentiac name in Western Argentina.

49 **Pangolins** PHOLIDOTA

The Pangolins are in some ways similar to the armadillos and ant-eaters, though not actually related to them. The head, back and upper part of the tail are covered in overlapping epidermal scales; they have an amazingly long tongue for insertion into anthills, in some cases nearly as long as the body, and a long prehensile tail. The name is from a Malayan word *peng-goling*, a roller, on account of the habit the animal has of rolling itself up to form a ball when in danger.

Subclass **EUTHERIA**
(see pages 460 and 477)

Order **PHOLIDOTA**
pholis (Gr) genitive *pholidos*, a horny scale; the head, back, and upper part of the tail are covered with horny scales.

Family **MANIDAE**   7 species
Derived from *manes* (L) in Roman religion the spirits of the dead, or ghosts; so named because of their nocturnal habits and peculiar appearance.

Giant Pangolin   *Manis gigantea*
Manis is an assumed Latin singular, coined from *manes* (see above) *gigas* (L) a giant, hence *giganteus* (L) big, gigantic. It inhabits the western part of Africa south of the Sahara.

Order
PHOLIDOTA
|
Family
MANIDAE
Pangolins

Short-tailed or Cape Pangolin   *M. temmincki*
Professor C. J. Temminck (1778–1858) was a Dutch zoologist, and at
one time Director of the Natural History Museum at Leyden in the
Netherlands. This pangolin lives in South Africa.

Small-scaled Tree Pangolin   *M. tricuspis*
tres, tria (L) three   *cuspis* (L) a point; the edges of the scales on the
young pangolin have three points, though these disappear in the
adult. Inhabiting the western part of Africa, south of the Sahara, and
ranging eastwards into Uganda.

Long-tailed Pangolin   *M. tetradactyla* (*tetradactyla* or *longicauda*)
tetra (Gr) four   *daktulos* (Gr) a finger, toe. The tail is about twice the
length of the head and body which is most unusual in mammals. It is
sometimes known as the Black-bellied Pangolin from the dark hair
on the underparts. Range as for *M. tricuspis*.

Chinese Pangolin   *M. pentadactyla*
pente (Gr) five   *daktulos* (Gr) a finger, toe. Inhabiting Nepal, China,
Hainan and Formosa.

Indian Pangolin   *M. crassicaudata*
crassus (L) thick, heavy   *cauda* (L) the tail   *-atus* (L) suffix meaning
provided with; it has a very broad heavy tail. It lives in southern Asia.

Malayan Pangolin   *M. javanica*
Inhabiting Burma, Java, Sumatra, Borneo, Celebes and southern
parts of the Philippines.

It should be noted that three or four subgenera are often recognised,
for example *Phataginus* and *Smutsia*; phatagen is an East Indian name
for the pangolin, and Johannes Smuts was an early nineteenth century
South African naturalist.

Until fairly recently the hares and rabbits were classified with the rodents, partly on account of the teeth, and also perhaps the shape of the head. However, they have four upper incisors instead of two, as in the rodents, and the upper and lower jaws oppose each other only on one side at a time, thus the motion of chewing is lateral instead of longitudinal. They are now placed in a separate order, Lagomorpha, and this includes the furry little animals known as pikas, sometimes called piping-hares on account of the shrill whistling and calling noises they make.

Domestic rabbits derive from the wild species *Oryctolagus cuniculus*, and by selective breeding in captivity many varieties of this one species have been produced.

Subclass EUTHERIA
(see pages 460 and 477)

Order LAGOMORPHA
lagōs (Gr) a hare *morphē* (Gr) the form, shape.

Family OCHOTONIDAE 14 species
ochotona (New L) derived from *ochodona*, a Mongolian name for the pika.

Russian or Steppe Pika *Ochotona pusilla*
pusillus (L) very small. The name pika originates from *piika* (Tun-

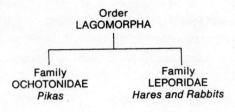

Order
LAGOMORPHA

Family	Family
OCHOTONIDAE	LEPORIDAE
Pikas	*Hares and Rabbits*

gusic). The Tungus are a people of eastern Siberia. Inhabiting Russia and Asia.

Large-eared Pika   *O. macrotis macrotis*
makros (Gr) long   *ous* (Gr), genitive *ōtos*, the ear. From Asia.

Mount Everest Pika   *O. m. wollastoni*
Named after Dr A. F. R. Wollaston (1875-1930), an English naturalist and explorer; it was found at an altitude of 5,300 m (17,500 ft). A subspecies of *O. macrotis* above.

Pallas's Pika   *O. pallasi*
Named after Professor Peter Simon Pallas (1741-1811), a German zoologist and explorer and a professor at St Petersburg University in Russia. He made contributions to most of the natural sciences. This pika inhabits the Volga district and the Ural Mountains in Russia.

American or Rocky Mountain Pika   *O. princeps*
princeps (L) a chief; a reference to an Amerindian name translated as 'Little Chief Hare'. Inhabiting North America.

Family   LEPORIDAE   about 50 species
lepus (L), genitive *leporis*, a hare.

Ryukyu Rabbit   *Pentalagus furnessi*
pente (Gr) five (in composition *penta-*)   *lagōs* (Gr) a hare; an allusion to the five pairs of upper cheek teeth instead of the usual six: 'Collected on the Liu Kiu Islands by Dr W. H. Furness and Dr H. M. Hiller on Feb 26th 1896.' No further information about these two collectors seems to be available. Inhabiting the Ryukyu Islands that lie to the south of Japan in the Pacific Ocean, it is jealously preserved as a 'Natural Monument'.

Natal Red Hare *Pronolagus crassicaudatus*
pronus (L) leaning forward; can mean belonging to what is before;
this hare has characters of an earlier form   *crassus* (L) thick, heavy
cauda (L) the tail of an animal. Inhabiting South Africa.

Volcano Rabbit *Romerolagus diazi*
Romero is a town in Mexico that lies in the volcanic belt that crosses
the country from east to west, and consisting of famous volcanos such
as Popocatapetl. However, the name is in honour of Don Matias
Romero (1837–1898) in recognition of his assistance to the Biological
Survey in Mexico; he was a Government Minister. The name *diazi*
commemorates D. A. Diaz de Leon and was given by Dr Jesus Diaz
de Leon. This rabbit is found only on the slopes of two volcanos lying
to the south-east of Mexico City.

Assam Rabbit *Caprolagus hispidus*
kapros (Gr) a wild boar   *lagōs* (Gr) a hare; probably an allusion to
the coarse, bristly fur   *hispidus* (L) rough, hairy. This rabbit is confined to a small area in Assam, and at one time was thought to be
extinct, but a few have been seen recently and it is now to be protected.

Cape Hare *Lepus capensis capensis*
-ensis (L) suffix meaning belonging to; it is not confined to the Cape
area, South Africa, and ranges through Europe and parts of Asia.

European or Brown Hare *L. c. europaeus*
The hare of the British Isles and Europe, formerly *Lepus europaeus*, is
now considered to belong to the Afro-Mediterranean species complex
of *L. capensis* (above). Thus, by the law of priority *L. capensis* named
by Linnaeus in 1758, must take precedence over *L. europaeus* named
by Pallas in 1778, 20 years later. This means it is now a subspecies,
Lepus capensis europaeus. It ranges through Europe, Africa and western
Asia.

Alpine Hare *L. timidus timidus*
timidus (L) afraid, timid. Living in the Alps of Europe and the
Scandinavian mountains, and ranging eastwards as far as Japan.

Scottish or Blue Hare *L. t. scoticus*
-icus (L) suffix meaning belonging to. This subspecies of *L. timidus*
lives in the mountains of Scotland and ranges south to northern
England and Wales; it is sometimes known as the Varying Hare
because the coat becomes white in winter.

Snowshoe Rabbit   *L. americanus*
-*anus* (L) suffix meaning belonging to. It is not a rabbit, but a hare;
stiff bristles grow on the feet in autumn to help when running on snow
and ice, hence the name 'snowshoe'. It inhabits North America.

White-tailed Jack Rabbit   *L. townsendi*
This is a hare with the typical long hind legs. The names 'hare' and
'rabbit' become rather indiscriminately mixed, so that some animals
known as rabbits are, in fact, hares. J. K. Townsend (1809–1851) was
an ornithologist and author who was exploring in the Rockies in 1834.
It inhabits the Rocky Mountains area in the north-west part of North
America.

Black-tailed Jack Rabbit   *L. californicus*
-*icus* (L) suffix meaning belonging to. Like some other hares, known
as a 'varying hare' because its coat becomes white in winter. This
'Jack Rabbit' is a hare inhabiting North America.

Cottontail   *Sylvilagus floridanus*
silva (L) a wood   *lagōs* (Gr) a hare; in spite of this name it is a rabbit
-*anus* (L) suffix meaning belonging to; of Florida. The tail is white
underneath, and looks like a ball of white cotton when raised; this is
used as an alarm signal and also in certain courting ceremonies. It is
not confined to Florida, and ranges through Canada, the USA,
Central America and the northern part of South America.

Pygmy Rabbit   *S. idahoensis*
-*ensis* (L) suffix meaning belonging to. This is the smallest rabbit, only
about 25 cm (10 in) long; it is found in Idaho and other parts of North
America.

Marsh Rabbit   *S. palustris*
paluster (L), genitive *palustris*, marshy, boggy. Inhabiting the south-
eastern part of North America.

Swamp Rabbit   *S. aquaticus*
aqua (L) water   -*icus* (L) suffix meaning belonging to, hence *aquaticus*
(L) living in water; it likes swampy conditions. Inhabiting the
southern part of North America.

European Rabbit   *Oryctolagus cuniculus*
oruktēr (Gr) a tool for digging   *lagōs* (Gr) a hare; 'a digging hare',
i.e. a rabbit, as hares do not make burrows   *cuniculus* (L) a rabbit;
can also mean an underground passage. This is the common rabbit

known throughout Britain and Europe. It is essentially a burrowing animal. Originally inhabiting Europe including the British Isles, and North Africa, it has now been introduced to many other countries.

Short-eared Rabbit   *Nesolagus netscheri*
nēsos (Gr) an island   *lagōs* (Gr) a hare; 'an island hare': named after E. Netscher, a naturalist who was at one time a member of the Council of Dutch East Indies (now Indonesia). This is a hare, but with remarkably short hind legs and short ears; it is sometimes known as the Sumatran Hare. A rare animal, found only in the tropical forest areas on the Island of Sumatra, Indonesia.

This is a large group, consisting of gnawing animals and containing over 1,500 species. It includes the mice, rats, guinea pigs, hamsters, squirrels, beavers, porcupines and many other less well-known rodents. They are, on the whole, vegetarians, and their teeth are specially adapted for such a diet. However, some will eat insects; the tooth formation is an infallible guide in identifying them as rodents. They have a single pair of continuously growing incisors in both upper and lower jaws, but have no canines.

For purposes of classification the order Rodentia has been divided into three suborders: Sciuromorpha, the 'squirrel-like'; Myomorpha, the 'mouse-like'; Hystricomorpha, the 'porcupine-like'.

Subclass **EUTHERIA**
Order **RODENTIA**

rodo (L) I gnaw.

Suborder **SCIUROMORPHA**

sciurus (L) a squirrel; also *skiouros* (Gr) a squirrel. This is derived from *skia* (Gr) shade, and *oura* (Gr) the tail; a 'shade-tail', on account of the way a squirrel holds his bushy tail over his back   *morphē* (Gr) form, shape, resemblance.

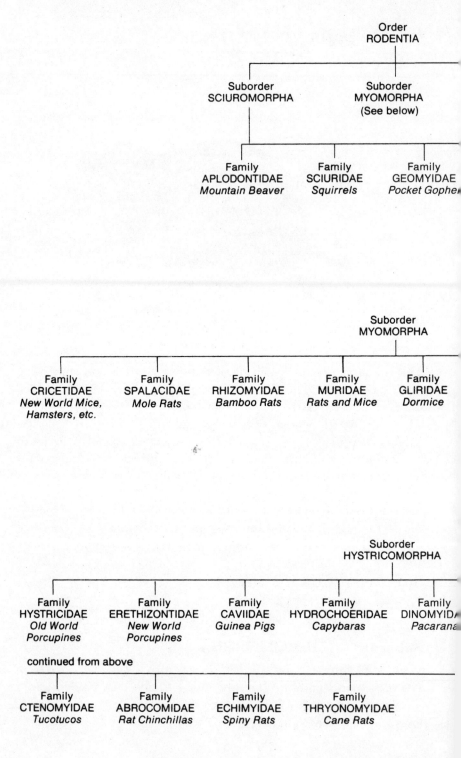

Order
RODENTIA

Suborder
SCIUROMORPHA

Suborder
MYOMORPHA
(See below)

Family
APLODONTIDAE
Mountain Beaver

Family
SCIURIDAE
Squirrels

Family
GEOMYIDAE
Pocket Gopher

Suborder
MYOMORPHA

Family
CRICETIDAE
New World Mice,
Hamsters, etc.

Family
SPALACIDAE
Mole Rats

Family
RHIZOMYIDAE
Bamboo Rats

Family
MURIDAE
Rats and Mice

Family
GLIRIDAE
Dormice

Suborder
HYSTRICOMORPHA

Family
HYSTRICIDAE
Old World
Porcupines

Family
ERETHIZONTIDAE
New World
Porcupines

Family
CAVIIDAE
Guinea Pigs

Family
HYDROCHOERIDAE
Capybaras

Family
DINOMYID/
Pacarana

continued from above

Family
CTENOMYIDAE
Tucotucos

Family
ABROCOMIDAE
Rat Chinchillas

Family
ECHIMYIDAE
Spiny Rats

Family
THRYONOMYIDAE
Cane Rats

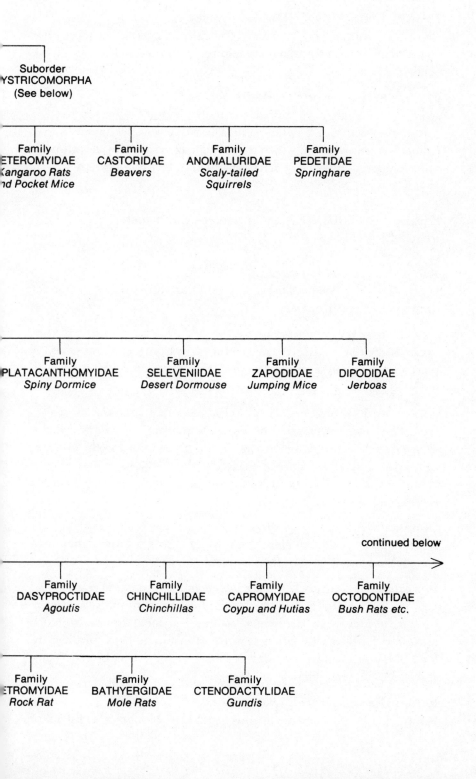

Suborder
HYSTRICOMORPHA
(See below)

Family
HETEROMYIDAE
Kangaroo Rats
and Pocket Mice

Family
CASTORIDAE
Beavers

Family
ANOMALURIDAE
Scaly-tailed
Squirrels

Family
PEDETIDAE
Springhare

Family
PLATACANTHOMYIDAE
Spiny Dormice

Family
SELEVENIIDAE
Desert Dormouse

Family
ZAPODIDAE
Jumping Mice

Family
DIPODIDAE
Jerboas

continued below

Family
DASYPROCTIDAE
Agoutis

Family
CHINCHILLIDAE
Chinchillas

Family
CAPROMYIDAE
Coypu and Hutias

Family
OCTODONTIDAE
Bush Rats etc.

Family
PETROMYIDAE
Rock Rat

Family
BATHYERGIDAE
Mole Rats

Family
CTENODACTYLIDAE
Gundis

Family **APLODONTIDAE**   1 species
aploos (Gr) simple, single   *odous* (Gr), genitive *odontos*, a tooth; the cheek teeth are simple; they have no roots and the molar teeth have single crowns.

Mountain Beaver or Sewellel   *Aplodontia rufa*
rufus (L) red, ruddy. In spite of the name, it is not a beaver, though it is fond of water and swims well. The name sewellel comes from a Chinook word *shewallal*, which means a cloak made from the animal's skin. There is only one species, and it inhabits the western side of North America.

Family **SCIURIDAE**   250 species, probably more
sciurus (L) see above.

Red Squirrel   *Sciurus vulgaris*
sciurus (L), see above   *vulgaris* (L) common, ordinary. Widespread in the forests of Europe and Asia.

Grey Squirrel   *S. carolinensis*
-ensis (L) belonging to; it is not confined to Carolina, from where it takes its name, and is widespread in North America, and latterly also in Great Britain.

American Red Squirrel or Chickaree   *Tamiasciurus hudsonicus*
tamias (Gr) a treasurer, one who stores; 'a hoarder'   *sciurus*, see above *-icus* (L) suffix meaning belonging to; it is not confined to the Hudson Bay area, and is widespread in North America. It should not be confused with the European Red Squirrel, *Sciurus vulgaris*. Chickaree is said to be an imitation of its cry.

Douglas's Squirrel   *T. douglasii*
Named after David Douglas (1798–1834) a botanist and explorer of North America, it lives in the California area.

Indian Palm Squirrel   *Funambulus palmarum*
funis (L) a rope and *ambulo* (L) I walk, hence funambulus (L) a rope dancer, a tight-rope performer   *palma* (L) the palm tree, hence *palmarius* (L) of palms; often seen in palm forests because they are fond of oil-palm nuts, they have other quite varied habitats. Widespread in southern Asia.

Indian Giant Squirrel   *Ratufa indica*
Ratuphar is a native name for this squirrel in Monghyr, a district of

Bengal   -*icus* (L) suffix meaning belonging to. It is a real giant among squirrels with an overall length of about 1 m (3 ft). Inhabiting southern Asia, Sumatra, Java and Borneo.

African Giant or Oil-palm Squirrel   *Protoxerus stangeri*
prōtos (Gr) first, primary; could indicate the first one to be discovered *xerōs* (Gr) dry, parched; so called from the character of the fur which is harsh and often spiny; the name can be misleading as it does not have the bristly hair of *Xerus erythropus*. Dr W. Stanger (1812–1854) was an English scientist and explorer. This squirrel inhabits the western part of Africa.

African Palm Squirrel   *Epixerus ebii*
epi- (Gr) upon, near; indicating it is anatomically like *Protoxerus* (above)   Ebo, a people of southern Nigeria. Inhabiting the western part of Africa.

Sun Squirrel   *Heliosciurus gambianus*
hēlios (Gr) the sun   *skiouros* (Gr) a squirrel; so called from its tropical habitat   -*anus* (L) suffix meaning belonging to; 'of Gambia'. A brightly coloured Sun Squirrel inhabiting Africa.

African Pygmy Squirrel   *Myosciurus pumilio*
mus (Gr) a mouse, genitive *muos*   *skiouros* (Gr) a squirrel   *pumilio* (L) a dwarf; it is only about 20 cm (4 in) long overall (c.f. *Ratufa indica*, above). Inhabiting the western part of Africa.

Hog or Long-snouted Squirrel   *Hyosciurus heinrichi*
hus (Gr), genitive *huos*, a pig   *skiouros* (Gr) a squirrel; a reference to the long nose   Gerd Heinrich (born 1896) was a collector in Celebes from 1930 to 1932. This squirrel inhabits Celebes.

Striped Ground Squirrel   *Xerus erythropus*
xerōs (Gr) dry, parched; a reference to the fur which is harsh and often spiny   *eruthros* (Gr) red   *pous* (Gr) a foot. Inhabiting savanna country of Africa from Mauritania to Uganda and Kenya.

South African Ground Squirrel   *Xerus (Geosciurus) inauris*
gē (Gr) earth, ground   *skiouros* (Gr) a squirrel   *in-* (L) prefix meaning not, without   *auris* (L) the ear; it has ears, but they are very small and usually hidden under the fur. Inhabiting South Africa. This is a subgenus of *X. erythropus* above.

Alpine Marmot   *Marmota marmota*   (*Marmota* formerly *Arctomys*)
marmotta (It) derived from *murmont* (Romansch, an Upper Rhine

dialect), which was derived from *mus* (L), genitive *muris*, a mouse; and *mons* (L), genitive *montis*, a mountain; 'mountain mouse'; it lives in the mountains, sometimes at an altitude of 2,500 m (8,000 ft) in central and north-eastern Europe.

Bobak *M. bobak*
bobak (Pol) a marmot. Inhabiting the Himalayas and other mountains in that part of Asia.

Hoary Marmot *M. caligata*
caliga (L) a boot, hence *caligatus* (L) wearing boots; the lower legs and feet are black, giving the appearance of boots. 'Hoary' indicates a greyish-white coat. Living in the mountains of Alaska and the western part of North America.

Woodchuck or Marmot *M. monax*
Monax is an American Indian name for the marmot; it lives high up in the mountains of Canada and the USA.

Black-tailed Prairie Marmot *Cynomys ludovicianus*
kuōn (Gr), genitive *kunos*, a dog *mus* (Gr) a mouse; sometimes known as the Prairie Dog on account of its sharp barking alarm call, it is not a dog: *ludovicianus* is an adjective connected with the name Louis; it indicates that the animal was found in Louisiana, a southern state in the USA, and it is the area where this marmot lives.

Antelope Ground Squirrel *Ammospermophilus harrisi*
ammos (Gr) sand, also a sandy place *sperma* (Gr) seed *philos* (Gr) loved, pleasing; 'seed-loving one of sandy places'; it lives in the hot sandy desert, and the diet is mostly seed and other plant life. It was named in honour of Edward Harris (1799-1863) who accompanied J. J. L. Audubon (1785-1851), the famous American ornithologist, on his Missouri River trip in 1843. The name, given by Audubon, commemorates their friendship, and several birds also bear his name, for example Harris's Woodpecker *Dryobates villosus harrisi*. The name 'antelope' derives from the tail which shows a white patch when raised, as with the Pronghorn Antelope. This squirrel inhabits desert areas of the USA in the south-west.

White-tailed Antelope Ground Squirrel *A. leucurus*
leukos (Gr) white *oura* (Gr) the tail. Inhabiting northern Mexico and Arizona, USA.

Rock Ground Squirrel *Otospermophilus beecheyi*
ous (Gr), genitive *ōtos*, the ear; the ears are more prominent than in
allied genera *sperma* (Gr) seed *philos* (Gr) loved, pleasing; the diet
is mostly seed and other plant life: named after Rear Admiral F. W.
Beechey (1796-1856) who was at one time President of the Royal
Geographical Society. This squirrel lives in the California area.

European Ground Squirrel or Souslik *Citellus citellus*
(*Citellus* = *Spermophilus*)
citellus (L) a ground squirrel suslik is a Russian name for the Ground
Squirrel. Inhabiting Europe and Asia.

Thirteen-striped Ground Squirrel *C. tridecemlineatus*
tria (L) three *decem* (L) ten *lineatus* (L) lined. Widespread in the
central area of North America.

Barrow Ground Squirrel *C. parryi*
Dr C. C. Parry (1823-1890) was an American botanist and explorer
who is commemorated in several plant names, for example the lily
Lilium parryi. This ground squirrel lives in Alaska, and takes its English
name from Barrow Point, on the northern coast of Alaska; this was
named by Sir John Barrow (1764-1848) the English Arctic explorer.

Eastern Chipmunk *Tamias striatus*
tamias (Gr) a treasurer, one who stores; 'a hoarder' *stria* (L) a
furrow, hence *striatus*, striped; it has stripes along the back. It inhabits
the eastern part of Canada and the USA.

Siberian Chipmunk *Eutamias sibiricus*
eu- (Gr) a prefix meaning well, nicely; usually used to indicate typical
tamias, see above *-icus* (L) belonging to; of Siberia.

European Flying Squirrel or Polatouche *Sciuropterus russicus*
sciurus (L) a squirrel *pteron* (Gr) a wing *-icus* (L) belonging to,
hence *russicus*, of Russia. The flying squirrels cannot fly in the true
sense of the word, but are able to glide a distance of 30 m (100 ft) or
more by means of a membrane of skin stretched between the forelegs
and the ankles. This squirrel ranges all the way from Scandinavia to
Japan. *Polatouche* (Fr) is derived from the Russian *poletusha*, a flying
squirrel.

Red and White Flying Squirrel *Petaurista alborufus*
petauron (Gr) a perch, a springboard *-ista* (L) suffix denoting ability,

one who practises; 'a springboard jumper' *albus* (L) white   *rufus* (L) red; it has reddish-brown rings round the eyes on a white face. Inhabiting eastern Asia.

American Flying Squirrel   *Glaucomys volans*
glaukos (Gr) silvery, gleaming; can mean grey   *mus* (Gr) a mouse: *volans* (L) flying; like all the flying squirrels, it cannot fly in the true sense of the word, but is able to glide quite long distances by means of a membrane of skin stretched between the forelegs and the ankles. Inhabiting Central and North America.

Family   GEOMYIDAE   about 38 species
gē (Gr) earth, ground   *mus* (Gr) a mouse; referring to the animal's subterranean mode of life.

Pocket Gopher   *Geomys bursarius*
bursa (Gr) the skin stripped off, a hide; giving rise to *bursa* (New L) a pouch, a pocket made of skin   *-arius* (L) suffix meaning belonging to; it has fur-lined pockets on the outside of the cheeks for carrying food. Inhabiting the western part of Canada, the USA and Central America.

Northern Pocket Gopher   *Thomomys talpoides*
thōmos (Gr) a heap   *mus* (Gr) a mouse; an allusion to the heaps of earth thrown out at frequent intervals along the line of the burrows *talpa* (L) a mole   *-oides* (New L) from *eidos* (Gr) apparent shape, resemblance; 'mole-like' because of its burrowing habits. Found in south-western Canada and north-western USA. The genus *Thomomys* has many subspecies.

Family   HETEROMYIDAE   70 species, possibly more
heteros (Gr) the other; can mean different from the usual   *mus* (Gr) a mouse; it is different from *Mus*.

Pocket Mouse   *Perognathus penicillatus*
pēra (Gr) a pouch, a pocket   *gnathos* (Gr) the jaw, the mouth; it has fur-lined pockets on the outside of the cheeks where it can store and carry food   *penicillus* (L) a painter's brush   *-atus* (L) suffix meaning provided with; the tail has a tuft at the end like a painter's brush. It inhabits the western part of Canada, the USA and Central America.

Kangaroo Mouse   *Microdipodops megacephalus*
mikros (Gr) small   *di-* from *dis-* (Gr) two, double   *pous* (Gr), genitive

podos, a foot   *opsis* (Gr) appearance; 'one that appears to have two small feet'; the front legs are very small and not used for moving about *megas* (Gr) big   *kephalē* (Gr) the head. Inhabiting North and Central America.

Giant Kangaroo Rat   *Dipodomys ingens*
di- from *dis-* (Gr) two   *pous* (Gr), genitive *podos*, a foot   *mus* (Gr) a mouse; *ingens* (L) vast, enormous; that is compared with other species. Inhabiting Mexico and ranging north to California.

Merriam's Kangaroo Rat   *D. merriami*
Dr C. Hart Merriam (1855–1942) was a zoologist who studied mammals and birds of America, and was Chief of the US Biological Survey in 1885. This kangaroo rat inhabits western and southern parts of North America.

Black-tailed Spiny Pocket Mouse   *Heteromys nigricaudatus*
heteros (Gr) different from the usual   *mus* (Gr) a mouse   *niger* (L) black   *caudatus* (L) having a tail. Inhabiting southern Mexico.

Family   CASTORIDAE   2 species
kastōr (Gr) the beaver.

European Beaver   *Castor fiber*
fiber (L) the beaver. Some authorities now consider there is only one species, but there are some small differences between the North American and the Eurasian type. This beaver inhabits Europe and Asia.

Canadian Beaver   *C. canadensis*
-ensis (L) suffix meaning belonging to. In addition to Canada, it is found in the northern part of the USA.

Family   ANOMALURIDAE   9 species, possibly more
anōmalos (Gr) uneven, irregular; here it is taken to mean 'strange' *oura* (Gr) the tail; an allusion to the scales, arranged in two longitudinal rows, on the under part of the basal part of the tail.

Pel's Scaly-tailed Squirrel   *Anomalurus peli*
H. S. Pel was Governor of Dutch Gold Coast, West Africa (now Ghana) from 1840 to 1850. The scaly-tails are squirrel-like animals resembling the flying squirrels; with the exception of *Zenkerella insignis*

(below), they too have a web of skin extending from the forelegs to the hind legs and out to the tail; this enables them to glide from tree to tree for distances of about 20 m (60 to 70 ft). Inhabiting western and central parts of Africa.

Beecroft's Scaly-tailed Squirrel   *Anomalurops beecrofti*
ops from *opsis* (Gr) aspect, appearance; it is very much like *Anomalurus* (above). John Beecroft was an Englishman who was made Governor by the Spaniards of their island Fernando Po, in 1844. Inhabiting western and central parts of Africa.

Pygmy Scaly-tailed Squirrel   *Idiurus zenkeri*
idios (Gr) one's own, private; can mean peculiar, distinct, hence strange   *oura* (Gr) the tail; this is a reference to the thinly haired tail, which has a number of rows of small scales on the under side near the base. G. Zenker (1855–1922) was a botanist and ornithologist who spent several years in West Africa from 1900 onwards. Inhabiting central Africa and Cameroun.

Non-gliding Scaly-tail   *Zenkerella insignis*
-ellus (L) the diminutive suffix which is sometimes used with a personal name to mean 'of Zenker' (see above)   *insignis* (L) remarkable, notable; of the nine species this is the only one that does not have the gliding membrane. Inhabiting Cameroun.

Family PEDETIDAE   1 species
pēdētēs (Gr) a dancer, a leaper.

Springhaas or Springhare   *Pedetes capensis*
-ensis (L) suffix meaning belonging to; taking its name from the Cape of Good Hope, South Africa. It also inhabits parts of eastern Africa; it has big powerful hind legs and moves by a series of huge leaps.

Suborder MYOMORPHA
mus (Gr) a mouse   *morphē* (Gr) form, resemblance; 'mouse-like'; many are much bigger than a mouse.

Family CRICETIDAE   about 570 species
cricetus (New L) derived from *criceto* (It) the hamster.

Coues's or Texan Rice Rat   *Oryzomys couesi*
oruza (Gr) rice   *mus* (Gr) a mouse. Dr E. B. Coues (1842–1899) was

an American surgeon and naturalist. This rice rat inhabits Belize in Central America; it may range northwards to Texas.

Black-eared Rice Rat *O. melanotis*
melas (Gr) black *ous* (Gr), genitive *ōtos*, the ear. Inhabiting Central and South America.

Alfaro's Rice Rat *O. alfaroi*
Dr A. Alfaro (1865–1951) was a zoologist, and at one time was Director of the Natural History Museum in Costa Rica. Inhabiting Central America.

Common Rice Rat *O. palustris*
palustris (L) marshy, boggy; can mean living in marshy places. Inhabiting southern USA, and south-eastern and northern parts of South America.

Jamaican Rice Rat *O. antillarum*
antillarum (New L) of the Antilles, a name applied to the islands of the West Indies, and which includes Jamaica. This rat is now very rare and may be extinct.

American Harvest Mouse *Reithrodontomys humulis*
rheithron (Gr) a stream, also a channel *odous* (Gr), genitive *odontos*, a tooth *mus* (Gr) a mouse; 'groove-toothed mouse'; there are grooves on the upper incisor teeth *humus* (L) the ground, hence *humilis*, on the ground. Inhabiting North America.

White-footed Mouse *Peromyscus maniculatus*
pēra (Gr) a pouch *mus* (Gr) a mouse *-iscus* (L) diminutive suffix; 'little pouched mouse'; it has cheek pouches like *Perognathus penicillatus*; *manus* (L) a hand, hence *manicula*, a little hand *-atus* (L) suffix meaning provided with. Inhabiting North and Central America.

White-footed Mouse *P. leucopus*
leukos (Gr) white *pous* (Gr) the foot. Range as above.

Cotton Rat *Sigmodon hispidus*
sigma (Gr) the letter Σ; *oudous* (= *odōn*) (Gr) a tooth; an allusion to the sigmoid pattern of the enamel of the molars when their crowns are worn down *hispidus* (L) hairy, shaggy; a reference to the harsh hair. Inhabiting southern USA and the northern part of South America.

Dusky-footed Woodrat   *Neotoma fuscipes*
neos (Gr) new   *tomos* (Gr) sharp, cutting; an allusion to the teeth
indicating a new genus of rodent   *fuscus* (L) dark-coloured   *pes* (L)
the foot. Inhabiting Canada and the coastal strip of western USA.

Bushy-tailed Woodrat   *N. cinerea*
cinis (L) ashes, hence *cinereus*, ash-coloured. Inhabiting British
Columbia and most of the western part of the USA.

Eastern Woodrat or Florida Packrat   *N. floridana*
-anus (L) suffix meaning belonging to. Named from Florida, but also
inhabiting other parts of south-eastern USA.

Fish-eating Rat   *Ichthyomys stolzmanni*
ikhthus (Gr) a fish   *mus* (Gr) a mouse; an allusion to its habit of eating
fish; it is notably modified for a semi-aquatic life;   named after Dr
Jean Stanislas Stolzmann (1854–1928), 'One of the best known and
most successful of Peruvian collectors; the discoverer of many new
mammals' (O. Thomas, 1893). He was Director of the Branicki
Zoological Museum in Warsaw from 1887 until his death in 1928.
This rat inhabits Peru.

Dwarf Hamster   *Phodopus sungorus*
phōs (Gr), genitive *phōdos*, a burn, a blister   *pous* (Gr) the foot; the
tubercles on the soles of the feet form a blister-like mass   *sungorus* is
from Dzungaria, a vast valley between the Altai Mountains and the
Tienshan Mountains in Sinkiang, China; the name was given by
Pallas in 1777. It also inhabits Siberia and Manchuria.

Common Hamster   *Cricetus cricetus*
cricetus (New L) derived from *criceto* (It) the hamster. It is found in
Russia, western Asia, Africa and parts of Europe.

Grey Hamster   *Cricetulus triton*
-ulus (L) diminutive suffix; a small hamster   Triton was a Greek sea-
god; the word is used here to indicate 'large'; 'A triton among the
minnows'. Although this genus consists of small short-tailed hamsters,
this species is the largest in the genus. Inhabiting eastern Europe and
northern Asia.

Golden Hamster   *Mesocricetus auratus*
mesos (Gr) middle   *cricetus* (New L) the hamster; indicating its intermediate position between *Cricetus* and *Cricetulus* (above)   *auratus* (L)
golden. This hamster, well known as a domestic pet, is descended

from a family of hamsters found in Syria in 1930; apart from this occasion it has never been seen in the wild, or at least any sightings have not been recorded.

Maned Rat *Lophiomys imhausii*
lophos (Gr) a crest   *mus* (Gr) a mouse; sometimes known as the Crested Hamster, it has a crest of erectile hairs along the back which are raised when the animal is alarmed. According to Dr Wilhelm C. H. Peters, 'a skull of the singular rodent lately described by M. Alphonse Milne-Edwards under the name *Lophiomys imhausii*, in the zootomical collection at Berlin, had been obtained by Dr Schweinfurth from the tombs of Maman, northward of Kassalá in Upper Nubia'   *imhausii* apparently refers to a Monsieur Imhaus, of Aden, who purchased a specimen that had been collected in north-east Africa; it is reported that this unusual rodent has also been seen in Uganda. Dr Wilhelm C. H. Peters (1815-1884) was a Professor of Zoology in Berlin. He was in East Africa from 1842 to 1848. Professor Alphonse Milne-Edwards (1835-1900) was a French zoologist. Dr G. A. Schweinfurth (1836-1925) was an author and naturalist who explored Africa during the nineteenth century.

Collared Lemming *Dicrostonyx hudsonius*
dikroos (Gr) forked, cloven   *stonux* (Gr) a sharp point   *onux* (Gr) a claw; in winter two claws on the forefeet become enlarged and prominent; the purpose is not established, but may have something to do with digging in the snow   *hudsonius*, of Hudson; named from Hudson Bay. It is widespread in the Arctic regions.

Norwegian Lemming *Lemmus lemmus*
lemmus (New L) the lemming, derived from the Norwegian *lemming*. This is the lemming that is occasionally involved in mass migration, which may take them into the sea where many are drowned. Other species live in northern North America and northern Asia.

Bank Vole *Clethrionomys glareolus glareolus*
(*Clethrionomys* formerly *Evotomys*)
klethron (Gr) a bolt or bar for closing a door   *mus* (Gr) a mouse; it is said this means 'bolt-toothed mouse', possibly because the enamel ridges on the teeth are rounded and without the angular ridges found in the Common Vole. Another reason could be that the molar teeth are rooted, i.e. 'bolted', and so differ from true voles which have constantly growing molars. It has also been suggested that the name

derives from *klethra* (Gr) an alder   *-ion* (Gr) diminutive suffix, and meaning 'alder-grove mouse', from its habitat   *glarea* (L) gravel *-olus* (L) diminutive suffix; actually it usually frequents earth banks rather than gravelly areas. It is widespread in Europe.

Jersey Bank Vole   *C. g. caesarius*
Caesarea is the Roman name for Jersey, one of the Channel Islands. The subspecific name indicates a subspecies of *C. glareolus*. The Bank Voles are sometimes known as Red-backed Voles; the fur on the back is a chestnut-red colour.

Skomer Bank Vole   *C. g. skomerensis*
Skomer is an island off the coast of Pembrokeshire, Wales. A subspecies; see above.

Raasay Bank Vole   *C. g. erica*
erice (L) heath, ling, from *ereikē* (Gr) heath, heather; 'a heath-vole'. Raasay is an island near the Isle of Skye, Inverness-shire, Scotland. A subspecies, see above.

Martino's Snow Vole   *Dolomys bogdanovi*
dolos (Gr) deceit   *mus* (Gr) a mouse; 'unter Anspielung auf die Bedeutung des Namens *Phenacomys'*—evidently on account of the puzzling affinities of the type species (see below). Named after Professor M. N. Bogdanov (1841-1888) of St Petersburg (Leningrad). A fossil of this vole was found in Hungary in 1898, and it was not until 1925 that a living specimen was found. First described by V. and E. Martino, and possibly given the name Snow Vole because it lives in the mountains, as high as 600 m (2,000 ft) in Yugoslavia.

Water Vole   *Arvicola terristris*
arvum (L) ploughed land, a field   *colo* (L) I till, I cultivate; can mean dwell in, inhabit   *terristris* (L) of the ground; it lives on river banks and is a good swimmer. Sometimes known as the Water Rat, it is a vole and not a rat. Widespread in Europe including the British Isles, but not Ireland. Also found in the Arctic, in North America, Russia and Siberia, and ranging southwards to parts of Asia Minor and Israel.

Muskrat   *Ondatra zibethica*
Ondatra is a North American Indian name for the muskrat   *zibetto* (It) the civet-cat; *zibethicus* (New L) civet-odoured. Muskrats were brought to Great Britain in 1929 and kept in fur-farms for the valuable fur, but many escaped and became established over large areas. They

were then registered as pests and exterminated by about 1940. Inhabiting North America.

Newfoundland Muskrat   *O. obscura*
obscura (L) dark, dusky. The muskrat is an aquatic animal and adapted for swimming; this species is found only on the island of Newfoundland.

Tree Mouse or Heather Vole   *Phenacomys intermedius*
phenax (Gr), genitive *phenakos*, a cheat   *mus* (Gr) a mouse; an allusion to the fact that 'the external appearance of the animal gives no clue to its real affinities'   *intermedius* (L) between, intermediate; probably a reference to the tail which is of medium length; other species have long tails. Inhabiting North America.

Common Field Vole   *Microtus agrestis*
mikros (Gr) small   *ous* (Gr), genitive *ōtos*, the ear   *agrestis* (L) belonging to the fields. A characteristic of the voles is that the ears are small and inconspicuous. Inhabiting Europe and northern Asia.

Steppe Lemming   *Lagurus lagurus*
lagos (Gr) a hare   *oura* (Gr) the tail; it has a short tail, similar to a hare. Inhabiting southern Russia and western Siberia.

Chinese or Yellow Steppe Lemming   *L. luteus*
luteus (L) the colour of the plant lutum, saffron-yellow. It lives in Mongolia and northern China.

American Steppe Lemming or Sagebrush Vole   *L. curtatus*
curtus (L) shortened   *-atus* (L) suffix meaning provided with; probably referring to the short legs and tail. Inhabiting the western part of the USA and Canada.

Pygmy Gerbil or Sand Rat   *Gerbillus gerbillus*
gerbil, and *jerboa*, derived from *jarbū* (Ar) a rodent   *-illus* (L) diminutive suffix; 'a little gerbil'. Inhabiting Palestine and ranging south to northern Africa.

Indian Gerbil   *Tatera indica*
Tatera, obscure; Professor Lataste says it is a euphonic name of unknown origin   *Indica* (L) Indian. Inhabiting sandy and semi-desert areas of India, Arabia and Africa. Professor Fernand Lataste was a French zoologist. His *Étude de la Faune des Vertébrés de Barbarie (Algérie, Tunisie et Maroc)*, published in 1885, was for many years a standard work on the animals of North Africa.

Mid-day Jird   *Meriones meridianus*
Mērionēs and Idomeneus were companions in arms in the Trojan War of Greek mythology. Since the early days of classification, Mērionēs has been used as the name of a genus of gerbils, and until fairly recently Idomeneus was the name of a subgenus   *meridianus* (L) at mid-day; there seems no obvious reason for the name. Jird is another name for the gerbil from the Berber name *gherda*. Living in desert areas of northern Africa.

Fat Sand Rat   *Psammomys obesus*
psammos (Gr) sand   *mus* (Gr) a mouse   *obesus* (L) fat; this is a kind of gerbil living in desert areas of eastern Europe, Africa and south-west Asia.

Great Gerbil   *Rhombomys opimus*
rhombos (Gr) a rhombus, a parallelogram having its sides equal and two angles oblique; i.e. lozenge-shaped; a reference to the enamel on the upper molars which shows a lozenge-shaped pattern   *mus* (Gr) a mouse   *opimus* (L) rich, fat. Inhabiting the deserts of Mongolia, Turkestan and Iran.

Family   SPALACIDAE   3 species
spalax (Gr), genitive *spalakos*, a mole.

Palestine Mole Rat   *Spalax ehrenbergi*
Dr C. G. Ehrenberg (1795-1876) worked in North Africa in 1820, where these mole rats live; they range north to Palestine and east to Russia.

Family   RHIZOMYIDAE   18 species
rhizōma (Gr) a root   *mus* (Gr) a mouse; they dig in the ground for roots, their main food.

African Mole Rat   *Tachyoryctes splendens*
takhus (Gr) fast, swift   *orussō* (Gr) I dig   *oruktēs* (Gr) one who digs; 'a fast digger'   *splendeo* (L) I shine; some have a shining black coat. They are very fast diggers, and eat roots and possibly cane. Inhabiting East Africa.

Family   MURIDAE   about 460 species
mus (L), genitive *muris*, a mouse or rat.

Marmoset Mouse   *Hapalomys longicaudatus*
hapalos (Gr) soft   *mus* (Gr) a mouse; this refers to the soft fur   *longus*
(L) long   *cauda* (L) the tail of an animal   *-atus* (L) suffix meaning
provided with. Inhabiting the Malay Peninsula, Thailand and Indo-
China.

Harvest Mouse   *Micromys minutus*
mikros (Gr) small   *mus* (Gr) a mouse   *minutus* (L) small, minute; it
only weighs about 7 g (¼ oz). Inhabiting Europe, including the British
Isles, eastern Europe and China.

Long-tailed Field or Wood Mouse   *Apodemus sylvaticus*
apodemus (Gr) away from home; in the fields; the name is given
to distinguish it from the House Mouse   *silva* (L) a wood, hence
silvaticus, of woods or trees; applied to plants and animals it usually
means wild. Inhabiting Europe, North Africa and parts of Asia.

Yellow-necked Mouse   *A. flavicollis*
flavus (L) yellow   *collum* (L) the neck. The range is similar to the
Long-tailed Mouse, above.

New Guinea Giant Rat   *Hyomys goliath*
hus (Gr), genitive *huos*, a pig   *mus* (Gr) a mouse   Goliath was the
Philistine giant in the Bible. This large rat, with a massive skull, can
be up to 76 cm (30 in) long, including the tail.

Striped Grass Mouse   *Lemniscomys striatus*
lemniscus (L) a ribbon; in this case the 'ribbons' are stripes   *mus* (L) a
mouse   *stria* (L) a groove, a furrow, hence *striatus* (New L) striped;
it has buff stripes on a dark brown coat. Living in Africa.

Zebra Mouse   *Rhabdomys pumilio*
rhabdos (Gr) a rod, a stick, so *rhabdōtos* (Gr) streaked, striped   *mus* (Gr)
a mouse; it has several light and dark stripes along the centre of the
back   *pumilio* (L) a dwarf; sometimes known as the Four-striped Rat,
it is only about the size of a house mouse. It lives in Uganda and
Kenya and ranges southwards into South Africa.

Black Rat   *Rattus rattus*   (*Rattus* formerly *Epimys*)
rattus (New L) a rat; sometimes known as the Ship Rat. This rat, and
the Brown Rat (below), probably originated in Asia, but can now be
found almost anywhere in the world, except the Arctic and Antarctic
regions.

Brown or Norwegian Rat   *R. norvegicus*
-icus (L) suffix meaning belonging to; of Norway; it does not originate
in Norway and is not confined to that country. Like the Black Rat it
probably originated in Asia. Now found throughout the world, and
probably even more common in the British Isles than the Black Rat.

Multimammate Rat   *Mastomys natalensis*
mastos (Gr) the breast   *mus* (Gr) a mouse   *-ensis* (L) suffix meaning
belonging to; found in Natal, South Africa, in 1843, it inhabits most
of Africa except desert areas. Multimammate refers to the fact that
it may have as many as sixteen pairs of teats.

Stick-nest Rat   *Leporillus apicalis*
lepus (L), genitive *leporis*, a hare   *-illus* (L) diminutive suffix; with its
large ears it is rather like a small hare   *apex* (L), genitive *apicis*, the
top, the tip   *-alis* (L) suffix meaning relating to; it has a white tip to
the tail. It builds large nests made of sticks. Inhabiting southern
Australia.

Giant Naked-tailed Rat   *Uromys caudimaculatus*
oura (Gr) the tail   *mus* (Gr) a mouse   *cauda* (L) the tail of an animal
macula (L) a spot   *-atus* (L) provided with; 'spotted tail'; the tail is
of two colours, and is hairless. A very big rat inhabiting New Guinea.

House Mouse   *Mus musculus*
mus (L) a mouse or a rat; under this name the Romans also included
animals like the marten and sable   *-culus* (L) diminutive suffix;
'little mouse'. It is found throughout the world wherever there are
people living, even in Arctic regions.

New Guinea Kangaroo Mouse   *Lorentzimys nouhuysii*
Dr H. A. Lorentz (1871–1944) was in the Dutch Consular Service,
in New Guinea during the period 1903 to 1910. Captain J. W. v.
Nouhuys (born 1869) of the Dutch Navy was also there at that time.
This mouse has strong hind legs and hops when moving about.

Australian Kangaroo Mouse   *Notomys mitchelli*
notos (Gr) the south   *mus* (Gr) a mouse   Sir Thomas Livingstone
Mitchell (1792–1855) was a Scottish surveyor and explorer, and was
the Surveyor General for New South Wales in 1827. Sometimes
known as Hopping Mice, they hop rather than walk or run; they are
in no way related to Kangaroos.

Fawn-coloured Kangaroo Mouse   *N. cervinus*
cervus (L) a stag, a deer, hence *cervinus*, tawny, like a deer. See notes
above; inhabiting Australia.

Cairo Spiny Mouse   *Acomys cahirinus*
akōkē (Gr) a sharp point   *mus* (Gr) a mouse; for self protection the
hair on the body of this mouse has evolved into spiny prickles. When
Cairo was founded in AD 968 it was named El-Kahira, meaning 'The
Victorious'; the name was gradually corrupted and became Cairo.
In addition to Africa, it inhabits southern Asia.

Cape Spiny Mouse   *A. subspinosus*
sub- (L) prefix meaning under; can also mean somewhat, slightly
spina (L) a thorn; spinosus, thorny, full of thorns; it is less prickly than
other spiny mice. Inhabiting South Africa.

Bandicoot Rat or Pig Rat   *Bandicota indica*
pandikokku (Telugu) a pig rat   Telegu is a language spoken in the
central and eastern parts of southern India   *Indica* (L) Indian. Sometimes known as the Malabar Rat, Malabar being a district in south-
west India. It is supposed to smell rather like a pig, and grunts like a
pig at times. Inhabiting southern India.

Bengal Bandicoot Rat   *B. bengalensis*
-ensis (L) suffix meaning belonging to. Sometimes known as a Mole
Rat, it is a good burrower, like a mole. First found in the Bengal area,
it is widespread in southern Asia, including the Malay Peninsula,
Sumatra and Java.

Giant Rat or Hamster Rat   *Cricetomys gambianus*
cricetus (New L) from *criceto* (It) the hamster   *mus* (Gr) a mouse   also
mus (L) a mouse or a rat   *-anus* (L) suffix meaning belonging to;
'of Gambia'. It is not a hamster, but a very large rat, sometimes
measuring more than 75 cm (30 in) overall. Inhabiting forested
tropical areas of West Africa.

African Climbing or Tree Mouse   *Dendromus mesomelas*
dendron (Gr) a tree   *mus* (Gr) a mouse   *mesos* (Gr) middle   *melas* (Gr)
black; it has a black stripe down the middle of the back. Widespread
in Africa.

South African Swamp Rat   *Otomys irroratus*
ous (Gr), genitive *ōtos*, the ear   *mus* (Gr) a mouse; T. S. Palmer does

not give any explanation for 'ear-mouse'   *irroratus* (L) moistened with dew, wetted; a reference to its damp marshy habitat. Inhabiting South Africa.

New Guinea Rat   *Mallomys rothschildi*
mallos (Gr) wool, a fleece   *mus* (Gr) a mouse; it has thick woolly fur; named after L. W. Rothschild, 2nd Baron, FRS (1868–1937) who founded the Zoological Museum, Tring, England in 1889; he also wrote much on zoology. Another large rat, up to 86 cm (34 in) from nose to tail; living in New Guinea.

Philippine Cloud Rat   *Phloeomys cumingi*
phloios (Gr) bark of trees, peel   *mus* (Gr) a mouse; 'suggested by the habit of the animal, which Mr Cuming states feeds chiefly on the bark of trees'. H. Cuming (1791–1865) was a sailmaker in Valparaiso, Chile, in 1819. He collected the first specimen of this gigantic rat in Luzon, in the Philippines. It is probably named Cloud Rat because of its mountain habitat. It is a very large mountain species inhabiting the Philippines.

Bushy-tailed Rat   *Crateromys schadenbergi*
krateros (Gr) strong, mighty   *mus* (Gr) a mouse; referring to the fact that it is about the largest and heaviest member of the Muridae. Named after Dr A. Schadenberg who sent the first specimens to Dresden Museum in 1894. He collected them on Mount Data, Luzon, in the Philippines. It is bushy like a Persian cat, the body as well as the tail and lives in the Philippine mountains.

Philippine Shrew Rat   *Rhynchomys soricoides*
rhunkhos (Gr) the snout, beak   *mus* (Gr) a mouse; referring to the very long snout   *sorex* (L) genitive soricis, a shrew-mouse   *-oides* (New L), from *eidos* (Gr) apparent shape, resemblance; the pointed snout is like that of a shrew. Inhabiting the Philippines.

Beaver Rat   *Hydromys chrysogaster*
hudōr (Gr) water; in composition the prefix *hudro-* is used   *mus* (Gr) a mouse   *khrusos* (Gr) gold   *gastēr* (Gr) the stomach; 'golden-bellied water-rat'. A good swimmer, and very common in New Guinea and Australia.

Shaw Mayer's Mouse   *Mayermys ellermani*
F. W. Shaw Mayer (born 1899) an Australian zoologist. He collected for the British Museum (Natural History) and in New Guinea in 1928

for the Zoological Museum at Tring. Sir John Ellerman (1909-1973) was the son of Sir John Reeves Ellerman, founder of the Ellerman Steamship Line. He was the author of a standard work on rodents. An interesting rodent, having only one molar tooth on each side of upper and lower jaws; this is unique among rodents. It lives in New Guinea.

Family GLIRIDAE   about 28 species

glis (L), genitive *gliris*, a dormouse; the English name probably comes from *dormio* (L) I sleep; they have a very long period of hiberation, and so have earned a reputation for being sleepy.

Edible or Fat Dormouse   *Glis glis*   (*Glis* formerly *Myoxus*)

It used to be eaten by the Romans; it becomes very fat before hibernating. It inhabits the southern part of Europe, Russia and ranges south to Syria. It has been introduced into the southern Midlands in England.

Hazel or Common Dormouse   *Muscardinus avellanarius*

Muscardinus (New L) from *muscardin* (Fr) a doormouse, from *muscadin*, a musk-scented lozenge; a reference to the odour of the animal   *avellana* or *abellana* (L) filbert or hazel nut, from Abella, a town in Campania, Italy, abounding in fruit-trees and nuts   *-arius* (L) belonging to; these nuts are the favourite food of this dormouse. Inhabiting Europe and Asia.

Garden Dormouse   *Eliomys quercinus*

eleios (Gr) a kind of dormouse   *mus* (Gr) a mouse   *quercus* (L) the oak   *-inus* (L) suffix meaning pertaining to; dormice like acorns and so would often be seen in oak trees; they also make nests in trees. Inhabiting Europe, north-west Africa and south-west Asia.

Forest Dormouse   *Dryomys nitedula*

drus (Gr), genitive *druos*, the oak (see above)   *mus* (Gr) a mouse   *nitedula* (L) a small mouse, or a dormouse. From Europe and Asia.

Japanese Dormouse   *Glirulus japonicus*

glis (L), genitive *gliris*, a dormouse   *-ulus* (L) diminutive suffix   *-icus* (L) belonging to; 'little dormouse of Japan'.

Mouse-tailed Dormouse   *Myomimus personatus*

mus (Gr), genitive *muos*, a mouse   *mimos* (Gr) an actor, a mimic   *persona* (L) a mask, as for drama   *-atus* (L) suffix meaning provided with; unlike other dormice this one does not have a bushy tail; 'one who imitates an ordinary mouse'. This is a rare species inhabiting the Russo-Iranian border, and surprisingly has recently been found in Bulgaria.

African Dormouse   *Graphiurus murinus*
grapheion (Gr) a pencil   *oura* (Gr) the tail; the tail has a pencil of hairs at the tip   *murinus* (L) mouselike. Inhabiting forested areas throughout the whole of Africa.

Family   PLATACANTHOMYIDAE   2 species
platus (Gr) wide, flat   *akantha* (Gr) a thorn, a prickle   *mus* (Gr) a mouse; they have flattened spines in the hair and are known as 'spiny dormice' because the hairs are sharp and prickly.

Spiny Dormouse   *Platacanthomys lasiurus*
lasios (Gr) hairy   *oura* (Gr) the tail; the tail is scaly at the base and has a bushy tip. Inhabiting southern India.

Family   SELEVINIIDAE   1 species

Desert Dormouse   *Selevinia betpakdalaensis*
This animal was first discovered in Central Kazakhstan, a southern desert region of the USSR, by a zoologist named W. A. Selevin, about the year 1938. It is rare and unique, and has to be placed in a family on its own. The name has also been given as *Selevinia paradoxa*, presumably because classifying it was something of a paradox. By the law of priority the original name is the only legitimate one. Bet-Pak-Dala is the name of a village in Kazakhstan close to where the animal was found.

Family   ZAPODIDAE   11 species
za- (Gr) a prefix with intensive meaning; much or very   *pous* (Gr), genitive *podos*, the foot; taken to mean 'strong feet' or 'big feet', as they have strong well developed back legs for jumping.

Northern Birch Mouse   *Sicista betulina*
Sikistan is a Tartar name meaning 'gregarious mouse'   *betula* (L) the birch   *-inus* (L) suffix meaning belonging to; 'of birch trees'. Inhabiting Scandinavia and Finland.

Southern Birch Mouse   *S. subtilis*
subtilis (L) slender, fine, not thick or coarse; could be interpreted as 'graceful'. Inhabiting southern Russia and Rumania.

Jumping Mouse   *Zapus hudsonius*
Zapus, see above. Named from Hudson Bay, Canada; it inhabits other areas of North America.

Szechuan Jumping Mouse   *Eozapus setchuanus*
ēos (Gr) the dawn; can also mean the east; meaning living in the east *zapus*, as above. Szechwan is a province in China. Inhabiting eastern Asia.

Family   DIPODIDAE   25 species, possibly more
di- from *dis-* (Gr) two   *pous* (Gr), genitive *podos*, the foot; they appear to have only two feet because the front legs are tiny and not used for moving about; they hop like kangaroos and the hind legs are long and well developed. Jerboa is from *jarbū* (Ar) a rodent.

Feather-footed Jerboa   *Dipus sagitta*
Dipus, see above   *sagitta* (L) an arrow; this suggests 'feather-tailed', referring to the tuft of hair at the tip of the tail, like the feathers on an arrow; it also has hairy feet. Inhabiting desert areas of Africa and Asia.

Egyptian Jerboa   *Jaculus jaculus*
jaculor (= *iaculor*) (L) to throw a javelin, hence *iaculus*, thrown, darting; it is only 12–18 cm (5–6 in) long but can leap a distance of nearly 2 m (6 ft). Inhabiting desert areas of North Africa.

Euphrates Jerboa   *Allactaga euphratica*
Alak-dagha is a Mongolian name for the jerboa   *-icus* (L) suffix meaning belonging to; 'of the Euphrates'. It inhabits the desert lands of Arabia.

Siberian Jerboa   *A. sibirica*
Inhabiting Siberia; this usually means the whole territory between the Ural Mountains and the Pacific Ocean.

Three-toed Dwarf Jerboa   *Salpingotus kozlovi*
salpinx (Gr), genitive *salpingos*, a war trumpet   *ous* (Gr), genitive *ōtos*, the ear; it has funnel-shaped or trumpet-shaped ears; named after General P. K. Kozlov (1863–1935) a Russian zoologist who explored Central Asia during the years 1899 to 1926. This remarkable little jerboa, measuring only 5 cm (2 in) head and body, has hairy tufts

under the three hind toes. Named only recently, in 1922, it inhabits the Gobi Desert in Mongolia.

Long-eared Jerboa   *Euchoreutes naso*
eu- (Gr) prefix meaning well   *khoreutēs* (Gr) a dancer; 'a good dancer'; a reference to the animal's mode of progression by leaps   *nasus* (L) the nose; it has a pointed nose and very long ears. Inhabiting Sikiang and Inner Mongolia.

Suborder   HYSTRICOMORPHA
hustrix (Gr), genitive *hustrikhos*, a hedgehog, a porcupine   *morphē* (Gr) shape, resemblance.

Family   HYSTRICIDAE   15 species

Great Crested Porcupine   *Hystrix cristata*
cristatus (L) crested. Inhabiting southern Europe, North and West Africa and south-western Asia.

Malayan Porcupine   *H. brachyura*
brakhus (Gr) short   *oura* (Gr) the tail. Inhabiting Malaya, Borneo and Sumatra.

Brush-tailed Porcupine   *Atherurus africanus*
athēr (Gr) the beard of an ear of wheat   *oura* (Gr) the tail; it has a tuft of hair on the tip of the tail. Inhabiting West Africa and part of western central Africa.

Rat Porcupine   *Trichys lipura*
thrix (Gr), genitive *thrikhos*, hair; the body has a lot of stiff hair mixed with spines and bristles   *leipō* (Gr) I am lacking, wanting   *oura* (Gr) the tail; the tail is shorter than in the allied genus *Atherurus*. Inhabiting the southern part of Malaysia, and Borneo and Sumatra.

Family   ERETHIZONTIDAE   11 species, possibly more
erethizō (Gr) I rouse to anger, I irritate.

Canadian Porcupine   *Erethizon dorsatum*
dorsum (L) the back   *-atus* (L) suffix meaning provided with; 'having a back that irritates'. Inhabiting large areas of northern North America, including Canada.

Brazilian Tree Porcupine   *Coendou prehensilis*
Coendou is a Brazilian native name for the porcupine   *prehenso* (L)

I lay hold of   -*ilis* (L) adjectival suffix denoting capability; 'able to hold'; it has a prehensile tail used in tree climbing. Inhabiting Central and South America.

Family   CAVIIDAE   15 species
cavia from *çaviá* (Port) now *savia* from the Tupi word *sawiya*, a rat.

Wild Guinea Pig or Restless Cavy   *Cavia porcellus*
porcus (L) a pig, hence *porcellus*, a little pig; it is no relation to the pig family, the Suidae, nor does it come from Guinea; a mistake for Guyana. Inhabiting South America.

Brazilian Cavy   *C. aperea*
aper (L) a wild boar, or pig.

Rock Cavy or Moco   *Kerodon rupestris*
keras (Gr) the horn of an animal   *odōn* (Gr Ionic dialect) a tooth; the reason is obscure; T. S. Palmer offers no explanation in his standard work *Index Generum Mammalium* (1904)   *rupes* (L) a rock, hence *rupestris*, living among rocks. Moco is a Tupi name for this cavy; it lives in Brazil.

Patagonian Hare or Mara   *Dolichotis patagona*
dolikhos (Gr) long   *ous* (Gr), genitive *ōtos*, the ear; it is not a hare, but has long legs and rather bigger ears than other members of the family   *mara* is American Spanish. Patagonia is a region in the southern part of Argentina.

Salt Desert Cavy   *Pediolagus salinicola*
pedion (Gr) a plain, level country   *lagōs* (Gr) a hare   *sal* (L), genitive *salis*, salt, hence *salinus* (New L) salty   *colo* (L) I inhabit; it is not known except in the salt desert areas of the southern part of South America. Some now give *Pediolagus* as a subgenus of *Dolichotis*.

Family   HYDROCHOERIDAE   2 species
hudōr (Gr) water; in composition the prefix *hudro-* is used   *khoiros* (Gr) a young pig.

Capybara or Carpincho   *Hydrochoerus hydrochaeris*
Sometimes known as the Water Hog, it is not a hog, but always lives near water and is a good swimmer and diver. It is the largest rodent and can weigh up to 45 kg (100 lb) and measure over 1 m (4 ft) in

length. Capybara and Carpincho are South American names derived from Tupi. Inhabiting a large area of South America except the very southern parts.

Panama Capybara   *H. isthmius*
isthmos (Gr) a narrow passage, also a neck of land between two seas; this capybara lives in the Panama area, so the name refers to the Isthmus of Panama.

Family   DINOMYIDAE   1 species
deinos (Gr) terrible, formidable   *mus* (Gr) a mouse; it has the appearance of an enormous guinea pig, with a head and body length of up to 76 cm (30 in) and a tail of 20 cm (8 in).

Pacarana or False Paca   *Dinomys branickii*
Named after Grafen Constantin Branicki. Jelski made zoological journeys of discovery in Surinam and Peru which were rich in results; he was able to do this owing to the splendid liberal support of Branicki. Professor K. Jelski (1838–1896) was a Polish zoologist and Curator of the Museum in Cracow from 1878 until his death in 1896. Grafen Constantin Branicki (1823–1884), after whom the animal was named, was a wealthy Polish nobleman whose son and nephew established the Branicki Zoological Museum in Warsaw in 1887, Dr J. S. Stolzmann being the Director (see page 550). *Paca* (Sp from Tupi) a South American rodent   *rana* (Tupi) false; so named because it is not a true paca of the family Dasyproctidae (below). It inhabits mountainous areas in Peru.

Family   DASYPROCTIDAE   17 species
dasus (Gr) hairy   *proktos* (Gr) the hindpart, the rump; the hair is not confined to the rump though it is longer there and usually of a different colour which makes it conspicuous.

Paca   *Cuniculus paca*   (*Cuniculus* formerly *Coelogenys*)
cuniculus (L) a rabbit; it is not a rabbit but more like a very large rat, with a short tail and a spotted coat   *paca* is a South American Spanish name derived from Tupi. Inhabiting a large area from Mexico south to the Argentine.

Mountain Paca   *C. taczanowskii*
Dr W. Taczanowski (1819–1890) was a Polish zoologist who made an

expedition to Peru in 1884. This paca lives high in the Andes Mountains, possibly up to about 3,000 m (10,000 ft).

Orange-rumped Agouti   *Dasyprocta aguti*
Aguti, or *acuti*, is a South American Spanish name; this Agouti lives in Brazil.

Acouchi   *Myoprocta acouchy*
mus (Gr), genitive *muos*, a mouse   *proktos* (Gr) hinder parts, the rump (from *Dasyprocta*, 'hairy buttocks', see above); the acouchi does not have the coloured rump hair of the agouti, but the hair is thick, and erected when fighting: acouchy is a native name. Inhabiting the north-western part of South America.

Family   CHINCHILLIDAE   6 species
Chinchilla is really a Spanish name for this animal, probably derived from the Quechua Indian.

Chincilla   *Chinchilla laniger*
lana (L) wool   *gero* (L) I carry; the coat is possibly the best and most highly valued of any animal; it is very soft and silky. Inhabiting the mountains in northern Chile.

Mountain Viscacha   *Lagidium peruanum*
lagos (Gr) a hare   *-idium* (New L) derived from *-idion* (Gr) a diminutive suffix   *peruanum*, of Peru; it lives in the mountains of Peru. *viscacha* (New L) from American-Spanish *vizcacha*.

Plains Viscacha   *Lagostomus maximus*
lagos (Gr) a hare   *stoma* (Gr) the mouth; 'the hare-mouthed one' *maximus* (L) the largest; a much larger animal than the chincilla. Inhabiting the southern part of South America.

Family   CAPROMYIDAE   11 species
kapros (Gr) a wild boar   *mus* (Gr) a mouse; Professor A. G. Desmarest (1784-1838) a French Professor of Zoology said the name was a result of the animal's resemblance to a wild boar in general appearance, character of hair, colour and manner of running.

Hutiacouga   *Capromys pilorides*
Known locally as the Pilori Rat, the name is probably of Arawak origin, an Indian people of South America formerly living in Cuba,

where this hutia lives   *-ides* (L) suffix indicating a relationship   *hutia* is a Spanish name for a rodent   *couga*, probably from *cougar*, a Guarani name for a type of cat. Inhabiting Cuba.

Hutiacarabali   *C. prehensilis*
prehenso (L) I take hold of   *-ilis* (L) suffix denoting capability; 'able to hold'; it has a prehensile tail used in tree climbing   *carabali*, obscure. Inhabiting Cuba.

Jamaican or Short-tailed Hutia.   *Geocapromys brownii*
gē (Gr) the earth, ground; a reference to its ground-dwelling habits, unlike other species of *Capromys* which live in trees; named after Patrick Browne (1720-1790) from whose *Civil and Natural History of Jamaica* J. B. Fischer (1730-1793) a German zoologist, took the account of this Hutia or Indian Coney. This animal is rather like a very large rat with no tail, for this is only a stump and hidden under the fur. It lives in the Blue Mountains area of Jamaica.

Dominican House Rat   *Plagiodontia aedium*
plagios (Gr) slanting, oblique   *odous* (Gr), genitive *odontos*, a tooth; a reference to the diagonal grooves in the upper molars   *aedes* (L) a room, a house   *-ium* (L) a suffix sometimes used to denote a place, as in 'aquarium'. Inhabiting Haiti, West Indies.

Coypu   *Myocastor coypus*
mus (Gr) a mouse   *kastōr* (Gr) a beaver; it is largely aquatic and lives on river banks, but it is not a beaver   *coypu* was originally a South American native name. Valued for its fur, known as nutria, it was brought to England and reared in captivity. Many animals escaped and became a threat to agriculture in East Anglia. However it is gradually being exterminated. Originally inhabiting southern South America.

Family   OCTODONTIDAE   8 species
octo (L) eight   *odous* (Gr), genitive *odontos*, a tooth; the grinding surfaces of the lower molars are shaped like a figure eight.

South American Bush Rat or Degu   *Octodon degus*
Degu is a South American native name for this rodent which looks like a large rat; it lives in the mountains of Peru and Chile.

Cururo   *Spalacopus cyanus*
spalax (Gr), genitive *spalakos*, a mole   *pous* (Gr) a foot; it is a burrow-

ing animal with feet adapted for this purpose   *kuanos* (Gr) dark-coloured, bluish   *cururo* is Spanish derived from the Araucan name *curi*; the Araucanians are an Indian people living in southern Chile and adjacent regions of Argentina. This animal looks rather like the Tucotuco (below); it inhabits South America.

Family   CTENOMYIDAE   27 species
kteis (Gr), genitive *ktenos*, a rake or comb   *mus* (Gr) a mouse; referring to the large fringes on the long claws of the hind feet, probably used to remove dirt from the fur and general grooming.

Tucotuco   *Ctenomys brasiliensis*
Tuco-tuco (Sp) an imitation of its cry   *-ensis* (L) suffix meaning belonging to; in Spanish Brazil is spelt 'Brasil'. A small greyish-brown burrowing animal up to 30 cm (12 in) long and inhabiting the southern part of South America.

Peruvian Tucotuco   *C. peruanus*
-anus (L) suffix meaning belonging to.

Banded Tucotuco   *C. torquatus*
torquatus (L) wearing a collar. Inhabiting southern Peru and ranging south to Tierra del Fuego.

Family   ABROCOMIDAE   2 species
abros (Gr) soft, luxurious   *komē* (Gr) hair; they have dense underfur which is extremely soft and silky.

Rat Chinchilla   *Abrocoma bennetti*
Named after E. T. Bennett (1797-1836) who was Secretary of the Zoological Society of London in 1831. It is not a rat, but something between the degu (page 566) and the chinchilla; it lives in Bolivia.

Family   ECHIMYIDAE   43 species, possibly more
echi (New L) derived from *ekhinos* (Gr) a hedgehog   *mus* (Gr) a mouse; they have sharp bristly fur.

Trinidad Spiny Rat   *Proechimys trinitatis*
pro (Gr) before, and *echimys* (see below)   *pro-* is used here to denote an allied form   *trinitatis*, 'of Trinidad'. Inhabiting South America, but now very rare.

Porcupine Rat   *Euryzygomatomys spinosus*
eurus (Gr) wide   *zugon* (Gr) a yoke, a joining together; the zygomatic
bone is situated in the upper part of the face, and forms the prominence
of the cheek   *mus* (Gr) a mouse; this could be interpreted as 'wide-
cheeked mouse'; it has a broad zygoma   *spina* (L) a thorn, so *spinosus*,
full of thorns. Inhabiting Brazil and Paraguay.

Spiny Rat   *Cercomys cunicularis*
kerkos (Gr) the tail   *mus* (Gr) a mouse; it has a rat-like tail, though
some species have no tail   *cuniculus* (L) a rabbit   *-aris* (L) suffix
meaning pertaining to. Widespread in South America.

Arboreal White-faced Spiny Rat   *Echimys chrysurus*
echi (New L) derived from *ekhinos* (Gr) a hedgehog   *mus* (Gr) a mouse;
it has sharp bristly fur   *khrusos* (Gr) gold   *oura* (Gr) the tail. A tree-
climbing rat with a white face and a pale yellow end to the tail.
Inhabiting north-eastern South America.

Family   THRYONOMYIDAE   2 species
thruon (Gr) a reed   *mus* (Gr) a mouse; they often inhabit reed beds.

Cane Rat or Cutting Grass   *Thryonomys swinderianus*
This cane rat was obtained by the original describer Temminck from
Professor Van Swinderen of Groningen, Netherlands. It is not a rat
or a mouse, but more like a coypu. It is not noticeably fond of cane,
but generally a vegetable grazer, hence 'cutting grass'. Inhabiting
Africa south of the Sahara.

Family   PETROMYIDAE   1 species
petra (Gr) a rock   *mus* (Gr) a mouse; it inhabits rocky areas and sleeps
in holes in rocks.

Rock Rat   *Petromus typicus*
tupos (Gr) a blow, the mark of a blow, an impression, like a wax seal,
hence *tupikos* (Gr) a type. A rare rat-like animal found only in the
southern part of Africa.

Family   BATHYERGIDAE   16 species, possibly more
bathus (Gr) deep   *ergō* (Gr) I work; they burrow and live almost
entirely underground.

Common Mole Rat *Cryptomys hottentotus hottentotus*
kruptos (Gr) secret, hidden *mus* (Gr) a mouse; seldom seen as they
spend most of their life underground; the Hottentots are a native
people of Namibia (South West Africa). The name is derived from
the Dutch for 'stutterer', on account of the peculiar Hottentot
language. This mole-like animal inhabits the western area of Cape
Province.

Damaraland Mole Rat or Blesmol *C. h. damarensis*
-ensis (L) suffix meaning belonging to; it inhabits Damaraland,
Namibia (South West Africa) and ranges eastwards to Rhodesia. The
name 'blesmol' comes from the Dutch *bles*, a blaze, and mole; there
is often a white patch on the head.

Lugard's Mole Rat *C. h. lugardi*
Named after F. J. D. Lugard, 1st Baron (1858-1945), a British soldier
and colonial administrator in Africa. Lugard and Speke (below) have
also had Nile steamers named after them, *SS Lugard* and *SS Speke*.
This mole rat inhabits eastern areas of Africa south of the Sahara.

Cape Mole Rat *Bathyergus suillus*
Bathyergus (see above under Family) *sus* (L), genitive *suis*, a pig
-illus (L) diminutive suffix, 'a small pig'. Inhabiting South Africa.

Naked Mole Rat or Sand Puppy *Heterocephalus glaber*
heteros (Gr) different *kephalē* (Gr) the head; probably a reference to
the bald head *glaber* (L) bald; the head and body are almost completely hairless. It lives in the hot Somaliland deserts in eastern Africa.

Family CTENODACTYLIDAE 4 species
kteis (Gr), genitive *ktenos*, a rake, a comb *daktulos* (Gr) a finger, a toe;
the inner two toes have horny combs on them used for grooming the
fur.

Gundi *Ctenodactylus gundi*
Gundi is Arabic and probably derived from Berber. Inhabiting the
northern part of Africa on the northern borders of the Sahara.

Speke's Pectinator *Pectinator spekei*
pecto (L) I comb, hence *pectinator*, one who combs (see above) Captain J. H. Speke (1827-1864) was the well-known explorer who
discovered Lake Victoria and the source of the Nile in 1857. This
gundi lives on the borders of the Somali Desert in eastern Africa.

52 Whales, Dolphins and Porpoises
CETACEA

This group consists of the whales, the dolphins and the porpoises. Strictly speaking, they are all whales, the dolphins and the porpoises being small whales. They are mammals that have become adapted to a life in the sea or rivers and have some remarkable features. Like all mammals they are warm-blooded, breath air, and suckle their young with the mother's milk.

The big examples, like the Blue Whale, can weigh well over 100 tonnes; the mouth is enormous and the tongue alone may weigh as much as 2 tonnes! Certainly whales are the largest living creatures, while at the other end of the scale the dolphins and porpoises may be only about 2 m (6 ft) long and weigh less than 45 kg (100 lb).

For purposes of classification the Order Cetacea is divided into two Suborders: Mysticeti, the 'moustached whales' and Odontoceti, the 'toothed whales'.

Subclass EUTHERIA
(see pages 460 and 477)

Order CETACEA
cetus (L) a large sea creature, the whale or dolphin.

Suborder MYSTICETI
mustax (Gr), genitive *mustakos*, a moustache   *kētos* (Gr) (= *cetus* (L)) a sea monster, a whale; this refers to the sheets of whalebone, or baleen plates, that hang down from the upper jaw, used for 'netting' the small marine creatures which are their food.

Family BALAENIDAE   3 species
balaena (L) a whale.

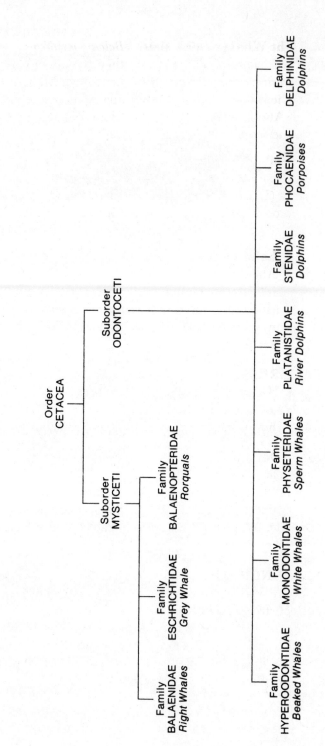

Greenland Right Whale or Bowhead *Balaena mysticetus*
Supposedly called 'right whales' because they were originally the
right whales to hunt, as they did not sink when killed. Modern tech-
niques enable other kinds of whales to be caught, though whaling is
now controlled. Also known as the Arctic Right Whale; 'bowhead'
is suggested by the shape of the enormous head.

Black Right Whale *Eubalaena glacialis*
eu- (Gr) prefix meaning well, nicely; in this case used to indicate
'typical' *balaena* (L) a whale *glacialis* (L) icy. The Right Whales
are not confined to Arctic seas and are found in other areas including
southern oceans.

Pygmy Right Whales *Caperea marginata*
(*Caperea* formerly *Neobalaena*)
capero (L) I am wrinkled; referring to the wrinkled appearance of the
tympanic bone in the skull *margo* (L), genitive *marginis*, the edge;
it has dark margins on the baleen plates. Inhabiting southern seas
around New Zealand and Australia.

Family ESCHRICHTIDAE 1 species
(formerly RHACHIANECTIDAE)
Named in honour of Daniel Fredrik Eschricht (1798–1863), a Dutch
zoologist, and author of several important papers on cetaceans.

Grey Whale or Devilfish *Eschrichtius gibbosus*
(formerly *Rhachianectes*)
gibbus (L) a hump, hence *gibbosus*, humped; in place of the dorsal fin
there is a row of nine or ten small humps. Sometimes known as the
Californian Whale, it lives in the northern Pacific and migrates to
southern seas in the winter. It is the only known species.

Family BALAENOPTERIDAE 6 species
balaena (L) a whale *pteron* (Gr) feathers or wings, or can mean a fin;
an allusion to the strong dorsal fin.

Lesser Rorqual *Balaenoptera acutorostrata*
acutus (L) sharp, pointed *rostrum* (L) the beak, snout; it has a more
pointed snout than other whales. The name rorqual is derived from
the Norwegian *royrkval*, meaning 'red whale'. However, the rorquals
are usually bluish-black or grey. Widespread throughout the world,
but mostly in Arctic or Antarctic seas.

Rudolphi's or Sei Whale   *B. borealis*
boreus (L) northern   *-alis* (L) suffix meaning pertaining to. Professor
K. A. Rudolphi (1771–1832) was a professor of anatomy in Berlin
sei is derived from *seje*, the name of a fish living off the coast of Norway
which forms part of this whale's diet. World-wide, moving to warmer
waters in winter for breeding.

Common Rorqual or Fin Whale   *B. physalus*
phusalis (Gr) a pipe, a wind instrument; this refers to the blow-hole
through which the whale breathes, in some cases it can make whistling
sounds   *rorqual*, see above: it has a high dorsal fin. Widespread
throughout the world.

Blue Whale or Sibbald's Whale   *B. musculus*
musculus (L) a little mouse; a word used by Pliny for a sea animal and
Linnaeus thought it was the blue whale, but more probably Pliny
meant it for a pilot-fish that was supposed to guide whales. Although
known as the Blue Whale, it is only a bluish grey, and can even be
grey to light grey and white underneath. It is the largest living creature
and can weigh over 100 tonnes. Sir Robert Sibbald (1641–1722) was
a Professor of Medicine in Edinburgh, and a plant genus *Sibbaldia* is
named after him. This whale is found throughout the world but tends
to inhabit the Arctic and Antarctic seas, migrating to warmer seas in
the winter.

Humpback Whale   *Megaptera novaeangliae*
megas (Gr) big   *pteron* (Gr) feathers, wings or a fin; it has unusually
long flippers which can be as much as one-third of the body length
novaeangliae of New England; the whale from which this species
received its name was found on the coast of New England. This is that
part of the USA comprising Maine, New Hampshire, Massachusetts
and adjacent states, and given the name New England by Captain
John Smith (1580–1631), an English explorer, in the year 1614.
World-wide distribution migrating to warmer seas in the winter.

Suborder   ODONTOCETI
odous (Gr), genitive *odontos*, a tooth   *ketos* (Gr) any sea monster, a
whale.

Family   HYPEROODONTIDAE (or ZIPHIIDAE)   18 species
hyperōe (Ionic Gr) the upper part of the mouth, the palate   *odōn*
(Ionic Gr), genitive *odontos*, a tooth; a misleading name, as they

usually have only two teeth, and these are in the lower jaw. However, on the palate there are small protuberances which were mistaken for teeth by the French zoologist Count de Lacépède (1756-1825) when he examined two of these whales stranded on the beach near Le Havre, France, in 1788. This whale inhabits the North Atlantic Ocean.

Bottle-nosed Whale *Hyperoodon ampullatus*
ampulla (L) a flask, a bottle *-atus* (L) suffix meaning provided with; a reference to the long snout in front of the bulging forehead. Widespread in the Northern Hemisphere, particularly the Arctic Ocean, and fairly common round British coasts.

Sowerby's Whale *Mesoplodon bidens*
mesos (Gr) middle, in the middle *oplon* (Gr) any tool or weapon *ōdon* (=*odous*) (Gr) a tooth *bidens* (L) having two teeth; there are usually only two teeth situated about the middle of the lower jaw. James Sowerby (1757-1822) was an English naturalist and artist; he was the first to describe this whale, in 1804. Sometimes known as the Beaked Whale or the Cowfish, it inhabits the North Atlantic and European waters.

True's Beaked Whale *M. mirus*
mirus (L) wonderful, extraordinary; this whale is remarkable for having the two teeth at the extreme tip of the lower jaw instead of in the middle, as *M. bidens* (above). F. W. True, an American zoologist, was an authority on whales; he was studying whales and writing about his findings during the years 1885 to 1913 and he described this whale in 1913. Less well known than most whales, it lives in the North Atlantic.

Stejneger's Beaked Whale *M. stejnegeri*
Dr L. Stejneger (1851-1943) was an American zoologist and author whose particular subject was taxonomy. This whale inhabits the northern Pacific Ocean.

Cuvier's Beaked Whale *Ziphius cavirostris*
ziphius (New L) derived from *xiphos* (Gr) a sword; this refers to the beak, or snout *cavus* (L) hollow *rostrum* (L) in animals, the snout; there is a hollow formation at the base of the snout. Baron Georges Cuvier (1769-1832) was the famous French comparative anatomist and Professor of Natural History. It has a world-wide distribution and is known round British coasts.

Family MONODONTIDAE 2 species
monos (Gr) single, alone *odous* (Gr), genitive *odontos*, a tooth.

White Whale or Beluga *Delphinapterus leucas*
delphis (also *delphin*) (Gr) the dolphin *a-* (Gr) prefix meaning not, or there is not *pteron* (Gr) a feather or wing; can mean a fin; this whale has no dorsal fin *leukos* (Gr) white; the adult whale is almost pure white *beluga* is said to derive from *belyi* (Russ) white. Inhabiting mostly Arctic seas.

Narwhal *Monodon monoceros*
Monodon, see above *keras* (Gr) the horn of an animal. They have a few rudimentary teeth when young but these never develop, except in the males which retain one canine; this grows to become an enormous horn, projecting forward to a length of over 2 m (7 to 8 ft) and twisted in an anticlockwise spiral. Narwhal comes from Old Norse, meaning 'corpse-whale'; their pallid colour is said to resemble that of a floating corpse. Inhabiting cold northern seas.

Family PHYSETERIDAE 3 species
phusētēr (Gr) a blow-pipe or tube; this refers to the blow-hole on the top of the head.

Sperm Whale or Cachalot *Physeter catodon* (or *macrocephalus*)
kata (Gr) down, below *odous* (Gr), genitive *odontos*, a tooth; there are up to 30 teeth on each side of the lower jaw, but few in the upper jaw and these are vestigial and non-functional *cachalot* (Fr) the sperm whale. Inhabiting all tropical and sub-tropical seas and occasionally colder areas.

Pygmy Sperm Whale *Kogia breviceps*
Kogia, a barbarous unmeaning name but it might be a tribute to a Turkish naturalist named Cogia Effendi who observed whales in the Mediterranean, in the early part of the nineteenth century *brevis* (L) short *ceps* (New L) derived from *caput* (L) the head; the head is short, and smaller than that of *Physeter catodon*, whose head is enormous. A rare species but probably world-wide distribution.

Dwarf Sperm Whale *K. simus*
simus (L) snub-nosed; it has a distinctly upturned snout. This rare species has been found in the Indian Ocean.

Family PLATANISTIDAE 4 species
platus (Gr) flat, hence *platanos* (Gr) the plane-tree, so called from its broad flat leaf *platanistēs* (Gr) or *platanista* (L) is a rare word, apparently only used by Pliny to mean some kind of fish in the Ganges. These dolphins have somewhat flattened beaks, used for digging in the mud of rivers.

Gangetic Dolphin or Susu *Platanista gangetica*
(*Platanista* or *Susu*)
-icus (L) suffix meaning belonging to *gangetica*, of the Ganges; it is also found in the Indus and other rivers of southern Asia. Susu is a Bengalese name for this dolphin.

La Plata Dolphin *Pontoporia blainvillei* (*Pontoporia* or *Stenodelphis*)
pontos (Gr) the sea *poreuō* (Gr) I carry (on land) or I ferry (on water) H. D. de Blainville (1777–1850) was a French Professor of zoology. This dolphin inhabits the Rio de la Plata in South America.

Amazonian Dolphin or Bouto *Inia geoffrensis*
Inia is a Bolivian name for a dolphin *geoffrensis* is after Etienne Geoffroy Saint-Hilaire (1772–1844), a French zoologist; the suffix *-ensis* (L) meaning belonging to, is usually a reference to a locality rather than a person. Bouto is a Portuguese name for this dolphin; it is found only in the Amazon and other rivers in that part of South America.

Chinese River or White Flag Dolphin *Lipotes vexillifer*
leipō (Gr) I am left behind, as in a race *-tes* (Gr) suffix meaning pertaining to; referring to this dolphin as an isolated relict species; it was not named until 1918 *vexillum* (L) a standard, a flag *fero* (L) I bear, I carry; this refers to the dorsal fin often seen above the surface of the water. Sometimes known as the Chinese Lake Dolphin, it is found in a lake several hundred kilometers up the Yangtze River in China, and also other rivers in South-East Asia.

Family STENIDAE 8 species
Named in honour of Dr Nikolaus Steno (1638–1687), a celebrated Danish anatomist, geologist and author.

Guiana River Dolphin *Sotalia guianensis*
Sotalia, a coined name of unknown origin *-ensis* (L) suffix meaning

belonging to; 'of Guiana'. Inhabiting the coast of north-eastern South America and some rivers in that area.

Plumbeous Dolphin   *S. plumbea*
plumbus (L) lead, so *plumbeus* (L) leaden, lead-coloured. Inhabiting the Indian Ocean.

Rough-toothed Dolphin   *Steno bredanensis*
Steno, see above under Family   *-ensis* (L) suffix meaning belonging to; named after a Monsieur Van Breda of Ghent, who sent a sketch of the skull of this dolphin to Georges Cuvier who attributed it to another species he had already named. However, it was a new species, and the present name was given by René Primevère Lesson (1794-1849) a French zoologist and surgeon, and best known as an ornithologist. He was the zoologist on the voyage of La Coquille during the years 1822 to 1825. The surfaces of the teeth of this dolphin are furrowed with ridges; it has a world-wide distribution.

Family   PHOCAENIDAE   6 species, possibly 7
phōkaina (Gr) a porpoise.

Common Porpoise   *Phocaena phocoena*
Porpoises, though similar to dolphins, have blunter snouts, lacking the beaks of dolphins. This porpoise is widespread in the waters of the Northern Hemisphere.

Burmeister's Porpoise   *P. spinipinnis*
spina (L) a thorn, can mean the spines of an animal, for example a porcupine   *pinna* (L) a feather, a wing, can mean a fin; 'spiky-finned'. Dr K. H. K. Burmeister (1807-1892) was a zoologist and at one time Director of the Zoological Museum at Halle University in Germany. This porpoise inhabits both the west and east coasts of South America.

Black Finless Porpoise   *Neophocaena phocoenoides*
(*Neophocaena* formerly *Neomeris*)
neos (Gr) new   *phōkaina* (Gr) a porpoise   *-oides* (New L) suffix derived from eidos (Gr) shape, form; can mean a species, a sort; the name indicates a new subdivision of porpoises. There is no dorsal fin; it inhabits the Pacific and Indian Oceans.

Dall's or White-flanked Porpoise   *Phocaenoides dalli*
Phocaenoides, see above   William H. Dall (1845-1927) was an Ameri-

can naturalist, and worked with the US Geographical Survey from 1884 to 1909. This porpoise inhabits the northern part of the Pacific Ocean.

Family   DELPHINIDAE   about 30 species
delphis, also *delphin* (Gr) the dolphin.

Blackfish or Pilot Whale   *Globicephala melaena*
globus (L) a round ball, a globe   *kephalē* (Gr) the head   *melas*, also *melaina* (Gr) black, dusky; the head is short and the forehead bulging, and the colour is black except for the white throat. It would be more correctly called a dolphin. They travel in large schools, with one apparently leading, hence the name 'pilot'. Inhabiting the Atlantic and Pacific Oceans and also European waters, including the British Isles.

Indian Pilot Whale   *G. macrorhyncha*
makros (Gr) long   *rhunkhos* (Gr) beak, snout. It is found in the Pacific, Atlantic and Indian Oceans.

Irrawaddy River Porpoise   *Orcaella brevirostris*
orca (L) a whale   *-ellus* (L) diminutive suffix; 'a small whale'; it should be remembered that although some of the animals in this order are known as dolphins or porpoises, because of certain differences in their structure, strictly speaking they are all whales   *brevis* (L) short *rostrum* (L) the beak or snout; the Irrawaddy is a river in Burma. Also found in the Indian and southern Pacific Oceans.

Killer Whale   *Orcinus orca*
orca (L) a whale, which gave rise to orc, or ork, meaning a mysterious sea monster of horrible aspect   *-inus* (L) suffix meaning like, pertaining to. It is known as the Killer Whale because it is an aggressive predator, killing and eating seals and porpoises as well as fishes. Sometimes known as the Grampus (see Risso's Dolphin, below) it has a world-wide distribution including the coast round the British Isles.

False Killer   *Pseudorca crassidens*
pseudēs (Gr) false   *orca* (L) a whale, the killer whale; 'false killer' because in some respects it resembles the killer whale   *crassus* (L) thick, heavy   *dens* (L) a tooth; it has large powerful teeth, but circular in cross-section instead of oval as in the killer. This whale was first described from a fossilised skull by Sir Richard Owen in 1846, which

was found in 'the great fen of Lincolnshire . . . near Stamford'; it resembled that of the Killer Whale. Present day knowledge of the False Killer has been obtained almost entirely from examination of the bodies of these whales that have become stranded ashore, sometimes in large numbers. This is because although they have a worldwide distribution, they are seldom seen. Sir Richard Owen (1804–1892) was a distinguished anatomist and Director of the British Museum (Natural History) from 1856 to 1883.

Northern Right Whale Dolphin   *Lissodelphis borealis*
lissos (Gr) smooth   *delphis*, also *delphin* (Gr) the dolphin; a 'smooth-skinned' dolphin   *boreus* (L) northern   *-alis* (L) suffix meaning relating to   *borealis* (L) northern (a rare Latin word). It inhabits the northern Pacific Ocean.

Commerson's Dolphin   *Cephalorhynchus commersonii*
kephalē (Gr) the head   *rhunkhos* (Gr) the snout, beak; the whole head is curved, or beaked, not the snout only as in most dolphins. Dr P. Commerson (1727–1773) was a naturalist who worked with the French world navigator Vice-Admiral Baron de Bougainville (1729–1811) during the years 1766 to 1769. This dolphin inhabits the Pacific and Atlantic coasts of South America.

Common Dolphin   *Delphinus delphis*
delphis, also *delphin* (Gr) the dolphin   *-inus* (L) suffix meaning like. As the English name implies it is frequently seen at sea, often leaping clear of the water and playing round ships' bows. World-wide distribution, particularly warm and temperate seas, including British waters.

Pacific or Baird's Dolphin   *D. bairdi*
Professor S. F. Baird (1823–1887) was an American zoologist who did a survey of the Pacific Ocean during the years 1857 to 1859. This dolphin inhabits the southern Pacific in the Australia and New Zealand area.

Risso's Dolphin   *Grampus griseus*   (*Grampus* or *Grampidelphis*)
The origin of the word grampus is probably Spanish *grande pez*, 'great fish', and has come to mean 'one who puffs'   *griseus* (New L) grey, from *gris* (Sp) grey; the general colour is grey, though lighter on the head, with possibly a white belly. A. Risso (1777–1845) was an Italian naturalist. World-wide distribution.

Bornean Dolphin   *Lagenodelphis hosei*
lagēnos (Gr) a flagon   *delphis* (Gr) the dolphin, 'bottle-shaped dolphin'   Dr C. Hose (1863–1929) was a naturalist who lived in Sarawak from 1884 to 1907; little is known of this dolphin but one skeleton was found in Sarawak.

White-sided Dolphin   *Lagenorhynchus acutus*
lagēnos (Gr) a flagon   *rhunkhos* (Gr) snout, beak, 'flagon-nosed' *acutus* (L) pointed; it has a short distinct beak about 5 cm (2 in) long; there is a light broad band along the flanks. Inhabiting the northern part of the Atlantic.

White-beaked Dolphin   *L. albirostris*
albus (L) white   *rostrum* (L) the snout, beak. Inhabiting the northern part of the Atlantic including European waters and fairly common in the North Sea.

Bridled Dolphin   *Stenella frontalis*   (*Stenella* or *Prodelphinus*)
stenos (Gr) narrow   *-ellus* (L) diminutive suffix   *frons* (L), genitive *frontis*, the forehead   *-alis* (L) suffix meaning pertaining to; a slender dolphin with a striped pattern on the head. This taxon has been called 'A genus in chaos' as the systematics are still very doubtful; about ten species have been described. This species inhabits the Atlantic and Indian Oceans.

Long-beaked Dolphin   *S. longirostris*
longus (L) long   *rostrum* (L) the beak, snout. Inhabiting the Pacific Ocean.

Bottle-nosed Dolphin   *Tursiops truncatus*   (*Tursiops* or *Tursio*)
tursio (L) a kind of fish resembling the dolphin; a name used by Pliny *opsis* (Gr) aspect, appearance   *ops* (Gr) eye, face; 'looking like a dolphin'   *trunco* (L) I shorten, I cut off, hence *truncatus* (New L) shortened; they have shorter beaks than other dolphins. This is the friendly dolphin, easily tamed and taught to perform in 'dolphinaria'; it is widespread in the Atlantic, and found in the Bay of Biscay, the Mediterranean, and round the coasts of the British Isles.

53 Dogs, Weasels, Lions and their kin

CARNIVORA

In this order we shall find many familiar animals such as dogs, cats, weasels, badgers, otters, bears; and of course the 'big cats' such as the lion and tiger.

Although the name Carnivora means flesh-eating, some of these animals have a partly vegetarian diet and some are almost complete vegetarians. As will be seen, they differ enormously in outward form and size—for example, from a small weasel to an animal the size of a bear or a lion. However, a study of their anatomy shows them to be all related.

For purposes of classification they are divided into two suborders: Canoidea, the 'dog-like', and Feloidea, the 'cat-like'.

Subclass EUTHERIA
Order CARNIVORA
caro (L), genitive *carnis*, flesh   *voro* (L) I devour.

Suborder CANOIDEA
canis (L) a dog   *-oides* (New L) from *eidos* (Gr) apparent shape, form; can mean a kind, sort. (Canoidea or Arctoidea.)

Family CANIDAE   37 or more species

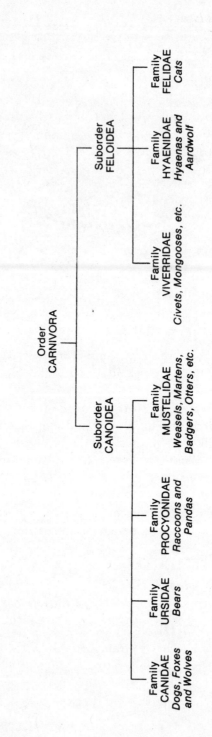

Grey Wolf *Canis lupus lupus*
canis (L) a dog *lupus* (L) a wolf. Inhabiting the wilder parts of
northern Europe, North America and Canada, and Asia.

Red Wolf *C. l. niger*
niger (L) black; a misleading name; it was given to a melanistic
individual, which by chance was black, or nearly black. (This phenomenon occasionally occurs, especially in mammals and birds, and is
called melanism, from the Greek *melas*, black.) Normally, the coat is
tawny. Inhabiting southern parts of North America, it is very rare and
practically extinct.

Coyote *C. latrans*
latro (L) I bark, hence *latrans* (L) a barker; coyote is derived from
coyotl, a Mexican name. Inhabiting North America.

Golden or Common Jackal *C. aureus*
aureus (L) golden. Inhabiting south-eastern Europe, Central and
northern Africa, the Middle East and a large part of southern Asia,
including Thailand but not Malaysia.

Black-backed Jackal *C. mesomelas*
mesos (Gr) middle *melas* (Gr) black. Living in Sudan and south to
the Cape, on the eastern side of Africa.

Side-striped Jackal *C. adustus*
aduro (L) I set fire to, so *adustus* (L) burnt, sunburnt; it is a brownish
colour with dark stripes on the sides. Inhabiting large areas of Africa,
including Congo, Kenya and south to the Cape.

Simenian Jackal *C. simensis*
-ensis (L) suffix meaning belonging to; it inhabits the Simien region
and mountains in the northern part of Ethiopia and is one of the
world's rarest animals.

Rüppell's Fox *C. rueppelli*
Dr W. P. E. S. Rüppell (1794–1884) was a German zoologist who
travelled in North Africa during the years 1822 to 1827. This fox
inhabits the Near and Middle East.

Dog *C. familiaris*
familiaris (L) domestic, belonging to a household, a servant; can mean
a familiar friend. World-wide.

European Red Fox   *Vulpes vulpes vulpes*
vulpes (L) a fox. Inhabiting North America, Europe, the northern part of Africa, Asia and India. (See Tautonyms, page 13.)

North American Red Fox   *V. v. fulva*
fulvus (L) tawny, yellowish-brown. A subspecies of *V. vulpes*.

Desert Fox   *V. leucopus*
leukos (Gr) white   *pous* (Gr) the foot. Inhabiting Iraq, Iran and the western part of India.

Arctic Fox   *Alopex lagopus*
alōpēx (Gr) a fox   *lagōs* (Gr) a hare   *pous* (Gr) the foot; a hare's foot is hairy; like the polar bear, this fox has hair on the soles of its feet, probably to prevent slipping on the ice and as a protection against the cold. The range is widespread in the Arctic Circle.

Fennec Fox   *Fennecus zerda*
Fenek is an Arabic word for a small fox   *zerda* (Arab) derived from *zardawa*, a fennec. Inhabiting northern Africa.

Grey Fox   *Urocyon cinereoargenteus*
oura (Gr) the tail   *kuōn* (Gr) a dog; a reference to the tail because it has a concealed mane of stiff hairs, without any soft fur intermixed *cinereus* (L) ash coloured   *argenteus* (L) silvery. Inhabiting southern Canada, the USA, Central America and the north-west corner of South America.

Island Grey or Beach Fox   *U. littoralis*
litoralis (L) of the shore; although in English it is 'littoral', in Latin there should only be one 't'. However, as the name has long been internationally accepted the incorrect spelling must remain (see page 15). This fox lives on some islands off the coast of California.

Raccoon Dog   *Nyctereutes procyonoides*
nux (Gr), genitive *nuktos*, night   *ereuna* (Gr) an enquiry, a search, hence *ereunētēs* (Gr) a searcher; 'a night-hunter'   *pro* (Gr) before, in front of   *kuōn* (Gr) a dog   *-oides* (New L) from *eidos* (Gr) apparent shape, form; indicating its remarkable likeness to the raccoon *Procyon* (page 589). They are night-hunters and plant food also forms part of their diet. They live in eastern Asia, including Vietnam and Japan, and have been introduced into Russia and Poland. They must not be confused with the common raccoon *Procyon lotor*, which lives in America.

Azara's or Western Dog *Dusicyon gymnocercus*
dusis (Gr) a setting of the sun; *dusis helion*, 'setting sun'; can mean the west, western *kuōn* (Gr) a dog *gumnos* (Gr) naked *kerkos* (Gr) the tail. Named after D. Felix de Azara (1746–1811), a zoologist who made expeditions to Paraguay during the years 1781 to 1801. This dog inhabits the forests of Paraguay.

Small-eared Fox or Zorro *Atelocynus microtis*
atelēs (Gr) imperfect *kuōn* (Gr) a dog; considered imperfect on account of the small ears *mikros* (Gr) small *ous* (Gr), genitive *ōtos*, the ear *Zorra* is Spanish for a fox. A rare animal, inhabiting South America.

Crab-eating Fox *Cerdocyon thous*
kerdō (Gr) the wily one, a fox *kuōn* (Gr) a dog *thōs* (Gr), genitive *thōos*, the jackal. Small mammals and birds are the main foods, and although it may occasionally eat crabs there is no reliable confirmation of this; the name dates from very early days of animal discovery. Inhabiting South America.

Maned Wolf *Chrysocyon brachyurus*
khrusos (Gr) golden *kuōn* (Gr) a dog; it has a handsome coat of long reddish-brown hair *brakhus* (Gr) short *oura* (Gr) the tail; it has a mane on the neck and shoulders which is erected when the animal is excited. Inhabiting South America.

Bush Dog *Speothos venaticus*
speos (Gr) a cave *thōs* (Gr) the jackal; they dig and live in holes in the ground *venaticus* (L) for the chase, belonging to hunting. Inhabiting the Amazonian forest area.

Dhole *Cuon alpinus*
kuōn (Gr) a dog *alpinus* (L) alpine; 'alpine' probably refers to the slopes at the foot of mountains because they live on the plains as well as in the hills. The anatomy of the dhole, for example 40 teeth instead of the usual 42, and the convex profile of the skull, separate it somewhat from typical Canidae. Dhole is an East Indian name. It inhabits India but not Sri Lanka and is widespread in south-eastern Asia, including parts of Russia, China and Korea.

Cape Hunting Dog *Lycaon pictus*
lukos (Gr) a wolf, so *lukaon* (Gr) a wolf-like animal *pingo* (L) I paint, hence *pictus* (L) painted; the coat looks as though it had been care-

lessly daubed with splodges of white and orange on a dark background. Inhabiting the southern half of Africa.

Bat-eared Fox   *Otocyon megalotis*
ous (Gr), genitive *ōtos*, the ear   *kuōn* (Gr) a dog; an allusion to the large ears   *megas* (Gr) big, wide. Inhabiting Africa.

Family   URSIDAE   8 species
ursus (L) a bear.

Spectacled Bear   *Tremarctos ornatus*
trema (Gr) a hole   *arktos* (Gr) a bear; a reference to an unusual hole in the humerus bone   *ornatus* (L) dress, equipment; it has yellow rings round the eyes giving the appearance of spectacles. Inhabiting the Andes Mountains of South America.

Moon Bear or Himalayan Black Bear   *Selenarctos thibetanus*
selēnē (Gr) the moon   *arktos* (Gr) a bear; it has a white mark on the chest like a crescent moon   *thibetanus*, of Tibet.

Brown Bear   *Ursus arctos*
ursus (L) a bear   *arktos* (Gr) a bear. It varies in colour from very pale brown to black. Inhabiting North America from Alaska south to northern Mexico, central Europe where there are forests and mountains and most of Eurasia.

American Black Bear   *U. americanus*
Occasionally brown, grey or even white, in colour, their range is widespread in North America.

Polar Bear   *Thalarctos maritimus*
thalassa (Gr) the sea   *arktos* (Gr) a bear   *maritimus* (L) of the sea. Living among the Arctic pack-ice.

Malayan or Sun Bear   *Helarctos malayanus*
helē (= *eilē*) (Gr) the heat of the sun   *arktos* (Gr) a bear; although according to Palmer 'Probably referring to its tropical habitat', it is more likely an allusion to a yellow mark on the chest said to represent the sun   *-anus* (L) suffix meaning belonging to; it is not confined to Malaya and is widespread in that area.

Sloth Bear   *Melursus ursinus*
mel (L) honey   *ursus* (L) a bear   *-inus* (L) suffix meaning like, belong-

ing to; they are fond of honey and tear open and eat ants' and bees' nests. Known as 'sloth bears' because they usually move with a slow, shambling gait. Inhabiting Sri Lanka and India.

Family PROCYONIDAE 18 species
pro- (Gr) prefix meaning before, in front of *kuōn* (Gr) a dog; this is the raccoon family; it is considered that the ancestors of the dogs were probably raccoons, hence 'before dogs'.

Cacomistle or Cacomixtle *Bassariscus astutus*
bassara (Gr) a fox; of Thracian origin with the same meaning as *alopex* (Gr) a fox *-iscus* (L) diminutive suffix *astutus* (L) cunning. Cacomixtle is a Mexican name for this animal; sometimes known as the Ringtailed Cat, it inhabits Mexico and ranges north to south-western USA as far as southern Oregon and Alabama.

Guayonoche or American Cacomistle *B. sumichrasti*
Professor F. E. Sumichrast (1828–1882) was a Mexican zoologist. Guayonoche is a Spanish name for this small fox-like animal, which inhabits the forest areas of southern Mexico and Central America.

North American Raccoon *Procyon lotor*
Procyon, see Family above *lotor* (New L) a washer, from *lavo* (L) I wash; when in captivity, this raccoon has been seen to wash its food before eating, but there are no reliable reports of this behaviour concerning raccoons in the wild. The reason for the behaviour seen in captive animals is not really established but is probably not for cleansing purposes. This raccoon inhabits North and Central America.

Crab-eating Raccoon *P. cancrivorus*
cancer (L) a crab *voro* (L) I devour; raccoons eat a wide variety of small animals and some plants, and the animal part of the diet includes shellfish. Inhabiting river valleys and coastal areas from Panama to the south of the Amazon.

Red or Ringtailed Coatimundi *Nasua nasua*
nasus (L) the nose; the nose is exceptionally long, flexible and almost like a trunk. Coatimundi is a Tupi name for this raccoon-like animal which inhabits the forests of Central and South America.

White-nosed Coati *N. narica*
naris (L) the nostril *-icus* (L) suffix meaning belonging to; it has a

long nose like that of *N. nasua*, above. A small raccoon-like animal living in southern Mexico and ranging south to Panama.

Nelson's Coati   *N. nelsoni*
Named in honour of Dr Edward Nelson (1855–1934), an American zoologist and explorer. This coati inhabits an island off the coast of Quintana Roo, Mexico.

Kinkajou   *Potos flavus*
Potos is derived from *Potto*, the native name for this animal   *flavus* (L) yellow; it has gingery-yellow fur   Kinkajour is a South American Indian name. It inhabits the forests of Central and South America. It is interesting to note that another animal, living in Africa, and known locally by the name Potto, is rather similar in appearance but completely unrelated; it is a primate (page 509).

Bushy-tailed Olingo   *Bassaricyon gabbii*
bassara (Gr) a fox; of Thracian origin with the same meaning as *alopex* (Gr) a fox   *kuōn* (Gr) a dog   Professor W. M. Gabb (1839–1878) was an American scientist who was in Costa Rica in 1876. This small raccoon-like animal, similar to the kinkajou, lives in Central and South America.

Beddard's Olingo   *B. beddardi*
Dr F. E. Beddard, FRS (1858–1925) was an English zoologist and at one time President of the Zoological Society of London. This olingo, known locally as the Cuataquil, inhabits South America; it may be the same animal as *Bassaricyon gabbii* (above).

Cat-Bear or Red Panda   *Ailurus fulgens*
aiolos (Gr) quick moving, wriggling   *oura* (Gr) the tail, hence *ailouros* (Gr) a cat; Liddell and Scott's Greek Lexicon says 'as expressive, not of colour, but of the *wavy motion of the tail* peculiar to the cat kind'; a 'tail-waver'   *fulgens* (L) shining, gleaming; it has a brightly coloured shiny coat   *panda* is originally an East Indian word. Sometimes known as the Lesser Panda, it inhabits Nepal, the Himalayas, northern Burma, Szechwan and Yunnan.

Giant Panda   *Ailuropoda melanoleuca*
ailouros (Gr) a cat (see Cat-Bear above)   *pous* (Gr), genitive *podos*, a foot; a reference to the likeness of the feet to those of *Ailurus*; *melas* (Gr) black   *leukos* (Gr) white. It is strange that this carnivore has a reputation for eating only bamboo shoots; in fact it will also eat small

animals such as rodents, birds and fishes. It inhabits the bamboo forests of Eastern Tibet and Szechwan in south-western China.

Family MUSTELIDAE 70 species
mustela (L) a weasel.

Stoat or Short-tailed Weasel *Mustela erminea erminea*
ermine (Old Fr) *hermine* (Fr) a stoat. Normally the coat is brown with white beneath; it may change to white in winter, when it is used for robes and known as ermine. Widespread in northern Europe and northern North America, including Canada and the Arctic.

Irish Stoat *M. e. hibernica*
Hibernia (L) Ireland -*icus* (L) suffix meaning belonging to.

Islay Stoat *M. e. ricinae*
ricinium (L) a small veil. Islay is a small island of the Inner Hebrides, Scotland, and the Romans gave it the name Ricina because it was always veiled in mist.

Weasel *M. nivalis nivalis*
nix (L), genitive *nivis*, snow, so *nivalis* (L) of snow, snowy; the weasel is a reddish-brown colour and white beneath, though a northern form turns quite white in winter. Inhabiting North America, Europe, North Africa and Asia.

Least Weasel *M. n. rixosa*
rixa (L) a fight, a quarrel -*osus* (L) suffix meaning full of, prone to; 'quarrelsome'. Inhabiting North America.

Black-footed Ferret *M. nigripes*
niger (L) black *pes* (L) a foot. A weasel very similar to the domesticated ferret of Britain and Europe; it inhabits North America.

American Mink *M. vison*
vison (Fr) the American mink. Vison is also used as an English name for this animal.

European Mink *M. lutreola*
lutra (L) the otter -*olus* (L) diminutive suffix, a small otter; it swims and dives well. Probably widespread in Russia but becoming scarce or even unknown in certain parts of Europe.

Asiatic or Russian Polecat *M. eversmanni*
Professor Dr E. Eversmann (1794–1860) was a German zoologist.

European Polecat   *M. putorius putorius*
putor (L), genitive *putoris*, a bad smell; polecats can make a foul-smelling discharge from glands under the tail.

Common Ferret   *M. p. furo*
fur (L) a thief, from which the English name 'ferret' is derived. This is the ferret that is used for hunting. It is probably a domesticated descendant of the Russian or Steppe polecat, possibly *M. eversmanni*. It is an albino, having an almost white coat and pink eyes, and is unknown except in captivity.

Stripe-bellied Weasel   *Grammogale africana*
gramme (Gr) a mark, a line   *galē* (Gr) a weasel or marten-cat; it has a chestnut-brown stripe along the belly   *africana* is misleading; it was first described as an African weasel by Professor A. G. Desmarest in 1818, but this was incorrect as it had not been collected in Africa. It was not until 1913 that Dr Angel Cabrera y Latorre (1879–1960), a Spanish zoologist, established that it was a Brazilian weasel, and even then the amendment was neglected by other mammalogists until 1937. However, by the rules of the International Commission on Zoological Nomenclature the specific name must remain *africana*; 'The first scientific name given to an animal after Jan 1st 1758 stands, even though it is not descriptively accurate.' This weasel inhabits the Amazon basin in Brasil.

Mottled or Marbled Polecat   *Vormela peregusna*
T. S. Palmer, in his standard work *Index Generum Mammalium* (1904) gives *Vormela* as Latin derived from the German, and quotes '*Animal cujus Agricola sub nomine* Vormelae *(Germanice Wormlein) mentionem fecit.*' (Pallas). This could be translated as 'An animal whom Agricola cited under the name of Vormela.' *peregusna* is a Latinised form of *pereguznya*, the Ukrainian name for the polecat, after Professor S. I. Ognev (1866–1951), a Professor at Moscow University. The Mottled Polecat inhabits south-eastern Europe and south-west and central Asia as far eastwards as the Gobi Desert.

Pine Marten   *Martes martes*
martes (L) a marten. Inhabiting Europe and central Asia.

Beech or Stone Marten   *M. foina*
foina (It dialect) a polecat. Inhabiting Europe, the USSR and the Himalayas.

Yellow-throated Marten   *M. flavigula*
flavus (L) yellow   *gula* (L) the throat. Inhabiting eastern Asia.

American Marten   *M. americana*
Inhabiting North America including Canada and Alaska.

Fisher or Pennant's Marten   *M. pennanti*
This refers to Dr T. Pennant, FRS (1726-1798), an English zoologist. This marten inhabits North America. The name 'fisher' could be misleading as it is not established that it catches live fish; it lives near water but is not aquatic. Exceedingly rare, and a protected species.

Tayra   *Eira barbara*
Eira is the Guarani name for the Tayra; the Guaranis are a people of Bolivia and Paraguay   *barbaros* (Gr) strange, foreign; originally meaning all those who were not Greek. The Tayra is a giant weasel of South America.

Grison   *Galictis vittata*
galē (Gr) a marten-cat or polecat   *iktis* (Gr) a marten-cat   *vittatus* (L) bound with a ribbon; can mean striped   *grison*, from *gris* (Sp) grey; the fur is blackish below and grey above. Inhabiting Central and South America.

Patagonian Weasel   *Lyncodon patagonicus*
lunx (Gr) a lynx   *odon* (= *odous*) (Gr) a tooth; a reference to the peculiarity of having only three pairs of molar teeth in each jaw. Like the lynx it has only three cheek teeth behind the canines in both upper and lower jaws, and so, like the lynx, has only 28 teeth; this is a small number for a member of the Carnivora   *-icus* (L) suffix meaning belonging to, 'of Patagonia'. An uncommon species inhabiting South America.

Striped Polecat   *Ictonyx striatus*
iktis (Gr) a marten-cat or polecat   *onux* (Gr) a claw; an allusion to the strong non-retractile claws on the fore feet   *striatus* (L) striped; it has a black coat with broad white stripes along the back. It is sometimes known as the Zorille or Zorilla, as *zorillo* is Argentine Spanish for the skunk, a related animal. The American Striped Skunk *Mephitis mephitis* is similar in appearance and behaviour. The Striped Polecat is widespread in Africa.

Libyan Striped Weasel   *Poecilictis libyca*

poikilos (Gr) many-coloured, dappled   *iktis* (Gr) a marten-cat or polecat   *libyca*, of Libya. Inhabiting northern Africa.

White-naped Weasel   *Poecilogale albinucha*
poikilos (above)   *galē* (Gr) a marten-cat or weasel   *albus* (L) white *nucha* (New L) the neck. Inhabiting central and southern Africa.

Wolverine or Glutton   *Gulo gulo*
gula (L) the throat, so *gulosus* (L) gluttonous; it probably eats no more than other carnivores, though it has a reputation for greed. Living in North America, northern Europe, the Arctic and northern Asia.

Ratel or Honey Badger   *Mellivora capensis*
mel (L), genitive *mellis*, honey   *voro* (L) I devour; a reference to its favourite food   *capensis*, of Cape Province, South Africa. The name Ratel was originally South African Dutch, possibly from *raat* (Du) a honeycomb. Inhabiting Africa and southern Asia.

Eurasian Badger   *Meles meles*
meles (L) a badger. Inhabiting Europe and Asia; it ranges from the British Isles to China.

Hog Badger or Sand Badger   *Arctonyx collaris*
arktos (Gr) a bear   *onux* (Gr) a claw; a reference to the long slightly curved blunt claws, as in the bear   *collum* (L) the neck   *-aris* (L) suffix meaning pertaining to; there is a white patch on the neck. Inhabiting southern Asia and Sumatra.

Teledu or Stink Badger   *Mydaus javanensis*
mudaō (Gr) I am wet, damp, from *mudos* (Gr) damp, decay; an allusion to the fetid skunk-like odour of the animal; like the skunk, it has a gland that secretes a foul-smelling liquid which it can expel a short distance   *-ensis* (L) suffix meaning belonging to; 'of Java'; it is also found in Sumatra and Borneo. Teledu is from the Malayan native name.

American Badger   *Taxidea taxus*
taxus (New L) a badger   *idea* (Gr) appearance; can mean kind, sort. Inhabiting North America.

Chinese Ferret Badger   *Melogale moschata*
meles (L) a badger   *galē* (Gr) a marten-cat or polecat   *moskhos* (Gr) musk, hence *moschatus* (New L) musky; it has a rather strong smell. Inhabiting southern Asia.

Javan Ferret Badger *M. orientalis*
-alis (L) suffix meaning relating to, thus *orientalis*, from the Orient.
It lives in Java and Borneo.

Striped Skunk *Mephitis mephitis*
mephitis (L) a noxious smell; skunks actually squirt a foul-smelling
liquid at an enemy with considerable accuracy. Inhabiting Hudson
Bay in the north and southwards through to Mexico.

Hooded Skunk *M. macroura*
makros (Gr) large, long *oura* (Gr) the tail. Living in Mexico and
Central America.

Spotted Skunk *Spilogale putorius*
spilos (Gr) a stain, a spot *galē* (Gr) a marten-cat or weasel *putor*
(L), genitive *putoris*, a bad smell. The range is from the southern part
of North America and south to Panama.

Hog-nosed Skunk *Conepatus leuconotus*
konis (Gr) dust *pateō* (Gr) I walk; can mean to frequent a place;
it lives in open country and deserts and it may also refer to its burrowing habits *leukos* (Gr) white *notos* (Gr) the back. Inhabiting North
America and Central America.

Eurasian Otter *Lutra lutra*
lutra (L) the otter. Inhabiting Europe including the British Isles,
North Africa, Asia, Java and Sumatra.

Canadian Otter *L. canadensis*
-ensis (L) suffix meaning belonging to; it inhabits North America
including Canada.

Indian Small-clawed Otter *Amblonyx cinerea*
(or subgenus of *Aonyx*)
amblus (Gr) blunt, point taken off *onux* (Gr) a claw; it has short
blunt claws *cinis* (L), genitive *cineris*, ashes, hence *cineria* ash-
coloured. Inhabiting southern Asia, Borneo, Java, Sumatra and the
Philippines.

Cape Clawless Otter *Aonyx capensis*
a- (Gr) prefix meaning not, or there is not *onux* (Gr) a claw; the
claws are small and rudimentary *capensis*, of Cape Province, South
Africa. It ranges north to the border of the Sahara.

Giant Otter or Saro   *Pteronura brasiliensis*
pteron (Gr) feathers; can mean wings   *oura* (Gr) the tail; this otter's tail has lateral flanges that make it appear something like a wing; 'the fin-like dilatation on each side of the hinder half of the tail' (Gray 1837)   *-ensis* (L) suffix meaning belonging to; also inhabiting Guyana and Surinam and not confined to Brazil; very rare.

Sea Otter   *Enhydra lutris*
enhudris (Gr) an otter, from *enhudro-bios*, living in the water   *lutra* (L) an Otter; this otter lives almost entirely in the water and rarely comes ashore. Inhabiting the shore areas of the northern Pacific, it is a rare animal and is protected by the governments of both the United States and Russia.

Suborder   FELOIDEA (or AELUROIDEA)
see *Ailurus fulgens* (page 590)
feles (L), genitive *felis*, a cat   *-oides* (New L) from *eidos* (Gr) apparent shape, form; can mean a kind, sort; 'cat-like'.

Family   VIVERRIDAE   about 82 species
viverra (L) a ferret; this is misleading as the animals in this family are not ferrets, they are more 'cat-like'.

African Linsang   *Poiana richardsoni*
Poiana, apparently from 'Po' of Fernando Po, the island off the coast of Cameroun, West Africa, where the species was found and described *-anus* (L) belonging to; named after Dr John Richardson, Inspector of the Naval Hospital at Haslar, by T. R. H. Thompson, RN, surgeon of 'the late Africa expedition' (1842).

Small Spotted or Feline Genet   *Genetta genetta*
genette (Fr) a genet or civet-cat. Inhabiting Africa and Spain.

Abyssinian Genet   *G. abyssinica*
-icus (L) suffix meaning belonging to; of Abyssinia; it also inhabits other parts of Africa. It can be tamed and domesticated and was probably the original 'cat' of ancient Egypt.

Blotched Genet or Tigrine Genet   *G. tigrina*
tigris (L) a tiger, so *tigrinus* tiger-like. Inhabiting Africa.

Rusty-spotted Genet   *G. rubiginosa*
rubor (L) redness, hence *rubiginosus* (L) rusty. From South Africa.

Giant or Victorian Genet   *G. victoriae*
This genet takes its name from Lake Victoria, Africa. It lives in a belt of country that surrounds this lake and is also found in the Ituri Forest, Zaire.

Small Indian Civet   *Viverricula indica*
viverra (L) a ferret (see under Family above)   *-culus* (L) diminutive suffix   *-icus* (L) suffix meaning belonging to; of India. It ranges from India through Sumatra and Java to Bali.

Water Civet   *Osbornictis piscivora*
This civet, discovered as recently as 1916 at Niapu, in northern Zaire, is named after Dr Henry F. Osborne (1857-1935), an American zoologist; no live specimens have been found since then   *iktis* (Gr) a marten-cat   *piscis* (L) a fish   *voro* (L) I devour. The extent of its range is not established.

African Civet   *Viverra civetta*
viverra (L) a ferret (see above under Family)   *civette* (Fr) the civet-cat; civet is the name of a highly valued perfume fixative obtained from this animal. Widespread in Africa.

Large Indian Civet   *V. zibetha*
zibetto (It) a civet-cat. Inhabiting southern Asia.

Banded Linsang   *Prionodon linsang*
priōn (Gr) a saw   *odōn* (Gr) a tooth; a reference to the simple and similar teeth which form a saw-like row   *linsang* is a Javanese name for this animal which lives in Java, Sumatra, Borneo and southern Asia.

Spotted Linsang   *P. pardicolor*
pardus (L) a leopard   *color* (L) colour; a reference to the leopard-like spots. Inhabiting Nepal, Assam and Vietnam.

West African or Two-spotted Palm Civet   *Nandinia binotata*
Nandine is a West African native name for the palm civet   *bi-* (L) two *notatus* (L) marked; it has two white spots on the shoulders.

Small-toothed Palm Civet   *Arctogalidea trivirgata*
arktos (Gr) a bear   *galē* (Gr) a marten-cat or weasel   *galideus* (Gr) diminutive of *galē*, a young weasel; head conical, nose compressed, broad forehead; seemingly bear-like   *tri* (L) three   *virgatus* (L) made of twigs; can mean striped; it has three dark stripes along the

back. The first of these civets discovered were frequently seen in palm trees and so became known as Palm Civets. They inhabit large areas of India and China, and Sumatra, Java and Borneo.

Common Palm Civet or Musang   *Paradoxurus hermaphroditus*
para (Gr) beside; can mean contrary to, against   *doxa* (Gr) opinion; *para doxan* (Gr) contrary to opinion   *oura* (Gr) the tail; T. S. Palmer, in his standard work *Index Generum Mammalium* (1904), gives this explanation 'from the mistaken idea that the tail was prehensile. Though the tail is not prehensile the animal has the power of coiling it to some extent, and according to Blanford "in caged specimens the coiled condition not infrequently becomes confirmed and permanent".' Hermaphroditos, in the Greek legend, was the son of Hermes and Aphrodite; a nymph from the fountain at Salmacis was so attracted by him that she prayed for perpetual union with him, and the resulting form had the characteristics of both sexes. This civet is not a hermaphrodite (though some animals are, e.g. the earthworm *Lumbricus terrestris*), but probably it was found difficult to distinguish one sex from another, which is a feature common to many carnivores. It inhabits southern Asia and the islands of the Philippines, Malaysia and Indonesia.

Masked Palm Civet   *Paguma larvata*
Paguma is a word coined evidently from puma; the name was given by Dr J. E. Gray in 1831 and he gave a number of other obscure names of animals   *larvatus* (L) bewitched; can mean masked; the body is a uniform grey-brown but the black face has a conspicuous white stripe down the middle, hence 'masked'. It lives in the forests of south-east Asia, Sumatra and Borneo.

Binturong   *Arctictis binturong*
arktos (Gr) a bear   *iktis* (Gr) a marten-cat; it is rather bear-like and sometimes known locally as the Bear-Cat   *binturong* is the Malayan native name. Inhabiting southern Asia including the Himalayas, Burma, Malaya, Java and Sumatra.

Fanaloka or Malagasy Civet   *Fossa fossa*
Fossa or *foussa* is a native name for this civet, and not to be confused with the Fossa (see page 601), a different animal although they both live in Madagascar. Fanaloka is another native name.

Banded Palm Civet   *Hemigalus derbyanus*
hēmi- (Gr) half; a prefix, in this case used to indicate like, or similar to

galē (Gr) a marten-cat or weasel; named after the Thirteenth Earl of Derby (formerly the Honourable E. S. Stanley) (1775-1851); he was President of the Zoological Society of London in 1831. This civet inhabits southern Asia, Sumatra and Borneo.

Owston's Banded Civet   *Chrotogale owstoni*

khrōs (Gr), genitive *khrōtos*, the skin; can mean colour of the skin, complexion   *galē* (Gr) a marten-cat or weasel; it has dark bands round the body. Alan Owston (1853-1915) was a zoologist who lived in Japan for many years. This civet inhabits Tonkin and Laos.

Otter Civet   *Cynogale bennetti*

kuōn (Gr), genitive *kunos*, a dog   *galē* (Gr) a marten-cat or weasel; it has an elongate face, much produced and compressed. Another rather obscure name by Gray (*see Paguma larvata*, page 598); it has an otter-like body, and is similar to an otter in its habits and diet. Gray gave the specific name for his friend Edward T. Bennett (1797-1836), zoologist and surgeon. He was prominent in starting the London Zoological Society and was the Secretary from 1831 until his death in 1836. He would have described the Otter Civet himself had he lived. It inhabits southern Asia, Sumatra and Borneo.

Falanouc   *Eupleres goudoti*

eu- (Gr) prefix meaning well   *pleres* (Gr) full; can mean complete; a reference to the full number of five toes on both fore and hind feet J. Goudot, a French naturalist, found this animal in Madagascar about 1830. Falanouc or *falanaka* is a Malagasy native name. Sometimes known as the Small-toothed Mungoose (sic), the teeth are so small that it was originally believed to be an insectivore. Inhabiting Madagascar.

Giant Falanouc   *E. major*

major (=*maior*) (L) greater, larger. Inhabiting Madagascar.

Ring-tailed Mongoose   *Galidia elegans*

galē (Gr) a marten-cat or weasel   *galideus* (Gr) diminutive of *galē*, a young weasel or kitten   *elegans* (L) neat, elegant. Living in Madagascar.

Broad-striped Mongoose   *Galidictis striata*

Galidia, see above   *iktis* (Gr) a marten-cat   *striatus* (L) striped. Inhabiting Madagascar.

Suricate or Slender-tailed Meerkat   *Suricata suricatta*

Surikate or *suricat* are South African native names   *meerkat* is South African Dutch, meaning 'lake cat'; although living in dry places in South Africa, it is fond of water.

Egyptian Mongoose   *Herpestes ichneumon*
herpestes (Gr) a creeper; an allusion to its stalking habits   *ikhneumon* (Gr) the tracker, an Egyptian animal of the weasel kind, or mongoose. Sometimes known as Pharaoh's Rat, it was highly regarded in ancient times because it hunted for crocodile eggs and ate them. Inhabiting Africa and southern Europe.

Small Indian Mongoose   *H. auropunctatus*
aurum (L) gold, golden   *pungo* (L) I puncture, thus *punctatus*, spotted as with punctures. Inhabiting southern Asia.

Crab-eating Mongoose   *H. urva*
Urva is a Nepalese name for this mongoose from southern Asia.

Dwarf Mongoose   *Helogale parvula*
helos (Gr) low ground by rivers, a marsh   *galē* (Gr) a marten-cat or weasel; a reference to its habitat   *parvus* (L) small, so *parvulus* (L) very small. Inhabiting Africa.

Marsh Mongoose   *Atilax paludinosus*
a- (Gr) prefix meaning not, or there is not   *thulax* (Gr) a pouch; Cuvier says 'par la considération de toute absence de poche à l'anus'. This refers to the alleged absence of the anal scent gland, but some say Cuvier was wrong when he named it as it does have the gland *palus* (L), genitive *paludis*, a marsh, a swamp   *-osus* (L) suffix meaning full of; i.e. 'marshy'. A good swimmer inhabiting central and southern Africa.

Banded Mongoose   *Mungos mungo*
Mungos is derived from *mangus* (Marathi); the Maratha are a people of the south central part of India. Although this mongoose inhabits East Africa the name derives from the Indian mongoose. It has about twelve dark transverse bands across the back.

Angolan Cusimanse (or Kusimanse)   *Crossarchus ansorgei*
krossoi (Gr) a fringe, tassels   *arkhos* (Gr) the rump, the hind parts; it has a bushy tail   Dr W. J. Ansorge (1850-1913), a naturalist and author, was in Angola in 1905. This mongoose lives in Angola and other parts of Central Africa. Cusimanse is from *kusimanse*, a native name, probably from Liberia.

White-tailed Mongoose   *Ichneumia albicauda*
ikhneumon (Gr) the tracker (see *Herpestes ichneumon*, page 600)   *albus* (L)
white   *cauda* (L) the tail of an animal. Inhabiting southern Arabia and
Africa.

Bushy-tailed Mongoose   *Bdeogale crassicauda*
bdeō (Gr) I break wind, I stink   *galē* (Gr) a marten-cat or weasel;
it can produce a foul-smelling glandular secretion similar to a skunk
crassus (L) thick, dense   *cauda* (L) the tail of an animal. Inhabiting
East Africa.

Meller's Mongoose   *Rhynchogale melleri*
rhunkhos (Gr) the snout, beak   *galē* (Gr) a marten-cat or weasel; a
reference to the pointed nose   Dr C. J. Meller (1836-1869) was
Superintendent of the Botanical Gardens in Mauritius in 1865; this
island in the Indian Ocean lies to the east of Madagascar. The mongoose, a very rare species, inhabits East Africa.

Yellow Mongoose   *Cynictis penicillata*
kuōn (Gr), genitive *kunos*, a dog   *iktis* (Gr) a marten-cat; this means
intermediate between, or connecting, the dogs and civets   *penicillus*
(L) a painter's brush; it has a bushy tail with a white tip. Inhabiting
South Africa.

Fossa or Foussa   *Cryptoprocta ferox*
kruptos (Gr) hidden   *prōktos* (Gr) hinder parts, tail; 'hidden in the
hinder parts'; it has an anal gland that produces an evil-smelling
liquid, like the skunk   *ferox* (L) brave, fierce; it has a reputation for
being unusually fierce but this is now considered to be exaggerated.
Fossa is a Malagasy name, and the animal must not be confused with
the Malagasy Civet which unfortunately has been given the Latin
name *Fossa fossa* (see page 598); quite a different animal but they both
live in Madagascar.

Family   HYAENIDAE   4 species
huaina (Gr) the hyaena, from *hus* (Gr) a hog, on account of the bristly
mane.

Aardwolf   *Proteles cristatus*
prōtos (Gr) first; can mean in front   *teleos* (Gr) complete, perfect;
'complete in front'; referring to the aardwolf having five toes on the
fore feet but only four on the hind feet   *cristatus* (L) crested; all

hyaenas have a crest or mane that normally lies flat. However, some say the aardwolf's mane is permanently erected, but there is disagreement about this among zoologists. Aardwolf is a Dutch name meaning 'earth-wolf'. Inhabiting southern and eastern Africa.

Spotted Hyaena   *Crocuta crocuta*
crocus (L) the crocus; also the colour of saffron, yellow   *utus* (L) suffix meaning provided with   *crocuta* is a rare Latin word originally meaning 'an unknown wild animal of Ethiopia, perhaps the hyaena'; it has no other meaning, but there is a Greek word *krokōtos* meaning saffron-coloured. This hyaena is tawny to yellow with dark spots; the range is widespread in Africa south of the Sahara.

Brown Hyaena   *Hyaena brunnea*
Hyaena, see above under Family   *brunneus* (New L) dark brown. A rare hyaena in danger of extinction, inhabiting South Africa.

Striped Hyaena   *H. hyaena*
It has six vertical black stripes on the flanks. Inhabiting Africa, the Near East and India.

Family   FELIDAE   about 36 species
feles (L), genitive *felis*, a cat.

European Wild Cat   *Felis silvestris*
silva (L) a wood, hence *silvestris* (L) belonging to woods. Inhabiting western Asia, the wild parts of Europe, and still found in Scotland.

African Wild Cat or Mu   *F. libyca*
libyca, belonging to Lybia. The Egyptians called this cat the *mu*, and trained it to hunt mice; it is not confined to northern Africa and also inhabits Asia.

Cat   *F. catus*
catta (New L) a cat; the domestic cat, common throughout the world, and sometimes known as *F. domestica*.

Serval   *F. serval*
serval (Port) a deer-wolf, transferred from another animal. Inhabiting Africa.

Sand Cat   *F. margarita*
Named after a Général Margueritte on duty in Algeria in the 1850's. From North Africa and south-west Asia.

Jungle Cat *F. chaus*
Chaus is an ancient name for a wildcat of Africa but the origin of the word is obscure. Inhabiting North Africa and Asia.

Leopard Cat *F. bengalensis*
-ensis (L) suffix meaning belonging to; it is not confined to Bengal and is widespread in south-eastern Asia.

Black-footed Cat *F. nigripes*
niger (L) black *pes* (L) a foot. Inhabiting South Africa.

Persian Lynx or Caracal *F. caracal*
caracal is from *karakulak* (Turk) meaning black ear; it has a fawn coat and large black ears with black tufts. It inhabits Africa, Arabia and India.

Northern Lynx *F. lynx lynx*
lynx (L) a lynx *lunx* (Gr) a lynx; from its light colouring and bright eyes and akin to Old English *lox*, and Old High German *luhs*, and Greek *leukos*, meaning white, bright. Inhabiting northern Asia, Europe and North America.

Canada Lynx *F. l. canadensis*
-ensis (L) suffix meaning belonging to; 'of Canada'.

Bobcat *F. rufa*
rufus (L) red, ruddy. Widespread in the USA and Mexico.

Pallas's Cat *F. manul*
manul (Mongolian) a small wildcat. Professor P. S. Pallas (1741-1811) was a German zoologist and explorer. This wild cat lives in the mountains of Mongolia, Siberia and Tibet.

Marbled Cat *F. marmorata*
marmor (L) marble, hence *marmoratus* (L) marbled. Inhabiting southern Asia, Sumatra and Borneo.

African Golden Cat *F. aurata*
aurum (L) gold, the colour of gold, so *aureatus* (L) adorned with gold. Inhabiting Africa.

Fishing Cat *F. viverrina*
viverra (L) a ferret *-inus* (L) suffix meaning like, pertaining to. Inhabiting southern Asia, Sumatra and Java.

Ocelot   *F. pardalis*
pardus (L) a panther, a leopard (see Leopard, below)   *-alis* (L)
suffix meaning relating to. Ocelot was originally French, from the
Nahuatl word *ocelotl*, a jaguar. Inhabiting Central and South America.

Tiger Cat   *F. tigrina*
tigris (L) a tiger   *-inus* (L) suffix meaning like, pertaining to. Inhabiting Central and South America.

Jaguarundi   *F. yagouaroundi*
Jaguarundi was originally a Tupi name, but this animal is not a jaguar
in spite of the name, and looks nothing like one. It is a rather unusual
member of the cat family and not even like a cat in appearance.
Inhabiting Central and South America.

Puma   *F. concolor*
concolor (L) of the same colour (as opposed to *discolor* (L) of different
colours). Adult pumas are plain greyish-brown with no conspicuous
markings. From Canada, USA and South America. It is also known
as the Cougar or Mountain Lion.

Clouded Leopard   *Neofelis nebulosa*
neos (Gr) new; can mean unexpected or strange; zoologists have not
found it possible to classify this leopard as either *Felis* or *Panthera*, and
so have given it a new genus *Neofelis*   *nebula* (L) cloud, hence *nebulosus*
(L) cloudy; the coat is mottled and striped with black on a brownish-
grey back-ground. Inhabiting south-east Asia, Borneo, Sumatra and
Java.

Lion   *Panthera leo*
panthera (L) a panther or a leopard (see Leopard)   *panthēr* (Gr) a
panther   *leo* (L) a lion. Widespread in Africa except in the north and
a few still remain in India under protection.

Tiger   *P. tigris*
tigris (L) a tiger. Now found only in Asia, Sumatra and Java.

Leopard   *P. pardus*
pardus (L) a panther or a leopard. It is easy to become confused by
these various names, so let us be quite clear that a leopard *is* the same
animal as a panther. The two names may have been used to distinguish between the various sizes and colours that occur in this animal,
and 'panther' is usually used for the black variety. Clearly the English

word leopard comes from the two Latin words *leo* and *pardus*. Widespread in Africa, apart from desert areas, and in Asia from the southwest to China and Korea.

Jaguar *P. onca*

onca   the Greeks were familiar with a moderate sized feline and called it *lynx* (see page 603). The Romans borrowed the Greek word and it became the Tuscan *lonza*. Later the word passed into French as *lonce*, and the initial *l* being mistaken for the article it was elided and the word became *once*, whence it passed into Spanish as *onca* and English as *ounce*. Finally, *onca* was Latinised to *uncia*, although this was Latin for a measure of weight and not an animal! The jaguar inhabits the forests of North, Central and South America.

Snow Leopard or Ounce *P. uncia*

uncia, see above. Inhabiting the mountains of Central Asia.

Cheetah *Acinonyx jubatus jubatus*

akaina (Gr) a thorn, a goad   *onux* (Gr) a claw; a reference to the non-retractile pointed claws; this is the recognised explanation but I suggest a possible interpretation would be *a-* (Gr) not   *kineō* (Gr) I move + claw; 'non-moving claws'   *jubatus* (=*iubatus*) (L) maned; young cheetahs have a crest or mane on the shoulders and back. Cheetah is from *cital* (Hind) from *chitraka* (Sanskrit), having a speckled body. Inhabiting Africa south of the Sahara and south-west Asia, it is rare and in danger of extinction.

Indian Cheetah *A. j. venaticus*

venaticus (L) belonging to hunting; for hundreds of years the human race has tamed and trained cheetahs to assist them in hunting. This subspecies is very rare.

54 Seals, Sea-Lions and the Walrus
PINNIPEDIA

This group, like the whales, consists of mammals that have become adapted to a life in the water. However, unlike the whales they are able to come ashore and move about on land, albeit in rather a slow and ungainly fashion. They are all carnivorous, and in addition to fish they eat a variety of other foods, such as sea birds, shellfish and other small marine life.

Subclass EUTHERIA
Order PINNIPEDIA
pinna (L) a feather or wing; can mean a fin *pes* (L), genitive *pedis*, a foot; 'fin-footed'; they walk on their fins when they come ashore.

Family OTARIIDAE 13 species
ous (Gr), genitive *ōtos*, an ear, thus *otarion*, a little ear; unlike the common seals, of the family Phocidae, they have external ear-flaps.

Cape Fur Seal *Arctocephalus pusillus*
arktos (Gr) a bear *kephalē* (Gr) the head; it has a bear-like appearance *pusillus* (L) very small. Fur Seals are so-named because the hide makes

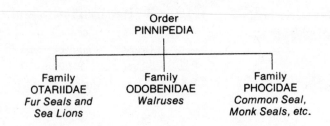

excellent fur coats; *pusillus* because the description was based on a picture of a young pup. Inhabiting the sea round The Cape of Good Hope, South Africa.

Forster's Fur Seal   *A. forsteri*
Named after J. G. A. Forster, FRS (1754-1794), an artist who accompanied Captain James Cook on his second voyage of exploration in *H.M.S. Resolution* during the years 1772 to 1775, and he made a drawing of this seal. It lives in the sea round the coast of New Zealand.

Australian Fur Seal   *A. doriferus*
dora (Gr) a skin, a hide; the word only applies when the skin is removed for the fur   *fero* (L) I bear. From the southern coasts of Australia and Tasmania, it has particularly fine fur.

Kerguelen Fur Seal   *A. tropicalis*
tropikos (Gr) tropical   *-alis* (L) suffix meaning relating to; the name was given mistakenly as it does not inhabit tropical seas. Kerguelen Island lies well to the south of the Indian Ocean; it also inhabits other sub-antarctic islands.

South American Fur Seal   *A. australis*
australis does not mean Australia; *auster* (L) the south   *-alis* (L) suffix meaning pertaining to   *australis* (L) southern. It inhabits most of the coastal sea round South America.

Guadalupe Fur Seal   *A. philippii*
Dr R. A. Philippi (1808-1904) was at one time Director of the Museum in Santiago in Chile. He collected the skull of this seal in 1864. It inhabits one small protected breeding colony on Guadalupe Island, off the coast of California.

Northern Fur Seal   *Callorhinus ursinus*
kallos (Gr) a beauty, a beautiful object   *rhinos* (Gr) the skin or hide

of an animal; 'beautiful hide'. T. S. Palmer and others have translated this as *rhinos* (Gr) the nose, and admit to not knowing the
explanation; the Greek for nose is *rhis*, genitive *rhinos*. The Greek
word *rhinos* also means the skin or hide of beasts and is frequently used
in this sense in Homer; I offer this as a more likely interpretation
ursus (L) a bear *-inus* (L) suffix meaning like; 'bear-like'; it is sometimes known as the Sea-Bear. Inhabiting Pribilof and other islands
in the Bering Sea area of the northern Pacific Ocean.

Californian Sea Lion *Zalophus californianus*
za- (Gr) an intensive prefix *lophos* (Gr) a crest; 'a high crest'; there
is a high sagittal crest on the male's skull. This is the sea lion which
you will have seen performing in a circus. It is not confined to the
Californian area; it also inhabits the western side of the Pacific north
of Japan.

Steller's Sea Lion *Eumetopias jubatus*
eu- (Gr) well, typical *metopion* (Gr) the forehead; it has a broad forehead *jubatus* (= *iubatus*) (L) having a mane; the male has a shaggy
mane. G. W. Steller (1709–1746) was a German zoologist who was
exploring in the northern Pacific Ocean in 1740. This sea lion inhabits coastal areas of both sides of the northern Pacific.

Southern Sea Lion *Otaria byronia*
Otaria, see under Family Commodore John Byron was serving with
HMS Tamar on a voyage of discovery in the South Seas from 1764 to
1766; he brought home a skull of this sea lion. It inhabits coasts of
South America and the Falkland Islands.

Australian Sea Lion *Neophoca cinerea*
neos (Gr) new *phōkē* (Gr) a seal; indicating a newly named sea lion
cinis (L) ashes, hence *cinerea* (L) ash-coloured. Inhabiting coasts of
southern Australia.

Hooker's Sea Lion *Phocarctos hookeri*
phōkē (Gr) a seal *arktos* (Gr) a bear; from the skull which is like that
of a bear. Sir J. D. Hooker (1817–1911) was a famous English naturalist and travelled widely as the botanist on expeditions in Australia,
Tasmania and New Zealand. This sea lion inhabits the southern
coast of New Zealand.

Family ODOBENIDAE 1 species and 1 subspecies
odous (Gr) a tooth *bainō* (Gr) I step, I walk; 'one that walks with its

teeth'; the walrus has been seen to drag itself along the ice using its tusks which are actually over-developed canine teeth.

Atlantic Walrus or Sea Cow   *Odobenus rosmarus rosmarus*
Rossmaal or *rossmar* are Scandinavian names for the walrus; it lives mostly in Arctic areas, from Canada to northern Russia.

Pacific Walrus or Sea Cow   *O. r. divergens*
di- (L) prefix, two; can mean apart   *vergo* (L) I turn   *divergens* (New L) turning apart; referring to the tusks. The Old English name was *horschwael*, meaning 'horse-whale'; it is also sometimes called the morse, from *mors* (Lappish) a walrus. This walrus lives in the area round Alaska, the eastern coasts of Siberia and the Bering Sea.

Family   PHOCIDAE   18 species
phoca (L) a seal, a sea-calf.

Common Seal   *Phoca vitulina*
vitula (L) a calf   *-inus* (L) suffix meaning like. Living on the shores of the northern oceans, including the British Isles.

Ringed Seal   *Pusa hispida*
Pusa is the Greenlandic name for a seal; it is probably a misspelling as the Greenlandic name for the harp seal is *puirse*; *hispidus* (L) hairy, bristly; a reference to the whiskers: the body is marked with oval white rings. Inhabiting Arctic coasts.

Baikal Seal   *P. sibirica*
-icus (L) suffix meaning belonging to; living in Lake Baikal, Siberia, a huge fresh-water lake some 640 km (400 miles) long; probably the only seal living entirely in fresh water.

Caspian Seal   *P. caspica*
-icus (L) suffix meaning belonging to; of the Caspian Sea.

Harp Seal   *Pagophilus groenlandicus*
pagos (Gr) anything hardened, ice, frost   *philos* (Gr) loving, fond of; they live among ice and snow   *-icus* (L) suffix meaning belonging to; of Greenland; they are not confined to this area, being widespread in the Arctic. They have unusual and distinct markings on the back which have given rise to the names 'Harp' and 'Saddlebacked'. It is also sometimes known as the Greenland Seal.

Grey Seal   *Halichoerus grypus*
halios (Gr) from, belonging to the sea   *khoiros* (Gr) a pig; 'a sea-pig'
grupos (Gr) hook-nosed; in profile the nose is distinctly rounded.
Inhabiting coastal areas of northern oceans including the British
Isles; it is said that in autumn there are more living round our shores
than all other areas put together. The other 'British' seal is the
Common Seal *Phoca vitulina*.

Crabeater Seal   *Lobodon carcinophagus*
lobos (Gr) a lobe   *odōn* (Ionic Gr) a tooth; the molars are compressed
and have lobes both in front and behind   *karkinos* (Gr) a crab
phagein (Gr) to eat; they probably eat more krill than crabs and this
may be the reason for the shape of the teeth. They live in the Antarctic
area.

Ross's Seal   *Ommatophoca rossi*
omma (Gr), genitive *ommatos*, the eye   *phoca* (L) a seal; it has exceptionally large eyes. Rear Admiral Sir James Ross (1800–1862), a
famous British Arctic and Antarctic explorer, also discovered Ross's
Gull *Rhodostethia rosea*. The gull lives in the Arctic whereas the seal
lives in the Antarctic.

Leopard Seal or Sea-Leopard   *Hydrurga leptonyx*
hudōr (Gr) water; in composition the prefix *hudro-* is used   *ergō* (Gr)
I work; a misspelling, or possibly derived from *urgeo* (L) I drive, I
urge; a reference to its aquatic life   *leptos* (Gr) slender, thin   *onux*
(Gr) a claw, nail. The skin is usually dull brown with light spots;
some have been caught with colouring remarkably like a leopard;
they are also ferocious predators. Inhabiting southern oceans.

Weddell's Seal   *Leptonychotes weddelli*
leptos (Gr) slender, thin   *onux* (Gr), genitive *onukhos* a claw, nail
-otēs (Gr) suffix denoting possession; referring to the rudimentary
claws of the hind feet. This seal was discovered by James Weddell,
the Scottish sealer, and named in his honour. He brought home a
drawing and a skeleton in 1824. The Weddell Sea, discovered in
1823, also bears his name; it is in the Antarctic, where this seal lives.

West Indian Monk Seal   *Monachus tropicalis*
monakhos (Gr) a monk, solitary; they are not always solitary and may
live in colonies in certain areas; they have rings of fat round the neck
which might suggest a monk's hood or cowl   *-alis* (L) suffix meaning

pertaining to; of the tropics. They live in the Caribbean Sea area and the Gulf of Mexico, but are now very scarce and may even be extinct.

Hawaiian Monk Seal   *M. schauinslandi*
Professor H. H. Schauinsland (1857–1937), a German zoologist and at one time Director of the Bremen Museum, discovered this seal on Laysan Island, Hawaii. There are probably no more than 1,000 to 1,500 still in existence.

Hooded Seal or Crested Seal   *Cystophora cristata*
kustis (Gr) a bladder; also a bag or pouch   *phoros* (Gr) carrying *cristata* (L) crested; the male has a peculiar pouch forming a crest on the nose which can be inflated like a bladder. Inhabiting large areas round Greenland in the Arctic.

Southern Elephant Seal   *Mirounga leonina*
mirounga (New L) derived from *miouroung*, an Australian native name for the seal   *leo* (L), genitive *leonis*, a lion   *-inus* (L) suffix meaning like; leonine. A very big seal which can weigh up to four tonnes, with a trunk-like snout; it can make a roaring noise said to be like that of a lion. Widespread in the southern oceans including the coast of Australia.

Northern Elephant Seal   *M. angustirostris*
This seal lives on the western coast of North America, and nowhere near Australia, but it takes the Latin name *Mirounga* from the Southern Elephant Seal (above)   *angustus* (L) narrow   *rostrum* (L) the snout; the male has a snout rather like an elephant's trunk and up to 600 cm (2 ft) long. However, it is narrower than the snout of the southern species *M. leonina*.

All by itself, this peculiar termite-eating mammal constitutes the order Tubulidentata, and the family Orycteropodidae. Why, the reader may ask, does it stand so isolated from other mammals? The reason is that the anatomical structure is quite unlike any other mammal, and gives no indication of the group in which it should be placed. The teeth are simple cylinders of dentine traversed from base to crown by hundreds of minute passages, or tubules. It has no immediate ancestors but probably stems from very early ungulates, as did the hyraxes.

Subclass   EUTHERIA
Order   TUBULIDENTATA

tubus (L) a pipe, a tube   *-ulus* (L) diminutive suffix   *dens* (L), genitive *dentis*, a tooth   *-atus* (L) suffix meaning provided with (see introductory note above).

Family   ORYCTEROPODIDAE   1 species
oruktēr (Gr) a tool for digging   *pous* (Gr), genitive *podos*, a foot.

Aardvark   *Orycteropus afer*
'One that has feet for digging'; it has powerful claws and is a remarkably fast digger   *Afer* (L) African. Aardvark was originally South

Order
TUBULIDENTATA
|
Family
ORYCTEROPODIDAE
The Aardvark

African Dutch, now known as Afrikaans, for earth-pig. Though rarely seen, being nocturnal, it is widespread in Africa south of the Sahara.

56 **Elephants** PROBOSCIDEA

There are now only two species of elephants, the African and the Indian. There is only one family; their exceptional tusks, trunk and many other characteristics set them apart from all other mammals. The trunk has evolved from a shorter proboscis such as that of the tapirs, and has become a specialised organ capable of picking up food, and drawing up water, and various other uses. It is thought that they may be distantly related to the manatees, and represent the only survivors of several hundred extinct species known from fossils.

Subclass **EUTHERIA**
Order **PROBOSCIDEA**

pro- (Gr) prefix meaning before, in front of; also used to express motive *boskō* (Gr) I feed, hence *proboskis* (Gr) a trunk or proboscis used for feeding as with elephants and certain insects *idea* (Gr) appearance; can mean a kind, a sort.

Family **ELEPHANTIDAE**   2 species
elephantus (L) the elephant.

Order
PROBOSCIDEA
|
Family
ELEPHANTIDAE
Elephants

African Elephant   *Loxodonta africana*
loxos (Gr) slanting   *odous* (Gr), genitive *odontos*, a tooth; the grinding surfaces of the teeth appear to be 'lozenge-shaped'.

Indian Elephant   *Elephas maximus*
elephas (Gr) the elephant   *maximus* (L) largest; this is misleading as the African species is usually the bigger. It is not confined to India, and inhabits Sri Lanka, forested areas of south-east Asia and Sumatra.

57 **Hyraxes** HYRACOIDEA

The hyraxes are small rabbit-like animals. They have always been a problem for zoologists as there seems to be no obvious group in which to place them for purposes of classification. For many years they were thought to be related to the elephants, but recently some zoologists have come to the conclusion that this may not be the case. Because of the unusual anatomical structure, particularly with regard to the teeth and the feet, they must be classed as a separate order. The upper cutting teeth, the incisors, are to some extent rodent-like, but the upper cheek teeth are like those of a rhinoceros, and the lower cheek teeth like those of a hippopotamus. The toes have small hoofs, and on the sole there is what appears to be a sort of suction pad which may account for their agility when climbing rocks and trees. They are the last survivors of about twelve extinct genera.

Subclass **EUTHERIA**
Order **HYRACOIDEA**

hurax (Gr), genitive *hurakos*, a mouse, a shrew-mouse   *-oidea* (New L), from *eidos* (Gr) form, sort, a particular kind; it is much bigger than a mouse.

Family PROCAVIIDAE   about 10 species (authorities differ)
pro- (Gr) before, hence *prōtos* (Gr) first   *cavia* (New L) from cabiai, a Brazilian word used in South America for a rodent, probably a 'guinea-pig'; *Procavia* could mean 'first guinea-pigs', suggesting that other similar animals are descended from these.

Order
HYRACOIDEA
|
Family
PROCAVIIDAE
Hyraxes

Tree Hyrax   *Dendrohyrax arboreus*
dendron (Gr) a tree   *hurax* (Gr) a mouse, a shrew-mouse; it is much
bigger than a mouse, being more like a large guinea-pig, or a rabbit
arbor (L) a tree, hence *arboreus* (L) relating to trees. Inhabiting most
of Africa south of the Sahara.

Beecroft's Hyrax   *D. dorsalis*
dorsum (L) the back   *-alis* (L) suffix meaning relating to; there are
scent glands on the back covered by a patch of white hair; the hair
stands erect when the animal is roused. John Beecroft was an English
naturalist who was made Governor by the Spaniards of their island
Fernando Po, in 1844; five years later he became the British Consul.
This hyrax was first found on the island about 1850, probably by
Beecroft. It inhabits the western part of Africa.

Yellow-spotted Hyrax   *Heterohyrax brucei*
heteros (Gr) different + hyrax; an allusion to the skull, which is like
that of *Dendrohyrax* except that the orbit, or eye-socket, is incomplete
behind; James Bruce (1730–1794), a naturalist who was in Ethiopia
(formerly Abyssinia) from 1768 to 1773, made some expeditions to
discover the sources of the Nile. At the time his amazing travel stories
about Abyssinia were not believed. This hyrax inhabits most of Africa
south of the Sahara.

Rock Hyrax or Dassie   *Procavia capensis capensis*
Procavia. see Family   *-ensis* (L) belonging to; it inhabits the Cape
Province area and Namibia. Dassie is the Afrikaans name for the
hyrax.

Syrian Hyrax   *P. c. syriacus*
-acus (L) suffix meaning relating to; this hyrax ranges through Syria
and Sinai and south to Saudi Arabia. It is the 'Coney' mentioned in
the Bible, but it is not a rabbit. A substance known as hyraceum, said
to be excreted by the hyrax and presumably its droppings, was used
as a medicine, a 'folk remedy', and was supposed to be good for
epilepsy and convulsions!

58 **Manatees and the Dugong** SIRENIA

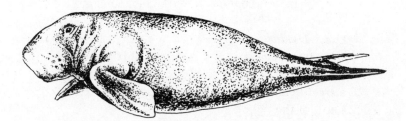

The Manatee is probably the origin of the mermaid legend, and this may have come about because they have a habit of standing up in the water with their head and shoulders above the surface. When doing this, the females often hold a baby manatee in one flipper to enable it to suckle at the breast.

The Manatee and the Dugong are a puzzle for the zoologists, as among the mammals, for anatomical reasons, it has been necessary to place them in a group by themselves. They are almost hairless and live entirely in the water, but unlike the whales they are completely herbivorous. The hind limbs have evolved into a fleshy horizontal paddle. The general opinion is that they are very distantly related to the elephants. They are the last survivors of about twenty fossil genera.

Subclass **EUTHERIA**
Order **SIRENIA**
seirēn (Gr) a siren; according to the legend the Sirens were nymphs who allured sailors by their sweet songs and then slew them.

Family **DUGONGIDAE** 1 species
Duyong is a Malayan name for this marine animal.

Dugong or Halicore *Dugong dugon*
The name halicore is from *hals* (Gr) the sea, and *korē* (Gr) a girl; like the manatee it is supposed to have given rise to the mermaid legend. It is found along the coasts round the Indian Ocean, and ranging south to Indonesia and northern Australia; it is becoming very rare.

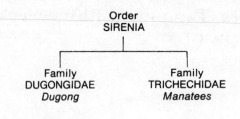

Family   TRICHECHIDAE   3 species
thrix (Gr), genitive *trikhos*, hair   *ekhō* (Gr) I have; they have hair on
the face and bristly moustaches.

North American or Caribbean Manatee   *Trichechus manatus*
manati (Sp) the manatee, derived from a native name in Haiti.
Inhabiting the coast of Haiti and other islands in the Caribbean Sea.

South American Manatee   *T. inunguis*
in- (L) prefix meaning not, without   *unguis* (L) a claw, a hoof; the
flippers are paddle-like and without claws. Inhabiting the Atlantic
coasts of both North and South America.

West African Manatee   *T. senegalensis*
-ensis (L) suffix meaning belonging to. It is not confined to the Senegal
area of West Africa and may also travel far inland up the rivers. These
strange animals are sometimes known as Sea Cows.

59 Horses, Tapirs and Rhinoceroses

PERISSODACTYLA

This group, together with the larger group comprising the antelopes, pigs, cattle and their kin, are known as ungulates, though the term is not usually employed in classification; it means hoofed, from the Latin *unguis*, a claw or hoof.

The members of the group we are now going to deal with have an uneven number of toes or hoofs and have been given the name Perissodactyla, meaning 'odd-toed'. They have either one or three toes, though the tapirs are an exception, having three toes on the hind feet and four on the front feet. The rhinoceroses have three toes on each foot.

In all cases, the central axis of the foot passes through the third digit, which may take most of the weight, and is large and symmetrical. In the case of the horses, asses and zebras, it is the only digit remaining, taking all the weight; the others have disappeared during evolution, a process occupying some 18,000,000 years, though rudimentary remains can still be seen. Thus, they walk and run on the middle finger!

For purposes of classification, the order is divided into two suborders, namely Hippomorpha, the 'horse-kind', and Ceratomorpha, the 'horned-kind'. The antelope group, having an even number of toes or hoofs, have been given the name Artiodactyla, meaning 'even-toed'; this will be dealt with in Chapter 60.

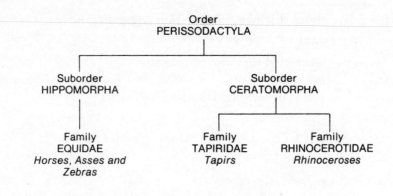

Order   PERISSODACTYLA
Suborder   HIPPOMORPHA

perissos (Gr) strange, unusual; of numbers odd, uneven   *daktulos* (Gr) a finger; can mean a toe.

hippos (Gr) a horse   *morphē* (Gr) apparent form, a kind, sort; the 'horse-kind'.

Family   EQUIDAE   about 6 species
equus (L) a horse

Wild Horse   *Equus przewalskii*
General N. M. Prjevalski (1839–1888) was a famous Russian explorer and naturalist who made several expeditions to Central Asia; he collected thousands of birds as well as mammals. Sometimes known as Prjevalski's Horse it is very rare, but still inhabits the Altai Mountains in Western Mongolia. It is the only true wild horse known to be still in existence.

Horse   *E. caballus*
caballus (L) a pack horse, a nag; this is the domestic horse.

Central Asian Ass   *E. hemionus hemionus*
hēmi- (Gr) prefix meaning half   *onos* (Gr) an ass   *hēmionos* (Gr) a mule; the name is misleading, as a mule is the offspring of a male ass mating with a mare, and it would be sterile and could not breed. In this case *hēmi* is meant to indicate 'like an ass' rather than 'half-ass'.

It has a variety of different local names depending on where it is found; in Mongolia it is known as the Dziggetai.

Onager or Persian Wild Ass   *E. h. onager*
onos (Gr) an ass   *agrios* (Gr) wild, savage   *onagrus* or *onager* (L) a wild ass. In India it is known as the *Ghor-kar*.

Kiang   *E. kiang*
Kiang or *Kyang* is a Tibetan native name; this wild ass lives in Tibet, western China. Some give it as a subspecies of *E. hemionus*.

African Wild Ass   *E. africanus*
Generally accepted as the ancestor of the domestic donkey (below). It inhabits northern parts of Africa but is becoming very rare.

Domestic Donkey   *E. asinus*
asinus (L) an ass; the name was given by Linnaeus in 1758.

Cape Mountain Zebra   *E. zebra zebra*
zebra is obscure, but could be from a Congolese or possibly an Abyssinian word *zibra*, meaning striped. Inhabiting the southern mountains of South Africa, it is very rare, and found only in certain protected areas.

Hartmann's Zebra   *E. z. hartmannae*
Premier Lieutenant Dr Hartmann discovered this zebra on the Kaoko Veld plateau, Namibia, and sent two skins to the Berlin Museum as a gift. It was named in 1898 by Dr Paul Matschie, of the Berlin Museum, in honour of Mrs Hartmann. Inhabiting the mountains of South West Africa.

Common Zebra   *E. burchelli*
Dr W. J. Burchell (1782–1863) was an English zoologist who made an expedition to Africa in 1811; this zebra is widespread in Africa south of the Sahara. The nominate subspecies, that is the animal originally described in 1824 as *E. burchelli*, from the northern Cape Province and the Orange Free State, and now known as *E. b. burchelli*, has been extinct for many years.

Chapman's Zebra   *E. b. antiquoram*
antiquorum (L) the ancients, of old times. It was named after James Chapman, an English naturalist who travelled in Africa in the nineteenth century. Sometimes known as the Damaraland Zebra, it inhabits the southern part of Africa, ranging from Angola to the Transvaal.

Selous's Zebra   *E. b. selousi*
F. C. Selous (1857–1917) was a big game hunter in Africa from 1871
to 1917. This zebra lives in the southern part of Rhodesia, Malawi,
and the area between.

Grant's Zebra   *E. b. boehmi*
Named after Colonel J. A. Grant (1827–1892) who explored Central
Africa during the years 1860 to 1862, and Dr R. Böhm (1854–1884),
a German zoologist who was in Africa during the years 1880 to 1884.
Probably the commonest zebra, it ranges from Sudan and Ethiopia
south to Uganda, Kenya, Tanzania and Zambia.

Grevy's Zebra   *E. grevyi*
Named after Francois P. J. Grevy (1807–1891), who was President
of the French Republic from 1879 to 1887. This zebra is easily
distinguished by its large rounded ears and unusually narrow stripes;
it is interesting to note that since November 1912 it has been given
five different generic or subgeneric names, and four of these during
the course of one year! It inhabits Ethiopia, Somaliland, and northern
Kenya.

Suborder   CERATOMORPHA

keras (Gr), genitive *keratos*, a horn   *morphe* (Gr) form, shape; can
mean kind, sort; the 'horned-kind'; however it is only the rhinoceroses
that possess the nose-horn.

Family   TAPIRIDAE   4 species
Tapir is from the Tupi word *tapira*; the Tupis are a tribe of South
American aborigines living in the Amazon area.

Malayan Tapir   *Tapirus indicus*
Indicus (L) of India; the name is misleading as this tapir does not live
in India. When Professor Desmarest named it in 1819, he probably
meant to indicate the East Indies, as it lives in the Malay Peninsula
and Sumatra.

Brazilian Tapir   *T. terrestris*
terrestris (L) of the earth, land-dwelling. Inhabiting a large area in
Brazil and also Venezuela and other areas in the north.

Mountain or Woolly Tapir   *T. pinchaque*
La Pinchaque was a large fabulous animal believed to live in the same
range of mountains as this tapir, which inhabits the Cordillera

Occidental, a range of mountains in the north-west of Colombia in South America.

Baird's Tapir   *T. bairdi*
W. M. Baird was an American naturalist who made an expedition to Mexico in 1843. In fact, previous accounts of the animal had been recorded by W. T. White, another American naturalist. It lives in Mexico, Central America, and ranges south to Ecuador.

Family   RHINOCEROTIDAE   5 species
rhis (Gr), genitive *rhinos*, the nose   *keras* (Gr), genitive *keratos*, a horn.

Indian Rhinoceros   *Rhinoceros unicornis*
unus (L) one   *cornu* (L), genitive *cornus*, the horn of an animal; there is only one horn. It inhabits limited areas in Bengal, Assam and Nepal, and is now very rare.

Javan Rhinoceros   *R. sondaicus*
-icus (L) suffix meaning belonging to; the Malay islands are known as the Sunda Islands and the Sunda Strait lies between Java and Sumatra, hence *sondaicus*. This rhino is very rare and is confined to a small area in Java.

Sumatran Rhinoceros   *Didermocerus sumatrensis*
di- from *dis* (Gr) two, double   *derma* (Gr) skin   *keras* (Gr) a horn; the horn is composed of keratin, a substance derived from the skin, like finger-nails; there are two horns   *-ensis* (L) suffix meaning belonging to; it is not confined to Sumatra and was once widespread in south-eastern Asia, from Assam to Malaya. Like the other rhinos from this part of the world, it is very rare.

White Rhinoceros   *Diceros simus*
Diceros, see above   *simum* (L) snub-nosed, flat-nosed; it has a flat, wide nose and lips adapted for grazing. The white rhino is not white, and the black rhino is not black; they are both grey; the name may be a corruption of the Dutch word *weit*, meaning wide, and referring to the nose and lips. The numbers of all species have been seriously reduced because they are hunted for the horn, which is supposed to be an aphrodisiac when ground into a powder. It inhabits two separate areas of Africa, one in the south, and one central, to the north-west of Lake Victoria.

Black Rhinoceros   *D. bicornis*

Diceros, see above   *bi-* from *bis* (L) twice, two   *cornu* (L), genitive *cornus*, the horn of an animal; so here it is first in Greek for the genus, and then in Latin for the species; 'a two-horned two-horn'. This rhino is dark grey, not black, and has a narrow rather pointed nose and lips, adapted for browsing; see notes re White Rhinoceros, above. The Black Rhinoceros ranges over a wide area in the southern half and the eastern side of Africa.

60 Pigs, Camels, Deer, Giraffes, Antelopes and their kin ARTIODACTYLA

This group, together with the smaller group comprising the horses, zebras, rhinoceroses and their kin, are known as ungulates, though the term is no longer in regular use in classification. It means 'hoofed', from the Latin *unguis* a claw or hoof. They have an even number of toes or hoofs and so have been given the name Artiodactyla, meaning 'even-toed'; the main axis of the foot passes between the third and fourth digits, which are capped with hoofs. In most cases, these two digits, forming the 'cloven hoof', take all the weight, the others having disappeared during the course of evolution. In some cases, however, such as the hippopotamuses and the chevrotains, four digits are still in use, or at least visible.

The horses and rhinoceroses, already dealt with in Chapter 59, have an uneven number of toes or hoofs and so have been given the

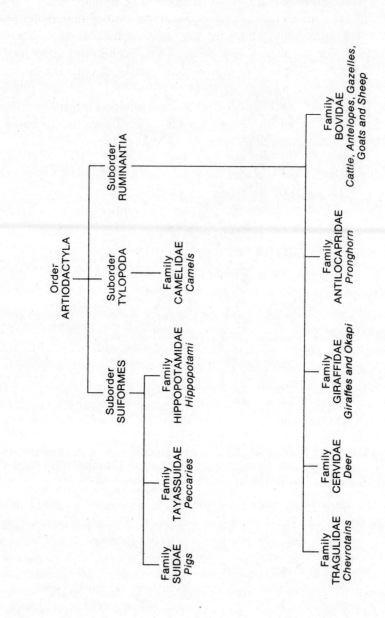

name Perissodactyla, meaning 'odd-toed'. All the animals in these two groups are related though very different in appearance.

Among the animals that follow will be found most of those that man has domesticated. Throughout the world, many millions of these are in use today for supplies of food, milk, wool, leather and other purposes. It is interesting to note that they are nearly all vegetarians and yet they are our main supply of meat.

For purposes of classification the order is divided into three suborders, namely Suiformes, meaning 'pig-like'; Tylopoda, meaning 'pad-footed'; and Ruminantia, meaning 'cud-chewers'.

Subclass EUTHERIA
Order ARTIODACTYLA
artios (Gr) complete, perfect of its kind; of numbers, even *daktulos* (Gr) a finger; can mean a toe.

Suborder SUIFORMES
sus (L), genitive *suis*, a pig *forma* (L) shape, sort, kind; the 'pig-kind'.

Family SUIDAE 8 or 9 species

African Bush Pig *Potamochoerus porcus*
potamos (Gr) a river *khoiros* (Gr) a pig; sometimes known as the River Hog, they are fond of wallowing in water in hot weather *porcus* (L) a pig or hog. Like most other pigs, it is not entirely vegetarian; the range is widespread in Africa south of the Sahara, and it is also found in Madagascar, where it is the only 'Ungulate'.

Giant Bornean Pig *Sus barbatus*
sus (L) a pig *barbatus* (L) bearded; it has a moustache, rather than a beard. Inhabiting Borneo, the Philippines and Malaya.

Eastern Wild Boar *S. verrucosus*
verruca (L) a wart *-osus* (L) suffix meaning 'full of'; it has three warts on each side of the head and muzzle. It inhabits Java.

European Wild Boar *S. scrofa*
scrofa (L) a breeding sow; this is the boar from which domestic pigs have been derived. It has a wide distribution, including Europe, North Africa, Asia, Sumatra, Java, Formosa and Japan.

Pygmy Hog *S. salvanius*
salvanius, belonging to Saul Forest in Nepal. The name was given by

Hodgson in 1847. Inhabiting Nepal, Sikkim and Bhutan. It was thought to be extinct but some specimens were found in 1971.

Wart Hog   *Phacochoerus aethiopicus*
phakos (Gr) a mole or wart on the body   *khoiros* (Gr) a pig or hog; it has warts of various sizes on both sides of the face   *aethiopicus* is often used to indicate Africa as a whole. This hog is widespread in Africa south of the Sahara.

Meinertzhagen's Giant Forest Hog   *Hylochoerus meinertzhageni*
hulē (Gr) a wood, forest   *khoiros* (Gr) a pig or hog, Colonel R. Meinertzhagen (1878–1967), the naturalist who discovered this hog, was in East Africa from 1902 to 1905. Like most other hogs it is not entirely vegetarian; it inhabits forested areas of the central part of Africa.

Family   TAYASSUIDAE   2 species
tajacu or *tayassu* is a Brazilian native name for the peccary.

Collared Peccary   *Tayassu tajacu*
Pecary was originally a Brazilian native (Tupi) name, and said to mean 'many paths through the woods', a reference to the animals habit. It has a narrow collar of light hair on the shoulders and is an agressive, fierce, hog-like animal with scimitar-like tusks. It will attack most living things and is not entirely vegetarian. Inhabiting Arizona, New Mexico, Texas, Mexico, Central America and the northern part of South America.

White-lipped Peccary   *T. albirostris*
albus (L) white   *rostrum* (L) the snout; it has white lips and a white moustache. Inhabiting an area similar to the Collared Peccary (above).

Family   HIPPOPOTAMIDAE   2 species
hippos (Gr) a horse   *potamos* (Gr) a river.

Hippopotamus   *Hippopotamus amphibius*
amphi (Gr) around, on both sides   *bios* (Gr) living, manner or means of living; 'living on both sides', i.e. land and water. In a strict sense, the hippo is not an amphibian as that sort of animal starts life equipped with gills, like a fish. The hippopotamus is widespread in the rivers and lakes of the southern half of Africa, but not now in the most southern part.

Pygmy Hippopotamus   *Choeropsis liberiensis*
khoiros (Gr) a pig or hog   *opsis* (Gr) aspect, appearance   *-ensis* (L)
suffix meaning belonging to; not confined to Liberia, it also inhabits
Guinea and Sierre Leone, but is now very rare.

Suborder   TYLOPODA

.tulē (Gr) a swelling or lump, a pad for carrying burdens   *pous* (Gr),
genitive *podos*, a foot; 'pad-footed' referring to the tough padded soles
of the feet.

Family   CAMELIDAE   6 species
camelus (L) a camel.

Arabian or One-humped Camel   *Camelus dromedarius*
A dromedary is really a domesticated camel bred for travel, for
running. The name comes from *dromeus* (Gr) a runner, and *-arius* (L)
a suffix meaning pertaining to, hence *dromedarius* (New L) a running
camel. It is unknown except in a domesticated state.

Bactrian Camel   *C. bactrianus*
Bactria is a province of the ancient Persian Empire   *-anus* (L) suffix
meaning belonging to. This is the two-humped camel, and also
domesticated, being almost unknown except in that state. However,
there may be some still living in the desert regions of China and
Central Asia.

Llama   *Lama peruana* or *L. glama*
Llama is a Peruvian name for this animal, which is closely related to
the camel, and probably originates from the Quéchua, an ancient
people of Peru   *glama*, a name given by Linnaeus, is a corruption of
llama. There are no llamas in the wild now, but large herds are
domesticated and kept by the South American Indians. Llamas are
used as pack animals and kept also for the hide, the fur for rugs and
clothing, and the flesh for food.

Guanaco   *L. guanicoe*
Guanaco is a Spanish name from the Quéchua people. This is a slightly
different animal from the llama and still survives in the wild, ranging
from Peru south to Patagonia in Argentina.

Alpaca   *L. pacos*
Paca and *pacos* are Peruvian names for the llama: *al* (Arabic, adopted
in Spanish) means 'the', so alpaca means 'the paca'; the word is used
in the English language to mean a cloth made from its wool. It no

longer survives in the wild but probably more than a million domesticated animals are kept in South America for their wool.

Vicuña   *Vicugna vicugna*
Vicuña is a Peruvian name for this animal; it still survives in the wild
and lives high up in the Andes in Peru, Bolivia, Chile and Argentina.
Like all the animals in this family, it is camel-like in many ways.

Suborder   RUMINANTIA
rumino (L) I chew the cud.

Family   TRAGULIDAE   4 species
tragos (Gr) a goat   *-ulus* (L) diminutive suffix.

Water Chevrotain   *Hyemoschus aquaticus*
hus (Gr), genitive *huos*, a hog   *moskhos* (Gr) musk; 'hog musk-deer',
referring to the type of skull and its pig-like habits   *aquaticus* (L) living
in water or near water. It inhabits Liberia and part of Nigeria.

Indian Chevrotain   *Tragulus meminna*
tragos (Gr) a goat   *-ulus* (L) diminutive suffix; it is not a goat, but
somewhat resembling a very small deer and structurally more akin
to pigs and camels   *memina* (Ceylonese) a small deer. Chevrotain is
from the French *chèvre*, a she-goat. It lives in the southern part of
Asia.

Larger Malay Chevrotain   *T. napu*
Napu is a native name for the chevrotain in Sumatra. Inhabiting
southern Asia, Sumatra and Borneo.

Lesser Malay Chevrotain   *T. javanicus*
-icus (L) suffix meaning belonging to; the range is similar to other
chevrotains but it is also found in Java. Sometimes known as the
Mouse Deer it is only about 30 cm (1 ft) high.

Family   CERVIDAE   between 30 and 40 species
cervus (L) a stag, a deer.

Musk Deer   *Moschus moschiferus*
moskhos (Gr) musk, *moschus* (New L) musk   *fero* (L) I bear, I carry;
the male has a pouch on the abdomen and the glands in this pouch
produce a strongly scented musk. It is widespread in Asia ranging
from the north down to southern China.

Chinese Water Deer *Hydropotes inermis*
hudōr (Gr) water, *hudr-* as a prefix *potēs* (Gr) a drinker; referring to
the animal's liking for marshy ground *inermis* (L) unarmed; it does
not have antlers. Inhabiting eastern Asia, and feral in Great Britain.

Indian Muntjac *Muntiacus muntjak*
Muntiacus (New L) from *muntjak*, the native name in the Sunda
language, in western Java. It has a wide range, from India to southern
China, Burma, Malaya, Sumatra, Borneo and Java; it is sometimes
known as the Javan Muntjac.

Chinese Muntjac *M. reevesi*
Named after John Reeves, FRS (1774–1856), a British naturalist who
was resident in China during the years 1812 to 1831. He studied the
natural history of the country and sent specimens to England. Reeves's
Long-tailed Pheasant *Syrmaticus reevesi* is also named in his honour.
This muntjac inhabits China, Burma and Thailand, and is feral in
Great Britain.

Black or Hairy-fronted Muntjac *M. crinifrons*
crinis (L) hair *frons* (L) forehead, brow. It inhabits south-east Asia.

Tenasserim Muntjac *M. feae*
Leonardo Fea (1852–1903) was an Italian zoologist who collected in
Burma. Tenasserim is a Division in Burma and there is also a town
and a river with this name. It inhabits south-east Asia.

Roosevelts' Muntjac *M. rooseveltorum*
The unusual ending of the specific name is the genitive plural, and
refers to the two sons of Theodore Roosevelt, at one time President
of the USA. It inhabits the southern part of Asia.

Tufted Deer *Elaphodus cephalophus*
elaphos (Gr) a deer *odous* (Gr) a tooth; referring to the large upper
canines of the male *kephalē* (Gr) the head *lophos* (Gr) a crest; the
male has a crest of hair on the forehead and around the base of the
antlers. Inhabiting China.

Fallow Deer *Dama dama*
dama (*damma*) (L) a general name for animals of the deer kind. Fallow
is from Old English *falu*, meaning brownish yellow. Originally from
the Mediterranean region of southern Europe and Asiatic Turkey,
it is now widespread in Europe, including the British Isles.

Persian Fallow Deer   *D. mesopotamica*
mesos (Gr) middle, between   *potamos* (Gr) a river; between two rivers, the Euphrates and the Tigris, hence Mesopotamia, now Iraq. It is found in a small area in Iran.

Chital or Axis Deer   *Axis axis*
Axis (L) is said to be Pliny's name for this deer, though some records only show it as 'an unknown wild animal in India'. It inhabits southern Asia including Sri Lanka. The name chital comes from *cital* (Hind) meaning spotted, variegated; it has conspicuous white spots on a reddish-brown coat. The name cheetah (*Acinonyx jubatus*, page   ) is from the same source.

Hog Deer   *A. porcinus*
porcus (L) a hog   *-inus* (L) suffix meaning like; referring to its supposed likeness to a hog. It is found in India, Assam, Burma, Thailand and Vietnam.

Kuhl's or Bawean Deer   *A. kuhli*
Dr H. Kuhl (1796–1821) was a German naturalist who was in the East Indies in 1820 and 1821. This deer is confined to Bawean Island, a small island in the Java Sea, between Java and Borneo.

Calamian Deer   *A. calamianensis*
This deer is found only on a group of islands in the Philippines, known as the Calamian Islands, between Mindoro and Palawan. The species is in danger of becoming extinct.

Thorold's Deer   *Cervus albirostris*
cervus (L) a stag, a deer   *albus* (L) white   *rostrum* (L) the snout; it has a white muzzle and lips. Dr W. G. Thorold obtained the second two specimens in Tibet in 1891. This deer was originally discovered by Prjevalski in 1879. Inhabiting part of China and Tibet.

Barasingha or Swamp Deer   *C. duvauceli*
A. Duvaucel (1796–1824) was a French naturalist. *Barasingha* is Hindi (a language of northern India) meaning 'twelve horns', from *bārah*, twelve, and *sig*, a horn. This deer usually has a total of twelve tines on the antlers. Inhabiting India and Assam.

Red Deer   *C. elaphus*
elaphos (Gr) a deer. Inhabiting Europe, including the British Isles, and part of northern Africa and Asia. It is estimated that there are over 150,000 living in the Highlands of Scotland.

Wapiti *C. canadensis*
-ensis (L) suffix meaning belonging to; it is not confined to Canada. *Wapiti* is an Algonkian name for this large deer; the Algonkin, an American Indian tribe, still exist today in some reserves in Ontario and Quebec. The Wapiti inhabits North America on the western side, and because of its large size it is known there as the Elk; this is zoologically incorrect and Wapiti is a better name.

Eld's Deer *C. eldi*
Named after Percy Eld who discovered this 'nondescript' species in Assam in 1842. Also known as the Thamin, this is a Burmese name for this deer. It is confined to a small area in southern Burma and in danger of becoming extinct.

Sika *C. nippon*
nippon (Jap) Japan *sika* (Jap) a small deer; the Sika is regarded as sacred by the Japanese. Now rare, but still found in southern Asia, Formosa (now Taiwan) and Japan, and feral in Great Britain.

Schomburgk's Deer *C. schomburgki*
Sir Robert H. Schomburgk (1804–1865) was the British Consul in Bangkok from 1857 to 1864. This deer is very rare and confined to a small area in Thailand; it may even be extinct.

Rusa Deer *C. timorensis*
-ensis (L) belonging to; it is not confined to Timor and inhabits Celebes, Java, Borneo and other islands in that area. *Rūsā* is the Hindi name for this deer.

Sambar *C. unicolor*
unicolor (L) of one colour; the coat is a rather dull brown throughout though there may be some white hair on the neck of the male. *Sambar* is another Hindi name for this deer; it is widespread in southern Asia.

Père David's Deer *Elaphurus davidianus*
elaphos (Gr) a deer *oura* (Gr) the tail; this refers to the unusually long tail, almost like that of a donkey. Père Armand David (1826–1900) was a missionary in China and a keen naturalist. He observed this deer in the Imperial Hunting Park near Peking. Originally from China it is unknown in the wild though there may be as many as 300 in zoos and wildlife sanctuaries.

Mule or Hollow-toothed Deer *Odocoileus hemionus*
odous (Gr) a tooth *koilos* (Gr) hollow; referring to the well-hollowed

teeth   *hēmionos* (Gr) a half-ass, a mule; it has very large ears like a mule. Inhabiting the western part of North America from Alaska in the north to Mexico in the south.

White-tailed Deer   *O. virginianus*
-*anus* (L) suffix meaning belonging to; it is not confined to Virginia and is widespread in the southern half of North America and the northern half of South America. In contrast to the black tail of *O. hēmionus* (above), it has a white tail and rump. The tail is raised as a danger signal when the animal is in flight from predators.

Roe Deer   *Capreolus capreolus*
caper (L) a goat   -*olus* (L) diminutive suffix   *capreolus* (L) the roe-buck. The Roe Deer is widespread in Europe including the British Isles and ranges to the east through China as far as Korea.

Moose or Elk   *Alces alces*
alces (L) the elk. It is the largest living species of deer and it has been said the name derives from the Greek *alkē*, meaning strength. Moose is derived from *musee*, an Algonkian Indian name. Usually known as a moose in America and an elk in the Old World, it inhabits northern parts of North America, northern Europe, and northern Asiatic Russia.

Reindeer   *Rangifer tarandus*
Old French *rangier*, a reindeer, and *ferus* (L) wild, untamed; also said to come from *ren* (Old Swed) a reindeer, and *ferus*; *tarandrus* (L) an animal of northern countries, and according to Cuvier the reindeer. It inhabits the tundras and northern woodlands of Europe, Asia, North America and Greenland. Sometimes known as the Caribou.

Marsh Deer   *Blastocerus dichotomus*
blastos (Gr) a shoot, or bud   *keras* (Gr) a horn; the horns are said to resemble a bud   *dikhē* (Gr) in two ways   *tomē* (Gr) cutting, sharp; the small antlers are doubly forked. Inhabiting central Brazil and south to northern Argentina.

Pampas Deer   *Ozotoceras bezoarticus*
ozōtos (Gr) branched, forked   *keras* (Gr) the horn of an animal; an allusion to the large, complex, forked antlers   *bezoar* (Sp) a wild goat -*icus* (L) belonging to (for meaning of bezoar see Wild Goat, page 650). The Pampas Deer inhabits central Brazil and south to Argentina.

Andean Deer or Guemal *Hippocamelus antisensis*
hippos (Gr) a horse *kamēlos* (Gr) a camel; the scientific name of this deer has been changed several times in the past. It has been described as a 'horse-camel'. It really bears no resemblance to either of these animals, but was considered intermediate between a horse and a llama. Antisana is the name of a 5,800 m (19,000 ft) peak in the Andes Mountains where this deer lives *-ensis* (L) suffix meaning belonging to *guemul* or *guemal* is the American Spanish name for this small deer.

Red Brocket *Mazama americana*
Mazame or *maçame* were Mexican names given to some species of deer in the seventeenth century *brocart* (Fr) a brocket *broc* (Old Fr) tine of a stag's horn. Widespread in Central America and the northern part of South America.

Little Red Brocket *M. rufina*
rufus (L) red *-inus* (L) like, pertaining to. Inhabiting South America.

Brown Brocket *M. gouazoubira*
Guazú-birá is a Guarani name in Paraguay for a brocket; a good example of a 'barbarism' (see page 15). Inhabiting Central and South America.

Pudu *Pudu pudu*
Pudu is Spanish from the Mapuche, a people of southern Chile. This very small deer, standing only about 30 cm (12 in) high, is now found only in the southern Andes in Argentina and on the nearby Chiloe Island.

Family GIRAFFIDAE 2 species
giraffa (New L) a giraffe, from the Arabic *zarāfah*, meaning 'one who walks swiftly'.

Okapi *Okapia johnstoni*
Okapi is a native name used by the pygmies in the Semliki Forest area, where this animal lives. It is related to the giraffe, though very different in appearance and habitat. Sir Harry H. Johnston (1858-1927), explorer and author, was in the Colonial Administration of British Central Africa when he discovered the okapi in 1900. A very strange animal, it was given the name *Equus ? johnstoni*, 'Johnston's Horse'. However, it soon proved to be nothing like a horse, and was related to the giraffe, and given the name *Okapia johnstoni*. Sometimes known

as the Forest Giraffe, it is found only in the Semliki Forest area of
Zaire.

Giraffe   *Giraffa camelopardalis*

camelus (L) a camel   *pardus* (L) a panther or a leopard   *-alis* (L)
relating to, like; this could be interpreted as 'a camel marked like a
leopard', but the name does not seem very apt. There is only one
species in the genus *Giraffa*, but there is so much variation of colour
and marking that a number of subspecies are recognised, most authors
giving as many as eight; four well known subspecies are given below.
The giraffe is found only in Africa.

Northern Giraffe   *G. c. camelopardalis*

Sometimes known as the Nubian Giraffe, it inhabits Nubia, in Sudan,
and parts of Ethiopia.

Reticulated Giraffe   *G. c. reticulata*

reticulum (L) a little net, hence *reticulatus* (L) reticulated, marked like
a net; probably the most handsome giraffe.

Masai Giraffe   *G. c. tippelskirchi*

E. L. von Tippelskirch (1774–1840) was a German General and
explorer. This giraffe inhabits the Masai Steppe in Tanzania.

Angola Giraffe   *G. c. angolensis*

-ensis (L) suffix meaning belonging to; inhabiting Angola and that
area on the western side of South Africa, it is now very rare.

Family   ANTILOCAPRIDAE   1 species

antholops (Late Gr) a horned animal, probably an antelope   *capra* (L)
a she-goat. Cuvier suggests that *Antilope* is a corruption of antholops
... 'which seems to refer to the beautiful eyes of the animal'   *anthos*
(Gr) a flower, and *ops* (Gr) the eye; '*Ce nom n'est pas ancien, il est corrompu
d'antholops . . . qui semble se rapporter aux beaux yeux de l'animal.*'

Pronghorn   *Antilocapra americana*

A solitary species and not one of the true antelopes, which are found
only in Africa and Asia. The horns are shed every year, as with deer,
but the bony core remains, as with antelopes. It is the last survivor
of an ancient family that lived only in North America. The name
pronghorn refers to the short forward-pointing prong on the upper
part of the horns. It is now found only in the western part of North
America.

Family   BOVIDAE

Zoologists do not agree about the number of species; estimates vary from 100 to 154; this is because some have raised certain subspecies to full specific rank, or have decided that certain species are really only subspecies; the latter are not counted when estimating the number of species in a family

bos (L), genitive *bovis*, an ox.

Red-flanked Duiker   *Cephalophus rufilatus*

kephalē (Gr) the head   *lophus* (Gr) a crest; referring to the tuft of hair on the head   *rufus* (L) red   *latus* (L) broad, wide; can mean the side, flank, of men or animals. It inhabits a large area in the central part of western Africa. Duikers are small antelopes, most of them only about 35-45 cm (14-18 in) high   *duiker* is Afrikaans for a diver and supposed to be because they 'dive' into the undergrowth when alarmed.

Blue Duiker   *C. monticola*

mons (L), genitive *montis*, a mountain   *colo* (L) I cultivate; can mean I dwell in a place, I inhabit, hence *monticola* (L) a dweller among mountains. This is the smallest species, being only about the size of a hare; it has a bluish grey coat. Inhabiting a large area of central Africa and ranging south, on the eastern side, down to South Africa.

Zebra Antelope or Banded Duiker   *C. zebra*

zibra (Congolese or possibly Abyssinian) a zebra; can also mean striped; it has a bright orange-coloured coat with black bands or stripes across the back. The rarest species and found only in Liberia and surrounding areas.

Red Duiker   *C. natalensis*

-ensis (L) suffix meaning belonging to; it is not confined to Natal and is fairly widespread on the eastern side of South Africa.

Black Duiker   *C. niger*

niger (L) black; can mean dark, dusky; the coat is dark grey. Inhabiting Liberia, Ivory Coast, Ghana and surrounding areas.

Grey Duiker   *Sylvicapra grimmia*

silva (L) a wood, a forest   *capra* (L) a she-goat; it is not a goat, of course, but a very small antelope like *C. niger* (above) with a blue-grey coat. It is named in honour of Dr Hermann Nicolas Grimm, a

German scientist, who described the duiker as early as 1686, and was given the name by Linnaeus in 1758.

Royal or Pygmy Antelope   *Neotragus pygmaeus*
neos (Gr) new   *tragos* (Gr) a he-goat; when named, in 1827, it was thought to be a new kind of antelope   *pugmē* (Gr) the fist, hence *pugmaios* (Gr) as small as a fist, dwarfish; probably the smallest antelope known, less than 30 cm (12 in) high. On account of its size it became known locally as the King of Hares, hence Royal Antelope. Inhabiting Ghana.

Bates's Dwarf Antelope   *N. batesi*
Named after George Latimer Bates (1883–1940), who settled as a farmer in Cameroun and devoted himself to collecting birds of West Africa. By all accounts he was a retiring and modest man but a most accurate recorder of zoological detail. This very small antelope lives in the Cameroun area of West Africa.

Zanzibar Antelope or Suni   *Nesotragus moschatus*
nesos (Gr) an island   *tragos* (Gr) a he-goat; originally from French Island, or possibly from Chapani Island, close to Zanzibar   *moschatus* (New L) musky; it has musk glands on the feet. It is not confined to any island and can be found in a wide area of territory along the east coast of Africa southward to Mozambique. *Suni* is a native name in south-eastern Africa.

Salt's Dik-dik   *Madoqua saltiana*
Madoqua, from the Amharic (an Ethiopian language) *medaqqwa* a small antelope. Sir Henry Salt (1780–1827) an explorer, was also British Consul General in Alexandria from 1815 to 1827. The name dik-dik is an imitation of the female's cry of alarm. It lives in northern Ethiopia.

Kirk's Dik-dik   *Rhynchotragus kirki* (or *Madoqua*)
rhynkhos (Gr) the snout   *tragos* (Gr) a he-goat; a reference to the trunk-like flexible nose. Sir John Kirk (1832–1922), a Scottish naturalist, was British Consul General in Zanzibar from 1880 to 1887 and was formerly physician and naturalist to Dr Livingstone on his second journey. This dik-dik inhabits East Africa and Namibia.

Klipspringer   *Oreotragus oreotragus*
oros (Gr), genitive *oreos*, a mountain   *tragos* (Gr) a he-goat   *klip* is Dutch for a rock; a 'rock-jumper'; it walks on the tips of its hoofs, and

its agility and sure-footedness on rocks is incredible. Living in mountainous rocky areas throughout Africa.

Beira Antelope *Dorcatragus megalotis*
dorkas (Gr) a gazelle or antelope *tragos* (Gr) a he-goat *megas* (Gr) big *ous* (Gr), genitive *ōtos*, the ear; the ears are unusually large and wide *beira* is from *behra*, the Somali name for this antelope. It is a rock-dweller in the mountains of Ethiopia and Somaliland.

Steinbok *Raphicerus campestris*
raphis (Gr) a needle, a pin *keras* (Gr) the horn of an animal; the horns are only short spikes *campestris* (L) level country, a plain. Steinbok is from *steen* (Du) a stone, and *bok* (Du) a buck. Inhabiting South Africa and part of East Asia.

Grysbok *R. melanotis*
melas (Gr) black *ous* (Gr) genitive *ōtos*, the ear. Grysbock is from *grys* (Du) grey, and *bok* (Du) a buck; there is white hair scattered in the coat. It is found only in the most southern part of South Africa.

Oribi *Ourebia ourebi*
The name *oribi* is a Hottentot word, of the Nama dialect. This small graceful antelope is widespread in Africa south of the Sahara.

Chousingha or Four-horned Antelope *Tetracerus quadricornis*
tetra (Gr) four *keras* (Gr) the horn of an animal *quadri-* (L) prefix meaning four *cornu* (L) the horn of an animal. *Chousingha* is a Hindustani name meaning 'four horns'. Inhabiting India.

Nilgai or Nilgau *Boselaphus tragocamelus*
bos (L) an ox *elaphos* (Gr) a deer *tragos* (Gr) a he-goat *kamēlos* (Gr) a camel; a peculiar mixture of names for this rather beautiful antelope with a pale blue coat. Nilgau is from the Hindustani word *nil*, meaning blue, and the Persian word *gaw*, a cow. Inhabiting India and Tibet.

Bushbuck *Tragelaphus scriptus*
tragos (Gr) a he-goat *elaphos* (Gr) a deer; in most cases tragos, a goat, is used in the sense of antelope *scribo* (L) I write, thus *scriptum*, something written; it has conspicuous markings of white stripes and spots on a reddish-brown coat. This is the small member of the genus *Tragelaphus* and it inhabits almost the whole of Africa south of the Sahara.

Sitatunga   *T. spekei*
Named after Captain J. H. Speke (1827–1864) the explorer of Central
Africa. He made expeditions with J. A. Grant and also Sir Richard
Burton. The name *sitatunga* is from an archaic Bantu language of
Rhodesia. Inhabiting a large area of Africa including Rhodesia,
Zaire, and ranging to the east and west coasts.

Nyala   *T. angasi*
George French Angas (1822–1886) was an English explorer, artist
and zoologist. He went to South Africa in 1846 and made drawings
of men and various animals. *Nyala* is a Swahili name for this bushbuck
inhabiting south-east Africa.

Mountain Nyala   *T. buxtoni*
The first specimen was found by a Mr Ivor Buxton in the Abyssinian
highlands in 1909. This nyala is found only in Ethiopia, formerly
Abyssinia.

Lesser Kudu   *T. imberbis*
imberbis (L) beardless; referring to the absence of the throat mane.
Inhabiting Somalia and north-eastern Kenya.

Greater Kudu   *T. strepsiceros*
strepho (Gr) I twist, so *strepsis*, a twisting   *keras* (Gr) the horn of an
animal; it has remarkable spirally twisted horns   *kudu*, or *koodoo*, is
the Hottentot name for this antelope. It inhabits Ethiopia, eastern
Africa and South Africa.

Common Eland   *Taurotragus oryx oryx*
tauros (Gr) a bull   *tragos* (Gr) a he-goat   *orux* (Gr) a gazelle or
antelope. This large antelope lives in the southern part of Africa.

Giant Eland   *T. o. derbianus*
-anus (L) suffix meaning belonging to. The Thirteenth Earl of Derby
(1775–1851) was formerly the Hon. E. S. Stanley, President of the
Zoological Society of London in 1831 (not to be confused with Sir
Henry Morton Stanley, famous for exploring Africa and his meeting
with Livingstone at Ujiji in 1871). This eland is even bigger than the
Common Eland (above), and inhabits Sudan, part of western Africa,
and the northern part of Zaire.

Bongo   *T. euryceros*
eurus (Gr) broad, widespread   *keras* (Gr) the horn of an animal; a

reference to the long and elegant horns: bongo is an African native name. It inhabits a belt of forest country from Guinea in the west, across Central Africa and northern Zaire, to Kenya.

South African Oryx or Gemsbok *Oryx gazella gazella*
orux (Gr) a gazelle or antelope *gazella* (New L) from *ghazāl* (Ar) a wild goat. Gemsbok is from *gemse* (Ger) the chamois, and *bok* (Du) a buck. Inhabiting the desert areas of South Africa.

East African Oryx or Beisa *O. g. beisa*
Beisa is from *bezā*, Amharic, an Ethiopian language from the Amhara district where this oryx lives.

Tufted Oryx *O. g. callotis*
kallos (Gr) beauty; can mean a beautiful object *ous* (Gr), genitive *ōtos*, the ear; a reference to the long tufts of hair on the ears. Inhabiting the southern part of Somaliland and part of northern Kenya.

Scimitar Oryx *O. dammah*
damma (L) a fallow deer, an antelope; also *dammar* (Ar) a sheep. Sometimes known as the Scimitar-horned Oryx, the horns are sabre-shaped and bent downwards. It inhabits the southern part of the Sahara.

Arabian or White Oryx *O. leucoryx*
orux (Gr) a gazelle or antelope *leukos* (Gr) white. This oryx is near to extinction and recently the Sultan of Muscat and Oman has prohibited their hunting. Some specimens have been taken to Arizona, USA, for protection. It inhabits the desert in the south-west of Saudi Arabia.

Addax *Addax nasomaculatus*
addax (L) a wild animal with crooked horns, derived from an African word *nasus* (L) the nose *macula* (L) a spot, a mark *-atus* (L) suffix meaning provided with; this refers to brown patches on the nose; it has spirally twisted horns. Becoming rare due to hunting, it is still found in parts of the Sahara.

Waterbuck *Kobus ellipsiprymnus*
kobus (New L) from *koba*, an African native name *ellipēs* (Gr) wanting, defective; an ellipse is a shape deviating from a circle, i.e. a 'defective circle' *prumnos* (Gr) the hind part; this refers to a conspicuous ellipse-shaped white ring on the rump. Widespread in most of Africa south of the Sahara, but not the extreme south of South Africa.

Defassa Waterbuck   *K. (e.) defassa*
defassa (New L) probably from a native name. A possible subgenus
of *K. ellipsiprymnus*. Widespread in areas not inhabited by the common
waterbuck, as far as the west coast of Africa south of the Sahara.

Buffon's Kob   *K. kob*
Named in honour of the great French naturalist Comte de Buffon
(1707-1788). Inhabiting Africa south of the Sahara and also part of
Angola.

Puku   *K. vardoni*
Named in honour of Major Frank Vardon, an English elephant
hunter, and a friend of Livingstone when in Africa about the year
1850; he wrote the first scientific paper on the Tsetse Fly. Puku is an
African native name. This waterbuck inhabits the Zambezi area.

Lechwe   *K. leche*
Lechwe is a name of Bantu origin meaning an antelope. Widespread
in the Zambezi area, and parts of Angola, Botswana and South West
Africa.

Mrs Gray's or Nile Lechwe   *K. megaceros*
megas (Gr) big, wide   *keras* (Gr) the horn of an animal. Named *Kobus
maria* in 1859 by Dr J. E. Gray FRS (1800-1875), keeper at the British
Museum, in honour of his wife Maria E. Gray (1787-1876); but the
name had to be replaced by the earlier *K. megaceros* (1855). Mrs Gray
was herself a talented artist of molluscs, and the popular English name
has survived. This lechwe is a rare species inhabiting the Bahr-el-
Ghazal and White Nile area in southern Sudan.

Common Reedbuck   *Redunca arundinum*
reduncas (L) bent backwards, curved; most people would describe the
horns as 'bent forward', though they do start at an angle backwards
from the head   *arundo* (*harundo*) (L) a reed, hence *arundinum*, pertaining to reeds; they are usually found in the vicinity of water. Inhabiting
the southern half of Africa.

Bohor Reedbuck   *R. redunca*
Bohor is an Ethiopian (Amharic) name for this reedbuck. Inhabiting
central parts of Africa from west coast to east coast.

Mountain Reedbuck   *R. fulvorufula*
fulvus (L) tawny   *rufus* (L) red, so *rufulus*, somewhat red, reddish;
it is usually more tawny than red. Not always found near water in its

mountain habitat, it lives in scattered areas in Nigeria, Cameroun, Kenya and eastern South Africa.

Impala   *Aepyceros melampus*
aipos (Gr) high, lofty   *keras* (Gr) the horn of an animal; referring to the long lyre-shaped horns of the male   *melas* (Gr) black   *pous* (Gr) the foot   *melampous* (Gr) black-footed; sometimes used in ancient Greece as a proper name, Blackfoot, but in this case an allusion to the tufts of black hair covering a gland on the heel of the hind legs. *Impala* is a Zulu name. It inhabits a large area of southern Africa extending as far north as Uganda.

Roan Antelope   *Hippotragus equinus*
hippos (Gr) a horse   *tragos* (Gr) a he-goat   *equus* (L) a horse and so *equinus*, relating to horses; the body is similar to that of a horse, with a reddish-brown coat. The range is widespread in Africa south of the Sahara.

Sable Antelope   *H. niger niger*
niger (L) black, dark coloured; the coat of both male and female is not jet-black, though in older bulls the coat becomes black; in heraldry sable means black. It inhabits the forests of the southern half of Africa but not found in the extreme south.

Giant Sable Antelope   *H. n. variani*
The type specimen was obtained in 1913 by H. F. Varian, chief engineer on the Benguela Railway, Angola. This large antelope is very rare, and only found in Angola.

Blesbok   *Damaliscus dorcas*
damalis (Gr) a young cow, a heifer   *-iscus* (L) diminutive suffix *dorkas* (Gr) an antelope or gazelle; a small antelope with a white patch on the forehead and nose. The name blesbok comes from *bles* (Du) a mark or blaze, and *bok* (Du) a buck, alluding to the white blaze on the forehead. A rare antelope found only in South Africa, and protected in some game reserves.

Hunter's Hartebeest   *D. hunteri*
H. C. V. Hunter (1861–1934) was a big game hunter and zoologist; he discovered this antelope in 1888 about 240 km (150 miles) up the Tana River in Kenya.

A rare hartebeest found only in a small area in southern Somalia and northern Kenya. The name is a combination of 'hart' and 'beast'; the hart is a male deer and the hind is a female.

Bontebok   *D. pygargus*
pugē (Gr) the rump, buttocks   *argos* (Gr) shining, bright   *pugargos* (Gr) white-rump; it has a conspicuous white rump and tail   *bont* (Du) particoloured   *bok* (Du) a buck. Found only in the south-west of South Africa, it is now extremely rare.

Korrigum or Topi   *D. korrigum*
Korrigum is from *kargum* (Kinuri) the language of a tribe living near Lake Chad, Niger   *topi* is a native name of Mande origin, akin to *ndope*, an antelope. Another rare antelope from West Africa.

Common Hartebeest   *Alcelaphus buselaphus*
alkē (Gr) the elk   *elaphos* (Gr) a deer   *bous* (Gr) a bullock or cow. For hartebeest, see above. Inhabiting a wide area in central Africa stretching from the west coast to the east.

Lichtenstein's Hartebeest   *A. lichtensteini*
M. H. C. Lichtenstein (1780–1857) was Director of Zoology at the Berlin Museum in 1815; he made an expedition to South Africa in 1804. This hartebeest lives in Tanzania, Mozambique and Zambia.

White-tailed Gnu or Wildebeest   *Connochaetes gnou*
konnos (Gr) the beard   *khaitē* (Gr) flowing hair, can mean a mane; a reference to the conspicuous beard and mane   *gnou* is a Hottentot name for these peculiar and rather ugly antelopes. This one is a very rare species, and probably now only existing in protected areas in Namibia and in zoos.

Southern Brindled or Blue Gnu   *C. taurinus taurinus*
taurus (L) a bull; taurinus, like a bull. Inhabiting the southern part of Africa.

White-bearded Gnu   *C. t. albojubatus*
albus (L) white   *juba* (= *iuba*) (L) the mane of an animal, hence *jubatus*, having a mane; the long hairs on the throat and beneath the neck of an animal are sometimes known as the mane. Inhabiting East Africa and parts of South Africa.

Indian Antelope or Blackbuck   *Antilope cervicapra*
antholops (Gr) an antelope: (see page 638)   *cervus* (L) a deer   *capra* (L) a she-goat. The only species in the genus *Antilope*, it is widespread in India.

Springbok *Antidorcas marsupialis*
anti- (Gr) against, opposed to; in composition can mean resemblance
to the word that follows *dorkas* (Gr) a gazelle; a reference to their
likeness to *Gazella*; *marsupium* (L) a pouch, a pocket *-alis* (L) suffix
meaning similar to, like; on the back there is a long pocket of skin
which can be opened at will, revealing a conspicuous mass of white
hair; this serves as a danger signal. Inhabiting semi-desert areas of
South Africa.

Goa or Tibetan Gazelle *Procapra picticaudata*
pro (L) before *capra* (L) a she-goat; suggesting the ancestral or original type of *Capra*; *pictus* (L) painted *cauda* (L) the tail of an animal
-atus (L) suffix meaning provided with; 'having a painted tail'; a
reference to its conspicuous white tail and rump; goa is from *dgoba*
(Tibetan) a gazelle. It inhabits Tibet and neighbouring areas of Asia.

Mongolian Gazelle *P. gutturosa*
guttur (L) the throat *-osus* (L) suffix meaning full of; the males have
an enlargement of the neck and throat during the mating season
which resembles a goitre. Inhabiting Mongolia and neighbouring
areas of Asia.

Grant's Gazelle *Gazella granti*
ghazal (Ar) a wild goat *-ellus* (L) diminutive suffix; named after
Colonel J. A. Grant (1827–1892) who explored central Africa during
the years 1860 to 1862, and on one occasion with J. H. Speke. This
gazelle inhabits Kenya and Tanzania.

Mountain Gazelle *G. gazella*
Inhabiting the mountains of North Africa and to some extent the
northern part of the Sahara; also Arabia and south-western Asia.

Dorcas Gazelle *G. dorcas*
dorkas (Gr) a gazelle. Inhabiting northern Africa including parts of
the Sahara, and Saudi Arabia.

Thomson's Gazelle *G. thomsoni*
Named after Joseph Thomson (1858–1895) who explored the area of
the Central African Lakes during the years 1879 to 1883. These
beautiful little gazelles live in that area, particularly in the Serengeti
National Park. They are always known locally by the popular name
'Tommies'.

Heuglin's Gazelle   *G. tilonura*
ptilon (Gr) a feather, down; also *tilos* (Gr) anything pulled or shredded, flock, down   *oura* (Gr) the tail; 'feathery-tailed'. M. T. von Heuglin (1824–1876) was a German zoologist who was in Africa from 1851 to 1864. This gazelle is becoming rare; it is still found in a small area in the eastern part of Sudan and northern Ethiopia.

Cuvier's Gazelle or Edmi   *G. cuvieri*
Baron G. L. Cuvier (1769–1832) was the famous French comparative anatomist and Professor of Natural History   *edmi* or *idmi* is an Arabic local native name. This gazelle is now very rare; it inhabits the northern part of the Sahara.

Pelzeln's Gazelle   *G. pelzelni*
A. von Pelzeln (1825–1891) was a zoologist and Custodian of the Vienna Museum from 1859 to 1883. This gazelle inhabits Somaliland and the eastern part of Ethiopia.

Goitred or Persian Gazelle   *G. subgutturosa*
sub (L) below   *guttur* (L) the throat   *-osus* (L) suffix meaning full of; the male has an enlargement of the neck and throat during the mating season which resembles a goitre. In fact, a goitre is a swelling of the throat caused by enlargement of the thyroid gland and has no connection with the enlargement of the throat of the gazelle. This gazelle inhabits a large area in central Asia including part of Iran at the western end of the range.

Clarke's Gazelle or Dibatag   *Ammodorcas clarkei*
ammos (Gr) sand, a sandy place   *dorkas* (Gr) a gazelle; a reference to its dry, sandy habitat; a specimen was obtained by T. W. H. Clarke in 1890 in the Marehan country, in the southern part of Somalia. Dibatag is from the Somali name *dabatag*. It inhabits Somaliland and Ethiopia.

Waller's Gazelle or Gerenuk   *Litocranius walleri*
lithos (Gr) stone   *kranion* (Gr) the upper part of the skull; 'stone-skull'; a reference to the skull which is almost solid bone at the base of the horns; named after the Rev. H. Waller (1833–1901) who was a missionary in Africa and a friend of the famous explorer Dr Livingstone. Gerenuk is from the Somali *garanug*, a long-necked gazelle. It inhabits Somaliland and part of neighbouring Ethiopia and Kenya.

Saiga Antelope *Saiga tatarica*
Saiga is a Russian name for an antelope   Tatary is an area in the
eastern part of European Russia   *-icus* (L) suffix meaning belonging
to. It has been in danger of extinction but can still be found in Kazakhstan and neighbouring areas. It has a dome-shaped nose, even larger
than that of *Pantholops* (below).

Tibetan Antelope or Chiru *Pantholops hodgsoni*
pas (neuter *pan*) (Gr) all   *antholops* (Gr) an antelope; a strange name.
T. S. Palmer, in his standard work *Index Generum Mammalium*, quotes
'The vulgar old name for the unicorn' (Hodgson). He goes on to
explain that when seen in profile the two horns appear like one, which
has given rise to the belief that the animal is the unicorn antelope
mentioned by the Abbé Huc. The Mr B. H. Hodgson FRS (1800–
1894) was a biologist who lived in Nepal during the years 1833 to
1843. Chiru is probably a local native name in Tibet. This goat-like
antelope has a peculiar dome-shaped nose, the purpose of which is still
under discussion among zoologists. It inhabits the high mountains of
Nepal and Tibet.

Grey Goral *Nemorhaedus goral goral*
nemus (L), genitive *nemoris*, a grove, a forest   *haedus* (L) a young goat,
a kid; an allusion to its habitat in mountainous and woody regions
goral is a native name from eastern India. This goat-antelope covers
a vast range including Afghanistan, Tibet, eastern Siberia and parts
of China and Korea.

Red Goral *N. g. cranbrooki*
The Fourth Earl of Cranbrook (born 1900) is a zoologist who was in
Burma in 1930, and was a Trustee of the British Museum of Natural
History in 1963. This goral has a shaggy coat of bright fox-red; it lives
in the mountains of northern Burma and Assam.

Maned Serow *Capricornis sumatraensis*
capra (L) a she-goat   *cornu* (L), genitive *cornus*, the horn of an animal;
a reference to the goat-like horns   *-ensis* (L) suffix meaning belonging.
It is not confined to Sumatra, and ranges over a wide area to the north
including Pakistan, Tibet and most of China. Serow is a name used
by the Lapchas, who inhabit Sikkim in the Himalayas, and now used
for the other goat-like animals in this genus.

Japanese Serow   *C. crispus crispus*
crispus (L) curly-headed. This serow is confined to Japan and neighbouring islands.

Formosan Serow   *C. c. swinhoii*
R. Swinhoe FRS (1836-1877) was at one time in the British Consular Service in China. This serow lives on the island of Formosa (Taiwan).

Rocky Mountain Goat   *Oreamnos americanus*
oros (Gr), genitive *oreos*, a mountain   *amnos* (Gr) a lamb. In the Rocky Mountains of Canada and the northern part of the USA, this aberrant goat-antelope, a relative of the Chamois, can still be found though it is becoming rare; some are protected in national parks.

Chamois   *Rupicapra rupicapra*
rupes (L), genitive *rupis*, a rock or cliff   *capra* (L) a she-goat. *Chamois* is French for a wild goat. This agile rock-climber lives high up in the mountains of Europe and western Asia.

Wild Goat   *Capra aegagrus*
capra (L) a she-goat   *aix* (Gr), genitive *aigos*, a goat   *agrios* (Gr) living in the fields; of animals, wild, hence *aigagros* (Gr) a wild goat. This goat is considered to be the ancestor of the domestic goat. It is sometimes known as the Bezoar Goat, bezoar being a stony substance found in the stomach of some ruminants such as goats and which used to be considered an antidote for all poisons. The word derives from *pādzahr* (Persian), from *pād*, protecting, and *zahr*, poison. Formerly widespread in Asia Minor and the Greek Islands, it is now found only in parts of Turkey, Georgia and Iran.

Alpine Ibex   *C. ibex ibex*
ibex (L) a kind of goat, a chamois. Inhabiting the mountains of southern Europe.

Siberian Ibex   *C. i. sibirica*
-icus (L) suffix meaning belonging to. Inhabiting parts of Siberia, Tibet and China.

Nubian Ibex   *C. i. nubiana*
-anus (L) suffix meaning belonging to; Nubia is a tract of country in northern Africa with no precise limits, lying between Egypt and Sudan. This ibex also ranges through Saudi Arabia and north to Syria.

Severtzow's Ibex *C. i. severtzovi*
Professor N. A. Severtzow (1827–1885) was a scientist who explored
Central Asia. This ibex lives in the western part of the Caucasus
Mountains.

Walia or Abyssinian Ibex *C. i. walie*
Walia was originally an Ethiopian native name. This ibex is in danger
of extinction and attempts are being made to preserve it. A fine
animal, it is confined to a small area in the mountains of the eastern
part of Ethiopia.

Caucasian Ibex or Tur *C. caucasica*
-icus (L) suffix meaning belonging to   *tur* (Russ) a Caucasian goat.
Inhabiting the Caucasus Mountains in southern Russia, along the
border with Georgia.

Pyrenees Ibex *C. pyrenaica pyrenaica*
This ibex is now extinct but is included because it is the nominate
subspecies.

Spanish Ibex *C. p. hispanica*
-icus (L) suffix meaning belonging to. A rare ibex in danger of extinction, it inhabits the Sierra Nevada mountains in Southern Spain.

Queen Victoria's Ibex *C. p. victoriae*
Named in honour of Queen Victoria Eugenie (formerly Princess Ena
of Battenberg) and the god-daughter of Queen Victoria; she married
King Alfonso XIII of Spain in 1906. Another very rare ibex in danger
of extinction, inhabiting Spain.

Markhor *C. falconeri*
Hugh Falconer (1808–1865) was a Scottish palaeontologist and
botanist in India. Markhor is a name that derives from the Persian
mār, a snake, and *khor*, eating. This is a strange name as goats are
vegetarians and no actual case of the markhor eating snakes has been
recorded, though they have been known to kill snakes. It inhabits the
mountains of Central Asia, originally including Afghanistan and
Persia, though it is now probably extinct in those countries.

There are varying opinions as to the number of species in *Capra* and
Ovis. The German systematist Th. Haltenorth reduces *Ovis* to one,
namely *ammon*, whilst we follow Desmond Morris and others with
seven. The Russian scientist Nabonov recognised as many as nine in
the Old World alone.

Barbary Sheep or Aoudad   *Ammotragus lervia*
ammos (Gr) sand   *tragos* (Gr) a goat; this is a reference to the colour
of its coat   *lervia*, from the wild sheep of northern Africa called
'Fishtall' or 'Lerwee' by the Rev T. Shaw in his *Travels and Observa-
tions* relating to several parts of Barbary and the Levant. It was named
by Pallas in 1777. *Aoudad* is a name used by the Berbers, a people of
northern Africa. It inhabits various parts around the Sahara, particularly Kordofan in Sudan, and Barbary, which is the old name for the
belt of land north of the Sahara stretching from Egypt to the Atlantic.
Berber is from the Arabic *Barbar*.

Himalayan Tahr   *Hemitragus jemlahicus*
hēmi (Gr) half   *tragos* (Gr) a goat; meaning 'something like a goat';
a reference to the absence of a beard and the animal having some of
the habits and characters of a goat   *jemlah*, probably from *hima*
(Sanskrit) snow, and *alaya*, an abode; hence also Himalaya   *-icus* (L)
suffix meaning belonging to. Tahr is from *thār*, the Nepalese name for
this wild goat inhabiting the Himalayas.

Arabian Tahr   *H. jayakari*
Named after Surgeon Colonel A. S. G. Jayakar; he collected in the
Persian Gulf, chiefly birds, from 1878. During the years 1885 to 1899
he presented collections to the British Museum. Inhabiting the Muscat
and Oman area of south-eastern Arabia.

Nilgiri Tahr   *H. hylocrius*
hulē (Gr) a wood, a forest   *krios* (Gr) a ram. The Nilgiri Hills are in
the south-eastern part of India.

Blue Sheep or Bharal   *Pseudois nayaur*
pseudēs (Gr) false   *ois* (Gr) a sheep; referring to the absence of facial
glands, and the character of the tail, which makes this genus resemble
the goats more than the sheep   *nayaur* is a native name for this wild
sheep, probably from the Nepali word *nahūr*. *Bharal* is a Hindi name.
The coat is grey, becoming more blue in winter in the case of juveniles.
It inhabits the Himalayas from India to China.

Sheep   *Ovis aries*
ovis (L) a sheep   *aries* (L) a ram. This is the domestic sheep, and there
are now over 400 different breeds. It was first domesticated many
thousands of years ago and the actual ancestry is uncertain.

Mouflon   *O. musimon*
musimo (L) an animal of Sardinia   *mouflon* (Fr) a Sardinian wild

sheep. Considered to be one of the ancestors of the domestic sheep, it is only found on the islands of Corsica and Sardinia.

Laristan Sheep *O. laristanica*
Lar, formerly Laristan, is a town and district in Iran, bounded on the south by the Persian Gulf. This sheep is confined to the southern part of Iran.

Urial or Red Sheep *O. orientalis*
-*alis* (L) suffix meaning relating to. Urial is from *hureāl* (Punjabi) a Himalayan wild sheep. The coat is a brown to red colour. Inhabiting the Himalayas in Afghanistan and ranging south to Baluchistan.

Argali *O. ammon*
Ammon or Amen was an Egyptian deity, usually represented in human form with a ram's head. *Argali* is a Mongolian name for this sheep; it covers a wide area throughout eastern Central Asia, including parts of Mongolia, Tibet and the Gobi Desert.

Rocky Mountain Sheep or Bighorn *O. canadensis*
-*ensis* (L) suffix meaning belonging to; it is not confined to Canada and ranges through mountainous areas as far south as Mexico; it has massive horns.

Dall's or White Sheep *O. dalli*
William H. Dall (1845-1927) was an American zoologist; this sheep was discovered in 1884 and named in his honour. The coat is grey to white and it inhabits Alaska and the western part of Canada.

Takin *Budorcas taxicolor*
bu (= *boo, bous*) (Gr) an ox *dorkas* (Gr) a gazelle; a gazelle-like ox *taxus* (New L) a badger; badger-coloured, a yellowish grey. *Takin* is a Tibeto-Burman name for this animal which is related to the musk-oxen; it inhabits Tibet and neighbouring areas.

Musk Ox or Musk Sheep *Ovibos moschatus*
ovis (L) a sheep *bos* (L) an ox; it has features in common with the ox and the sheep *moschatus* (New L) musky; it has preorbital glands that secrete a musky odour. The heavy long-haired coat probably constitutes the best 'warmth preserver' among land-living animals; very necessary for this ox for protection against the intense cold of Greenland, northern Canada and Alaska.

Tamarau *Anoa mindorensis*
Mindoro is an island in the Philippines -*ensis* (L) suffix meaning

belonging to *anoa* is the Celebes native name for this buffalo. Tamarau (Tagalog) is from the language of a people of Luzon, in the Philippines.

Anoa   *A. depressicornis*
de (L) down from   *presso* (L) I press   *cornu* (L) a horn; the horns are short and depressed backwards. Inhabiting Celebes, Indonesia.

Water Buffalo   *Bubalus arnee*
boubalos (Gr) a buffalo   *arnee* is from the Hindi native name *arnā*. Although domesticated there are still some wild herds in Borneo, Malaya, Thailand and India.

African Buffalo   *Syncerus caffer caffer*
sun (Gr) together   *keras* (Gr) the horn of animals; an allusion to the horns which are close together at the base   *cafer* (New L) of Caffraria (or Kaffraria), the country of the Kaffirs. It is widespread in the southern half of Africa but not found now in the extreme south.

Forest or Dwarf Buffalo   *S. c. nanus*
nanus (L) a dwarf; this small buffalo lives in forested areas of western Africa including Congo, Cameroun and Nigeria and ranging west to Guinea.

Banteng   *Bos javanicus*
bos (L) an ox   *-icus* (L) suffix meaning belonging to; it inhabits Java, Borneo and part of southern Asia. *Banteng* is a Malayan name.

Gaur   *B. gaurus*
Gaur is the Hindustani name for this large ox inhabiting India, Burma, and other parts of southern Asia.

Kouprey   *B. sauveli*
Only recently discovered and first described by Professor A. Urbain, the French zoologist, in 1937. He had previously seen, for the first time, the horns of this ox in the home of Dr Sauvel, a veterinary surgeon in Cambodia. Thus, he named it in his honour. Kouprey is the native name in Cambodia. A rare animal, restricted to an area in the Mekong River valley in Cambodia, and part of Laos and Vietnam.

Aurochs   *B. primigenius*
primigenus (L) original, primitive; for many years there has been discussion about the probable ancestors of domestic cattle. Zoological

research in recent years shows that they are probably descended from this ox. The name aurochs is derived from Old High German. It is now extinct, but was known in Europe up to the sixteenth century.

Zebu (Domestic) *B. indicus*
-icus (L) suffix meaning belonging to; 'of India'; originally from India it exists now only in the domestic form and is widespread in the east. Zebu is a French name and first adopted at a French fair in 1752.

Western Cattle (Domestic) *B. taurus*
taurus (L) a bull. Almost world-wide in farms, ranches, etc.

Wild Yak *B. mutus*
mutus (L) dumb, unable to speak; they cannot 'moo' like normal cattle, but only grunt. Inhabiting high altitudes in Tibet.

American Bison *Bison bison bison*
bison (L) a bison *bisōn* (Gr) a species of wild ox, the humpbacked ox, bison. In the eighteenth and early nineteenth centuries, there were many millions of these fine animals in North America, but gradually as the white settlers invaded the country and began to build railways the beasts were systematically destroyed for food. Now there are probably none in existence, except in wildlife parks and game reserves. (See Tautonyms, page 13.)

Wood Bison *B. b. athabascae*
The type was named from a region known as Athabaska, in North West Canada, south of the Great Slave Lake. There is a Lake Athabaska and a River Athabaska and a large area is now established as Wood Buffalo National Park, where over ten thousand animals are preserved. This subspecies is larger and darker, and has longer thinner horns.

European Bison *B. bonasus*
bonasus (L) a kind of buffalo. It is sometimes known as the Wisent, from the German word *wisunt*, a bison. Like the American Bison, it is now unknown in the wild state, but many are protected in wildlife parks in Poland and other countries in Europe, and in the Caucasus Mountains.

Appendix

Transliteration of Greek Alphabet

Greek		Name	Modern System	Latin System
A	α	alpha	a	a
B	β	bēta	b	b
Γ	γ	gamma	g	g
Δ	δ	delta	d	d
E	ε	epsīlon	e	e
Z	ζ	zēta	z	z
H	η	ēta	ē	e
Θ	θ	thēta	th	th
I	ι	iōta	i	i
K	κ	kappa	k	c
Λ	λ	lambda	l	l
M	μ	mū	m	m
N	ν	nū	n	n
Ξ	ξ	xī	x	x
O	o	omīcron	o	o
Π	π	pī	p	p
P	ρ	rhō	r	r
Σ	σ ς	sigma	s	s
T	τ	tau	t	t
Y	υ	upsīlon	u	y
Φ	φ	phī	ph	ph
X	χ	chī	kh	ch
Ψ	ψ	psī	ps	ps
Ω	ω	ōmega	ō	o

Bibliography

Barnes R.D. (1987) *Invertebrate Zoology*. 4th Edition. Holt Saunders. Philadelphia.

Benton M.J. (1993) *The Fossil Record 2*. Chapman & Hall. London.

Blackwelder R.E. (1967) *Taxonomy: A Test and Reference Book*. Wiley & Sons Inc. London.

Darwin C. (1859) *The Origin Of Species*. Rowman & Littlefield. New Jersey.

Halliday T. & Adler K. (1986) *The Encyclopaedia of Reptiles and Amphibians*. Allen & Unwin. London.

Jaeger E.C. (1978) *A Source Book of Biological Names and Terms*. 3rd Edition. Charles C. Thomas. Illinois.

Jeffrey C. (1989) *Biological Nomenclature*. 3rd Edition. Edward Arnold. London.

Lawrence E. (1989) *Henderson's Dictionary of Biological Terms*. 10th Edition. Longman Scientific & Technical. London.

Macdonald D.M. (1985) *The Encyclopaedia of Mammals*. Allen & Unwin. London.

Margulis L. & Schwartz K.V. (1987) *Five Kingdoms. An Illustrated Guide to the Phyla of Life on Earth*. 2nd Edition. Freeman & Co. New York.

Mayr E. & Ashlock P.D. (1991) *Principles of Systematic Zoology*. 2nd Edition. McGraw-Hill. New York.

Natural History – the Magazine of the American Museum of Natural History. New York.

Nowak R.M. (1991) *Walker's Mammals of the World*. Vols I & II, 5th Edition. Johns Hopkins University Press. Baltimore.

Parker S.P. (1982) *Synopsis & classification of Living Organisms*. Vols 1 & 2. McGraw-Hill. New York.

Pearse V., Pearse J., Buchsbaum M. & Buchsbaum R. (1987) *Living Invertebrates*. Boxwood Press. California.

Perrins C.M. & Middleton A.L.A. (1985) *The Encyclopaedia of Birds*. Allen & Unwin. London.

Sibley C.G. & Monroe B.L. Jr (1990) *Distribution and Taxonomy of the Birds of the World*. Yale University Press. New Haven.

Skelton P. (1993) *Evolution. A Biological and Palaeontological Approach*. Addison-Wesley. Massachusetts.

Indexes

The Indexes of English names contain the English names of all the reptiles, birds and mammals in this book. Normally, the page number of the main reference is given, but where there is an additional reference of a general or less important nature, this page number is given in *italic* type.

The Indexes of Latin names similarly contain the Latin names of all the reptiles, birds and mammals in this book and subordinate references are similarly given in *italics*. The Phylum, Subphylum, Class, Subclass, Order, Suborder and Family are in SMALL CAPITALS.

Reptiles: English Names

Reptiles: Latin Names

Birds: English Names

669

Birds: Latin Names

MAMMALS: English Names

Mammals: Latin Names

Acknowledgements

The executors of the late A. F. Gotch wish to thank the following people for help given to the author:

Dr E.N. Arnold, British Museum (Natural History); Professor A. Bellairs; Gordon Bennett; Stuart Booth of Blandford Press; L.R. Conisbee; Dr H.B. Cott; Howard Crellin; Mrs Ann Datta, British Museum (Natural History); Mr R. Fish, The Zoological Society of London; M.J. Gardener; J.D. Gotch; J. Edwards Hill, British Museum (Natural History); Lawrence Jones; Dr K.A. Joysey; Norman Kearney; L.G. Kelleway, Department of Zoology, University College of Swansea; Lesley Lowe; Dr A. Maraspini; John Maries; Dr J.F. Monk; Dr Ernest Neal MBE; Dr J.G. Sheals; Michael Tweedie; Mrs Vale, Mr D.K. Read and Mr I.C.J. Galbraith, Tring Museum; Ferelyth and Bill Wills; Father B. Wrighton.